THE END OF ALZHEIMER'S

THE END OF ALZHEIMER'S

The Brain and Beyond

SECOND EDITION

THOMAS J. LEWIS, PhD
RealHealth Clinics, Founder
Jefferson City, TN, United States

CLEMENT L. TREMPE, MD
Beth Israel Deaconess Medical Center Boston
Boston, MA, United States
Harvard Medical School (Retired)
Boston, MA, United States

ACADEMIC PRESS

An imprint of Elsevier

Academic Press is an imprint of Elsevier
125 London Wall, London EC2Y 5AS, United Kingdom
525 B Street, Suite 1800, San Diego, CA 92101-4495, United States
50 Hampshire Street, 5th Floor, Cambridge, MA 02139, United States
The Boulevard, Langford Lane, Kidlington, Oxford OX5 1GB, United Kingdom

Library of Congress Cataloging-in-Publication Data
A catalog record for this book is available from the Library of Congress

British Library Cataloguing-in-Publication Data
A catalogue record for this book is available from the British Library

ISBN: 978-0-12-812112-2

For information on all Academic Press publications visit our website at
https://www.elsevier.com/books-and-journals

Working together
to grow libraries in
developing countries

www.elsevier.com • www.bookaid.org

Publisher: Mara Conner
Acquisition Editor: Natalie Farra
Editorial Project Manager: Pat Gonzalez
Production Project Manager: Lucía Pérez
Designer: Matthew Limbert

Typeset by Thomson Digital

Dedication

This work is dedicated to my father, "Papa," who passed from Alzheimer's disease a decade ago. It is also dedicated to my mother, Cecelia (Neema, Momalou).

My mother is very "old school," having lived through the Great Depression and WWII. She married my father at the age of 24 and was completely committed to family. She never could envision abandoning him to someone else as he slipped into dementia. My mother so completely and selflessly managed my father and the home, that my siblings and I were insulated from his true condition. We did learn later that his behavior was somewhat typical of Alzheimer's patients in that during his severe episodes, he would lash out and become violent. She often explained bruises as being caused by her clumsiness.

Regardless of my father's behavior and the prompting of his doctors, my mother disregarded any suggestions to place him in full-time care. She had vowed, at the time of their wedding, to be there for him for better and for worse and in sickness and in health. She was not one to compromise on her promise. What I neglected to consider was that, since his fate was sealed, my efforts were not for my father but rather to help my mother. She was always so strong and capable, so I assumed that she could and would handle anything.

Twelve years after the passing of my dad, my mom, at the age of 93, is doing heroically well. Thanks to God.

Thomas J. Lewis, PhD

Contents

About the Authors

Dr. Lewis holds a PhD in Chemistry from MIT. He served in various research capacities prior to starting a scientific consulting business in 1997. He is an entrepreneur and healthcare professional with expertise in toxic substances, drug development, biotechnology, health technology, and medical protocol development. In 2005, after the passing of his father from Alzheimer's disease, he has dedicated his time and career to finding a solution to this disease. After finding a unique and remarkable clinician, Dr. Trempe, with a profound understanding of Alzheimer's diagnosis and treatment, Dr. Lewis spent the past several years verifying the findings of Dr. Trempe using the vast medical and scientific literature available. His research over the past several years culminates in this book.

Dr. Trempe received his MD degree from Ottawa Medical School, Canada. He furthered his studies at Harvard's Schepens Eye Research Institute (SERI) and Massachusetts Eye and Ear Infirmary, Boston. He has been on staff at Harvard Medical School teaching hospitals since the 1970s. He is the author of hundreds of medical scientific papers.

Dr. Trempe didn't set out to solve the Alzheimer's conundrum; he did set out to treat eye diseases in a different way, however; he and many others recognize that a sick eye does not reside in a healthy body. A sick eye is, for the most part, is a sick body. Treat the causes of the sick body and the health of the eye will also improve. Dr. Trempe is one of very few Ophthalmologists who take this approach. Why? Because eye doctors treat eye diseases, cardiologists treat heart diseases, neurologist treat brain diseases, and so on. These specialties seldom collaborate. Each medical discipline has its own set of diagnostics and drugs for their special ailments.

When Dr. Trempe started diagnosing and treating his "eye" patients for systemic (whole body wide) diseases, back in the 1980s, their eyes did indeed get better. In fact, they got much better and stayed much better compared to people who were treated as if their eyes existed in isolation from the rest of the body.

Most importantly, many patients with serious diseases beyond the eye reported back to Dr. Trempe that these other conditions improved upon his "eye" (whole body) treatments. One of those conditions that improved was Alzheimer's disease.

tjlphd@gmail.com
tlewis@realhealthclinics.com
cltrempe@gmail.com

Foreword

BY JACK C. DE LA TORRE, MD, PhD

It's time to face facts. Suppose Dr. Alois Alzheimer came back from his grave to see how the disease that bears his name has progressed in the last 100 years since its discovery in 1907. He would be amazed to learn how much innovative research has been done to uncover the cellular, molecular, and biochemical mechanisms of the disease but only where animals and test tubes are concerned. It is my guess that Dr. Alzheimer would also be totally perplexed and disheartened at the fact that after a century of research and over 100,000 scientific papers written on the subject, patients presently diagnosed with Alzheimer's disease are no better off now than they were in 1907. This fact alone invites the troubling question, are we on the right track to finding a way to help Alzheimer's patients?

To search for an answer to this consequential question, one needs to read "The End of Alzheimer's" by Dr. Thomas Lewis and Dr. Clement Trempe who write about this disquieting problem and possible ways to solve it.

It is important to recall how research works, both at the basic and clinical levels. Clinical research is generally an off-shoot of basic research. Basic research to a problem usually involves a hypothesis, experimentation, and evidence to prove or disprove the hypothesis. If experimentation repeatedly fails to support a hypothesis, scientists usually move on to seek another hypothesis. This is not the case with the Abeta hypothesis, the reigning paradigm of Alzheimer's disease

whose concept of clearing amyloid plaques from the brains of Alzheimer's victims has entirely failed to help them in reported clinical trials held so far. Common sense dictates that when you discover you are riding a dead horse, the best strategy is to dismount.

Having said that, one assumes that although many basic researchers are quite smart, they are also totally dependent on funding to do their research. No funding, no research. Even the most brilliant hypothesis can lay in the corner of the laboratory gathering dust if funding is not obtained. Who provides the funding? The main funders are the pharmaceutical industry, the government (NIH) and private foundations, mostly in that order of money-giving generosity.

Government and private foundations rely on a panel of "experts" to advise the bureaucrats whether a research project is worthy of funding. Often, a conflict of interest arises from these supposedly impartial advisors who more often than not, opt to fund their friends or research projects close to their hearts. They are in essence, the keepers of the gate. Pharmaceutical-derived funding is more businesslike. They prefer to fund research projects that will bring them money by the truckload. Alzheimer's disease is a disorder that affects over 5 million people in the United States and 36 million worldwide so it has become an excellent target of investment.

To find even a negligible benefit to Alzheimer's patients, a patented drug sponsored by pharmaceutical money, can mean, as Drs. Lewis and Trempe correctly pointed out in their book, the mother lode of return investment reaching billions of dollars annually. This is what Dr. Alzheimer would find callous and mean-spirited, should he return from the grave.

Since it is axiomatic that most scientists with an intellectual or financial stake in a theory tend to ignore the facts that may undercut their views, it is not surprising that the Abeta hypothesis has survived this long. To survive, the Abeta hypothesis has creatively morphed into a 9-headed Hydra whose heads, like the mythical monster, can regrow after being cut-off. Thus, each time sharp evidence cuts off one of its heads, the monster hypothesis survives by quickly growing another head. In this fashion, each clinical trial failure greeted by jury of vested scientists whose chorus is, "it didn't work, BUT…" and thus, another head on the Hydra is regrown to fight another day. Consequently, the continued re-invention of these anti-Abeta compounds continue to be retested on Alzheimer's in multi-million dollar clinical trials.

Why do these pharmaceuticals persist in clinically retesting the same failed concept over and over again and expecting a different result? In the case of the Abeta hypothesis, the answer is, money. This point is fluently discussed by Drs. Lewis and Trempe. They offer a compelling argument that while the Abeta hypothesis is dying from an absence of supporting clinical evidence, millions of dollars continue to be poured into these single-minded Abeta projects by the greedy pharmaceutical companies. They hope to tap into this billion-dollar industry if one of their drugs is approved for any positive action on Alzheimer's disease, no matter how clinically inconsequential.

Tragically, research avenues not dealing with anti-Abeta therapy are ignored by these same pharmaceuticals who have decided, at least for the moment, not to hedge their bets with several promising concepts that may help prevent or control Alzheimer's onset.

Drs. Lewis and Trempe also discuss the important issue concerning how the start of Alzheimer's disease can be significantly prevented or controlled by early identification and detection of offending risk factors in both healthy and mildly symptomatic individuals. Such a strategy involves treating the modifiable precursors to Alzheimer's dementia will also ensure their control and prevention. This approach will not only result in a better mental health outlook for the patient but also will significantly lower the exponentially growing incidence of this devastating dementia and the explosive impact from its socio-economic consequences.

Drs. Lewis and Trempe have written a mind-opening, well-informed, and intelligent account of the history, present and future interventions and distillation of keen thinking on the subject of Alzheimer's disease. This book will be the focus of many prospective and pivotal discussions on how medical research will eventually govern this mind-shattering disorder.

Jack C. de la Torre MD, PhD
Professor of Psychology
University of Texas, Austin
Senior Editor, Journal of Alzheimer's Disease
Austin, TX, United States

BY KIMER S. MCCULLY, MD

In their brilliant and comprehensive analysis of Alzheimer's disease, Drs. Lewis and Trempe present an innovative strategy for prevention and treatment of this devastating disease. By understanding the underlying cause of the disease, rational measures are used to arrive at the correct diagnosis, which is the key to successful management of the disease. In this analysis, ophthalmological observation and thorough determination of general health are used to assess the potential for the development of dementia in the individual patient. By using the results of medical research available on the internet, a successful strategy can be developed from the "Trillion Dollar Conundrum" as published in scientific articles worldwide. The Trillion Dollar Conundrum refers to the two million research studies of Alzheimer's disease and other diseases, funded to the extent of $500,000 each that are published in the medical literature each year. In the conventional wisdom of the cause of Alzheimer's disease, the medical establishment, and more importantly, the pharmaceutical industry commit immense sums of money to development of drugs to counteract the amyloid cascade hypothesis. In their analysis, most of these efforts have proven to be fruitless, and the new approach of Drs. Lewis and Trempe, based on scientific understanding, is presented to guide therapy and prevention.

In the years since 1906, when neuropathologist Dr. Alois Alzheimer introduced the concept of tangles and plaques in the brain as a cause of early-onset dementia, the disease has been found to be closely related to vascular disease in arteries of all organs of the body. The conclusion of a century of medical research is that vascular dementia and dementia associated with tangles and plaques in the brain are closely related to and associated with aging, declining oxidative metabolism, and infections. A further conclusion is that inflammation and the immune system are participants in the initiation and progression of dementia observed in Alzheimer's disease, Parkinson's disease, amyotrophic lateral sclerosis, and other neurodegenerative diseases. These diseases are associated with inflammation of the brain, and two molecular markers of inflammation in the blood, homocysteine and C-reactive protein, are especially useful in following the inception, progress, and treatment of these diseases.

Homocysteine is a four-carbon amino acid containing sulfur in the form of a sulfhydryl group. Homocysteine was discovered in 1932 by the eminent American chemist Vincent DuVigneaud by heating the amino acid methionine in concentrated sulfuric acid. In contrast to methionine, homocysteine does not occur in the peptide linkages of proteins, even though the molecule differs from methionine, an important sulfur amino acid of proteins, only by a methyl group. The importance of the methyl group and its relation to the biochemistry of sulfur were explored in animals by DuVigneaud and many other investigators in the 1930s and 1940s. However, the importance of homocysteine in human disease was totally unknown until 1962, when cases of the disease homocystinuria were discovered in children with arterial and venous thrombosis, mental retardation, and other disturbances of the central nervous system. Analysis of vascular disease occurring in cases of homocystinuria caused by different inherited enzymatic abnormalities of methionine metabolism, revealed the atherogenic effect of homocysteine in causing arteriosclerotic arterial plaques. This concept is termed the homocysteine theory of arteriosclerosis, since many important aspects of atherogenesis occurring in the general population are attributed to the effect of homocysteine on the cells and tissues of the arteries.

Homocysteine became an important factor in understanding the cause and treatment of Alzheimer's dementia in 2002, when investigators at the Framingham Heart Study demonstrated that participants with elevated blood homocysteine levels are at greatly increased risk of developing Alzheimer's dementia when followed for a decade. This observation corroborated the hundreds of published studies documenting elevation of blood homocysteine as an independent, potent risk factor for atherosclerosis in the general population.

A further development in understanding the origin of atherosclerosis and dementia occurred when investigators demonstrated remnants of microorganisms in arterial plaques in subjects with atherosclerosis and in the brains of subjects with Alzheimer's disease. The pathogenesis of vulnerable atherosclerotic plaques was attributed to obstruction of vasa vasorum of artery walls, where inflammation and deposition of lipids is first observed in atherosclerosis, by aggregates of lipoproteins, micro-organisms, and homocysteinylated lipoproteins. These aggregates become trapped in vasa vasorum because of high tissue pressure of artery walls and because elevated blood homocysteine causes endothelial dysfunction, narrowing the lumens of capillaries and arterioles. Obstruction of vasa vasorum by these aggregates causes ischemia, death of arterial wall cells, hemorrhage, and rupture into the intima creating a microabscess, the vulnerable plaque.

In a similar process in the brain, spirochetes from the oral cavity invade the nerves of the nasopharynx, and olfactory tract, spreading to the brain, where inflammatory reaction and deposition of A-beta amyloid creates the plaques and tangles of Alzheimer's disease, as shown by the eminent Swiss neuropathologist, Judith Miklossy. The analysis of Drs. Lewis and Trempe takes advantage of these observations by showing that treatment of chronic intracellular infections by organisms, such as Chlamydia pneumoniae, Mycoplasma pneumoniae, Helicobacter pylori, Rickettsiae, Borrelia burgdorferi (of Lyme disease), and Archaea has the potential for arresting the pathogenesis of dementia by enhancement of immune system function through optimal nutrition and nutritional supplements and by elimination of sources of further infection by meticulous oral hygiene.

As the pathophysiological processes of aging, atherosclerosis, and dementia are characterized by elevation of blood homocysteine, an explanation of the origin of these systemic processes is related to biosynthesis and metabolism of homocysteine. Two decades ago a new theory of oxidative metabolism was introduced to explain the observations of oxidative stress and aerobic glycolysis in atherosclerosis, cancer, autoimmune diseases, and other degenerative diseases of aging. According to this theory, oxidative phosphorylation is dependent upon thioretinaco ozonide, the complex formed from retinoic acid, homocysteine thiolactone, cobalamin, ozone, and oxygen. This theory also explains the coordination of reduction of oxygen by electrons from electron transport particles of mitochondria with the polymerization of phosphate with a precursor of adenosine diphosphate (ADP) to produce adenosine triphosphate (ATP) and the proton gradient across mitochondrial membranes.

A recent development of this theory implicates nicotinamide adenine dinucleotide (NAD+) as a precursor of ADP, leading to the active site of oxidative phosphorylation, thioretinaco ozonide oxygen NAD+ phosphate. This theory explains the origin of elevated blood homocysteine in aging, atherosclerosis, and dementia, because this active site complex is consumed by microorganisms occurring in vulnerable plaques of the arteries and plaques and tangles of the brain in Alzheimer's disease. This active site of ATP synthesis is also the precursor of the important co-enzyme adenosyl methionine, the precursor of methylation reactions and the allosteric regulator of the enzymes of homocysteine metabolism. Adenosyl methionine and NAD+ within cells both decline in aging, and nicotinamide riboside, a precursor of NAD+, activates sirtuins which regulate mitochondrial function in aging. The antiaging properties of nicotinamide riboside are attributed to increased synthesis of

NAD+ and thioretinaco ozonide, molecules which both decline in aging.

The brilliant strategy by Drs. Lewis and Trempe takes advantage of revolutionary new concepts for guiding enhancement of immune function and treatment of chronic infections in prevention and treatment of Alzheimer's disease. The diagnosis of mild cognitive impairment by psychological testing, combined with assessment of ophthalmological abnormalities and determination of health status through thorough testing of biochemical markers related to infection and inflammation, are necessary for improving the prognosis and reducing the risk of dementia. The implications of this strategy for the individual and for the population are enormous. Control of dementia, atherosclerosis, and degenerative diseases of aging by the insights of Drs. Lewis and Trempe has the potential for revolutionizing management of chronic disease in the general population.

Kilmer S. McCully, MD
Chief of Pathology and Laboratory Medicine
US Department of Veterans Affairs Medical Center
VA Boston Healthcare System
Pioneer of the Homocysteine Theory
Boston, MA, United States

Preface

BY DR. LEWIS

The seeds of great discovery are constantly floating around us, but they only take root in minds well prepared to receive them.
— *Joseph Henry*

Globally, almost one trillion dollars ($1,000,000,000,000—one million times one million dollars) is spent annually on medical and related research. Are we getting what we pay for? Yes and no.

When you search through and read the medical literature, the depth and breadth of the information is almost beyond comprehension. I use http://www.scholar.google.com for most searches, and this engine allows for a fair number of inputs including searching for keywords in the body or the title of articles. The amount of research in the area of Alzheimer's is mind-boggling. If you want to know the association between Vitamin D and Alzheimer's, at least 20,000 titles are found. The number drops to 20 when the search is "title only." A search for "amyloid and Alzheimer's" yields over 123,000 records. Beta-amyloid is considered one of the two most important biological "hallmarks" of Alzheimer's disease (AD).

Pick an association you might think is important about Alzheimer's disease (AD) and to be sure, a search will yield many articles. A rule of thumb is that each article costs approximately $500,000 to produce considering researchers, their time, laboratories involved, meetings, and all ancillary items associated with performing research and creating a finished technical document, complete with a novel thesis. Thus there are about two million research articles published each year, give or take.

Translation of pure research into clinical practice is a big problem and rears many ugly heads. From a patient's perspective, it simply takes too long for the information obtained by researchers to reach the clinic. Some may estimate that the time lag between discovery and clinical application is 10 years but I believe it is at least, on average, 20 years. Compare this to other industries such as information technology. The time from discovery to the shelf is often less than one year and we, the consumers, demand that new technologies are at our fingertips immediately. It is likely that the lag time between discovery and clinic will only lengthen. This is in complete contradiction to essentially every other enterprise.

Consider the book *The Singularity Is Near: When Humans Transcend Biology* by Dr. Raymond Kurzweil. Four central postulates of the book are as follows:

1. A technological-evolutionary point known as "the singularity" exists as an achievable goal for humanity.
2. Through a law of accelerating returns, technology is progressing toward the singularity at an exponential rate.
3. The functionality of the human brain is quantifiable in terms of technology that we can build in the near future.
4. Medical advancements make it possible for a significant number of this generation (Baby Boomers) to live long enough for the exponential growth of technology to intersect and surpass the processing of the human brain.

Do you see any signs of #4 emerging anywhere? Medicine appears to be stagnant or even going backward compared to other "technologies." We are holding even on cancer and heart disease and losing ground in diabetes, Alzheimer's, and other neurodegenerative diseases.

What is the problem and solution? It is actually quite simple: "translational medicine." Consider this simple example: according to *U.S. News* in 2010, Harvard Medical School was ranked *first* in medical research globally. That same year, Massachusetts General Hospital, a Harvard Medical School teaching hospital, was ranked 57th. Yet these two institutions are connected. Mass General is part of Partners Health Care, and Partners is affiliated with Harvard Medical School. Most of the doctors at the hospital hold Harvard Medical School appointments. Why is there such a large discrepancy, first in research yet 57th in clinical delivery? There is an apparent lack of translation between research and patient care even within the same organization! Researchers perform research (mainly on animals that have artificially induced disease, thus have little correlation to actual disease in humans) and clinicians treat humans, and the two groups do not talk (and experimental ideas must pass over 10 years of FDA muster).

The entire medical industry is incredibly segmented into tight verticals, and there is little cross-pollination. Shrinking research dollars leads to research groups being very protective of their novel ideas, which exacerbates this. Also, doctors are busier than ever trying to care for patients while earning a decent wage as both Medicare and commercial insurance reimbursement are diminished. Are you aware that major hospitals are training their doctors to make a 10-minute visit feel like 30 minutes? (Private communication between TJL and attending clinicians.) Yes,

medicine has decayed to that point, far away from the house call.

I work with a very fine doctor, Dr. Clement Trempe, who, now in his seventies, should be retired. However, his love of patient care and medicine keeps him in the office daily. And, he has a slight financial issue. He frequently spends hours (2–5) with patients and follow-up tests, recommendations, phone calls, entry of electronic medical records, and a myriad of other new requirements. He often is only reimbursed $65 for an office visit under Medicare, for patients over 65 years of age. So, if he spends 3 hours with a patient and is reimbursed only $65, isn't he better off working his way up the ladder at Dunkin' Donuts?

I first learned from him what the "Trillion Dollar Conundrum" (as I now call it) is all about. It is illustrated by way of a simple story. He frequently goes to Avenue Louis Pasteur (to the Harvard Medical School auditorium) to attend lectures by prominent researchers. He told me, "I know I'm the only clinician who attends these lectures because I'm the only one wearing a tie. All the other attendees are in sneakers and jeans. They are all PhDs. When the lecture is over, they go back to their lab. I go back and see patients."

My father taught me long ago that, when something doesn't make sense, money is involved. I believe the same holds true in modern medicine. There are plenty of medications and even supplements that work to prevent and/or treat AD, but they never get notoriety. Why? Who is going to spend the money to test and promote generic drugs or even vitamins for this purpose? Yes, there is some degree of testing, but marketing drives our world, and drugs or products without a strong potential for financial reward have no backers. The drugs that are pushed are those that are "on patent" because the drug companies and their tremendous marketing machines have the financial impetus to drive these to the doctor's office. Many good drugs that are or become generic (and no longer have patent protection) just fade away from use in clinical practice because young medical students are not taught about them. Why? These medications do not make drug companies money thus young doctors are not taught about their value. To exacerbate this problem, since about 1980, drug companies have been allowed to sponsor medical school curriculum, and that "education" focuses on new "on patent" drugs, which are controlled and marketed by the pharmaceutical companies. (Wilson, Duff. "Harvard Medical School in ethics quandary." The New York Times 2, 2009.)

The point is a simple one. There is more than enough research, even for a disease like Alzheimer's. There are a myriad of options for both early detection and treatment of patients who already have AD, contrary to what the Alzheimer's Association and other pundits continue to say. These organizations constantly send the message that there is no cure or even a way to slow the progression of Alzheimer's, thus more research money is needed.

This book provides a thorough review of the trillion dollars of annual medical and scientific literature. Based on that review, a case is made for a differential diagnosis process for Alzheimer's and related disorders. We believe you will arrive at the conclusion that there is a way to slow the progression of, or even reverse, AD based on a proper and thorough diagnosis that goes well beyond Alzheimer's.

What does "differential diagnosis" mean? There are many medical definitions of differential diagnosis. We use the following:

A detailed diagnositic process that assesses every EVERY aspect of a patient's whole-body physiology, and pathology to determine a causes or causes of disease.

Here are some more classically worded definitions you will find online:

1. determination of the nature of a cause of a disease.
2. a concise technical description of the cause, nature, or manifestations of a condition, situation, or problem.
3. medical diagnosis based on information from sources, such as findings from a physical examination, interview with the patient or family or both, medical history of the patient and family, and clinical findings as reported by laboratory tests and radiologic studies.

In essence, our differential diagnosis is a broad and deep look into your personal health to answer the question, "why are you ill." Alzheimer's, as a diagnosis is NOT a differential diagnosis. More importantly, it, as a diagnosis, does not give a doctor any notion as to proper treatments. The current treatments, as you likely know, do not change the course of the disease. In our quest to determine what treatable causes your "Alzheimer's" state has, we follow the teaching of Claude Bernard, the father of experimental medicine who, in the 19th century stated,

If you do not understand a patient's disease, you have not look hard enough because there is only one science of health.

What does a differential diagnosis of Alzheimer's do for you? Again, a short story provides an ample illustration. When I talk to doctors about Alzheimer's and infer that there are ways to prevent, slow the progression, and even reverse the course of the disease, one hundred percent of the time the doctor will ask, "What is the treatment?" I always provide a terse answer, "The question should not be: 'what is the treatment?' The question should be: 'what is the diagnosis?'" It may seem like a diagnosis of Alzheimer's is a death sentence. However, a differential diagnosis that delves deeply and broadly into the patient, their environment, physiology, and all the things that makes a person a person, may arrive at a diagnosis that has bona fide treatment options.

Consider this description for the disease Typhus:

> Typhus is any of several similar diseases caused by Rickettsia bacteria. The name comes from the Greek typhos (τῦφοζ) meaning smoky or hazy, describing the state of mind of those affected with typhus.

Do Alzheimer's patients sometimes have a "smoky or hazy" state of mind? Yes. Could Rickettsia bacteria be the cause? Maybe. Has your neurologist tested for Rickettsia? No. Is Rickettsia disease, misdiagnosed as AD, potentially treatable? Yes.

Stop hoping for modern medicine to save you. It could if it were not for the way the industry is constructed, based on verticals, profit motives, and general lack of translation from research into the clinic where the information can benefit you. The good news is that you can save yourself. The Internet is not structured into verticals. It costs nothing except for a monthly subscription to get online, and you can translate the information for your own health and well-being. This book offers a detailed translation for you.

I hope you find the information I've translated for you compelling.

Good luck.
You can beat Alzheimer's disease.
Stay well,
Thomas J. Lewis, PhD

BY DR. TREMPE

First, I want to thank Dr. Lewis for putting together this book that explains what I have been doing for years. Yes, there are more comprehensive diagnosis and treatment for AD and other neurodegenerative diseases related to aging that is provided in clinical practice today. I am a clinician and have never taken any money from any drug companies. The point is that, I am free from bias that financial influences inevitably control.

I did not begin my career in medicine with the goal of helping people with Alzheimer's. However, I have been blessed with the opportunity to learn about disease from something far greater than a test tube in some laboratory. I learned from my patients. They dictated my career path. I am an Ophthalmologist and am also very curious. I also believe in the Hippocratic Oath and am true to its pledge, to do no harm and to help the body health itself.

In the 1980s I gave up a lucrative practice of treating eye diseases as diseases isolated to the eye only. Back then, surgery and laser treatment was the way to go. I soon realized that my patients with eye diseases were always sick in many ways. I'm a doctor so how could I ignore this fact? Can treating the eye with a laser or surgery "cure" the reason why my patients had the eye problems and were otherwise ill? Of course not. And, by reading the medical literature it was becoming clear at that time, that the eye disease was the symptom of a broader condition of poor general health.

My practice changed 30 years ago to be one where I used the eye and eye diseases as a biomarker for broader systemic (whole body) disease. The eye is quite unique for detecting disease. Using simple ophthalmic tools, eye doctors are able to perform disease "biopsy" simply by looking into the eye. Our tools magnify the tissue in the eye and some more advanced tools are able to map tissue very precisely. We are able to "see" disease happening at its earliest stages. I know your cardiologist would benefit greatly in their diagnosis by opening up your chest and peering in at the tissue. Clearly you would not approve of that just for the purpose of diagnosis. However, optometrists and ophthalmologist can do the same thing, but noninvasively. The eye contains both blood vessels and nervous system tissue. We can "open" a window into your health by simply having you, our patients, open your eyes. We all have one circulatory system and one nervous system. What is happening in your eyes is, for the most part, the same thing that is happening in your heart and your brain. This is a much underappreciated and under utilized part of medicine.

Modern medicine is seeking the holy grail of early detection through biomarkers and billions of research dollars are being spent to find biomarkers and develop expensive drugs to treat disease. They are looking for the one thing (it is never one thing) that causes the major disease of our society; cardiovascular disease, diabetes, Alzheimer's, and cancer. The answer to their quest is staring them right in the eye.

I am not aware that a simple eye examination was included in any of the more than 200 failed prospective drug studies done by pharmacological companies in their quest for a new Alzheimer's treatment. During my more than 40 years of practice on the Harvard University staff I had the opportunity to see patients that were in many such studies. On many occasions patients with memory disorders participating in those studies had no evidence of neurodegenerative changes in their eye and their memory problem were due to other causes, such as severe vitamin B12 deficiency, drug induced transient memory loss, or other causes not related to AD. Many of the memory problems related to aging are not related to AD and this could contribute failure of those 200+ studies. The early neurodegenerative changes in the eye are related to the future possibility of AD and not to other multiple causes of memory problems.

After more than 200 failed studies we have to change things. You know what they say about people that keep doing the same thing over and over yet expect different results. (Do not forget that the people involved in those studies are among the smartest in the country).

In future AD studies, patients should be recruited based on evidence of finding early ocular neurodegenerative diseases changes related to possible future development of AD.

A sick eye is, for the most part, in a sick body. Treat the causes of the sick body and the health of the eye will also improve. I'm one of very few Ophthalmologists who takes this approach. Why? Because eye doctors treat eye diseases, cardiologists treat heart diseases, neurologist treat brain diseases, and so on. These specialties seldom collaborate. Each medical discipline has its own set of diagnostics and drugs for their special ailments. But it should not be that way. We should all work together and face the facts that diseases overlap and are often connected.

When I started diagnosing and treating eye patients for systemic (whole body) diseases, back in the 1980s, their eyes did indeed get better. In fact they got much better and stayed much better compared to people who were treated as if their eyes existed in isolation from the rest of the body. Most importantly, many patients with serious disease beyond the eye reported back to me that these other conditions improved upon with whole body treatments. One of those conditions that improved was AD.

I also learned from my patients what does not work. I never use my patients as a laboratory but medicine, as a science, is constantly evolving and new ideas are the norm. One such idea was the value of antioxidants. Major National Institutes of Health studies promoted the use of antioxidants. However, when I suggested patient take, for example, vitamin E, they reported back to me that their eye got worse. Sure enough, when I examined these patients, they did show more bleeding, swelling, and scarring. When I removed them from the vitamin, their eye problem resolved. We can learn so much from patients. Dr. Alzheimer for whom AD is named taught us that medical development should start with patients in the clinic, followed by laboratory research to understand why. Today we have it backward as drug companies start in the test tube and hope their results will extrapolate into humans. Few, if any, major advances in medicine have occurred using this method.

Medical researchers have unequivocally proven that glaucoma, like Alzheimer's, is a neurodegenerative disease. The eye is an extension of the brain and the death of retinal ganglion cells in the back of the eye leads to glaucoma. The same or similar process happens in the brain of Alzheimer's patients where neurons die. It makes sense that these diseases are connected because we have one circulatory system, one central nervous system, and one lymphatic system. All these systems are interconnected. It is almost physiologically impossible for a disease, especially a slowly incubating chronic disease, to live in complete isolation from the whole body.

It is time for a new model for disease management. Two-thirds of disease is chronic in nature and accounts for almost $2 trillion dollars of healthcare spending annually in the United States alone. How does healthcare currently manage these diseases? By reacting to them once they are already impacting the patient's health. This is wrong and does not abide by the Hippocratic Oath. These diseases do not just suddenly strike a patient. Even cardiovascular diseases including heart attacks do not just suddenly happen without warning signs. A person who experiences a heart attack has a sick heart that got there through a slowly progressive decay over years or even decades. This is true for all the chronic diseases. It is time to institute new measures to evaluate the so-called "well person" before they have clinical symptoms of disease. Don't be fooled, just because you do not have symptoms does not mean you are illness free. Our bodies are both resilient and redundant and is often able to function well even when our health is partially compromised. It is at this stage patients are most receptive to treatments. But how do we inform people about their chronic subclinical disease?

The eye provides the answer for people interested in learning about their current and future potential for chronic disease. The beautiful part of the eye is that those most at risk already had the tests. That is, the answers to your current and future health condition is already done and it's free. How so? If you had an eye examination, then your eye doctor has the information you need to appreciate your health condition and risks. There are 50,000 eye doctors in the United States and each sees roughly 1000 patients each year. Thus eye doctors are examining and evaluating 50,000,000 United States patients each year. What if each of those patients were informed of their results as it related to chronic disease? You can be sure chronic disease would not be epidemic in American and the world as it is today. Here is a short list of eye diseases and their relationship to chronic diseases:

- Nuclear cataract: Associated with increased risk of cardiovascular disease.
- Cortical cataract: Associated with AD.
- Glaucoma: Now considered AD of the eye.
- Macular degeneration: Those with this disease are at increased risk of both cardiovascular disease and Alzheimer's.

- Loss of visual acuity: Sudden or steady vision loss is associated with increased risk of all cause mortality.

The issue you still have is finding someone willing to explain the meaning of the results. This is a big challenge we face in medicine today. The eye doctors, for the most part, understand the results and your risks but they are unwilling to share the information with you because it is "not their job." The fracturing of medicine has caused this. Doctors "pass the buck" from one specialist to the next and no one really takes charge of the information. Dr. Charles Mayo, the founder of the Mayo Clinic, used the concept of "Grand Rounds" to bring all the specialists together to confer on clinical cases. It worked and made Mayo famous. Today, the modern Mayo Clinic no longer uses this technique, it is too expensive. Instead, the patient is shuttled to each specialist who works in apparent isolation.

At your last routine eye examination did your eye doctor tell you that you have evidence of an early neurodegenerative disease process going on in your eyes? If you have a certain type of cataract, early evidence of macular degeneration, or glaucoma you have evidence of an early neurodegenerative process related to the possible future development of AD? This is if you survive another 10–15 years. All those eye diseases are associated with significant increase mortality and only the lucky survive long enough to have a chance of developing AD. I know this sounds counterintuitive but the average lifespan of Americans is less than 80 years yet many Alzheimer's patients are in their 80s and 90s. They somehow outlived the average, albeit with a serious degenerative disease.

Sadly, I am not aware of a single eye doctor that discussed the overall health consequences of eye diseases with their patients. I have trained over 200 fellows of ophthalmology and none of them have the courage to

go beyond a diagnosis of an eye disease with their patients. You have to ask your eye doctor if you have early evidence of any of those diseases after every eye examination and ask what should be done to control the chronic systemic inflammatory process related to those diseases.

As a doctor who always put my patients first, I find the big medical industrial complex aligned against the patient. The Alzheimer's Association, for example, proves to me that they are not interested in a cure for the disease. They continue to support researchers pursuing a failed approach to the disease (Chapter 2). And big pharma will never produce a pill that will "cure" Alzheimer's and other major chronic diseases. The human body is too complex for that "monotherapy" approach. Pills make money and treat symptoms, but seldom cure disease. People cure disease by taking good care of their health and seeking treatments as a last resort. Their eyes are important because it exposes diseases early. The instruments used to measure disease in the eye are very accurate and precise so I am able to show my patients how their lifestyle changes, and in some case medications, have improved their eyes and their overall health. This is motivating to most of my patients.

I hope you read, understand, and enjoy this book. It explains the pitfalls of modern medicine but it also shows you the bright side as well. There are researchers from all over the globe doing interesting and beneficial work to show why disease happens. These researchers paint a very clear picture that diseases like Alzheimer's do have treatments that work. That is, there are ways to prevent, slow, and reverse Alzheimer's. The key is to detect the disease early. This is where the eye comes in because the tests are quick, simple, non-invasive, low (or no) cost and provide a great deal about your current and future health.

As I come closer to retirement I hope I can leave a legacy of ways to improve my patient's and your health. I have worked with other doctors but with limited success. They are too busy keeping their heads about water. However, the people with the most to gain are people like you. Maybe if more people like you become informed about ways to protect your health you will demand this type of approach from medicine. You are our hope for what I consider the right and proper change to medicine.

Be Well, Sincerely,
Clement Trempe, MD

Image Credit: A profile of a human head, with a question mark, by EdwardRech (Own work). Available at Wikimedia Commons: https://commons.wikimedia.org/wiki/File:Lost_or_Unknown.svg

Is it Alzheimer's Disease?

If you stop at, and accept, a diagnosis of Alzheimer's disease (AD), all hope is lost. However, AD does have treatments if you (the patient or family) do not tolerate "Alzheimer's" as a final diagnosis. Ask healthcare professionals to spend extra time on you or your loved one to obtain a better understanding of the root cause(s) of the disease. You will find that a broader and deeper diagnostic approach, available today but seldom practiced, will yield information about effective treatments for AD. Medical and scientific research, through its nearly $1 trillion annual budget, has already revealed enough information to slow down, halt, or even reverse this disease, even in late stages, but this information is not filtering into the clinic where you are diagnosed and treated.

Everything presented in this book is "evidence-based," supported by millions (and in some cases, billons) of dollars of medical research, published in prestigious medical journals, from research groups all over the world. This information is intentionally scientific (but still readable) to provide you with the tools to help yourself or your loved one with "Alzheimer's." You need the backing of researchers who are part of the medical establishment in your quest for a cure. That is exactly who is referenced and quoted in this book. None of the information is from what might be considered "fringe science"; instead, it's from researchers at the most prestigious universities like Harvard Medical School, Stanford, MIT, and other top medical universities and research institutes from around the globe.

The goal of this book is to present to you the key research that points to causes of AD as well as possible treatments. We hope you experience an "ah ha!" moment in which you say, "This makes sense!" There are a lot of factors that contribute to this disease and many of these are well studied and published in medical literature. Many of the causes, thus preventions, are within your ability to control. Other factors that contribute to the disease are more complicated, but solutions certainly are within the current knowledge of medicine that your doctor can implement on your behalf. There is hope for all of us!

The term "Alzheimer's disease" is really for a constellation of symptoms associated with the loss of cognitive function. The label "Alzheimer's" does not give an indication about the cause(s) of the disease. The purpose of this book is to convince and empower you to go deeper than a diagnosis of Alzheimer's. We have combed through the medical research literature, and it reveals a clear path to advanced diagnoses and treatments that can stop and reverse so-called "Alzheimer's disease."

The End of Alzheimer's. http://dx.doi.org/10.1016/B978-0-12-812112-2.00001-X

So why are we stuck at a diagnosis of Alzheimer's and a belief that the disease is untreatable? This is a complex question whose answer lies in the fact that medicine today is big business. Thought leaders from leading university researchers to the heads of organizations like the Alzheimer's Association use the media to deluge us with the narrative that AD has no cure nor is there any way to even slow the course of the disease. This statement is part of a marketing strategy, because the Alzheimer's "industry" is competing against all other medical disciplines for shrinking research dollars and donations. Even researchers within the discipline compete against one another. The more urgent the message, the more likely dollars will flow their way. One could then assume that members of the medical profession do not have an understanding of the causes of this disease. However, as you progress through this book, you will see that researchers from around the world are closing in on real, treatable causes for Alzheimer's. This is very good news.

Dr. Alois Alzheimer himself (for whom the disease is named) understood the disease well and, if he were alive today, could give valuable guidance to clinicians assigned to diagnose and treat Alzheimer's patients. Dr. Alzheimer offered an educated guess about the cause in the early 20th century. Current studies are proving him substantially correct. However, medicine today is ignoring the evidence. We provide you with answers to true causes of Alzheimer's in subsequent chapters, as it is not one simple thing.

There is no disputing that the term "Alzheimer's" is appropriate because the disease is complex and based on many factors. Thus this "catch-all" disease name is aptly used in honor of Dr. Alzheimer who first characterized a patient with this relatively unknown form of senility or dementia well over 100 years ago. One factor holding back medicine from a cure is that to be paid, doctors must follow the prescriptive diagnostic and treatment codes created by insurance carriers (including Medicare) and their actuaries, accountants, lawyers, and lobbyists. "Alzheimer's" is an expedient landing point for what should be a much more rigorous diagnostic process. But doctors don't get paid to go further. Sadly, we have a health system that confuses health insurance with actual health care. They are not synonymous! The health insurance "tail" is wagging the healthcare dog and many of us suffer as a consequence.

Doctors and scientists have less and less input into the design of healthcare delivery as time moves forward. What does this mean to a patient "diagnosed" with AD? They are stuck in a cycle of treatment allowed under the "standard-of-care." That is, once you are diagnosed, your doctor goes to his or her codebook and determines what are reimbursable procedures and/or medications according to the patient's insurance. That is what you get regardless of where you go for diagnosis and treatment (almost). And these treatments are the ones you probably already know about. They fulfill the expectations that have been drubbed into us because they do not work. Thus the "Alzheimer's" diagnosis, from your neurologist within the standard medical delivery model, is a slow and degrading death for the sufferer and an equally slow and miserable emotional and financial decay for caregivers and family.

Does a diagnosis of AD have to be a dead end? No. What is stopping a solution for your loved one who is suffering from Alzheimer's? My dad (who passed away from Alzheimer's a decade ago) told me when I was quite young: "Son, if there are things in this world that just do not make any sense, then big money is involved." Yes, big money is involved in preventing known solutions for AD from being brought to the public. However, this is not necessarily a deliberate or malicious action. It is more a result of a complex system that does not always follow a logical path from brilliant ideas to clinical treatments. Finances all too often override science in this process.

The modern approach to medical development stifles clinical innovation. Dr. Alzheimer stated that outcomes and observations in the clinic should drive medical research. Today, just the opposite is true. Medical research is driven by the development of new drugs, which are first tested on animals that are brought to the clinic. There are many issues with this approach that are discussed in subsequent chapters. Very important points are that clinical discoveries by doctors go relatively unnoticed because these generally involve old drugs (or combinations) that do not have financial sponsors. Drugs have a short (20-year) patent life when the owners, the pharmaceutical companies, have the greatest financial interest to market these medications heavily. This forces "big pharma" (the 10 biggest pharmaceutical and biotech companies, including Pfizer, Merck, GSK, and the other big names we see on TV daily) to constantly produce new drugs that have an "on patent" status. The drug approval process is a critical part of a drug company's exclusive rights to a new drug, but it presents a tremendous bottleneck to delivery of new drugs and innovation.

There are an enormous number of medication and treatment ideas that never make it into the drug development pipeline. Why? There is a choke point created by cost and resource limitations that control the drug pipeline. Only the 10 big pharmas have the financial and technical resources to spend $1 billion and 10 years developing a drug. Do scientists and doctors have the final say on what candidate drugs will be developed? No. So who has the say? Well, actually, it is you! If you are a shareholder of any of these companies and are watching quarterly earnings reports, then you hold some responsibility for the action of the companies.

The CEOs of the "big pharma" have more to say about drug development projects compared to an individual like you, who owns 100 shares of Pfizer, for example. However, the point is that the scientists and doctors have little say about the drugs that enter into the FDA process. It is a business decision, and many groups within each of the big companies essentially compete for their project to be chosen. In essence, just 10 people determine what new treatments you will get. And they base their decisions on the business of medicine, not the realities of medical development or the best care of patients like you.

The truth about drug development helps define a term that puts our health care and the management of AD into perspective—it's the "Trillion Dollar Conundrum." This concept is derived from the nearly $1 trillion spent annually on medical and related research around the globe. One trillion dollars buys us approximately 2 million scientific and medical journal articles each year, at an approximate cost of $500,000 for each. The cost includes research time, professor and researcher salaries, benefits, and all the steps necessary to perform research, assemble the data, write the paper, and get it published. These 2 million ideas worthy of publication funnel down into a bottleneck of drug development that yields tens, not millions, of new treatments. This is the Trillion Dollar Conundrum. It is in essence that some of our (arguably) best medical ideas never leave the research laboratory and are not "translated" to the clinic where they can improve the health of patients suffering with disease. The good news is this overlooked research is published and available to all of us through the Internet.

The Trillion Dollar Conundrum offers a strong message of hope. Why? Because it tells us that medicine is offering us only a tiny fraction of the ideas circulating in the minds (and research articles) of medical researchers. The small percentage that enters into drug development may not even represent the best and brightest of the ideas. Why is that? Many great ideas from history came from clinicians who, while practicing medicine, discovered something beneficial to their patients. These ideas infrequently offer a drug company the patent protection they seek, thus are largely ignored. See Appendix 6 for poignant examples.

This book distills many of the best ideas, some of them theoretical, but many practical, that can, or eventually will, lead you to a health improvement, or even a cure, if you have the affliction currently called "Alzheimer's disease."

NEURODEGENERATIVE DISEASES AND DEMENTIA

Alzheimer's is broadly a neurodegenerative disease and, more specifically, a type of dementia. Neurodegenerative disease is the umbrella term for the progressive loss of structure or function of neurons, including the death of neurons. These diseases include dementias, Parkinson's, Alzheimer's, and Huntington's, to name a few. As research progresses, many similarities appear that relate these diseases to one another on a subcellular level. Discovering these similarities offers hope for therapeutic advances that certainly will ameliorate many diseases simultaneously.

According to the Mayo Clinic, dementia isn't a specific disease. Instead, dementia describes a group of symptoms affecting intellectual and social abilities severely enough to interfere with daily functioning. Memory loss generally occurs in dementia, but memory loss alone doesn't mean you have it. Dementia indicates problems with at least two brain functions, such as memory loss and impaired judgment or language. It can make you confused and unable to remember people and names. You also may experience changes in personality and social behavior. However, some causes of dementia are treatable and even reversible. We will show in Chapters 8 and 9 that reversible vascular risk factors and infection, respectively, are but a few of these causes.

There are many types of dementias, most prominently Alzheimer's, which are presumed to be irreversible. All dementias reflect dysfunction in the cerebral cortex, or brain tissue. Some disease processes damage the cortex directly; others disrupt subcortical areas that normally regulate the function of the cortex. When the underlying process does not permanently damage the cortical tissue, the dementia may sometimes be stopped or reversed. Many believe that the extent of the damage to the vascular network limits the possibility of recovery. For example, in a stroke, a large section of the brain is destroyed due to lack of oxygen. In that area, blood vessels can no longer access and service the diseased and damaged tissue. It is like putting out a fire in a building from the outside. The water has limited reach, so the core of the building continues to burn.

In dementias, this "burning" happens gradually. A "rule of thumb" is that every cell must be within three cells from a functioning capillary to survive. If the capillaries are damaged, then the tissue they support will deteriorate. If the extent or degree of capillary deterioration is not too broad, then the damaged tissue has a chance to be repaired or new tissue can be produced to replace that which was damaged, assuming treatment can repair the bad vessels.

TRUE DEMENTIAS

The word "démence" existed in the French language as far back as 1381. The Egyptians and Greeks of the period 2000–1000 BC were well aware that old age was associated with disorders of the memory. The Chinese used the words "Zhi Dai Zheng" for dementia and "Lao Ren Zhi Dai Zheng" for senile dementia, which was described basically as a disease of old people characterized by muteness, lack of response, and craziness. The Romans, that is, A.C. Celsus and Claudius Galen of the first and second centuries AD, referred to chronic mental disorders known to produce an irreversible impairment of higher intellectual functions. The Ayurvedic physicians of India used the Sanskrit term "Smriti Bhransh" as early as AD 800 to describe loss of memory.

Even William Shakespeare refers to the effects of old age on the mind in his plays, "*As You Like It*," "*Macbeth*," and "*King Lear*." The latter, written in about 1606, aptly describes what is known even today as dementia:

> LEAR: "Pray, do not mock me: I am a very foolish fond old man, Four score and upward, not an hour more nor less; And, to deal plainly, I fear I am not in my perfect mind. Me thinks I should know you, and know this man; Yet I am doubtful: for I am mainly ignorant What place this is; and all the skill I have Remembers not these garments; nor I know not Where I did lodge last night. Do not laugh at me; For, as I am a man, I think this lady To be my child Cordelia."

Dr. Philippe Pinel, the founder of modern psychiatry, first used the word "dementia" in 1797. The term "dementia" refers to a chronic, often progressive problem of cognition and other aspects of intellectual capability. Intellectual capability is a complex function consisting of many individual "components," such as memory, problem solving, calculation, speech, and the ability to find the way and analyze problems. So, dementia affects memory and almost always affects judgment, decision making, and relationships with others. The components of intellectual capacity are as follows:

- memory and learning
- attention, concentration, and orientation
- thinking (e.g., problem solving, abstraction)
- calculation
- language (e.g., comprehension, word finding)
- geographic orientation

Neuropsychologists have developed tests for each of the components of intellectual function. Dementia often leads to deterioration in all these components of intellectual function.

ALZHEIMER'S DISEASE

Alzheimer's disease was discovered by, and named for, Dr. Alois Alzheimer. He was a famous German pathologist who described a patient who had died of an unusual mental illness in 1906. The patient, a 56-year-old woman, showed unreasonable jealousy toward her husband as the first noticeable sign of the disease. Soon a rapidly increasing loss of memory was noticed. She could not find her way around in her own apartment. Concomitant, irrational behavior occurred. For example, she carried objects back and forth and hid them, and at times she would begin shrieking loudly.

Her ability to remember was severely disturbed. In early tests by Dr. Alzheimer, he would point to objects that she was able to identify correctly, but immediately afterward she would forget them. When reading, she went from one line to another, reading the letters or reading with senseless emphasis. When writing, she repeated individual syllables several times, left out others, and quickly became stranded. When talking, she frequently used perplexing phrases and some convoluted expressions (milk-pourer instead of cup, e.g.). She would frequently become confused by simple questions, becoming stuck and frustrated. She seemed no longer to understand the use of basic objects. The generalized dementia progressed with time and toward the end, the patient was completely in a stupor, lying in bed with her legs drawn up under her, and in spite of all precautions she acquired bedsores. After 4.5 years of suffering, death occurred.

"Organic Brain Syndrome," an early term for AD, was the first term used in the *Diagnostic and Statistical Manual* of the American Psychiatric Association, which is a worldwide reference source first published in 1952. Organic Brain Syndrome indicated mental abnormalities that were associated with chronic brain disorders. In 1968, the second edition of the *Diagnostic and Statistical Manual* introduced the expressions "presenile" dementia and "senile" dementia, which was unfortunate, insofar as it implied that cases with onset of disease before 60 years of age (presenile) had one disease called "Alzheimer's disease," whereas cases with onset after 60 years of age (senile) had another disease called "senile dementia." It is now well accepted that, regardless of the age of onset, presenile and senile dementias are manifestation of mainly one disease, AD.

AD is the most common cause of dementia, at least based on diagnosis of the living, which is recognized as suspect. It accounts for more than half of all presumed cases of dementia. People with AD also have lower-than-normal levels of brain chemicals called neurotransmitters that control important brain functions. It is not considered reversible, unless of course you obtain a differential diagnosis and a result that points toward treatable targets and hope.

AD disease is sometimes referred to as primary degenerative dementia. It is known as "degenerative" because the brain cells wither away and die. This disrupts the production and distribution of certain chemicals called neurotransmitters that carry messages within the brain. Under the assumption that there is no known cause for AD, it is referred to as a "primary" disorder, which in medical terms implies "without cause." This book may convince you that Alzheimer's is not a primary disorder.

There are two primary types of AD: sporadic and familial. Familial AD (FAD) or Early Onset Familial AD (EOFAD) is an uncommon form that usually strikes earlier in life, defined as before the age of 65 (usually between 50 and 65 years of age, but can be as early as 15) and is inherited in an autosomal dominant fashion, identified by genetics and other characteristics such as the age of onset. It accounts for approximately half of the cases of early onset AD. But in terms of total Alzheimer's cases, it adds up to fewer than 2% and some experts believe the number is less than 1%. FAD requires the patient to have at least one first-degree relative with a history of AD. Nonfamilial cases of AD are referred to as "sporadic" AD, where genetic risk factors are minor or unclear. This is the type of Alzheimer's we think about when we hear that diagnosis.

ALZHEIMER'S VERSUS DEMENTIAS

Neurologists currently diagnose AD using specific criteria (Appendix 2). AD and the different types of dementias are likely to have a common, or at least overlapping, set of root causes even though the neurological symptoms vary widely, but this idea remains somewhat speculative. There is a wealth of information in the medical literature pointing to common causes for a variety of neurodegenerative diseases that we now call "related," like Alzheimer's, dementias, multiple sclerosis, and other neurodegenerative diseases including glaucoma. (Yes, glaucoma is a neurodegenerative disease. It is a disease of the nerve of the retina, and the retina is part of the brain [1].)

If two diseases have the same or very similar root causes, why do they manifest differently? That is, why do they have different symptoms? The answer to the question is likely unknown but is extremely important. One explanation is that one disease does not share all the same root causes compared to another disease even though many aspects of the diseases overlap. Another explanation is that each of us has a different "phenotype" leading to the disease "expressing" somewhat uniquely with each individual. A simpler way to put this is to say that we are as different on the inside, in terms of physiology, as we are on the outside, like our height, gender, and all our other physical attributes.

Definition of Phenotype: A phenotype is the composite of an organism's observable characteristics or traits, such as its morphology, development, biochemical or physiological properties, phenology, behavior, and products of behavior. Phenotypes result from the expression of an organism's genes as well as the influence of environmental factors and the interactions between the two. [https://en.wikipedia.org/wiki/Phenotype]

Your phenotype, simply put, is everything that contributed and continues to contribute to who you are, as an individual today. And today, you are somewhat different compared to the other approximately 6 billion people on Earth at this moment. No wonder you respond to the world, from a health and disease perspective, differently compared to your neighbor. A component of phenotype is the genotype. Many people believe that their health destiny is more controlled by genetics compared to the other aspects of phenotype. This is not true. A prominent Harvard Medical School Professor and geneticist says in his book, *Super Brain*, that environment is more important compared to genetics, and that those with strong genetic disposition to AD comprise less than 2% of all sufferers of the disease [2].

Another strong case against a genetic "crutch" comes from *National Geographic Magazine*. In the May 2013 issue, the article titled "New clues to a long life" suggests that a baby born today will live to 120 [3]. The authors went all over the world to examine people and populations that live into their hundreds today. They attempt to make a strong case for genetics being the key factor that determines longevity, and, by inference, the lack of diseases like Alzheimer's.

However, they conclude that genetics actually plays a small part (25%) in longevity. The other 75% they attribute to environment and luck! Based on their findings, it is clear that our free will is responsible for at least three-quarters of our longevity. Our environment that we create for ourselves in which to exist and thrive determines our health. Our health is determined by diet, exercise, daily habits, exposures, and all the circumstances that contribute to our phenotypes, excluding our genotypes. That is our "environment." What about the term "luck?" If you accept the thesis of this book, then "luck" represents just a lack of understanding that we hope diminishes as you turn each page. Luck is eliminated through a process of well-being and, if you do become ill, a differential diagnosis that solves the fundamental problem and does not simply mask the symptoms.

We may not understand why disease occurs in a given part of the body, but the good news is, if we are correct, that many dementias and related diseases have common root causes. Then why the disease manifests in a certain way becomes far less important than knowing the causes. What is most important is the accuracy and completeness of the diagnosis that then guides the treatments.

MULTIFACTORIAL ALZHEIMER'S

Research performed over the past decade shows that AD is multifactorial. The definition of multifactorial disease is "pertaining to or characteristic of any condition or disease resulting from the interaction of many factors, specifically the interaction of several genes, usually polygenes, with or without the involvement of environmental factors" (http://medical-dictionary.thefreedictionary.com/multifactorial). In the context of Alzheimer's then, no two patients are likely to have the same set of factors, or at least the same degree of each factor, that causes their disease. However, in the standard-of-care all patients with a diagnosis of Alzheimer's are treated in essentially the same way. How can we expect people with different underlying disease "factors," who are treated in the same manner, to get well? More importantly, today, none are getting well. Chapter 2 explains that prospects for the future are bleak as more than 300 randomized trials on new drugs, sponsored by the largest drug companies on earth, have all failed. These companies are focused on one target and continue hoping to develop a single pill to treat AD. Since it is a multifactorial disease, this approach is not likely to work, ever.

Associations between newly discovered causes of AD and potential effects of emerging treatments are becoming clearer. Even the research and clinical failures of big pharma provide insight into cause and effect. This implies that our understanding of the disease, thus our ability to effectively treat the disease, is likely to advance rapidly. However, changes occur very slowly in medicine, as it is a very conservative discipline and insurance companies and regulatory agencies must approve the changes. This means that meaningful treatments, known to researchers today, will not be available to AD sufferers for up to a decade or more.

There is a major impediment to your hope for an Alzheimer's solution beyond the Trillion Dollar Conundrum. Both the practice of medical research and the practice of clinical delivery of medicine occur in silos. That is, there is little communication and cooperation between different disciplines within medicine, and this is clearly true for the clinical care of patients with AD. Just like in any business, turf battles occur, and medicine is no different. AD has largely been considered a disease of the brain only. Thus, most (if not all) cases are turned over to neurology. What should happen instead is that those afflicted with symptoms that currently receive the blanket label "Alzheimer's" should be assessed by a broader range of medical professionals who must look beyond the brain for answers to causes and treatments of AD. This is essentially not done today. However, this is what Dr. Trempe does on a daily basis. His patients improve TODAY. This can happen to you or your loved one if you take the time to investigate what we present in this book and find a practitioner like Dr. Trempe. There are many like him. This is where hope can emerge.

WHAT IF ALZHEIMER'S IS NOT ISOLATED TO THE BRAIN?

One of the major "hallmarks" of AD is beta-amyloid 1-42 protein (abbreviated beta-amyloid, Ab, Aβ, β-amyloid). This protein is associated with chronic diseases and is found in many tissues outside the brain including the heart and the muscles. Chapter 8 explores amyloid diseases that occur outside the brain. These diseases are all interconnected by abnormal beta-amyloid protein formation. Therefore, is it naive to view AD as "brain only?" Neurology so far has proven to be the wrong medical specialty for hope. To find a solution to AD, you will have to find enlightened practitioners who are willing to go beyond AD as a diagnosis and go beyond the brain for answers. To do this you must remove yourself from the influences of neurology that are stuck on Alzheimer's as a "brain-only" disease. Once you do this, you will be able to find doctors who will present to you bona fide personal treatment options.

Today, medical experts agree that a true diagnosis of Alzheimer's cannot be accomplished until death, at autopsy. Therefore the diagnosis is a guess based on brain atrophy, beta-amyloid burden (if it can be detected), and cognitive testing that assesses if the brain is malfunctioning (Appendix 2). The issue with this approach to a diagnosis is that:

- Many processes can lead to brain atrophy, not just AD.
- Beta-amyloid, the so-called "hallmark" of Alzheimer's, turns up in patients without AD and sometimes does not appear in patients with AD (Chapter 2).
- Oral and written cognitive tests for memory and function are very nonspecific (Chapter 3).

The danger of the blanket term "Alzheimer's," as a diagnosis, is that all people with it receive the same treatment, within reason. Is it any surprise that treatments are not working? What is surprising is that the treatments never work; thus the current methods available to clinicians clearly do not address the cause(s) of the disease, or, some may argue, even the symptoms. New treatments are in development by the large pharmaceutical companies. Unfortunately, most, if not all, are failing in the clinical trial process, and some companies are even considering ending their Alzheimer's drug development programs [4]. It is essential that diagnosis be sufficiently specific to afford patient treatments that provide real hope, because in the world of medicine today, the diagnosis dictates the treatment.

DIFFERENTIAL DIAGNOSIS

The title of this book includes the term "differential diagnosis." The most basic definition of differential diagnosis is the following: "the determination of which one of several factors or diseases may be producing the symptoms." The error in this definition is the word "one." There may be one cause with simple diseases but complex diseases usually have complex and multiple causes. During the past 20 years or so, medicine has put all its Alzheimer's efforts into one basket—beta-amyloid. Many of the big pharmaceutical companies have pursued beta-amyloid as "the" cause and have lost approximately $100 billion and 20 years in the process, with no progress gained toward a cure. That is the risk of viewing complex disease in a relatively simplistic way. A differential diagnosis, carried out across many medical disciplines, reveals the multifactorial aspect of Alzheimer's along with solutions that enable clinicians to reverse this disease.

Another definition of differential diagnosis is "a systematic diagnostic method used to identify the presence of an entity where multiple alternatives are possible. The process may be termed differential diagnostic procedure and may also refer to any of the included candidate alternatives, which may also be termed candidate conditions. This method is essentially a process of elimination or at least of obtaining information that shrinks the 'probabilities' of candidate conditions to negligible levels." This definition came from Wikipedia (http://en.wikipedia.org/wiki/Differential_diagnosis), where the article's authors show recognition for the complexity of disease. In all cases, they used the plural to describe possible causes.

In essence, our differential diagnosis is a broad and deep look into your personal health to answer the question, "why are you ill." Alzheimer's, as a diagnosis is NOT a differential diagnosis. More importantly, it, as a diagnosis, does not give a doctor any notion as to proper treatments. The current treatments, as you likely know, do not change the course of the disease. In our quest to determine what treatable causes your "Alzheimer's" state has, we follow the teaching of Claude Bernard, the father of experimental medicine who, in the 19th century, stated:

> If you do not understand a patient's disease, you have not look hard enough because there is only one science of health.

Here are some steps to a differential diagnosis process:

- First, the physician should gather all information about the patient and create a symptoms list. This process is not a 10-minute exercise; instead, it involves a broad diagnostic assessment. Certain tests, including those of the blood, may require the physician to perform additional tests as each result provides more clarity.
- Second, the physician should make a list of all possible causes (also termed "candidate conditions") of the symptoms.
- Third, the physician should prioritize the list by placing the most urgently dangerous possible cause of the symptoms at the top of the list.
- Fourth, the physician should rule out or treat the possible causes beginning with the most urgently dangerous condition and working his or her way down the list. "Rule out" practically means to use tests and other scientific methods to render a condition of clinically negligible probability of being the cause.

Making observations and using tests can remove diagnoses from the list.

From the Alzheimer's Association website (https://www.alz.org/professionals_and_researchers_13507.asp):

> Alzheimer's disease accounts for between 50 and 70 percent of all cases of dementia. Many researchers believe Alzheimer's is caused by the accumulation of protein plaques in the brain. The plaques interfere with communication between brain cells and cause the cells to die, leading to memory loss, changes in judgment and other behavioral changes characteristic of Alzheimer's. Physical changes in the brain can cause other forms of dementia as well. Diagnosis may be complicated by coexisting conditions or when symptoms and pathologies of various dementias overlap. Making an accurate diagnosis helps patients receive the treatment and support services appropriate for their condition and maintain the highest possible quality of life.

This is not a proper definition of differential diagnosis because the focus is just on the brain. Interestingly, their definition supports many of the points made thus far about the multifactorial aspect of AD and the need for a detailed diagnosis. Unfortunately the Alzheimer's Association continues to espouse the dogma that this is a brain-only disease.

If so many in medicine believe that Alzheimer's is multifactorial, why isn't diagnosis and treatment multifactorial? Neurology is stuck testing the patient's cognitive behavior related to the brain simply to rule out other brain-only syndromes. Medical practitioners are not using the differential diagnosis process to its fullest potential by testing as broadly and deeply as the patient deserves, especially based on the prognosis of this disease.

Albert Einstein provided us with another definition that appears to fit quite nicely when it comes to describing modern AD diagnosis: "doing the same thing over and over again and expecting different results." Of course this is his definition of insanity. Keep reading and you will soon have the knowledge and the power to impact change that can only come about through a thorough diagnosis that will lead to better treatments for you.

I happen to work with a very talented physician who is clearly improving the health of his Alzheimer's patients. As I attempt to spread the word, doctors and laypeople alike are very quick to ask the question: "What is the treatment?" My response is always: "The most important part of medicine is the diagnosis. If you have the right diagnosis, you don't need to ask about the treatment. You know what to do (if you are a doctor)."

Doctors do currently perform some tests outside of the brain to diagnose AD. They draw blood, for example. However, these are not AD-specific tests using the best that modern research has to offer. In this book we explore the diagnosis of Alzheimer's in great detail. Whether you are a physician or a layperson, this process will lead you to a better understanding of the disease. Clearly, some of the steps to a differential diagnosis are complex and need to be performed and interpreted by qualified medical professionals. However, there are plenty of symptoms and other early warning signs that portend AD, about which this book will make you keenly aware, even though it is likely that you are currently not apprised about their connection to AD.

To give modern medicine its due, root causes are seldom straightforward. Clearly the causes of Alzheimer's have evaded the greatest medical minds of the past two centuries. We endeavor to show that this disease (more accurately, set of diseases and/or causes) can be explained by a combination of theories presented over 150 years ago by medical giants including Louis Pasteur and Claude Bernard. And many researchers are championing new studies based on these classic thought leaders. Indeed, Dr. Alzheimer had a hypothesis about microorganisms' linkage to the senile plaques of AD. This is an important extension of so-called "Germ Theory" postulated by Pasteur and others back in the mid-1800s [5].

BETA-AMYLOID AND ALZHEIMER'S

Beta-amyloid protein is considered the hallmark of Alzheimer's. Chapter 2 delves into the efforts of research and the pharmaceutical industry to advance the "Amyloid Cascade Hypothesis." Many argue beta-amyloid (Ab, Aβ, β-amyloid) is a root cause of AD, the proof being a tremendous effort to develop drugs to eliminate this material from the brain. In fact, the major pharmaceutical companies have focused almost completely on beta-amyloid at the expense of every other theory. It is becoming quite clear that this one target is not the entire AD story. In fact, beta-amyloid may be a symptom, not a cause of the disease that leads to the destruction of brain neurons. Often, something that is found and presumed to be a cause may actually be a symptom or a marker, and may not be detrimental. In fact, emerging science shows that beta-amyloid is present in healthy people and sometimes is not presented in those diagnosed with presumed Alzheimer's.

Beta-amyloid studies continue to be heavily funded by the NIH and other governmental and private organizations. However, as this approach fades into ignominy, a "new" target is emerging. Here, the emphasis is on tau protein that, when diseased, forms neurofibrillary tangles. These formations are essentially only in the brain. Dr. Alzheimer, in the early 1900s, identified both beta-amyloid (he called them senile plaques) and disrupted tau, or neurofibrillary tangles. So if you are led to believe tau is new, it's not. In fact, Dr. Alzheimer suggested that tau is likely more of a perpetrator in Alzheimer's than beta-amyloid (senile plaques). Again, he was probably right. Finally there is a shifting from amyloid (in the brain) to tau (in the brain). Tau is an interesting target, although it is likely to prove to be a marker, not a cause, but time will tell. There is a broad range of potentially responsible targets that are discussed throughout this book that medical research is showing to be much more important than beta-amyloid and even tau.

DISEASES THAT OCCUR WITH ALZHEIMER'S

Alzheimer's patients are not healthy. Sure their brains are deteriorating, but often, so is the rest of their body and not necessarily by AD. There is a broad range of diseases that researchers are showing to be related to AD or are simply appearing alongside AD. Alzheimer's zealots may argue that some of these diseases are coincidental and are just confounding and confusing the diagnosis of AD. But other researchers argue that the diseases are connected by common risk factors and/or root causes and are not just coincident. Since the medical community does not know the true root cause(s), who is to say?

Diseases that occur alongside AD are probably related to it at a root cause level. There are several diseases that are now known to frequently occur at the same time or before AD. A short list is type 2 diabetes, macular degeneration, glaucoma, and cardiovascular diseases. These diseases are likely caused by the same underlying processes, as is AD. Furthermore, due to the complexity of the human body and the multifactorial nature of AD, we can only really talk statistically (or in generalities) even after a detailed differential diagnosis is performed. However, wouldn't you rather have the hope of "statistically" getting 50% better, rather than be condemned to your current fate if you have AD?

An example of a disease that is often concurrent with (comorbid), or occurs before, Alzheimer's in the same person is macular degeneration. Do all macular degeneration patients get AD? No. Do all AD patients have macular degeneration? No. However, there is enough statistical and other scientific evidence to infer a root cause link between the two diseases. These data may imply that the two diseases share some of, but not all of, the same root causes. Will your doctor refer you for an Alzheimer's assessment if you have macular degeneration? Certainly not! Should screening for macular degeneration become part of an early assessment method for people without symptoms or with the early signs of cognitive impairment? You bet! This and many other connections are investigated based on a significant amount of medical research in Chapters 6–8.

We strive to make a diagnosis of AD a distant memory. You may know or remember the diseases: scurvy (updated diagnosis: a vitamin C deficiency), or, better yet, Pellagra (updated diagnosis: vitamin B3 deficiency), but these terms are fading into history. Today we understand these diseases more by their cause than by their name. The same will (hopefully) be true for Alzheimer's in the near future. However, compared to scurvy, which is caused by one deficiency, AD is much more complex. Each patient will need to be classified by their differential diagnosis and the term "Alzheimer's" will seldom be adequately descriptive. Patients need to be placed in many "baskets." Targeted treatment will then emerge that is different and more appropriate from basket to basket, depending upon the diagnosis. We hope that in the future doctors are more specific about the "type" of Alzheimer's at a much more fundamental level than they are today, based on a more comprehensive differential diagnosis.

IMPACT OF ALZHEIMER'S

About 4–5 million people in the United States have some degree of dementia, and that number is growing quite rapidly due to our aging population and other emerging factors. The number is expected to nearly triple by 2050 when more than 13.2 million people will experience the frustration of memory loss and distortions in their personalities, according to a recent study in the journal *Archives of Neurology* [6]. Statistics on AD are provided in Appendix 4. Of the major diseases, Alzheimer's is the one that is increasing fastest at 60% over the past 5 years while diseases like cardiovascular disease have been on the decline over the same period. At the turn of the 19th century, when Dr. Alzheimer identified the specific type of "senility" that is named in his honor, dementia, and certainly AD, was quite rare. The dramatic increase in the number of Alzheimer's diagnoses suggests that this is not a genetic disorder. Genetic changes require many generations for their impacts to affect the large population now impacted by AD. This strongly suggests that environmental factors are likely at the root cause, rather than genetics.

Dementia affects about 1% of people aged 60–64 years and as many as 30%–50% of people older than 85 years. It is the leading reason for placing elderly people in institutions such as nursing homes because this disease, like none other, affects the ability of an individual to perform the basic actions necessary for survival. Thus many people with dementia eventually become totally dependent on others for their care. The sad aspect of dementia is, although people with disease typically remain fully conscious, the loss of short- and long-term memory is universal.

People with dementia often experience declines in any or all areas of intellectual functioning, for example, use of language and numbers; awareness of what is going on around him or her; judgment; and the ability to reason, solve problems, and think abstractly. These losses not only impair a person's ability to function independently but also have a negative impact on quality of life and relationships. Dr. Rudolph Tanzi of Harvard Medical School says that dementia and Alzheimer's rob sufferers of their personality.

Baby boomers may be in the toughest situation of all—that is, in the middle. Some are providing care to parents with Alzheimer's while at the same time trying to take care of their own families. And they wonder about the future. Most Alzheimer's patients' first caregivers are their spouses, who must deal with the person's wandering, confusion, violence, and sleepless nights. Patients often get to the point where they need care 24 hours a day. Placing a loved one in a care facility can be extremely difficult and plague those making the decision with guilt. But it is an act of necessity when primary caregivers, including family members, can often no longer maintain their own health and safety while ensuring the well-being of the person they love.

Even nursing care is feeling the pinch of Alzheimer's, with staff at care facilities stretched to their limit. Because of the difficulty of dealing with Alzheimer's patients, turnover is high among caregivers. The job of providing care is particularly difficult as it is often hard to figure out what the patients want or need. They cannot communicate well, often feeling afraid without knowing why. Additionally, many Alzheimer's patients develop violent behavior, likely triggered by their own unexpressed frustration, making care all the more difficult.

According to the Alzheimer's Association, half of all nursing home residents suffer from AD or a related dementia-type disorder. Many facilities have waiting lists and employ a limited number of licensed care providers. Alzheimer's is the disease that, for the first time, will cause our society to deal with the human and financial stress of an inverted pyramid as far as our workforce and our retired individuals. We will have many more elderly needy than we have resources to manage those complicated and continuing needs. After all, Alzheimer's patients are often in a state of high assisted and full-time need for as much as a decade. And the United States is not alone facing this problem. Many of the European nations, with socialized medicine systems, are in a significant dilemma when it comes to caring for their elderly, and particularly those afflicted with AD, as it is the most expensive of all the diseases of aging.

The most rigorous studies to date of how much it costs to care for Americans with dementia found that the financial burden is at least as high as that of heart disease or cancer, and is probably higher [7]. And both the costs and the number of people with dementia will more than double within 30 years, skyrocketing at a rate that rarely occurs with a chronic disease. A landmark report on the "Global Economic Impact of Dementia" finds that AD and other dementias are exacting a massive toll on the global economy, with the problem set to accelerate in coming years. Professor Anders Wimo of the Karolinska Institutet, Stockholm, Sweden, and Professor Martin Prince, Institute of Psychiatry, King's College London, jointly authored

"The World Alzheimer Report 2010"—issued on World Alzheimer's Day by Alzheimer's Disease International (ADI). It provides the most current and comprehensive global picture of the economic and social costs of the illness. According to the Chairman of the ADI, world governments are woefully unprepared for the social and economic disruptions this disease will cause.

"The World Alzheimer's Report 2010" urges the global community to take the following immediate actions:

- Governments worldwide should act urgently to make AD a top priority and develop national plans to deal with the social and health consequences of dementia. Several countries have moved forward to develop national plans, including France, Australia, and England. It is critical for other governments to follow suit.
- Governments and other major research funders must increase research funding to a level more proportionate to the economic burden of the condition. Recently published data from the United Kingdom suggest that a 15-fold increase is required to reach parity with research in heart disease, and a 30-fold increase to achieve parity with cancer research.
- Governments worldwide must develop policies and plans for long-term care that anticipate and address social and demographic trends and have an explicit focus on supporting family caregivers and ensuring social protection of vulnerable people with AD and other dementias.

The Alzheimer's Association published "Changing the trajectory of Alzheimer's disease: a national imperative" [8]. They detail the benefits of a humble goal: to delay the onset of AD by 5 years. With a 5-year delay, half of those who would have gotten Alzheimer's will pass for some other reason, and most likely they will pass in a much more compassionate, quicker, and less costly way. More importantly they will experience up to 5 more years of a quality life, enjoying grandchildren and the simple pleasure of full awareness. In addition, if the disease can be delayed for 5 years, a person's life savings may be spared and the Medicare/Medicaid system would save hundreds of thousands of dollars per person. (Four million sufferers multiplied by $100,000 per year equals $400 billion in savings annually.)

Many of us hope to pass on some financial help to our children but, due to the nature of medical coverage of long-term care, many of those with Alzheimer's likely die stripped of any savings. The cost of AD is not just tied up in care; Alzheimer's patients use nearly three times the amount of prescription drugs compared to a non-Alzheimer's patient, further burdening family, insurance, and the healthcare system. That's because Alzheimer's patients typically take medications not only for the disease but also for behavioral problems and other medical conditions that occur alongside of, or possibly because of, the disease.

Congress recently passed legislation such that a family must shelter assets at least 5 years before the expenses of Alzheimer's set in. It doesn't take long for a stay in a nursing home to drain the savings of a couple, even if they have substantial means. Once that happens, a person with Alzheimer's will likely qualify for care paid by Medicaid, a government program for the poor. Rules vary by state, but generally the healthy spouse can continue to live at home and is allowed to keep some assets that are not counted in determining Medicaid eligibility.

Not all aging memory disorders have the same impact as AD, and memory loss trajectory varies between AD cases. However, an aging person fears loss of memory, regardless of cause, because it threatens their freedom. They cannot find their glasses or remember someone's name, or otherwise have seemingly more frequent "senior moments." These very common problems are most often due to other processes not really related to dementia. Medical professionals call this "benign senescent forgetfulness," or "age-related memory loss." Although these types of conditions can be worrisome, they do not impair a person's ability to learn new information, solve problems, or carry out everyday activities, as dementia does.

The good news is that memory loss, even that assigned to AD, may be treatable and partially or fully reversible. We cannot stress enough that it is all in the rigor of the diagnosis. In the context of this book, we agree that AD is not treatable or reversible. But what if your diagnosis of Alzheimer's is incomplete or a misdiagnosis? We will spend a chapter exploring the frequency of misdiagnosis and medical mistakes in a later chapter. For now, though, be aware that you should not accept a diagnosis unless the attending medical professionals have exhaustively ruled out any number of potential causes discussed here.

One way to prevent overlooked diseases is to seek advice from more than one specialty. For example, do not necessarily get a second opinion from a second neurologist. Seek the advice of an internist, cardiologist, endocrinologist, and/or infectious disease specialist. Amalgamate the data, possibly with the help of your primary care provider, to create a clear picture of your situation. Will insurance cover all these providers? Unlikely! Are you worth it (i.e., is your health worth paying out of pocket for your own care in situations like this)? I hope your answer is yes.

TREATABLE DEMENTIAS AND "SENIOR MOMENTS"

Alzheimer's can sometimes be mistaken for common diseases that impact the brain. These afflictions are not the subject of this book as we focus on what is believed to be Alzheimer's. However, for example, up to 5% of Americans diagnosed with AD may actually have a treatable condition called "normal pressure hydrocephalus." We assume that if you or a loved one is suffering from diagnosed Alzheimer's, your medical team (PCP, internist, and neurologist) have ruled out other conditions that can look like AD.

It is important to be aware of the many possible misdiagnoses associated with AD. When you understand the scope and breadth of the possibilities, it helps support the specific ideas about cause/effect of Alzheimer's as presented in subsequent chapters. A simple Google search reveals many cases of people with presumed Alzheimer's who actually have some other affliction. However, some of the conditions cited here might constitute a triggering event for AD. This is quite plausible because many studies show clear evidence that Alzheimer's-type brain atrophy starts decades before a person has clinical symptoms of the disease.

Head Injury

This refers to brain damage from accidents and other forms of trauma. Depending upon the force, location, and other factors, recovery is clearly possible. Also your current states of health play a major role in recovery. Head injury creates acute (immediate) inflammation. People with more bodily inflammation at the time of injury are more likely to recover more slowly. We know that head trauma can catalyze dementia. Since dementias, and Alzheimer's-type dementia, are known to involve an inflammatory process, producing acute inflammation via trauma is logically something that can exacerbate the chronic inflammation associated with these diseases. Science is showing that acute and chronic inflammation of the brain triggers the same physiology, at least in part.

Acute Infections

Infections of brain structures, such as meningitis and encephalitis, are primary causes of dementia. Other infections, such as HIV/AIDS and syphilis, can affect the brain in later stages. In all cases, inflammation in the brain is part of a process that damages cells. Here is a case where the root cause is clearly identified: an infectious species. Many infections, even of the brain, have clear treatments. Again, since this is an inflammatory process, a well-cared-for body will be more resistant to both the infection and the negative impacts of the infection and resulting inflammation.

Normal Pressure Hydrocephalus

The brain floats in a clear fluid called cerebrospinal fluid. This fluid also fills internal spaces in the brain called cerebral ventricles. If too much fluid collects outside the brain, it causes hydrocephalus. This condition raises the fluid pressure inside the skull and compresses brain tissue from outside. It may cause severe damage and death. If fluid builds up in the ventricles, the fluid pressure remains normal ("normal pressure hydrocephalus"), but brain tissue is compressed from within. Many of the estimated 375,000 Americans who have NPH don't even know it, because its symptoms are strikingly similar to dementia, mental health specialists say. As a result, thousands of NPH patients never benefit from surgical procedures that can correct and reverse the condition.

Simple Hydrocephalus

Simple hydrocephalus may cause typical dementia symptoms or lead to coma. In normal pressure hydrocephalus, people have trouble walking and become incontinent (unable to control urination) at the same time they start to lose mental functions, such as memory. If normal pressure hydrocephalus is diagnosed early, putting in a shunt may lower the internal fluid pressure. This can stop the dementia, the gait problems, and the incontinence from getting worse.

Brain Tumors

Tumors can cause dementia symptoms in a number of ways. A tumor can press on structures such as the hypothalamus or the pituitary gland, which control hormone secretion. It can also press directly on brain cells, damaging them. Treating the tumor, either medically or surgically, can reverse the symptoms in some cases. The space around tumors is often inflamed. Consider the "environment" of your body so that it is resistant to some of the negative forces associated with tumors.

Toxic Exposure

People who work around solvents or heavy metal dust and fumes (mercury and lead especially) without adequate protective equipment may develop dementia from the damage these substances can cause to brain cells. Some exposures can be treated with chelating and other agents, and avoiding further exposure can prevent further damage.

Metabolic Disorders

Diseases of the liver, pancreas, or kidneys can lead to dementia by disrupting the balances of salts and other chemicals in the blood. Often, these changes occur rapidly and affect the person's level of consciousness. This is called delirium. Although the person with delirium, like the person with dementia, cannot think well or remember, treatment of the underlying disease may fully reverse the condition. If the underlying disease persists, however, brain cells may die, and the person may progress to a dementia. Diabetes is a type of metabolic disorder and is now believed to be caused by inflammation and the origins of inflammation. Is the profound connection between dementia, inflammatory disease, and other disorders that also involve inflammation becoming clearer? New research points to potentially reclassifying AD as type 3 Diabetes.

Hormone Disorders

Disorders of hormone-secreting and hormone-regulating organs such as the thyroid gland, the parathyroid glands, the pituitary gland, or the adrenal glands can lead to hormone imbalances, which can cause dementia if not corrected.

Poor Oxygenation (Hypoxia)

People who do not have enough oxygen in their blood may develop dementia because the blood brings oxygen to the brain cells, and brain cells need oxygen to live. The brain, although only 2% of body mass, consumes over 20% of all oxygen inhaled; therefore processes that impact oxygen transport and utilization also impact the brain significantly. The most common causes of hypoxia are lung diseases such as emphysema or pneumonia. These limit oxygen intake or transfer of oxygen from the airways of the lungs to the blood. Cigarette smoking is a frequent cause of emphysema. It can worsen hypoxic brain damage by damaging the lungs and also by increasing the levels of carbon monoxide in the blood. Heart disease leading to congestive heart failure may also lower the amount of oxygen in the blood. Sudden, severe hypoxia may also cause brain damage and symptoms of dementia. Sudden hypoxia may occur if someone is comatose or has to be resuscitated. Ischemia that impacts blood flow to the brain can also create symptoms akin to AD.

Blood Pressure

A blood pressure that is too low acts the same (is the same) as poor oxygenation. Many seniors are routinely put on blood pressure–lowering medication in an attempt to make their

BP "normal." An 80-year-old probably should not have the same BP as a 25-year-old because the older person's vessels are more occluded (constricted) and rigid. What regulates blood pressure in your body? Your brain! Why? It needs the oxygen for metabolism. The brain consumes 20% of the body's energy but is 2% of the body's mass. Therefore, look at the brain as the turbocharged part of your body. When it lacks sufficient oxygen for its high-octane needs, it triggers the vascular system to deliver more. How? It increases blood pressure. If your mother or father is on blood pressure–controlling medicines, ask your doctor(s) some uncomfortable questions. Many seniors on blood pressure–reducing medicine experience dizziness, especially when they first awake. They may get dizzy and fall, breaking a hip or other bone. This is usually the death knell for seniors and may be caused by the prescription drugs ordered by those attending to their care. This is also the time when they may experience repeated "senior moments."

Drug Reactions, Overuse, or Abuse

Some drugs can cause temporary problems with memory and concentration as side effects in elderly people. Misuse of prescription drugs over time, whether intentional or accidental, can cause dementia. The most common culprits are sleeping pills and tranquilizers. The FDA in their "Consumer Updates" website discuss reports of memory loss associated with statin use [9]. Other drugs that cause dry mouth, constipation, and sedation ("anticholinergic side effects") may cause dementia or dementia symptoms. Illegal drugs, especially cocaine (which affects circulation and may cause small strokes) and heroin (which is very anticholinergic), may also cause dementia, especially in high doses, if taken for long periods, or in older people. The withdrawal of the drug usually reverses the symptoms.

Nutritional Issues

On one side of the coin are deficiencies. Deficiencies of certain nutrients, especially B vitamins, can cause dementia if not corrected. On the other side are overdoses. Excess calcium, a chronic issue among modern women, may have a significant impact on Alzheimer's. This will be covered in more detail later. However, do note that women have higher incidences of AD than men. A quick Google search reveals many interesting articles on the connection between calcium and Alzheimer's. The original collection of works on this topic was published by the New York Academy of Sciences and is titled "The calcium hypothesis of dementia." Vitamin deficiency, excesses, and the lack of micronutrient homeostasis (vitamin and mineral balance) are, in our view, major preventable causes of dementia and AD. The solution is a proper diet. Much more on this topic is provided in a later chapter.

Sodium Deficiency

Salt is an important mineral that plays a role in maintaining fluid balance within cells. This mineral also helps support the health of nerve and muscle functions. Although most Americans consume excessive amounts of sodium, some people experience low levels of blood sodium, a condition known as hyponatremia. Older adults have a higher risk of developing hyponatremia. Conditions that can deplete the amount of sodium in your body include severe vomiting, diarrhea, kidney failure, heart failure, hypothyroidism, cirrhosis of the liver, and Addison's disease. Certain medications, such as antidepressants and diuretics, as well as some pain medications, may also lead to low sodium levels. Hyponatremia can cause drowsiness, fatigue, and confusion. This condition can also slow down your thinking and cause problems with memory.

Chronic Alcoholism

Dementia in people with chronic alcoholism is believed to result from other complications such as liver disease, cirrhosis, and nutritional deficiencies such as vitamin B1 (thiamine).

OTHER NEURODEGENERATIVE DISORDERS

Alzheimer's is clearly the most well known of the neurodegenerative disorders. Here are a list and a brief explanation of some of the other diseases that are similar to AD.

Vascular Dementia

This is the second most common cause of dementia, accounting for as many as 40% of cases. This dementia is caused by atherosclerosis, or "hardening of the arteries," in the brain. Deposits of fats, dead cells, and other debris form on the inside of arteries and partially (or completely) block blood flow. These blockages cause multiple strokes, or interruptions of blood flow, to the brain. Because this interruption of blood flow is also called "infarction," this type of dementia is sometimes called multiinfarct dementia. One subtype whose origin is not well understood is Binswanger's disease. Vascular dementia is related to high blood pressure, heart disease, diabetes, and related conditions. All these conditions point to chronic inflammation at the root. Treating those conditions can slow the progress of vascular dementia, but functions may not come back once they are lost, again, depending upon the extent of disruption to the vascular system.

Parkinson's Disease

People with this disease typically have limb stiffness, which causes shuffling when they walk, speech problems, and tremors (shaking at rest). Dementia may develop late in the disease, but not everyone with Parkinson's disease (PD) has dementia. Reasoning, memory, speech, and judgment are most likely to be affected.

Lewy Body Dementia

This is the disease my dad had based on autopsy. Abnormal microscopic deposits of protein, called Lewy bodies, which destroy nerve cells, cause this. These deposits can cause symptoms typical of PD, such as tremor and muscle rigidity, as well as dementia similar to that of AD. Lewy body dementia affects thinking, attention, and concentration more than memory and language, however, despite the diagnosis that was not the case for my dad. Like AD, Lewy body dementia is presumed irreversible and without cure.

Huntington's Disease

This apparently inherited disease causes wasting of certain types of brain cells that control movement as well as thinking. Dementia is common and occurs in the late stages of the disease. Personality changes are typical. Reasoning, memory, speech, and judgment may also be affected.

Creutzfeldt–Jakob Disease

This rare disease occurs most often in young and middle-aged adults. Infectious agents called prions invade and kill brain cells, leading to behavior changes and memory loss. The disease progresses rapidly and is fatal.

Pick's Disease (Frontotemporal Dementia)

This is another rare disorder that damages cells in the front part of the brain. Behavior and personality changes usually precede memory loss and language problems.

CONCLUSIONS

The premise of this book is that the diagnosis of Alzheimer's is inadequate. Through a differential diagnosis process, a more root cause diagnosis of AD is obtained. In some respects, without a differential diagnosis process, AD may be considered a misdiagnosis. However, with a broader and deeper diagnostic methodology compared to what is practiced universally by healthcare professionals today, treatable targets for presumed Alzheimer's may be identified and the disease path may be able to be changed. This book looks at potential root causes that are emerging but not well understood, and which are certainly not being diagnosed in the clinical environment anywhere today. Before delving into a detailed discussion about new theories, it is important to help you put a dagger into the heart of the prevailing conventional wisdom so that your mind is prepared to accept a new argument.

Remember:

> The seeds of great discovery are constantly floating around us, but they only take root in minds well prepared to receive them.

—*Joseph Henry*

References

[1] Dowling JE. The retina: an approachable part of the brain. Cambridge, MA: Harvard University Press; 1987.
[2] Chopra D, Tanzi RE. Super brain: Libera il potere esplosivo della tua mente per raggiungere salute, felicità e benessere. Segrate, MI: Sperling & Kupfer Editori; 2013.

[3] Hall SS. New clues to a long life. National Geographic Magazine, vol. 223, issue 5; 2013. p. 28-47. Available from: http://ngm.nationalgeographic.com/2013/05/longevity/hall-text

[4] Drug companies 'giving up' on Alzheimer's treatment after series of expensive failed trials. Mail Online; September 18, 2012. Available from: http://www.dailymail.co.uk/news/article-2205339/Leading-pharmaceutical-firms-giving-Alzheimers-treatment-series-expensive-failed-trials.html

[5] Pasteur L. On the extension of the germ theory to the etiology of certain common diseases. C R Acad Sci 1880:1033–44. Available from: https://ebooks.adelaide.edu.au/p/pasteur/louis/exgerm/ [Ernst HC, Trans., translated from French].

[6] Hebert LE, Scherr PA, Bienias JL, Bennett DA, Evans DA. Alzheimer disease in the US population: prevalence estimates using the 2000 census. Arch Neurol 2003;60(8):1119–22.

[7] Hurd MD, Martorell P, Delavande A, Mullen KJ, Langa KM. Monetary costs of dementia in the United States. N Engl J Med 2013;368:1326–34.

[8] Changing the trajectory of Alzheimer's disease: a national imperative. Chicago, IL: Alzheimer's Association; 2010. Available from: www.alz.org/documents_custom/alz_medicarecosts.pdf

[9] FDA expands advice on statin risks. US Food & Drug Administration; June 25, 2014. Available from: http://www.fda.gov/ForConsumers/ConsumerUpdates/ucm293330.htm#2

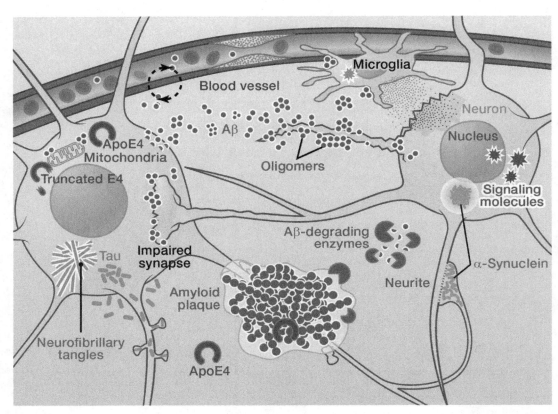

Image Credit: Multifactorial Basis of Alzheimer's Disease Pathogenesis, from Huang Y, Mucke L. Alzheimer Mechanisms and Therapeutic Strategies. Cell 2012;148(6):1204–22. Available at: http://www.cell.com/fulltext/S0092-8674(12)00278-4

The Amyloid Cascade Hypothesis

The Amyloid Cascade Hypothesis (or simply the Amyloid Hypothesis) continues to be the most widely espoused theory on Alzheimer's disease (AD). This hypothesis has driven the majority of the diagnosis and treatment research and medical trials on AD since its introduction in 1992 [1].

The goal of science, at its most fundamental level, is to construct hypotheses and thereafter to devise experiments that either disprove the hypothesis or, by not disproving it, provide additional support for the hypothesis. Therefore, this book provides information compiled by researchers from all corners of the globe that shows the Amyloid Cascade Hypothesis is one that is adequately disproven. Hopefully the information is provided objectively for you to reach your own conclusion. Our own government provides us with the impetus to continue to challenge this hypothesis because they continue to fund research in this area at very generous levels, at the sake of other worthy theories.

The amyloid hypothesis states that a buildup of deposits of beta-amyloid 1-42 protein (Ab, Aβ, β/A4, β-amyloid) is the fundamental cause of AD. It is a compelling theory because a gene associated with this form of amyloid is located on chromosome 21, and people with an extra copy of this gene (those with Down Syndrome) almost universally exhibit AD by 40 years of age.

Another genetic clue is the so-called APOE4. If a person has APOE4, it is considered a major genetic factor for AD, as it is often associated with excess amyloid buildup in the brain before AD symptoms arise. Thus, it is known and often true that beta-amyloid buildup precedes clinical AD. Further evidence comes from the finding that mice with a mutant form of the human genes associated with amyloid develop Alzheimer's-like brain pathology. Unfortunately mice have proven to be an extremely poor model for evaluating human AD.

Right out of the shoot came controversy surrounding the Amyloid Cascade Hypothesis. In 1993, Armstrong posited [2]:

> In Alzheimer's disease (AD), the '[Amyloid] Cascade Hypothesis' proposes that the formation of paired helical filaments (PHF) may be causally linked to the deposition of β/A4 protein. Hence, there should be a close spatial relationship between senile plaques and cellular neurofibrillary tangles in a local region of the brain. In tissue from six AD patients, plaques and tangles occurred in clusters and individual clusters were often regularly spaced along the cortical strip. However, the clusters of plaques and tangles were in phase in only four of thirty-two cortical tissues examined. Hence, the data were not consistent with the '[Amyloid] Cascade Hypothesis' that β/A4 and PHF are directly linked in AD.

The End of Alzheimer's. http://dx.doi.org/10.1016/B978-0-12-812112-2.00002-1

Evidence that potentially weakens the Amyloid Hypothesis is that an experimental vaccine was found to clear the amyloid plaques in early human trials, but it did not have any significant effect on dementia. And there are much more data that dispute an association between AD and amyloid as a cause. Much of the evidence regarding this hypothesis is considered here.

The amyloid hypothesis gained strength in part due to a correlation between amyloid load and loss of cognition. In mice, lowering amyloid load (through a variety of strategies) has been associated with improved cognitive function. However, there are some troubling aspects to the theory. Notably, there are many well-documented instances of autopsied brains that were full of amyloid plaques but came from the skulls of persons with high cognitive function at advanced ages. Perhaps this can be explained by timing. It is possible that the plaques build, after which there is a period of progressive brain damage due to the presence of the amyloid, and then cognition declines. This explanation is plausible and would explain why we sometimes find cognitively vital people with lots of amyloid in their brains.

New research from Saint Louis University suggests a more complex relationship between amyloid and the brain [3]. In a study on mice, researchers demonstrated that amyloid was associated with improved learning and memory, the exact opposite of what might have been expected. Keep in mind that this is a mouse study; thus extrapolation to humans is likely as misguided when the results are favorable as they are when they are unfavorable. The point made by the mouse studies at least suggests that viewing all amyloid as bad may be too simplistic. Perhaps having either too much or too little is the real problem, and therapeutic strategies should be refined to "regulate" amyloid as opposed to "eliminate" this important protein. This is how diabetes is managed today with regard to blood sugar regulation. The problem with this analogy is that diabetes is not a disease of sugar regulation. The root cause is associated with inflammation, and the consequences are elevated sugar. It may be that a truism is found between the sugar/diabetes and amyloid/Alzheimer's connection. Both treatment targets (sugar and amyloid) are spectators and consequences of a more profound underlying root cause. Each day we find another answer and pose another question to this complex puzzle but slowly our understanding of how to diagnose and treat the disease is becoming much more comprehensive and meaningful to patients.

Much presented in the remainder of this chapter is information taken directly and/or interpreted from the vast body of medical literature that casts strong doubt on the Amyloid Cascade Hypothesis. There is no dispute that amyloid exists; it clearly does. However, it is most likely not involved directly in the degeneration of brain neurons, although this may be secondary to its function. In this respect, it could trigger an autoimmune reaction of sorts. Emerging research shows Aβ has important protective characteristics and possible deleterious effects. Thus beta-amyloid may be both a foe and a friend.

Medical studies show beta-amyloid is not an appropriate target for therapy. This infers that something (or things) much deeper than amyloid is (are) the true cause (causes) of Alzheimer's. These other things are likely present in our body alongside of beta-amyloid. The amyloid actually provides us a clue to look further and deeper. So what are we left with? Aβ is a strong indicator that certain pathologies (disease processes), currently referred to as AD, are ongoing in a person's brain. Thus amyloid burden is a reasonably useful biomarker for the disease. Again, since AD is multifactorial, a single biomarker like amyloid does not make for an accurate diagnosis.

EVIDENCE AGAINST THE AMYLOID CASCADE HYPOTHESIS

The literature supporting the Amyloid Cascade Hypothesis is vast, while the literature contrary to the hypothesis is limited. Yet clinical trials are now proving to be the final nail in the coffin for this theory. In 2001, Professor Campbell of the University of California published a provocative paper titled "β-amyloid: friend or foe?" [4]. Ten years after its publication, I called Dr. Campbell to see how her amyloid research had progressed. I hadn't noted any new papers by her in the ensuing 10-year period. It took me a while to track her down at a new post in Utah. She is a gracious and professional woman. She explained to me that there was no follow-up research because there was no grant money for research opposing beta-amyloid. Whether or not you believe this (true) story, it does raise questions about the processes behind research funding.

Let's take a look at "β-amyloid: friend or foe?" by Dr. Campbell [4]. The summary of this remarkable paper is as follows:

> The function of the amyloid precursor protein (APP) and its product, b-amyloid, (Aβ) is at present unknown. The deposition of Aβ in senile plaques as well as meningeal and cerebral vessels has led many researchers to discount the possibility of a beneficial protective function for the protein. Thus it is generally believed that the aberrant processing of APP leads to increased b-amyloid secretion that in turn leads to subsequent plaque formation and Alzheimer's disease. Here, a hypothesis is presented that the protein may indeed be protective and that a potential role for beta-amyloid in innate immunity may exist.

Dr. Campbell makes several points as she develops a thesis that beta-amyloid is much more of a friend and guards against the adverse effects of some other root causes of AD. She states that cultured cells from the central nervous system produce beta-amyloid, and soluble forms of the protein are found in both normal cerebrospinal fluid (CSF) and CSF collected from patients with AD [5]. Furthermore, she states, "A concept of a role for beta-amyloid as a mediator of innate defense mechanism (that is, beta-amyloid is part of our immune system) of the cell is strengthened by the observation that in many circumstances beta-amyloid appears to act in a cytokine-like manner and promote inflammatory events." Cytokines are an important part of our immune defense system. "It is possible that this effect of the protein is protective and limits the entry of potentially harmful compounds and pathogens. Glial activation (brain immune system) and subsequent secretion of beta-amyloid … may be an integral part of an attempted defense mechanism of the central nervous system (CNS, including the brain, spinal cord, and eye) against pathogens."

Dr. Campbell goes on to suggest that formation of senile plaques (these contain the beta-amyloid) seems to be a component of natural aging when the integrity of blood vessels and possibly the barrier between the rest of the body and the brain (the blood–brain barrier) deteriorate. Her inference to microbial agents is quite interesting as Dr. Alzheimer hypothesized that microorganisms may be the cause of the disease that subsequently was named in his honor. She believes that the root cause of these processes is environmental toxins such as aluminum. It is plausible that such toxicity could lead to the proliferation of infectious species. Certain infectious species are thought to cause metal dysregulation so these two concepts may be working together to damage the brain.

Let's fast-forward from 2001 to 2010 and review the work of an esteemed Professor of Neurology from Harvard Medical School. Dr. Rudolph Tanzi holds the Joseph P. and Rose F. Kennedy

Professorship of Neurology, Harvard Medical School/Massachusetts General Hospital. It is hard to imagine a more prestigious title than that. He is a geneticist and has dedicated his life to disease of the brain in both the young and the old. Specific research that led to a 2010 publication discusses how the beta-amyloid protein of AD functions to protect, not harm, the brain. The title of this fine paper is "The Alzheimer's disease-associated amyloid b-protein is an antimicrobial peptide" [6]. Contributing authors were from Mass General Institute for Neurodegenerative Disease and Department of Neurology, Massachusetts General Hospital, Charlestown, MA; Department of Anatomy and Neurobiology, Boston University School of Medicine, Boston, MA; Department of Pathology, Beth Israel Deaconess Medical Center, Boston, MA; Department of Public Health/Geriatrics, Uppsala University, Uppsala, Sweden; and Boston University Alzheimer's Disease Center, Boston University, Boston, MA.

The "conclusion/significance" of this research is as follows:

> Our findings suggest Aβ is a hitherto unrecognized AMP (antimicrobial peptide) that may normally function in the innate immune system. This finding stands in stark contrast to current models of Aβ-mediated pathology and has important implications for ongoing and future AD treatment strategies.

What this conclusion means in simplified terms is that our own body produces Aβ like it does other components of our immune system. Furthermore, Aβ has a specific function, to attack hostile microorganisms. Dr. Tanzi et al. indicate that their work "stands in stark contrast to current models" that imply that Aβ is the fundamental cause of brain neuron degradation. This group from Harvard, BU, and Uppsala are diametrically opposed to the long-standing conventional wisdom about the causes of AD. Also, the Tanzi work nicely corroborates the hypothesis of Campbell published 9 years earlier.

This paper from the Harvard group is sufficiently important that we have called out some of the key findings as follows:

- "Aβ is generated in the brain and the peripheral tissue." This simple statement is extremely important to understanding AD. Aβ is clearly (but not always) connected to Alzheimer's. That Aβ appears in "peripheral" (nonbrain) tissue infers that the same processes that lead to the formation of Aβ in the brain are also occurring in these "peripheral" tissues. Simply put, the process(es) that involve brain decay associated with Aβ may also be occurring in these other tissues such as muscle tissue, eye tissue, and vascular tissue. AD is not a "brain-only" disease.
- "Aβ are similar to those of a group of biomolecules collectively known as 'antimicrobial peptides' (AMPs) which function in the innate immune system. AMPs (also called 'host defense peptides') are potent, broad-spectrum antibiotics that target Gram-negative and Gram-positive bacteria, mycobacteria, enveloped viruses, fungi, protozoans, and in some cases, transformed or cancerous host cells"; thus Aβ is produced by our bodies as natural antibiotics.
- "AMPs are also potent immunomodulators that mediate cytokine release and adaptive immune responses." Interpretation: Aβ is part of what the immune system produces to protect us from disease.
- "Here we show that Aβ is active against at least eight common and clinically relevant microorganisms." Apparently Dr. Tanzi and collaborators believe that microorganisms may play a role in the formation of Aβ and thus AD. Interestingly Dr. Alois Alzheimer

hypothesized in 1907 that microorganisms played a role in the disease named after him.

- "The in vitro (outside the body—an experiment done in a 'test tube') antimicrobial activity of Aβ matched, and in some cases, exceeded, that of LL37, an archetypical human AMP." Interpretation: Aβ of Alzheimer's is a strong or a stronger antibiotic compared to a well-known and well-characterized antimicrobial also produced by the body.
- "More recently, in a clinical trial of the Aβ lowering agent tarenflurbil, patients receiving the drug have significantly increased rates of infection." This too is an extremely important finding. This result takes us beyond the theoretical and into reality. Treatment of actual patients with a drug designed to lower the amount of Aβ in the body results in these patients becoming ill due to infection. Did they get AD? We don't know because it is well known that Alzheimer's progresses slowly and takes years to decades before clinical symptoms appear. This result does imply that infection may be involved in Alzheimer's, and strategies to attack (lower) Aβ in the body are likely to cause more, not less, disease.
- "Recent studies have shown that while the adaptive immune system has limited access to the brain, the central nervous system can still mount a robust response to invading pathogens via antimicrobial peptides and the innate immune system." This statement indicates that the antimicrobial peptides are extremely important for brain protection. We must now consider Aβ to be a very important antimicrobial peptide of the brain.

The researchers postulate that three mechanisms for Alzheimer's are reasonable based on the new finding that Aβ is antimicrobial. The one that makes the most sense based on an emerging understanding of Alzheimer's is the following:

> Persistent sub-acute CNS infection may drive chronic activation of the innate immune system. A number of studies have reported that the central nervous system (CNS) of AD patients is infected with pathogens including *Chlamydia* pneumonia [7], *Borrelia* spirochetes [8], *Helicobacter pylori* [9], and HSV [10]. Deposition of b-amyloid has also been reported for acquired immunodeficiency syndrome patients with brain HIV infection [11]. Consistent with the increase risk of AD associated with the e4 variant of the apolipoprotein E gene [12], carriers of the e4 allele are reported to have higher rates of CNS infection for several of these pathogens [13] ….

Simply put, this mechanism proposed by the Harvard team further asserts that infection may play a key role in AD, and the Aβ plays a key role in regulating infection.

Dr. Tanzi is certainly considered an insider in medical research. Interestingly, the grapevine disclosed that the publication of this paper was extremely controversial. Is it because it is contrary to the prevailing theory? It is more likely that the controversy stemmed from the fact that Amyloid Cascade Hypothesis, dying as it is, continues to generate millions of dollars for researchers all over the world.

Dr. Craig Atwood, a former colleague of Dr. Tanzi at Harvard and now at the University of Wisconsin at Madison, is also an expert on beta-amyloid. He, too, offers a view that this misfolded protein is not at the root of Alzheimer's. Dr. Atwood offers his proof through a variety of research papers, often with very interesting titles. Some of those are captured here:

- Amyloid-β: a vascular sealant that protects against hemorrhage? [14]
- Amyloid-β: a chameleon walking in two worlds: a review of the trophic and toxic properties of amyloid-β [15]

- The role of beta-amyloid in Alzheimer's disease: still a cause of everything or the only one who got caught? [16]
- The state versus amyloid-β: the trial of the most wanted criminal in Alzheimer's disease [17]
- Amyloid-β, tau alterations and mitochondrial dysfunction in Alzheimer's disease: the chickens or the eggs? [18]

Finally, a personal favorite title that is all about AD even though it is not obvious through the provocative title:

- Living and dying for sex [19]

Who said science and medicine are boring?

Dr. Atwood did not corner the market with creative titles. A group from Italy wrote a chapter titled, "Amyloid-β peptide: Dr. Jekyll or Mr. Hyde?" [20]. The content of their abstract includes the following thoughts:

> Amyloid-β Peptide is considered a key protein in the pathogenesis of Alzheimer's disease because of its neurotoxicity and capacity to form characteristic insoluble deposits known as senile plaques … Amyloid-β peptide has been widely considered as a 'garbage' fragment that becomes toxic when it accumulates in the brain, resulting in impaired synaptic function and memory. Amyloid-β peptide is produced and released physiologically in the healthy brain during neuronal activity. In the last 10 years, we have been investigating whether Amyloid-β peptide plays a physiological role in the brain. We first demonstrated that picomolar concentrations (very low levels) of a human Amyloid-β peptide (type 42) preparation enhanced synaptic plasticity and memory in mice. Next, we investigated the role of endogenous Amyloid-β peptide in healthy murine brains and found that treatment with a specific antirodent Amyloid-β peptide antibody and a siRNA against murine Amyloid-β peptide protein precursor impaired synaptic plasticity and memory. The concurrent addition of human Amyloid-β peptide rescued these deficits, suggesting that in the healthy brain, physiological Amyloid-β peptide concentrations are necessary for normal synaptic plasticity and memory to occur.

Here is some more food for thought, highlighted by a paper from *The American Journal of Pathology* that has yet another interesting title, "Amyloid-β vaccination: testing the amyloid hypothesis? Heads we win, tails you lose!" [21]. The authors of this 2006 research state:

> In the field of Alzheimer's disease (AD), the predominant hypothesis is the Amyloid Cascade Hypothesis, the original version of which posited that insoluble fibrillar amyloid-beta is central to disease pathogenesis [1]. In support of this hypothesis, Aβ fibrils have been found to be toxic in vitro (outside of the body, in a test tube) [22] and, since then, considerable effort has been and continues to be made on developing therapeutic modalities that target Aβ fibrils.
>
> [T]he original Amyloid Cascade Hypothesis underwent a slight modification in which the emphasis switched to oligomeric, rather than fibrillar, forms of Aβ [23]. Today, oligomeric Aβ is viewed, almost universally in the field, as the most toxic and, therefore, the most important species [24].
>
> While the two mainline amyloid theories consider amyloid as toxic, a second alternate role for Aβ in AD must also be considered. We predicted originally that vaccination strategies were likely to fail [25,26], not because of an increase in oligomeric amyloid, but because we suspect that Aβ is a protective consequence of the disease, not a cause of disease [27,28]. Our Alternate Amyloid Hypothesis posits that Aβ serves as a protective antioxidant and that its removal will exacerbate, rather than treat, disease [29].

Here is a summary from a 2011 research article titled "The pathogenesis of Alzheimer's disease: a reevaluation of the 'amyloid cascade hypothesis'" [30]. The same author who wrote against the theory back in 1993 writes this [2]. This is a very important paper because it addresses the interrelationship between the two "hallmarks" AD first presented by Dr. Alois Alzheimer in the early 20th century. These are the senile plaques, now known to contain beta-amyloid and neurofibrillary tangles. Dr. Alzheimer was sure that the senile plaques were not the cause of the disease. Interestingly, of the two original cases described by Alzheimer's, both had numerous senile plaques but only one of the cases had significant numbers of neurofibrillary tangles [31].

> The most influential theory to explain the pathogenesis of Alzheimer's disease (AD) has been the "Amyloid Cascade Hypothesis" (ACH) that was first formulated in 1992. The ACH proposes that the deposition of β-amyloid (Aβ) is the initial pathological event in AD leading to the formation of senile plaques (SPs) and then to neurofibrillary tangles (NFTs), death of neurons, and ultimately dementia. This paper examines two questions regarding the ACH: (1) is there a relationship between the pathogenesis of SPs and NFTs, and (2) what is the relationship of these lesions to disease pathogenesis? …. It was concluded that senile plaques and neurofibrillary tangles develop independently and may be the products rather than the causes of neurodegeneration in AD. A modification to the ACH is proposed which may better explain the pathogenesis of AD, especially of late-onset cases of the disease.

Note that "late onset," also known as sporadic Alzheimer's, is the most common form of the disease accounting for more than 98% of all cases. The early onset version of Alzheimer's is called "familial AD," and has a very strong genetic component compared to the sporadic form of the disease.

The authors propose a "Modification of the ACH," and this modification is summarized in the subsequent text. They appear to be substantially correct, based on new information coming out of drug trials in humans, but do not go far enough in describing the true root causes of AD and the real reason for the formation of the senile plaques (Aβ) and neurofibrillary tangles (NFTs). These true root causes are ferreted out by the research of Dr. Tanzi [6], but primarily by the work of many others that will be presented in later chapters. What is being discovered is that the true cause of Alzheimer's is accelerated, age-related immune deterioration (immunosenescence) through a range of mechanisms followed by propagation of bacteria that are able to enter the brain and create sustained, chronic neuroinflammation.

> In this modified hypothesis, the essential trigger to the development of AD is ageing of the brain and associated risk factors such as head trauma, vascular disease, and systemic disease, collectively referred to as the "allostatic load [32]." In this modified hypothesis, genetic factors, rather than initiating disease, indirectly influence the formation and composition of peptides formed when neurons degenerate [33].

The authors make some valuable points that you will see as recurring themes of this book. They mention an aging brain. Well, the oldest persons known, a woman from Scandinavia and one from England, recently died at the age of 115. They had essentially full mental faculties until the time of their death. In fact, many people maintain their cognitive abilities well into their 80s and beyond. The authors might have stated their hypothesis slightly differently by saying "the development of AD is an abnormal aging of the brain." They do state associated risk factors such as head trauma. Most importantly they indicate that AD is likely a vascular disease and a systemic disease. Both of these points are extremely important and will be discussed in much greater detail in subsequent chapters.

The concept of "systemic disease" is critical to all sufferers of neurodegenerative diseases because the current standard-of-care almost exclusively focuses on the brain and ignores the other body "systems." Many researchers are making the case that the genesis of the disease stems substantially (but not exclusively) from the vascular system. However, to fully understand the development of AD, you will have to begin to adopt a new and emerging set of root causes of vascular diseases that does not focus on cholesterol.

The evidence presented here, so far, is certainly compelling to say that the Amyloid Cascade Hypothesis, as originally stated and as widely accepted, is wrong. Not only is the Amyloid Cascade Theory wrong, it is potentially 180 degrees off base. However, keep in mind that several of the largest companies in the world, with multimillion dollar research departments and an army of PhDs, MDs, and consulting professors from the top universities in the world, press on by trying to develop drugs to remove beta-amyloid as a treatment or cure for AD. They clearly believe that this substance is deleterious to the brain.

RESEARCH OPPOSED TO THE AMYLOID CASCADE HYPOTHESIS

Sadly, despite much compelling information about beta-amyloid, most of the drugs being developed to treat AD are those that reduce beta-amyloid. Does that sound like the right strategy based on what you have read? Don't give up hope; there is a small faction of researchers pursuing nonamyloid approaches to Alzheimer's therapy. However, since there is such a strong ingrained belief system in favor of the Amyloid Cascade Hypothesis, it is important to support the case against such a strong idea with as much information as possible. Here are a few more papers that infer that the hypothesis is based on a flawed interpretation of the readily available science. A little later in this chapter, data from drug company trials, testing the hypothesis in real patients, put the nail in the coffin.

- 2007: "Amyloid-beta in Alzheimer's disease: the null versus the alternate hypothesis" [34]. Researchers from the University of Maryland, Asahikawa Medical College of Japan, The University of Texas at San Antonio, and Case Western Reserve University compiled this article. The authors state, "An increasingly vocal group of investigators are arriving at an 'alternate hypothesis,' stating that Aβ, while certainly involved in the disease, is not an initiating event but rather is secondary to other pathogenic events. Furthermore and perhaps most contrary to current thinking, the alternate hypothesis proposes that the role of Aβ is not as a harbinger of death but rather a protective response to neuronal insult."
- 2008: "Amyloid cascade hypothesis: is it true for sporadic Alzheimer's disease" [35]. This paper came from the Laboratory of Molecular Neuropharmacology, Department of Pharmacology, Medical School & Croatian Institute for Brain Research, University of Zagreb, Croatia. "These findings suggest that development of insulin resistant brain state precedes and triggers Aβ pathology in sporadic AD, challenging thus the Amyloid Cascade Hypothesis when sporadic AD is concerned." This is a very interesting finding, that insulin resistance, in other words, diabetes-type pathology in the brain, triggers AD. Indeed many researchers are espousing a new definition of Alzheimer's by calling it type 3 diabetes. This connection between diabetes and AD is explored later.
- 2012: "Alzheimer's disease and the amyloid cascade hypothesis: a critical review" [36]. Published by a researcher at Columbia University, this author did not make a

call to completely throw the Aβ "baby" out with the bathwater. She did, however, make a couple of important points. "Randomized clinical trials that tested drugs or antibodies targeting components of the amyloid pathway have been inconclusive." That is a nice, polite way of saying they failed. She goes on to say, "Finally, it makes sense to pursue other targets beyond Aβ as there is substantial evidence for additional potential pathways increasing disease susceptibility, among these lipid metabolism and inflammatory processes" [37]. Thus the author is not really hedging her bets and suggests that we move beyond the current approach to treat AD. Note that this paper has 107 references! Thus we are presenting only a small sampling of the available scientific and medical literature opposed to the current belief structure on beta-amyloid and AD.

- 2013: "Amyloid fibrils composed of hexameric peptides attenuate neuroinflammation" [38]. This paper was published by researchers at Stanford Medical School. "The amyloid-forming proteins tau, αB crystallin, and amyloid P protein are all found in lesions of multiple sclerosis (MS). Our previous work established that amyloidogenic peptides from the small heat shock protein αB crystallin (HspB5) and from amyloid β fibrils, characteristic of AD, were therapeutic … reflecting aspects of the pathology of MS. To understand the molecular basis for the therapeutic effect, we showed … amyloid β A4 … to be anti-inflammatory and capable of reducing serological levels of interleukin-6 (this tracks almost exactly with C-reactive protein, the most general marker for both acute and chronic inflammation in the body) and attenuating paralysis in EAE (Endotoxin Autoimmune Encephalitis) … Amyloid fibrils thus may provide benefit in MS and other neuroinflammatory disorders." These "other" neuroinflammatory disorders include AD.
- 1998: "Alzheimer's disease: a re-examination of the amyloid hypothesis" [39]. Harvard Medical School and Mount Sinai School of Medicine in New York researchers published this article. "This (the Amyloid Hypothesis) is a controversial theory, however, primarily because there is a poor correlation between the concentrations and distribution of amyloid depositions in the brain and several parameters of AD pathology, including degree of dementia, loss of synapses, loss of neurons and abnormalities of the cytoskeleton." The authors go on to state, "AD probably is a multifactorial disease that should be approached from many perspectives." If only we would.
- 2006: "Amyloid b and neuromelanin—toxic or protective molecules? The cellular context makes the difference" [40]. Columbia University researchers and collaborators from India and Italy contributed to this article. These authors connect Parkinson's disease (PD) with AD. The authors state that "neuromelanin (NM), whose role in PD is emerging …" has properties similar to Aβ. They conclude: "A careful analysis of these parallel effects of Aβ and NM, including their seemingly paradoxical ability to participate in both cell death and protection, may lead to an improved understanding of the roles of these molecules in neurodegeneration and also provide insights into possible parallels in the pathological mechanisms underlying AD and PD." Thus an effective treatment for AD may also be an effective treatment for PD. There is little doubt, based on emerging research, that these two diseases are connected at some basic root cause or causes.

- 2012: "Aberrant action of amyloidogenic host defense peptides: a new paradigm to investigate neurodegenerative disorders?" [41]. The British are leaders in many facets of Alzheimer's research. Here these Brits confirm the work at Harvard that describes beta-amyloid as an antimicrobial peptide. "Host defense peptides (HDPs) are components of the innate immune system with activity against a broad range of microbes. In some cases, it appears that this activity is mediated by the ability of these peptides to permeabilize microbial membranes via the formation of amyloid associated structures. Recent evidence suggests that the naturally occurring function of the Aβ, which are causative agents of Alzheimer's disease, may be to serve as amyloidogenic HDPs."

Thus we see another elegant piece of medical research that associates Aβ with antibiotic properties. Could microbes (bacteria, fungi, and/or viruses) be at the root of AD? And if so, what triggers their growth?

In 2004, a research team at Case Western Reserve University published a paper titled "Challenging the amyloid cascade hypothesis" [42]. These researchers predicted the future (which is now our present) very well, as you will soon see. Here are some highlights from their paper:

> Ever since their initial description over a century ago, senile plaques and their major protein component, amyloid-β, have been considered key contributors to the pathogenesis of Alzheimer's disease. However, counter to the popular view that amyloid-β represents an initiator of disease pathogenesis, we herein challenge dogma and propose that amyloid-β occurs secondary to neuronal stress and, rather than causing cell death, functions as a protective adaptation to the disease. ... Although controversial, a protective function for amyloid-β is supported by all of the available literature to date and also explains why many aged individuals, despite the presence of high numbers of senile plaques, show little or no cognitive decline. [...] Our arguments supporting Aβ as a crucial antioxidant defense mechanism are extremely relevant to current pharmacological efforts targeted at either removing Aβ or lessening Aβ production.
>
> Removing amyloid will likely leave neurons without one of their fundamental compensatory responses to aging and disease and, therefore, we would expect that current pharmacological strategies to lower amyloid levels will actually serve to worsen the disease.

Read on and you will see that this is exactly what happened about 6 years later.

In 2002, Case Western Reserve University published a paper titled "Predicting the failure of amyloid-beta vaccine" [43], by Craig S. Atwood and colleagues. Their correspondence to *The Lancet* is reproduced in its entirety:

> Sir—After the news of the (anti-amyloid) trial suspension, the rush to test a vaccine for Alzheimer's disease has proven ill-fated. Indeed, although it is no surprise that the inappropriate deposition of protein in the normal mouse brain because of massive overexpression of Aβ protein precursor modifies function, nor that its removal can then restore function, there is not, nor ever was, any evidence that interventions designed to remove or alter the deposition of Aβ would benefit patients with Alzheimer's disease.
>
> Treatment strategies based on removal of a naturally occurring endogenous (inside the body) substance harkens back to the time of leeches and exorcism for the removal of bad humour and spirits to restore function. Thankfully, such concepts died with the understanding of homoeostatic balance that defines modern biology. Or did they?
>
> Therapeutic strategies arguing for the removal of Aβ beg the question of why deposits of this substance develop with age in the first place. Amyloid develops in many long-lived mammalian species and, in human beings, most people older than 40 years have amyloid in their brain. Removal of this Aβ from the aged or diseased brain, we argue, is more likely to destabilize age-related or disease related compensations and be harmful. Unfortunately for patients who have Alzheimer's disease, this scenario seems true.

DRUG COMPANIES COMMITTED TO THE AMYLOID HYPOTHESIS

From the previous section you see that there is a significant amount of research that suggests Aβ is not the appropriate target for the treatment of AD. That is, Aβ is not the root cause of the disease, nor is it even in the "cascade" of events that is detrimental to brain neurons and thus the pathogenesis of the disease. To be fair, the large body of data that supports the Amyloid Cascade Hypothesis is not presented here. Feel free to search the literature yourself. Searching Google clearly directs you to plenty of sites and articles that argue for and against any given theory.

How, then, can a scientific bias become resolved, assuming that there are data to support opposing views? The best way to remove all doubt is to look at the "translation" of the research. That is, what happens when the relatively theoretical ideas are then put into clinical practice with real people, real patients? We, in the United States, have a very well-established mechanism for testing medical theory in a real setting: the FDA drug approval process.

The mission of the FDA, specific to drug development, is to ensure that those entities proposing new drugs for human use are offering such substances that are both safe and effective. The process of "promoting" a new drug entity from theory to the marketplace now costs roughly $1 billion and takes 8–10 years of rigorous testing and evaluation. And a drug can fail due to safety or efficacy at any stage of the trial, including the final stage after the $1 billon dollar investment. The very nature of the drug approval process makes "changing course" very difficult for the big drug companies. A company cannot invest 5 or more years on a drug concept and spend half a billion dollars and then just willy-nilly switch to a new concept unless the evidence is so overwhelming in support of the new concept and absolutely damning to the concept being pursued. In other words, once a pharmaceutical company enters a drug into the FDA approval process, it must be fully committed for the duration of the trial, unless it becomes an abject failure somewhere along the process. So abandoning a drug in FDA trials in favor of a different drug is like changing the course of the Titanic when it is at full throttle. As you read about the results of drug trials on testing the Amyloid Cascade Hypothesis, you are likely to conclude that the Titanic has already hit the iceberg.

MedPage Today is very topical and informative (http://www.medpagetoday.com). Their mission is to put breaking medical news into practice. Toward the end of 2012 they published an article titled "Alzheimer's disease: amyloid 'proponents' soldier on" [44]. In this article, MedPage Today reviewed the outcomes of several antiamyloid drug development programs sponsored by the largest pharmaceutical companies in the world. The authors stated, "Here's what's happened with drug development for AD since we published the first 2012 piece on what appeared to be the demise of beta-amyloid as a drug target for symptomatic disease." In the next subsections is a brief review of what happened to the Amyloid Cascade Hypothesis drug clinical trials up through 2012.

Eli Lilly

Eli Lilly announced that both trials of its antiamyloid (monoclonal antibody) drug solanezumab had failed to show a significant benefit when tested on patients showing clear signs of cognitive impairment. However, the MedPage article infers that all onlookers realized that solanezumab was a failure but Lilly itself refused to accept their own data. The company still believes that solanezumab has a future in treating symptomatic AD but only (or maybe) if the therapy is applied earlier in the development of the disease.

The Eli Lilly Amyloid Cascade Hypothesis drug trial team is "hanging on to the life raft" but they are in 28° water.

Lilly justified their decision to move forward by claiming that "a pre-specified secondary analysis of pooled data across both trials showed statistically significant slowing of cognitive decline in the overall study population of patients with mild-to-moderate AD. [...] In addition, pre-specified secondary subgroup analyses of pooled data across both studies showed a statistically significant slowing of cognitive decline in patients with mild AD, but not in patients with moderate AD" [45]. It is quite clear that the impact of their drug is marginal at best.

Lilly continues to pursue their drug development program based on the Amyloid Cascade Hypothesis because of their mammoth investment in both time and money, not because of the results of the tests. Even though the science presented so far appears to oppose the success of their compound, the height of the "bar" established to measure the success of a drug to treat Alzheimer's is so low that anything that shows any level of impact on the disease may get approved. This is likely Lilly's angle. The sad part for Alzheimer's patients of the world is the concept of "opportunity cost." That is, Lilly is making a tremendous effort on a dubious approach, and that effort is then being taken away from methods and drug candidates that have a much higher likelihood of success and thus a better chance to help people. It is doubtful that solanezumab will really curb the misery of AD. But Lilly "soldiers on."

Johnson & Johnson and Pfizer

Their monoclonal antibody drug, bapineuzumab, had also shown no clinical benefit in a large trial. The companies announced that one group of a Phase II clinical trial of bapineuzumab failed to show that the drug slowed the progression of the disease. Irish pharmaceutical manufacturer Elan Corporation, in conjunction with Wyeth, developed bapineuzumab. Elan Corporation still retains a significant ownership interest in the drug [46].

A Wall Street analyst wrote, "We interpret yesterday's news as a definitive end to bapineuzumab, and any glimmer of hope is all but extinguished" [47]. "What's really surprising about bapineuzumab's failure is that expectations were as high as they were. And what's shocking is that Johnson & Johnson spent more than a $1 billion to invest in Elan to get one-quarter of the drug, and that Pfizer (or, rather, Wyeth, which Pfizer bought) and Elan chose to push the drug into broad clinical trials despite a single, uncomfortable fact that Bapineuzumab failed in earlier studies, and Wyeth and Elan decided to plunge forward anyway" [47]. The *Forbes* article continues by explaining the motivation behind the drug companies' failed efforts [47]. "The logic behind going forward probably went something like this: Alzheimer's is one of the world's biggest health problems and any drug that can impact it would be simply huge. Even if bapi (bapineuzumab) were not that effective, it could have generated $5 billion in annual sales, easy. It would be crazy not to try, right?" [47].

Bristol-Myers Squibb

Many news outlets presented Bristol-Myers's failed trail on avagacestat as "Another Alzheimer's drug hits the dust." This drug entered into FDA clinical trials with much promise (except for the fact that its target was beta-amyloid). Avagacestat is a γ-Secretase inhibitor (GSI) with potential to modify the progression of AD based on its ability to regulate amyloid-β (Aβ) accumulation. BMS-708163 (avagacestat) is an oral GSI designed for selective inhibition of Aβ synthesis currently in development for the treatment of mild to moderate and predementia AD. Avagacestat produced up to 190-fold greater selectivity for Aβ synthesis in preclinical studies and was expected to produce less toxic adverse events than other less selective compounds.

Mission accomplished, as avagacestat trials were not halted due to toxicity. They were halted because stopping beta-amyloid doesn't work. Bristol-Myers Squibb confirmed that conclusion in a 2013 announcement at the Alzheimer's Association International Conference in Paris. Study patients in a midstage trial showed hints of "negative cognitive effects." Dr. Craig Atwood was right; patients get worse on antiamyloid therapy. Is it possible to conclude, then, that beta-amyloid is protective if reducing it causes patients with beta-amyloid and Alzheimer's to get worse?

Surprisingly (maybe not surprisingly), Bristol-Myers says it will continue the trials but just at the lower drug dose. A company official said, "Reducing the maximum dose is expected to limit potential safety or tolerability issues while still testing a dose range with potential for clinical efficacy." Trying different doses is called a ladder study. It is likely, based on their data trends, that a dose of zero will provide the most benefit.

Corporate PR departments carefully select the language of clinical trial announcements. Is it plausible to conclude that these trial therapies actually made people become sicker faster compared to placebo? The lack of any positive language infers a negative result. This type of information is important because it could be used to contribute to the argument that Aβ is protective. Interestingly, the Bristol-Myers folks say, "Amyloid remains an important target for Alzheimer's research and B-MS continues to test the amyloid hypothesis with an investigational gamma secretase modulator in Phase I development."

Pfizer and Medivation

Their drug Dimebon did not show improvement in patients, and the project was shut down in early 2012. As a result Pfizer wrote off $725 million dollars after the final Phase III failure. The last flickering hope that Dimebon could help AD patients was thus extinguished. The biotech company announced that a 12-month study of the drug failed to register significant improvements for patients, mirroring two shorter Phase III studies in which Dimebon failed to outperform a sugar pill (it probably underperformed the sugar pill but we do not have full access to the drug company's results). Dimebon, also known as Latrepirdine, was thought to operate through multiple mechanisms of action, both blocking the action of neurotoxic beta-amyloid proteins and inhibiting L-type calcium channels.

Merck

They recently announced that it was taking MK-8931 into a large trial with patients with symptomatic disease. This is an oral inhibitor of beta-secretase or BACE, one of the enzymes (the other principal one is gamma-secretase) that cleaves beta-amyloid protein from a larger precursor molecule. By blocking this enzyme, beta-amyloid production should be greatly diminished. As with solanezumab, the idea is that formation of insoluble plaques will be diminished as well. A Phase I study showed that MK-8931 reduced beta-amyloid protein levels in cerebrospinal fluid by more than 90% in healthy individuals. The world anxiously awaits the results of this study.

Another gamma-secretase drug, this one by Lilly, was proven ineffective in 2013. The drug, called semagacestat, was designed to block an enzyme called gamma-secretase that makes beta-amyloid. Animal studies and early human trials had suggested that the drug did what it was designed to do, but in a test of the medication in more than 1500 patients with mild to moderate Alzheimer's, those taking semagacestat actually declined faster on thinking tests than those who took a dummy pill. And those on the drug experienced more serious side effects, including skin cancers and infections. The study results were published in the July 25, 2013, issue of *The New England Journal of Medicine* [48]. Dr. Craig Atwood should be the highest paid Alzheimer's consultant because he predicted all these results and could have saved drug companies billions of dollars and shortcut the path to a true treatment.

> "You've got a very clear look at some not good results here," said study author Dr. Rachelle Doody, a neurologist with the Baylor College of Medicine in Houston speaking to Healthday News. "It clearly leads to the conclusion that targeting gamma secretase as a way to reduce amyloid simply doesn't work," said Steven Ferris, director of the Alzheimer's Disease Center at NYU's Langone Medical Center in New York City. "It doesn't appear that this is a promising target for treatment."

Researchers hope that studying why drugs like this fail may help drug manufacturers learn from their mistakes and pave the way for future successes. Eli Lilly stopped the trial early in 2010 after preliminary analysis of the data suggested safety problems with the medication. The company then turned over all its study data to the Alzheimer's Disease Cooperative Study, a project of the US National Institutes of Health, for independent analysis.

> "In the Alzheimer's field, that's unprecedented," Doody said. "There's never been an industry group that's turned over their data and said, 'We're giving up our rights to publish the data.' Those on the drug did worse, and that's a very important thing to get out to the public," she said.

Smaller pharmaceuticals are also pursuing the beta-amyloid hypothesis.

> "Kareus Therapeutics Announces Phase I Trial for Product in Alzheimer's Disease." Kareus has developed a pipeline of novel molecules targeting diseases of the central nervous system based on its proprietary KARLECT chemistry and drug discovery platforms that target dysfunctional energy production in neurons. KU-046 targets bioenergetics pathways upstream from the increased Aβ (beta-amyloid) peptide production found in Alzheimer's disease. It has demonstrated significant improvement in cognition in a number of preclinical models (these studies involve animal models only).

Today, even staunch advocates of the amyloid hypothesis in AD are finally changing their thinking. However, some still cling to the notion that maybe some facet of beta-amyloid has been overlooked and deserves revisiting. Some top scientists are suggesting that, by the time symptoms appear, beta-amyloid has already done its damage, and that is the reason for the drug failures. Instead, they advocate antiamyloid drugs would be effective only if introduced much earlier in the disease process, before plaques have become extensive and before neurodegeneration has really taken hold. That might be a plausible claim if the clinical trials yielded the slightest positive outcomes.

Our government, not deterred by the abysmal track record of beta-amyloid–based drug development, continues to put money into this approach. Here is a headline and story from 2013 [49]:

Funding approved for Alzheimer's trial.

"Long-awaited federal funding has been approved for a first-of-its-kind, Boston-led study to test whether drugs can hold off Alzheimer's disease in people who have no symptoms of the illness, but who have an abnormal protein in their brain believed to be a marker of the disease." The National Institutes of Health announced that the clinical trial, to be held at Brigham and Women's Hospital, is one of four that will be funded this year to find treatments for the disease.

Brigham's will receive the lion's share of the money, roughly $36 million. The federal biomedical research agency said it will give the four trials a total of $11 million this year and as much as $55 million over five years. The study will give half of the participants a drug designed to clear amyloid plaques and the others a placebo, and researchers will track the rate of cognitive decline in both groups.

This will be a very long study because the subjects do not have the disease at the outset of the study. The head researcher said that if the researcher's hypothesis is right, the group that receives the amyloid-clearing medications would have a 25%–35% slower rate of decline than those receiving a placebo. This appears to be extremely optimistic.

"Advocates, who have been frustrated by the pace of funding for Alzheimer's studies hailed the announcement." "This is a message that the government is really serious about coming up with meaningful treatment for this very tough disease that is overwhelming society," said James Wessler, chief executive of the Alzheimer's Association of Massachusetts and New Hampshire Chapter.

"The NIH money is just a piece of the needed funding, but the researcher expects to hear soon on funding from industry and charitable foundations to finalize the support for the $140 million study."

What about the other three studies that NIH plans on funding?

The other three NIH-funded studies include one that will test whether exercise—often widely recommended to maintain physical function and reduce age-related declines—is effective in slowing further cognitive losses and brain atrophy in people with mild cognitive impairment, a condition that often leads to Alzheimer's disease. Exercise has not been shown in a longer-term clinical trial to improve cognition or alter the course of Alzheimer's disease.

THE ANIMAL MODEL FOR ALZHEIMER'S AND BETA-AMYLOID THERAPY

Models for AD in animals are where researchers obtained hope for the presumed efficacy of their beta-amyloid–reducing strategies. Before a drug is tested in humans, extensive animal testing is required, with a major emphasis on safety. This testing is done in the phase called "pre-IND," where IND stands for Investigational New Drug. Are there animal models for AD? It is widely held that AD incubates over a long period of time—up to 25 years. Thus, is it possible to simulate such a human disease in a mouse that lives but a few years? Arguably not, as drug companies and contract research organizations are required to perform animal testing and thus develop animal models for even long incubating chronic diseases such as AD. It is simply a mandatory requirement of the FDA approval process.

Evaluating the concept of animal testing of potential human drugs is well beyond our scope. However, please consider the following excerpt from a communication by a thought leader, Patrick McGeer, on the association between AD and inflammation. The article is titled "Amyloid-beta vaccination for Alzheimer's dementia," from 2008 [50]. He makes the case that the amyloid hypothesis is flawed science and that the mouse model(s) for Alzheimer's is part of the reason researchers were led astray. McGeer states, "The results are quite different in transgenic mouse models of Alzheimer's disease (compared to in humans). Human amyloid-β deposits are not powerful activators of mouse complement (a component of the immune system), and assembly of the mouse membrane attack complex is not seen in these mice. This crucial difference between patients with Alzheimer's disease and transgenic mice has not been properly appreciated" [51].

Who is to blame? No one really needs to harbor too much fault because the animal studies, through the course of history, have provided valuable insights and have protected human health. However, researchers need to, and probably do, realize that animal data do not always translate into humans, particular with regard to efficacy. A rule of thumb is that half of the results from animals reflect what will happen in human studies, at least regarding efficacy.

In yet another example of "clinging to the amyloid life raft," consider this paper from 2010 published by Brigham and Women's Hospital researchers in Boston and University of San Diego, California: "Can Alzheimer's disease be prevented by amyloid-β immunotherapy?" [52]. Here is the abstract:

Alzheimer's disease (AD) is the most common form of dementia. The amyloid-β (Aβ) peptide has become a major therapeutic target in AD on the basis of pathological, biochemical, and genetic evidence that supports a role for this molecule in the disease process. Active and passive Aβ immunotherapies have been shown to lower cerebral Aβ levels and improve cognition in animal models of AD. In humans, dosing in the phase II clinical trial of the AN1792 Aβ vaccine was stopped when ~6% of the immunized patients developed meningoencephalitis. However, some plaque clearance and modest clinical improvements were observed in patients following immunization. As a result of this study, at least seven passive Aβ immunotherapies are now in clinical trials in patients with mild to moderate AD. Several second-generation active Aβ vaccines are also in early clinical trials. On the basis of preclinical studies and the limited data from clinical trials, Aβ immunotherapy might be most effective in preventing or slowing the progression of AD when patients are immunized before or in the very earliest stages of disease onset.

The researchers appear to be shedding the most positive spin possible on the efficacy of an Aβ vaccine. All the recent data indicate that patients on Aβ-reducing therapies get worse. As a result, researchers are working backward to treat people with the most minimal of symptoms. Their mantra is the following: the antiamyloid treatment failed because the disease is just too far progressed. However, there are just too much data against the amyloid hypothesis for this to be plausible. Let's assume that this most recent study is valid. Would you, as a patient with the earliest signs of cognitive impairment that may not even progress, be willing to take a drug that causes meningoencephalitis and who knows what other side effects?

An irony about this particular group is that two of the researchers, those in Boston, work in Avenue Louis Pasteur, part of the Harvard Medical School campus. Why the irony? Dr. Louis Pasteur is considered one of the main developers of the Germ Theory of disease. That theory posits that certain germs (bacteria, virus, fungi) cause specific diseases. Examples include

Tuberculosis, lung disease, *H. pylori*, and stomach ulcers. As already stated, Dr. Alzheimer suggested "microorganisms" (germs) as a cause of AD over 100 years ago. Germs and AD may emerge as a more important connection compared to beta-amyloid.

Although not an antiamyloid therapy, the recent results on an immunoglobulin therapy failure are worth noting. Baxter International announced in 2013 that it would discontinue research on its experimental drug Gammagard, a much-anticipated drug that failed to slow the progression of AD after tests showed it did not stabilize or slow dementia in patients who received 18 months' worth of the blood-product compound. The study included patients with mild to moderate Alzheimer's.

In the past 10 years there have been almost 100 compounds that have failed in phase 3, just short of FDA drug approval. Many (but not all) of these drugs targeted beta-amyloid. Only a handful of drugs have been approved for Alzheimer's and dementia since the 1990s; none of them target beta-amyloid, and none of them work. Since then, billions of dollars have been allocated to experimental therapies, also with little to show for it. There is concern in the scientific community that drug companies may stop funding the research into Alzheimer's if it isn't paying off. Meanwhile, the cost of caring for Alzheimer's patients is already taking its toll as the number of those affected continues to rise. According to the World Health Organization, the cost to care for people with dementia around the world is currently over $604 billion (USD) per year.

IF NOT AMYLOID, WHAT NEXT?

Today the evidence from past antiamyloid drug trials has shown convincingly that, by the time symptoms have developed, it's too late to reverse them by shutting down further production of beta-amyloid protein. Some amyloid proponents argue that further trials could provide much useful data, and they continue to be excited about other studies set to get under way that will test antiamyloid agents in a "preclinical" AD population. That is, patients without clear signs of disease but based on a number of tests and observations may be likely to develop the disease. The diagnosis to qualify patients for this trial will be challenging.

Fortunately, skeptics of the Amyloid Cascade Hypothesis appear to be gaining ground. And there are a number of researchers who are pessimistic but hold that beta-amyloid is a factor in the pathology of AD—that is, it is part of the cause. But there are too many other factors that also play a role in the disease to justify singling out beta-amyloid as the point of therapeutic attack. Those who hold this glimmer of hope in the face of mounting evidence that Aβ plays more of a protective role rather than a detrimental role are slowly moving on to other targets. The key emerging target is one that Dr. Alzheimer identified at the turn of the past century that is neurofibrillary tangles. These contain so-called tau proteins that are "hyperphosphorylated" and apparently form toxic structures in the brain in AD. Tau will likely be the alternative target to Aβ especially if more of these antiamyloid studies fail.

The evidence in the broader medical literature suggests that tau is not a proper target either, but may yield better results compared to amyloid. Since AD is multifactorial, future efforts by the large drug development companies must include a thorough and objective review of the scientific and medical literature followed by pursuing multiple therapeutics together that might make an impact on Alzheimer's. Considering that the disease is multifactorial, a novel approach would clearly be to look at multiple factors simultaneously. This approach is considered "unscientific" because, if there is an effect, it is not possible to quickly discern which factor is truly important. But one must consider that human physiology is extremely complex and that a bona fide "cure" may be missed because the "curative" agent only functions in synergy with others. Simply put, there is no single "silver bullet" medication that will cure AD. The solution lies in a highly integrative approach. Currently, clinical trial design requirements, including approaches approved by the FDA, do not encourage multicomponent studies.

A scientist at Cleveland Clinic said, "Let's not think of Alzheimer's as one disease, but as many diseases, like breast cancer. When you have a breast cancer patient, you ask the question, is she HER2-positive? Then you treat her with Herceptin. Is she ER-positive? Then you use tamoxifen. You don't treat all the breast cancer patients the same way. A monotherapy is most likely not going to work. Lifestyle changes, which recent studies have shown can be effective in reducing Alzheimer's disease risk and perhaps in reversing some symptoms, will certainly be a component of future treatment strategies" [44]. Alzheimer's is likely to create yet a newer paradigm whereby, after a complete differential diagnosis, doctors will say you have this, that, and the other thing, you are deficient in X, Y, and Z, and the following are out of balance. Doctors will need to treat all of these factors at once and together if you are to overcome Alzheimer's!

WHAT USE DOES AMYLOID SERVE? DIAGNOSTICS

Going forward, does the Amyloid Cascade Hypothesis hold any merit when it comes to developing a solution to AD? Indeed. Beta-amyloid is one very important component of a differential diagnosis of AD. Based on information presented in this chapter, beta-amyloid does not occur definitively in Alzheimer's patients. However, its presence does provide an interesting piece to the Alzheimer's diagnostic puzzle.

Since beta-amyloid is not a significant part of the answer to Alzheimer's, we explore a differential diagnosis process to find the causes of and treatments for AD. The goal of this process is to:

- Establish the credibility of the science behind methods that outright confirm targets and processes that are contributing to AD and/or neurodegeneration in general.
- Show targets and processes that strongly infer what is likely contributing to AD and/or neurodegeneration in general.
- Present the science regarding other diseases that are often considered "confounding" to AD (in other words, interfering with diagnosis due to overlap). These diseases are more likely part of a systemic (system-wide) process that is impacting the health of tissue in a way that may be at the root of AD and neurodegeneration.
- Begin a discussion on the science on human homeostasis (balance within the body) as impacted by micronutrients (diet) that is a very strong "environmental" factor that appears to strongly impact the development of AD and other diseases that show up as accelerated diseases of aging.

At the end of this process, we hope to provide you with the means to prevent Alzheimer's from attacking you and your loved ones. We also hope to convey tools that allow you to recognize signs in your own body that may be telling you that a disease is happening in your body without you being aware. Our thesis remains that proper diagnosis is everything in medicine. Armed with the right information on cause and effect, you will be able to prevent the disease from occurring, or, in the unfortunate circumstance that you or a loved one has the disease, your doctor will be empowered with true root cause information from which they can design and implement a disease management program that hopefully will change the course of the disease in your favor.

TAU—NEUROFIBRILLARY TANGLES: THE "OTHER" HALLMARK

In AD, neurofibrillary degeneration results from the aggregation of tau proteins within neurons. The presence of these tangles (lesions) is very well correlated with cognitive deficits. Tau proteins are proteins mostly seen in neurons. There are six forms of tau protein in the adult human brain. These proteins play a role in the polymerization and stability of the microtubules (the supply rail system of the brain). This function is regulated by a chemical reaction called "phosphorylation" of the tau proteins.

In many neurodegenerative diseases (grouped together under the term "tauopathies"), abnormally phosphorylated tau proteins are aggregated in filaments that characterize the disease. In AD, neurofibrillary degeneration is initially found in the entorhinal cortex and the hippocampal formation, sequentially affecting neuronal subpopulations of the isocortex. It then appears in the associative polymodal regions, followed by the associative unimodal regions, and finally by the primary and secondary sensory-motor regions. In other neurodegenerative diseases, there is an aggregation of the tau proteins not only in neurons but also in glia cells (supporting tissue intermingled with the essential elements of nervous tissue especially in the brain, spinal cord, and ganglia). Activation of the glial cells is seen in all forms of neurological and neurodegenerative diseases including: AD, dementia, PD, amyotrophic lateral sclerosis (ALS), traumatic injury, inflammatory diseases, glaucoma, and macular degeneration.

Excess or hyperphosphorylation of tau and the change in the ratio between the different tau protein isoforms may be of importance in the formation of the toxic neurofibrillary tangles. The modifications leading to the aggregation of the tau proteins are phosphorylation and variations in the splicing of the tau protein. These modifications would appear to be the cause of a change in protein conformation leading to their accumulation and thus disease.

The aggregation of the tau proteins upsets neuronal functioning. Alteration of axonal transport constitutes the principal disturbance. Axonal transport is the simultaneous movement of proteins and other materials from the cell body of the neuron to the nerve fiber terminals and from the nerve fiber terminals to the cell body. There are other, as of yet poorly understood consequences of tau aggregation, such as the deficit in proteins that induce the survival, development, and function of neurons and agents that promote the transmission of signals between neurons. All these processes, when interrupted, could contribute to AD, either separately or together. Also, what is still very uncertain is tau at the root of the disease, or a marker for some other underlying cause(s).

As you will learn, there are a variety of ways to assess both beta-amyloid burden and neurofibrillary tangle burden. New tests are being developed that can probe the brain and other tissues for this information in a noninvasive way. This information is extremely important for characterizing the extent of disease and following the progression or regression of the disease in response to treatment. However, even though both are subject to therapeutic consideration, it may be that neither is truly associated with the root cause of AD. Measuring these materials may be useful in addressing the symptoms of the disease, but maybe not the root cause(s). That being the case, identification of either is useful from a diagnostic perspective.

Proposed Tau Therapies

Tau-based treatments for AD have become a point of increasing focus and current and previous investigational therapies can be grouped into four categories including tau-centric active and passive immunotherapeutics, microtubule-stabilizing agents, tau-protein kinase inhibitors, and tau-aggregation inhibitors (TAIs). Among different tau-directed approaches in AD, small-molecular-weight compounds developed to inhibit formation of tau oligomers and fibrils by blocking tau–tau aggregation have already been tested in humans [53]. In cell-based and/or in vitro screening assays, several classes of agents that may act to prevent tau aggregation have been identified, including but not limited to polyphenols, porphyrins, phenothiazines, benzothiazoles/cyanines, N-phenylamines, thioxothiazolidinones (rhodanines), phenylthiazole-hydrazides, anthraquinones, and aminothienopyridazines [54]. However, the efficacy for inhibiting tau aggregation in vivo for many TAIs has not yet been tested. On the other hand, several TAIs have toxic profiles that would preclude their use in vivo.

Currently, TAIs fall into two mechanistic classes depending on their way to interact with tau protein, that is, covalent and noncovalent molecules. Covalent TAIs can attack any or all species in an aggregation pathway, but appear to be especially efficacious modifiers of tau monomers. Natural polyphenols are covalent TAIs, such a as oleocanthal, a natural product aldehyde reacting with epsilon amino groups of lysine residues, oleuropein aglycone, abundant in the extra virgin olive oil, or the green tea–derived (−)-epigallocatechin gallate. Other redox-active compounds, including the nonneuroleptic phenothiazine methylene blue (MB) [methylthioninium chloride (MTC), Rember, TRx-0014, TauRx Therapeutics, Singapore, Republic of Singapore] can also modulate cysteine oxidation when incubated in the absence of exogenous reducing agents. In general, covalent mechanisms of tau-aggregation inhibition in AD are predicted to have low utility in vivo [55].

Hyperphosphorylated Tau Occurs Naturally in Healthy Animals

Hyperphosphorylated tau appears in healthy animals. When? Interestingly, this "Alzheimer's" state of tau appears in hibernating animals! What is different between an animal in hibernation and one that is active? Metabolism is greatly slowed in the hibernating animal. It is logical that there are physiological changes to an animal during hibernation that allow it to survive under severe conditions—the main one being a lack of oxygen (hypoxia). Hyperphosphorylation may be one such change. Is this hyperphosphorylation protecting both the Alzheimer's and the hibernating brain? One article on the subject is titled

"Physiological regulation of tau phosphorulation during hibernation" [56]. Here is an excerpt from the abstract:

> The microtubule-associated protein tau is abnormally hyperphosphorylated in the brains of individuals with Alzheimer's disease and other tauopathies and is believed to play a critical role in the pathogenesis of these diseases. While the mechanisms leading to abnormal tau phosphorylation remain elusive, the recent demonstration of reversible tau phosphorylation during hibernation provides an ideal physiological model to study this critical process in vivo (in a body rather than a test tube). In this study, Arctic ground squirrels (AGS) during hibernation were used to study mechanisms related to tau hyperphosphorylation. Our data demonstrate that tau is hyperphosphorylated at all six sites examined in hibernating AGS. Interestingly, only three of these sites are dephosphorylated in aroused animals, suggesting a reversible phosphorylation at selective sites. Summer-active AGS demonstrated the lowest tau phosphorylation at all these sites.

"Reversible phosphorylation" is an important concept obtained by studying hibernating animals. Let's presume that hyperphosphorylated tau plays an important role in the cause of AD. If it is reversible in humans as it is in hibernating animals, then there is certainly hope that the disease process may be slowed, halted, or reversed.

Based on this study of tau hyperphosphorylation/dehyperphosphorylation in animals in the wild, we predicted, in 2014, that any Alzheimer's therapy that simply reduces the phosphorylated tau without consideration for mechanism will fail to help, and likely will hurt, Alzheimer's sufferers. Two years later, our prediction came true as publicized at the Toronto International Alzheimer's Association meeting, July 2016. The July 27, 2016, *New York Times* provided the following summary of the failure:

> A new type of drug for Alzheimer's disease failed to slow the rate of decline in mental ability and daily functioning in its first large clinical trial. There was a hint, though, that it might be effective for certain patients.
>
> The drug, called LMTX, is the first one with its mode of action—trying to undo so-called tau tangles in the brain—to reach the final stage of clinical trials. So the results of the study were eagerly awaited. The initial reaction to the outcome was disappointment, with perhaps a glimmer of hopefulness.
>
> Over all, the patients who received LMTX, which was developed by TauRx Therapeutics, did not have a slower rate of decline in mental ability or daily functioning than those in the control group.
>
> Claude Wischik, a founder and the chief executive of TauRx, said in an interview. He spoke from Toronto, where the results were being presented at the Alzheimer's Association International Conference. Dr. Wischik said a second clinical trial sponsored by the company, whose results will be announced later, found the same phenomenon. He said the company planned to apply for approval of LMTX to be used by itself.
>
> Dr. Rachelle Doody, director of the Alzheimer's disease and Memory Disorders Center at Baylor College of Medicine, agreed. "To present it to the public now as a promising approach seems unjustified," she said.

Earlier in this chapter, we explained how many of the major drug companies continue to pursue the Amyloid Cascade Hypothesis in the face of overwhelming evidence against its efficacy. Could history be repeating itself? Here is more from the *New York Times* article on the failure of this first Tau therapy:

> Still, the failures of the amyloid drugs so far have prompted companies, including AbbVie, Biogen and Roche, to begin looking more at tau, another protein in the brain. When it becomes abnormal, it aggregates into tangles that kill neurons and can spread through the brain. Some studies suggest that levels of tau are more closely correlated with cognitive decline than levels of amyloid.
>
> "There is increasing evidence that tau is more proximal to the onset of disease symptoms," said William Jagust, Professor of Public Health and Neuroscience at the University of California, Berkeley.

The results of the LMTX trial do not necessarily spell doom for all tau drugs, because others might work differently. LMTX is "not the be-all and end-all for tau targeting," said Harry M. Tracy, publisher of *NeuroPerspective*, a newsletter that follows companies developing neurology drugs.

There is now a drive by top clinicians and researchers from around the world to illustrate the connection between infection and AD [57]. However, the mainstream Alzheimer's researchers continue to ignore the data behind the Alzheimer's/infection connection. In Chapter 9 we present strong research that shows infections of certain kinds promote formation of beta amyloid. Here we explain that the hyperphosphorylated Tau is most likely a manifestation of hypoxia. Infection is well known to create hypoxic physiological environments. Why do the researchers with strong ties to drug companies not pursue the infection/AD connection? We know that these scientists and doctors are not ignorant of the new research. We hypothesize that there is not sufficient monetary remuneration possible if the treatment for Alzheimer's winds up being antibiotic/antiinflammatory/immune system–boosting therapies. You can tell that the existing Alzheimer's programs that the drug companies are following are based on business decisions because the drugs they are producing are biologics—specifically monoclonal antibodies (note the "mab" at the end of the scientific name for these drugs). These drugs are extremely expensive—and unfortunately—ineffective.

Tau, just like with beta-amyloid, provides little or no therapeutic value. Both are hallmarks of Alzheimer's and thus have some utility in diagnosis.

> Every single possible cause must be explored, even those seemingly unrelated to AD. Because AD is so poorly defined, we cannot afford to rule out any possibility or overlapping factors. Differential Diagnosis is the key to successful treatment.
>
> —*Clement Trempe, MD*

Tau is Dying - Amyloid is dead! Long live Amyloid (and Tau)!

References

[1] Hardy JA, Higgins GA. Alzheimer's disease: the amyloid cascade hypothesis. Science 1992;256(5054):184–5.
[2] Armstrong RA, Myers D, Smith CUM. The spatial patterns of plaques and tangles in Alzheimer's disease do not support the 'cascade hypothesis'. Dementia 1993;4(1):16–20.
[3] Erickson MA, Farr SA, Niehoff ML, Morley JE, Banks WA. 95: antisense directed against the amyloid precursor protein reduces cytokine expression in the brain and improves learning and memory in the Tg2576 mouse model of Alzheimer's disease. Brain, Behavior, and Immunity 2012;26(1):S27.
[4] Campbell A. β-Amyloid: friend or foe. Med Hypotheses 2001;56(3):388–91.
[5] Chung H, Brazil MI, Soe T, Maxfield FR. Uptake, degradation, and release of fibrillar and soluble forms of Alzheimer's amyloid b-peptide by microglial cells. J Biol Chem 1999;274:32301–8.
[6] Soscia SJ, Kirby JE, Washicosky KJ, Tucker SM, Ingelsson M, Hyman B, et al. The Alzheimer's disease-associated amyloid β-protein is an antimicrobial peptide. PLoS One 2010;5(3):e9505.
[7] Itzhaki RF, Wozniak MA, Appelt DM, Balin BJ. Infiltration of the brain by pathogens causes Alzheimer's disease. Neurobiol Aging 2004;25:619–27.
[8] Miklossy J, Kis A, Radenovic A, Miller L, Forro L, et al. β-Amyloid deposition and Alzheimer's type changes induced by *Borrelia* spirochetes. Neurobiol Aging 2006;27:228–36.
[9] Kountouras J, Tsolaki M, Gavalas E, Boziki M, Zavos C, et al. Relationship between *Helicobacter pylori* infection and Alzheimer's disease. Neurology 2006;66:938–40.
[10] Itzhaki RF, Lin WR, Shang D, Wilcock GK, Faragher B, et al. Herpes simplex virus type 1 in brain and risk of Alzheimer's disease. Lancet 1997;349:241–4.
[11] Youngsteadt E. Virology. Alzheimer's risk factor also aids HIV. Science 2008;320:1577.

[12] Strittmatter WJ, Saunders AM, Schmechel D, Pericak-Vance M, Enghild J, et al. Apolipoprotein E: high-avidity binding to β-amyloid and increased frequency of type 4 allele in late-onset familial Alzheimer's disease. Proc Natl Acad Sci USA 1993;90:1977–81.

[13] Urosevic N, Martins RN. Infection and Alzheimer's disease: the APOE epsilon4 connection and lipid metabolism. J Alzheimers Dis 2008;13:421–35.

[14] Atwood CS, Bishop GM, Perry G, Smith MA. Amyloid-β: a vascular sealant that protects against hemorrhage? J Neurosci Res 2002;70(3):356.

[15] Atwood CS, Obrenovich ME, Liu T, Chan H, Perry G, Smith MA, et al. Amyloid-β: a chameleon walking in two worlds: a review of the trophic and toxic properties of amyloid-β. Brain Res Brain Res Rev 2003;43(1):1–16.

[16] Verdile G, et al. The role of beta amyloid in Alzheimer's disease: still a cause of everything or the only one who got caught? Pharmacol Res 2004;50(4):397–409.

[17] Rottkamp CA, et al. The state versus amyloid-β: the trial of the most wanted criminal in Alzheimer's disease. Peptides 2002;23(7):1333–41.

[18] Smith MA, et al. Amyloid-β, tau alterations and mitochondrial dysfunction in Alzheimer's disease: the chickens or the eggs? Neurochem Int 2002;40(6):527–31.

[19] Bowen RL, Atwood CS. Living and dying for sex. Gerontology 2004;50(5):265–90.

[20] Puzzo D, Arancio O. Amyloid-B peptide: Dr. Jekyll or Mr. Hyde? In: Perry G, editor. Alzheimer's disease: advances for a new century. Amsterdam: IOS Press; 2013.

[21] Lee H-G, et al. Amyloid-β vaccination: testing the amyloid hypothesis?: Heads we win. Tails you lose! Am J Pathol 2006;169(3):738.

[22] Lorenzo A, Yankner BA. Beta-amyloid neurotoxicity requires fibril formation and is inhibited by Congo red. Proc Natl Acad Sci USA 1994;91:12243–7.

[23] Hardy J. The amyloid hypothesis of Alzheimer's disease: progress and problems on the road to therapeutics. Science 2002;297:353–6.

[24] Lesne S, Koh MT, Kotilinek L, Kayed R, Glabe CG, Yang A, et al. A specific amyloid-beta protein assembly in the brain impairs memory. Nature 2006;440:352–7.

[25] Perry G, Nunomura A, Raina AK, Smith MA. Amyloid-beta junkies. Lancet 2000;355:757.

[26] Smith MA, Atwood CS, Joseph JA, Perry G. Predicting the failure of amyloid-beta vaccine. Lancet 2002;359:1864–5.

[27] Lee HG, Casadesus G, Zhu X, Takeda A, Perry G, Smith MA. Challenging the amyloid cascade hypothesis: senile plaques and amyloid-beta as protective adaptations to Alzheimer's disease. Ann N Y Acad Sci 2004;1019:1–4.

[28] Rottkamp CA, Atwood CS, Joseph JA, Nunomura A, Perry G, Smith MA. The state versus amyloid-beta: the trial of the most wanted criminal in Alzheimer's disease. Peptides 2002;23:1333–41.

[29] Lee HG, Zhu X, Nunomura A, Perry G, Smith MA. Amyloid-beta: the alternate hypothesis. Curr Alzheimer Res 2006;3:75–80.

[30] Armstrong RA. The pathogenesis of Alzheimer's disease: a reevaluation of the amyloid cascade hypothesis. Int J Alzheimers Dis 2011;2011:630865.

[31] Graeber MB, Kösel S, Egensperger R, Banati RB, Müller U, Bise K, et al. Rediscovery of the case described by Alois Alzheimer in 1911: historical: histological and molecular genetic analysis. Neurogenetics 1997;1(1):73–80.

[32] Carroll BJ. Ageing, stress and the brain. In: Novartis Foundation symposium, vol. 242; 2002. p. 26–45.

[33] Styczynska M, Strosznajder JB, Religa D, et al. Association between genetic and environmental factors and the risk of Alzheimer's disease. Folia Neuropathol 2008;46(4):249–54.

[34] Lee H-G, et al. Amyloid-β in Alzheimer's disease: the null versus the alternate hypotheses. J Pharmacol Exp Ther 2007;321(3):823–9.

[35] Šalković-Petrišić M. Amyloid cascade hypothesis: is it true for sporadic Alzheimer's disease. Periodicum Biologorum 2008;110(1):17–25.

[36] Reitz C. Alzheimer's disease and the amyloid cascade hypothesis: a critical review. Int J Alzheimers Dis 2012;2012:369808.

[37] Noble W, Planel E, Zehr C, et al. Inhibition of glycogen synthase kinase-3 by lithium correlates with reduced tauopathy and degeneration in vivo. Proc Natl Acad Sci USA 2005;102(19):6990–5.

[38] Kurnellas MP, et al. Amyloid fibrils composed of hexameric peptides attenuate neuroinflammation. Sci Transl Med 2013;5(179):179ra42.

[39] Neve RL, Robakis NK. Alzheimer's disease: a re-examination of the amyloid hypothesis. Trends Neurosci 1998;21(1):15–9.

[40] Rao KSJ, et al. Amyloid β and neuromelanin—toxic or protective molecules? The cellular context makes the difference. Prog Neurobiol 2006;78(6):364–73.

[41] Harris F, Dennison SR, Phoenix DA. Aberrant action of amyloidogenic host defense peptides: a new paradigm to investigate neurodegenerative disorders? FASEB J 2012;26(5):1776–81.

[42] Lee H-G, et al. Challenging the amyloid cascade hypothesis: senile plaques and amyloid-β as protective adaptations to Alzheimer's disease. Ann N Y Acad Sci 2004;1019(1):1–4.

[43] Smith MA, et al. Predicting the failure of amyloid-β vaccine. Lancet 2002;359(9320):1864–5.

[44] Gever J. Alzheimer's disease: amyloid 'proponents' soldier on. MedPage Today; December 20, 2012. Available from: http://www.medpagetoday.com/Neurology/AlzheimersDisease/36562.

[45] Eli Lilly and Company announces top-line results on solanezumab phase 3 clinical trials in patients with Alzheimer's disease. PRNewswire; August 24, 2012. Available from: https://investor.lilly.com/releasedetail.cfm?ReleaseID=702211.

[46] Pollack A. Alzheimer's drug fails its first big clinical trial. The New York Times; July 23, 2012. Available from: http://www.nytimes.com/2012/07/24/business/alzheimers-drug-fails-its-first-clinical-trial.html.

[47] Herper M. How a failed Alzheimer's drug illustrates the drug industry's gambling problem. Forbes; August 8, 2012. Available from: http://www.forbes.com/sites/matthewherper/2012/08/08/how-a-failed-alzheimers-drug-illustrates-the-drug-industrys-gambling-problem/#514232dc3aae.

[48] Doody RS, et al. A phase 3 trial of semagacestat for treatment of Alzheimer's disease. N Engl J Med 2013;369(4):341–50.

[49] Lazar K. Major Boston-led study to test drugs to delay Alzheimer's disease gets funded. Boston Globe; January 15, 2013. Available from: http://archive.boston.com/whitecoatnotes/2013/01/14/major-boston-led-study-test-drugs-delay-alzheimer-disease-gets-funded/pcpWpXGn2b23899RgrcQxL/story.html.

[50] Holtzman JL. Amyloid-beta vaccination for Alzheimer's dementia. Lancet 2008;372(1381):1381–2.

[51] Erickson Dunning LM, Holtzman JL. The effect of aging on the chaperone concentrations in the hepatic, endoplasmic reticulum of male rats: the possible role of protein misfolding due to the loss of chaperones in the decline in physiological function seen with age. J Gerontol Biol 2006;61A:435–43.

[52] Lemere CA, Masliah E. Can Alzheimer's disease be prevented by amyloid-β immunotherapy? Nat Rev Neurol 2010;6(2):108–19.

[53] Wischik CM, Harrington CR, Storey JM. Tau-aggregation inhibitor therapy for Alzheimer's disease. Biochem Pharmacol 2014;88(4):529–39.

[54] Cisek K, Cooper GL, Huseby CJ, et al. Structure and mechanism of action of tau aggregation inhibitors. Curr Alzheimer Res 2014;11(10):918–27.

[55] Schafer KN, Cisek K, Huseby CJ, et al. Structural determinants of tau aggregation inhibitor potency. J Biol Chem 2013;288(45):32599–611.

[56] Su B, et al. Physiological regulation of tau phosphorylation during hibernation. J Neurochem 2008;105(6): 2098–108.

[57] Itzhaki R, et al. Microbes and Alzheimers disease. J Alzheimers Dis 2016;51(4):979.

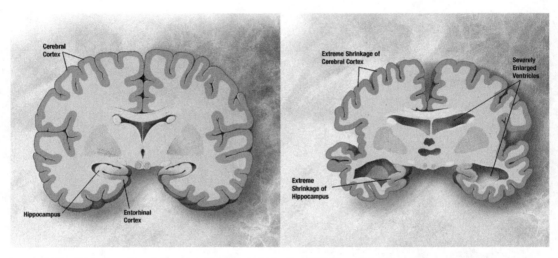

Image Credit: Diagrams of normal versus Alzheimer's brain, by ADEAR - Alzheimer's Disease Education and Referral Center, a service of the National Institute on Aging. Available at Wikimedia Commons: https://commons.wikimedia.org/wiki/File:Alzheimer%27s_disease_brain_comparison.jpg

Diagnosis of Alzheimer's— Standard-of-Care

Diagnosing complex diseases like Alzheimer's requires a fresh approach. In some ways this fresh approach existed a century ago when Dr. Charles Mayo developed the "Grand Rounds" concept at the inception of the Mayo Clinic. Grand Rounds triggers a differential diagnosis because it includes input from top professionals in each medical specialty. Thus the concept of a differential diagnosis is not new, but for diseases (syndromes really) like Alzheimer's disease (AD) that are so poorly understood, the diagnostic process must transcend all disciplines of medicine. Currently some may argue that a differential diagnosis process is used. However, if it is restricted to one medical vertical like neurology, then is it a truly differential process?

Modern medicine has a myriad of tools and techniques to diagnose AD. The existing tests provide tremendous value to patients and doctors and should be maintained within any enhanced model. Being from the camp that Alzheimer's is multifactorial means the diagnosis MUST be the same—as broad and deep as possible. No information should be discarded or overlooked.

Medicine does suffer from an affliction known as "the standard-of-care." Within the standard-of-care, patients receive approximately the same level of care from the Johns Hopkins or your local country doctor. Why? Health care has evolved to be confused with health insurance. That is to say, everything within health care is codified to meet the criteria established by the payer (insurance company, Medicare, HMO). Thus, whether at Mayo, MD Anderson, or your local doctor, a coded diagnosis is performed followed by a prescriptive treatment based on the diagnosis. There is very little flexibility allowed within this now rigid system, unless, of course, you are willing to pay out of pocket. How is this different compared to the inspection of your automobile where a "code" comes up from a computer and the technician changes an oxygen sensor?

The standard-of-care diagnostic process for a person with deteriorating mental cognitive functioning almost inevitably results in a diagnosis of Alzheimer's or a form of dementia. Once a patient obtains this diagnosis, essentially all hope for that patient is eliminated. Chapters 6–12 build a strong case that there are options beyond a standard-of-care diagnosis that is based on billions of dollars of medical research. Standard-of-care diagnosis is presented here as a reference for comparison to a true differential diagnosis presented in the later chapters.

The End of Alzheimer's. http://dx.doi.org/10.1016/B978-0-12-812112-2.00003-3

ALZHEIMER'S DIAGNOSIS: STANDARD-OF-CARE

Assessment of a patient with AD disease usually involves doctors, working independently within their own silo of specialization, consisting of a neurologist, a psychiatrist, and a neuropsychologist, assisted by a radiologist and a pathologist. Assessment involves a series of steps and can take several office visits to complete. There is no single test that confirms or excludes AD. Within the standard-of-care, most diagnostic efforts target the brain and minimal "systemic" testing is performed other than nonspecific blood tests.

Definition of Systemic: Pertaining to or affecting the body as a whole.

The Alzheimer's Association website has a section titled "Tests for Alzheimer's disease and dementia." They state, "There is no single test that proves a person has Alzheimer's." Their next statement is quite forward thinking. They say, "A diagnosis is made through a complete assessment that considers all possible causes." They present a litany of tests you can expect, as a potential Alzheimer's sufferer. Feel free to compare these tests to the ones proposed in Chapters 10 and 11. As you read through these tests, ask yourself if any of these diagnostics procedures help identify a root cause or causes. Also, ask yourself if treatment options become apparent based on these tests.

Medical History: During the medical workup, your health-care provider will review your medical history. He or she will want to know about any current and past illnesses, as well as any medications you are taking. The doctor will also ask about key medical conditions affecting other family members, including whether they may have had AD or related dementias.

Physical Exam and Diagnostic Tests: During a medical workup, you can expect the physician to:
- Ask about diet, nutrition, and use of alcohol.
- Review all medications. (Bring a list or the containers of all medicines currently being taken, including over-the-counter drugs and supplements.)
- Check blood pressure, temperature, and pulse.
- Listen to the heart and lungs.
- Perform other procedures to assess overall health.
- Collect blood or urine samples for laboratory testing.

Blood tests are a starting point to establish a baseline of systemic health of a patient. In the standard-of-care, blood tests are performed to do just that, to be able to compare standard testing parameters against norms. Clinicians are missing a tremendous opportunity to get to the root of Alzheimer's because, as you will see, markers in the blood are very telling about the presence and cause of Alzheimer's. The blood test performed in your doctor's office merely looks at sugar levels and indications of kidney and liver problems but little else. Also, the "so-called" normal ranges upon which doctors rely to judge health lack a solid scientific basis. This issue is also examined in more detail later.

While there is no such thing as a "standard blood test," there are several common blood test panels that doctors generally order. In many cases, blood tests rule out illnesses before having to resort to more invasive testing.

- Complete Blood Count: A complete blood count, or CBC, is part of routine testing and is often part of an annual exam. A CBC measures the amounts of red blood cells, white blood cells, and platelets in your blood. This test can detect anemia, inflammation, infections (but seldom the specific species without further testing), and bleeding disorders. A CBC can also help determine your response to medications and if dosage and type need to be adjusted.
- Blood Chemistry Tests: These tests are also known as a basic metabolic panel, or a BMP. This is a series of tests that is run on the plasma of your blood. These tests measure how much glucose and calcium is in your blood, and whether you have the right amount of electrolytes and minerals in your blood. A BMP can detect diseases such as diabetes, cancer, bone disease, kidney disease, and other disorders. Some tests require that you fast beforehand, and others don't.
- Blood Enzyme Tests: Enzyme tests can show whether you have damage or disease in various organs of the body. A creatine kinase (or CK) analysis can help determine if you've had a heart attack. A troponin test is another way to indicate a heart attack. Creatinine measures kidney function. Liver enzyme assays can rule out or diagnose liver disease. These also help to determine whether a medication may need to be adjusted or eliminated if it is causing liver damage.

Note how these tests used in Alzheimer's diagnosis are standard tests that are often obtained during a routine physical exam. There is no new blood testing parameters designated for Alzheimer's specifically. Thus, these tests are not designed to assert a diagnosis of Alzheimer's. Instead they focus more on ruling out common or standard diseases for which these tests are designed. In other words, the blood tests are not investigative and look only to establish general wellness in the patient.

According to the Alzheimer's Association, information from a physical exam and laboratory tests can help identify health issues that can cause symptoms of dementia. Conditions other than Alzheimer's that may cause confused thinking, trouble focusing, or memory problems include anemia, depression, infection, diabetes, kidney disease, liver disease, certain vitamin deficiencies, thyroid abnormalities, and problems with the heart, blood vessels, and lungs.

Genetic Testing (Not Normally Conducted): Researchers have identified certain genes that increase the risk of developing Alzheimer's and other rare "deterministic" genes that directly cause Alzheimer's. Although genetic tests are available for some of these genes, health professionals do not currently recommend routine genetic testing for AD.

- Risk genes: While there is a blood test for APOE-e4, the strongest risk gene for Alzheimer's, this test is mainly used in clinical trials to identify people at higher risk of developing AD. Carrying this gene mutation indicates only a greater risk; it does not indicate whether a person will develop Alzheimer's or whether a person has Alzheimer's. Genetic testing for APOE-e4 is controversial and should be undertaken only after discussion with a physician or genetic counselor.
- Deterministic genes: Testing is also available for genes that cause autosomal dominant AD (ADAD) or familial AD, a rare form of Alzheimer's that accounts for less than 5% of all cases. ADAD runs strongly in families and tends to begin earlier in life. Many people in these families do not wish to know their genetic status, but some get tested

to learn whether they will eventually develop the disease. Some ADAD families have joined clinical studies to help researchers better understand Alzheimer's.

Delving into the genetic component of disease is beyond our scope. However, the impact of genes is often discussed throughout this book with regard to specific diseases. A recent article in *National Geographic Magazine* appears to create a proper appreciation for the impact of genes on disease. In the May 2013 issue, the author explains [1]:

> But genes alone are unlikely to explain all the secrets of longevity … Passarino made the point while driving back to his laboratory after visiting the centenarians in Molochio. "It's not that there are good genes and bad genes," he said. "It's certain genes at certain times. And in the end, genes probably account for only 25 percent of longevity. It's the environment too, but that doesn't explain all of it either. And don't forget chance."

The purpose of the article was to interview populations of centenarians and evaluate their genetic makeup. That the author concluded genetics play 25% of role really indicates that the actual value is substantially less. Nongenetic environmental factors play the biggest role. Translated, that means YOU HAVE CONTROL over your health. And, by the way, do forget chance and replace it with a differential diagnosis.

Neurological Exam: During a neurological exam, the physician will closely evaluate the person for problems that may signal brain disorders other than Alzheimer's. The doctor will look for signs of small or large strokes, Parkinson's disease (PD), brain tumors, fluid accumulation in the brain, and other illnesses that may impair memory or thinking. The physician will test:

- reflexes
- coordination, muscle tone, and strength
- eye movement
- speech
- sensation

The neurological exam may also include a brain imaging study.

Mental Status Tests

Mental status testing evaluates memory, ability to solve simple problems, and other thinking skills. Such tests give an overall sense of whether a person:

- is aware of symptoms;
- knows the date and time, and where he or she is;
- can remember a short list of words, follow instructions, and do simple calculations.

The Mini-Mental State Examination (MMSE) and the Mini-Cog test are two commonly used tests.

Mini-Mental State Examination

During the MMSE, a health professional asks a patient a series of questions designed to test a range of everyday mental skills. The maximum MMSE score is 30 points. A score of 20–24 suggests mild dementia, 13–20 suggests moderate dementia, and less than 12 indicates severe dementia. On average, the MMSE score of a person with Alzheimer's declines about two to four points each year.

Mini-Cog

During the Mini-Cog, a person is asked to complete two tasks:

- remember and a few minutes later repeat the names of three common objects;
- draw a face of a clock showing all 12 numbers in the right places and a time specified by the examiner.

The results of this brief test can help a physician determine if further evaluation is needed.

Mood Assessment

In addition to assessing mental status, the doctor will evaluate a person's sense of well-being to detect depression or other mood disorders that can cause memory problems, loss of interest in life, and other symptoms that can overlap with dementia.

Brain Imaging

A standard medical workup for AD often includes structural imaging with MRI or CT; these tests are primarily used to rule out other conditions that may cause symptoms similar to Alzheimer's but require different treatment. Structural imaging can reveal tumors, evidence of small or large strokes, damage from severe head trauma, or a buildup of fluid in the brain.

Imaging technologies have revolutionized our understanding of the structure and function of the living brain. Researchers are exploring whether the use of brain imaging may be expanded to play a more direct role in diagnosing Alzheimer's and detecting the disease early on.

Objective Evaluation: Standard-of-Care

Let's objectively evaluate the testing performed on potential Alzheimer's patients based on today's standard-of-care. Medical history and physical exams for AD patients are no different than tests performed on a healthy person during a routine physical exam. There is nothing extraordinary performed and thus nothing unusual likely to be found regarding AD. The genetic tests, if performed, indicate only a predisposition to disease susceptibility. However, it is well proven that those with the predisposition are not much more likely to have the disease because nongenetic environmental factors are far more important. The next set of exams, including brain imaging, are strictly neurological and show response and structure. There is nothing physiological that could lead to a treatment evaluated within any of these tests.

What does the standard-of-care testing for Alzheimer's and dementia do? It raises a white flag over the patient. That is, the best these tests can do is establish a baseline for the extent of disease that the doctor(s) then track, over time, to see the regression of the patient. There is nothing within the testing/diagnostic protocol that leads to new or novel treatment. This is just a method to watch your decline. As the Alzheimer's Association says, if you have Alzheimer's, expect your mini-mental score to decline by two to four points each year.

Hope is on the horizon, however, because the big interests in Alzheimer's got together and published a series of peer-reviewed articles on the future of Alzheimer's and dementia diagnosis. Let's take a look into the future and see what help and hope we can expect. After all, diagnosis is the most important aspect of medicine, the results of which drive treatment decisions.

Experts from the Mayo Clinic, Massachusetts General Hospital, The National Institutes of Health, Johns Hopkins University, the Alzheimer's Association, and other prestigious and authoritative organizations combined their brainpower and wrote a series of recommendations on Alzheimer's diagnoses. The results of their work were published in four papers in *The Journal of the Alzheimer's Association* in May 2011 [2–5]. Each of these papers addresses a different aspect of the AD process. Here is a look into each one of these articles as the experts address the promise of a bona fide disease characterization and diagnosis.

Article 1: "Introduction to the recommendations from the National Institute on Aging–Alzheimer's Association workgroups on diagnostic guidelines for Alzheimer's disease" [1].

This is the first of the papers. It sets the tone for the guidelines that follow in papers 2–4. In this article, the team sets the stage for how they approach the setting of new guidelines. Here is the abstract:

Background: Criteria for the clinical diagnosis of Alzheimer's disease (AD) were established in 1984. A broad consensus now exists that these criteria should be revised to incorporate state-of-the-art scientific knowledge.

Methods: The National Institute on Aging (NIA) and the Alzheimer's Association sponsored a series of advisory round table meetings in 2009 whose purpose was to establish a process for revising diagnostic and research criteria for AD. The recommendation from these advisory meetings was that three separate workgroups should be formed, each with the task of formulating diagnostic criteria for one phase of the disease: the dementia phase; the symptomatic, pre-dementia phase; and the asymptomatic, preclinical phase of AD.

Results: Two notable differences from the AD criteria published in 1984 are the incorporation of biomarkers of the underlying disease state and the formalization of the different stages of disease in the diagnostic criteria. There was a broad consensus within all three workgroups that much additional work is needed to validate the application of biomarkers for diagnostic purposes. In the revised NIA–Alzheimer's Association criteria, a semantic and conceptual distinction is made between the AD pathophysiological processes and the clinically observable syndromes that result, whereas this distinction was blurred in the 1984 criteria.

Conclusions: The new criteria for AD are presented in three documents. The core clinical criteria of the recommendations regarding AD dementia and mild cognitive impairment (MCI) due to AD are intended to guide diagnosis in the clinical setting. However, the recommendations of the preclinical AD workgroup are intended purely for research purposes.

Sadly, the preclinical workgroup information is just for research purposes. Indeed these patients are asymptomatic (they do not show clinical signs of an ailing brain). However, these are the people most easily treated. They are relatively healthy. Chapter 6 shows that simple tests are available and show the possibility of disease in asymptomatic people. Medicine needs to screen our populations for early biomarkers of disease and then treat the patients rather than relegate a tremendously beneficial opportunity to "research."

Note it took 2 years just to create the recommendations. The big news is that new diagnostic criteria will contain biomarkers "of the underlying disease." If you read Chapter 2, you are in a good position to guess the biomarkers for the "underlying disease." Yes—beta-amyloid and tau. Tau will be the next big quest in drug research. Beta-amyloid is dead as a therapeutic, and it has limited value as a diagnostic. Yet the best and brightest in the field of Alzheimer's will make a case that we need to do the tests outlined in the standard-of-care and add a test for beta-amyloid. Since beta-amyloid is NOT at the root, what does this do for an AD sufferer?

Here is an excerpt from a section titled "3. Biomarkers of AD":

Evidence suggests that although both Aβ deposition and elevated tau/phosphorylated tau are hallmarks of AD, alterations in these proteins are seen in other neurological disorders. Because elevations in Aβ seem to be more specific than alterations in tau, it was decided to divide the biomarkers into two major categories: (1) the biomarkers of Aβ accumulation, which are abnormal tracer retention on amyloid PET imaging and low CSF Aβ42, and (2) the biomarkers of neuronal degeneration or injury, which are elevated CSF tau (both total and phosphorylated tau).

Let's take a closer look at the pedigree of the researchers involved in this workgroup who published these recommendations. This is taken from the acknowledgment section of the first article:

Clifford Jack serves as a consultant for Eli Lilly, Eisai, and Elan; is an investigator in clinical trials sponsored by Baxter and Pfizer Inc.; and owns stock in Johnson and Johnson. Marilyn Albert serves as a consultant to Genentech and Eli Lilly and receives grants to her institution from GE Healthcare. David Knopman serves on a Data Safety Monitoring Board for Lilly Pharmaceuticals; is an investigator for clinical trials sponsored by Elan Pharmaceuticals, Forest Pharmaceuticals, and Baxter Healthcare; and is deputy editor of *Neurology* and receives compensation for editorial activities. Guy McKhann serves on a Data Safety Monitoring Board for Merck. Reisa Sperling has served as a site investigator and/or consultant to several companies developing imaging biomarkers and pharmacological treatments for early AD, including Avid, Bayer, Bristol-Myers-Squibb, Elan, Eisai, Janssen, Pfizer, and Wyeth. Maria Carrillo is an employee of the Alzheimer's Association and reports no conflicts. Bill Thies is an employee of the Alzheimer's Association and reports no conflicts. Creighton Phelps is an employee of the US Government and reports no conflicts.

Note that drug companies pay all the esteemed professors. These are the same companies developing drugs for AD based on the biomarkers in the proposal. Are these the right people making decisions about the diagnostic process for your loved one suffering from Alzheimer's? The professors are all either neurologists or radiologists; thus they are not suggesting a truly differential diagnosis.

Article 2: "The diagnosis of dementia due to Alzheimer's disease: recommendations from the National Institute on Aging–Alzheimer's Association workgroups on diagnostic guidelines for Alzheimer's disease" [3].

What the workgroup is trying to do is place dementia sufferers into different baskets of "diagnosis" based on cognitive performance. What appears to be lacking in their efforts is a fundamental understanding of how the world truly works. The world is not linear; rather, it is asymptotic. What that means, by way of an example, is that it is relatively easy to ride a bike 20 mph, but very difficult to go 25 mph. Think about that curve in math—the "X-squared" curve. It starts off shallow and then rises precipitously toward infinity. That is our world and it (almost) always holds true, whether it describes grades, the ability to make a billion dollars, be the top football player, or the expansion of the universe. In their analysis, they are trying to determine when the "inflection" of the "X-squared" curve occurs. To make matters even more complicated, even these experts have admitted, in one forum or another, that Alzheimer's is multifactorial. Therefore, they are trying to characterize the onset and progression of a disease with many overlapping and intersecting "X-squared" curves.

Why is there such a necessity to put people into "baskets" of disease levels? Because, based on the Amyloid Cascade Hypothesis, the only hope for this therapy is to characterize people early in the disease process and hope this approach has therapeutic value. If there was a true root cause(s) understanding of the disease, then the extent of the disease would be irrelevant, and tests would be performed to determine causes and treatments administered. That being said, as a scientist, clearly the better a person (patient) is characterized, the more targeted potential treatments will be and the more insight medicine will gain from any cause/effect from treatment. In that respect, the efforts of the workgroup should be lauded.

Now let's take a look at Article 2 in some detail. Here is an excerpt from the abstract:

> On the basis of the past 27 years of experience, we made several changes in the clinical criteria for the diagnosis. We also retained the term possible AD dementia, but redefined it in a manner more focused than before. Biomarker evidence was also integrated into the diagnostic formulations for probable and possible AD dementia for use in research settings. The core clinical criteria for AD dementia will continue to be the cornerstone of the diagnosis in clinical practice, but biomarker evidence is expected to enhance the pathophysiological specificity of the diagnosis of AD dementia. Much work lies ahead for validating the biomarker diagnosis of AD dementia.

The workgroup states, "The core clinical criteria for AD dementia will continue to be the cornerstone of the diagnosis in clinical practice." That is to say, all the efforts of this workgroup really will not add to the 1984 definition; thus there will be no real improvement in the diagnostic process for Alzheimer's going forward. It will remain one that assesses cognitive function primarily.

Article 3: "The diagnosis of mild cognitive impairment due to Alzheimer's disease: recommendations from the National Institute on Aging–Alzheimer's Association workgroups on diagnostic guidelines for Alzheimer's disease" [4].

The abstract is reproduced here:

> The National Institute on Aging and the Alzheimer's Association charged a workgroup with the task of developing criteria for the symptomatic pre-dementia phase of Alzheimer's disease (AD), referred to in this article as mild cognitive impairment due to AD. The workgroup developed the following two sets of criteria: (1) core clinical criteria that could be used by healthcare providers without access to advanced imaging techniques or cerebrospinal fluid analysis, and (2) research criteria that could be used in clinical research settings, including clinical trials. The second set of criteria incorporate the use of biomarkers based on imaging and cerebrospinal fluid measures. The final set of criteria for mild cognitive impairment due to AD has four levels of certainty, depending on the presence and nature of the biomarker findings. Considerable work is needed to validate the criteria that use biomarkers and to standardize biomarker analysis for use in community settings.

One way to obtain biomarkers is to obtain cerebral spinal fluid through a spinal tap. This is the proposed method to assess tau and amyloid, along with very expensive imaging technology for beta-amyloid. Some time ago when the NIH (or NIH-funded organization) proposed a clinical trial that involved spinal taps, there were few volunteers. As you will see in subsequent chapters, there are much less costly and less invasive means to evaluate both tau and beta-amyloid. There may be a conflict of interest among one of the authors of the workgroups. According to the acknowledgment, one of the authors "has served as a site investigator and/or consultant to several companies developing imaging biomarkers and pharmacological treatments for early AD, including Avid, Bayer, Bristol-Myers-Squibb, Elan, Eisai, Janssen, Pfizer, and Wyeth." Sure, this person is an expert on imaging but the workgroups appear to be underrepresented in broader disciplines.

Article 4: "Toward defining the preclinical stages of Alzheimer's disease: recommendations from the National Institute on Aging–Alzheimer's Association workgroups on diagnostic guidelines for Alzheimer's disease" [5].

The abstract posits an interesting question. How do you detect a disease before it shows clinical symptoms? Many studies show that Alzheimer's progresses over decades and thus has a long, yet potentially detectable, incubation period. This is the stage when Alzheimer's must be diagnosed. MRI studies at Mass General and other medical centers show that there is brain atrophy in patients with normal cognition that either do or are likely to progress to cognitive impairment and dementia. The cost of screening the nation with MRI is cost prohibitive. However, there are bona fide low-cost ways to obtain the same type of data provided by MRI. It is clear from article 4 that these types of tests, discussed later in this book, are completely ignored. Also, specific blood tests are able to portent dementias, and our experts ignore these. Let's take a look at their proposal:

The pathophysiological process of Alzheimer's disease (AD) is thought to begin many years before the diagnosis of AD dementia. This long "preclinical" phase of AD would provide a critical opportunity for therapeutic intervention; however, we need to further elucidate the link between the pathological cascade of AD and the emergence of clinical symptoms. The National Institute on Aging and the Alzheimer's Association convened an international workgroup to review the biomarker, epidemiological, and neuropsychological evidence, and to develop recommendations to determine the factors which best predict the risk of progression from "normal" cognition to mild cognitive impairment and AD dementia. We propose a conceptual framework and operational research criteria, based on the prevailing scientific evidence to date, to test and refine these models with longitudinal clinical research studies. These recommendations are solely intended for research purposes and do not have any clinical implications at this time. It is hoped that these recommendations will provide a common rubric to advance the study of preclinical AD, and ultimately, aid the field in moving toward earlier intervention at a stage of AD when some disease-modifying therapies may be most efficacious.

A year later, a truly who's who of Alzheimer's research, but sadly from mainly just pathology and neurology, followed up with "National Institute on Aging–Alzheimer's Association guidelines for the neuropathologic assessment of Alzheimer's disease: a practical approach" [6].

We present a practical guide for the implementation of recently revised National Institute on Aging–Alzheimer's Association guidelines for the neuropathologic assessment of Alzheimer's disease (AD). Major revisions from previous consensus criteria are: (i) recognition that AD neuropathologic changes may occur in the apparent absence of cognitive impairment, (ii) an "ABC" score for AD neuropathologic change that incorporates histopathologic assessments of amyloid β deposits (A), staging of neurofibrillary tangles (B), and scoring of neuritic plaques (C), and (iii) more detailed approaches for assessing commonly co-morbid conditions such as Lewy body disease, vascular brain injury, hippocampal sclerosis, and TAR DNA binding protein (TDP)-43 immunoreactive inclusions. Recommendations also are made for the minimum sampling of brain, preferred staining methods with acceptable alternatives, reporting of results, and clinico-pathologic correlations.

Hopefully you will find the practical approach presented in this book not only more practical but also more comprehensive and science-based compared to what you just read.

THE FIRST ALZHEIMER'S DIAGNOSIS

The first use of the term "Alzheimer's disease" (and thus the first official Alzheimer's diagnosis) occurred in 1910 by a Germany doctor, Kraepelin, who knew Dr. Alois Alzheimer well. He used the term to describe cases with the features provided by Alzheimer himself. At the time, before the use of the term "Alzheimer's disease," the relatively few cases appearing in the elderly were classified as senile dementia or senile psychosis.

In 1906, Alois Alzheimer performed an autopsy on the brain of a 56-year-old woman with a history of progressive mental deterioration. This woman, Auguste D, became the first patient diagnosed with yet-to-be-named "Alzheimer's disease." In her cerebral cortex, the part of the brain responsible for reasoning and memory, were strange fiber bundles, which he termed "neurofibrillary tangles" (NFTs), and accumulations of cellular debris, or senile plaques (SPs), which together define AD.

Dr. Alzheimer was working on an understanding of memory and cognitive impairment before the turn of the century. In 1898, he published a significant paper on senile dementias of various causes with emphasis on those related to atheromatous (plaque-containing) vascular disease. Alzheimer's unique contribution to the disease that bears his name was the demonstration of the structures that are now called neurofibrillary tangles and the recognition that they were important markers of the disease process. Most important, he showed a connection between system-wide disease (vascular disease) and this newly described dementia.

The other hallmark of the disease, the senile or neuritic (amyloid as it's now known) plaque, had first been reported in the brains of old people affected by epilepsy in 1892. Fischer published a fuller description of these formations in 1907. Alzheimer disagreed with the hypothesis that the plaques were the most important structure associated with the disease. He concluded that they were not the cause of senile dementia but an accompaniment of a special case of senile aging of the nervous system.

Although Alzheimer performed some utterly progressive research, he considered himself primarily a clinician and always saw the lab as providing support and service to the clinic. Somehow we have the cart before the horse, and in our modern focus, the lab and drug development supersedes good medicine and translating the knowledge that we already have. The crux of the problem is that medical research is now an independent stand-alone industry that is no longer designed to support clinical medicine.

STANDARD-OF-CARE DIAGNOSIS AT AUTOPSY

The definitive diagnosis of AD, according to the current recognized standards of care, is based on the observation of characteristic brain lesions (usually found during a postmortem examination), first described by Drs. Alzheimer and Fischer: senile plaques and neurofibrillary tangles. The weight and the volume of the brain are reduced on average. Cortical areas playing a role in the memory functions are the first to lose volume, followed by the regions of the cortex implicated in such functions as language, the complex analysis of visual or auditory impulses, or the programming of voluntary movements.

The accumulations of beta-amyloid peptide and tau protein assume different shapes and structures. The deposits of beta-amyloid peptide are dense and spherical in form. It is also deposited in the vessel walls (amyloid angiopathy). Neurofibrillary tangles correspond to the aggregation of tau protein in the cellular body of the neuron. This protein also accumulates in the axons surrounding deposits of beta-amyloid peptide, forming the crown of the senile plaque that is the hallmark of AD identification. The senile plaque is thus made up of a deposit of beta-amyloid peptide surrounded by a crown of axons enriched in tau protein.

The neurofibrillary tangles are due to the accumulation in the neuron of a naturally present protein, the tau protein. This protein plays a role in the polymerization of the microtubules (microtubules are fibrous, hollow rods that are a cellular "rail" system for providing needed supplies to, and removing waste from, brain cells), while the beta-amyloid is characterized by the extracellular (outside the cell) accumulation of a protein, which is normally present in low concentrations. The normal function of this peptide and of its precursor remains unknown, but very recent research is suggesting that beta-amyloid might actually be part of an immune response (as discussed in Chapter 2). It may be both friend and foe.

Thanks to the analysis of a large number of cases, of varying age and severity, it has been possible to trace the space and time evolution of the lesions, and to describe the different stages. The neurofibrillary tangles accumulate in different parts of the brain, successively with the entorhinal (stages I and II), hippocampal (stages III and IV), and neocortical (stages V and VI) regions. Each stage adds a new affected structure to those affected at the previous stage. The same applies to the five "phases" describing the evolution of the beta-amyloid peptide deposit that occurs successively and additively in the neocortex, in the entorhinal area, in the hippocampus, in the subcortical nuclei, in the brain stem, and finally in the cerebellum.

Interestingly (and possibly a clue), the progression of the neurofibrillary tangles in the cortex (entorhinal cortex, then hippocampus, and lastly neocortex) corresponds to the progression of the symptoms. On the other hand, the deposits of beta-amyloid are less well correlated with the symptoms. It is quite common to find, in elderly subjects considered to be intellectually normal, diffuse deposits of beta-amyloid. These beta-amyloid lesions appear constant in the brain of centenarians on whom a postmortem examination has been performed. Their frequency is an indication that they could remain stable and that they solely represent physiological cerebral aging. According to another hypothesis, these lesions, even if they are without clinical consequence, could signal the presence of an as of yet asymptomatic AD. However, plaques are found in young persons. For example, 20% of young people who died by accident and who were autopsied showed some amyloid buildup.

COGNITIVE TESTS AND HUMAN MEMORY

The standard-of-care in diagnosis includes written and oral cognitive tests that probe into the functioning of different capacities of the brain. There are a myriad of psychological tests that can be administered to a potential Alzheimer's patient but, as indicated at the beginning of this chapter, usually only one or two are performed. The range of these tests is designed to determine the extent of overall brain impact and potentially point to specific damage to different components of the brain. These tests are also designed to place a patient into one of the three emerging classifications for Alzheimer's/dementia, and this is important to help manage care and possibly treatment. Although these tests speak to only symptoms (the "what" of the disease) and not the causes (the "why" of the disease), they have been available for a while and thus there is a body of information that makes the conclusions derived from these tests useful. They are also relatively easy to administer.

The considerable downside is that those patients in the early stages of the disease find these tests terribly demeaning and discouraging. The patient is essentially given a test that a 6-year-old can pass. For example, ask your loved one what 2 + 2 equals. They will not be happy with you if they assume you are serious.

Professionals in neurology contend that psychological evaluations highlight and characterize the various dementia disorders. Also, these tests are suggested to be able to differentiate, for example, AD from the other neurodegenerative diseases, or depressive syndromes and simple age-induced decline of certain cognitive capacities. The hope is this type of evaluation plays an important part in revealing those mental capacities, thus portions of the brain, that have been preserved.

It has been shown that the onset of the disease can take several forms; the first signs are usually related to memory disorders. Olfactory sensation is another early change. However, ocular disturbance arguably provides the first clues as to the underlying brain irregularity. The memory comprises several components, or memory systems, that are not affected in the same way. Episodic memory stores memories of personally experienced events, situated in the temporal spatial context. Disorders of episodic memory are central to AD and are characterized by difficulties in acquiring new information and in retrieving memories, particularly those relating to recent events. Such disorders can readily be distinguished from the decline in memory linked to increasing age, in both degree and kind, since they concern the different stages of memorization: encoding, storage, and retrieval of information. Retrieval disorders are less specific since they are observed in numerous diseases. This is a key area where these so-called mental functioning tests are useful.

Isolated disorders of memory episodes are characteristic of mild cognitive impairment with some component of amnesia. Most of the patients show impaired scores in tests of episodic memory: learning of lists of words, primacy effect (remembering the first words in the list), recognition of words, and remembering a story or a geometrical figure. The most sensitive and at the same time the most specific measurement would appear to be the delayed recall of a list of related words. This might be explained by the patients' difficulty in organizing the items to be memorized by categories of meaning (semantic memory).

Episodic memory is usually examined by means of tasks of learning words or remembering stories. One test in particular is now commonly used in memory consultations. In AD, there is a deficit in the free recall of information, and there is scarcely any improvement in performance with cued remembering (e.g., "What was the name of the flower?"). This type of recall problem points to difficulties in encoding and storing information.

Semantic memory, which stores words, concepts, and knowledge about the world, as well as personal semantics (general knowledge about oneself), may be disrupted in early Alzheimer's, while it stands up very well to the effects of age, thus suggesting a degenerative disease. Disturbances of semantic memory have regularly been highlighted in mild cognitive impairment patient groups and would appear to be among the best predictive indices of subsequent cognitive decline.

Semantic memory disorders can be revealed through questionnaires focusing on knowledge of concepts or famous persons. These disorders have a greater impact on specific knowledge than on general knowledge and are expressed by constant errors from one moment to another and from one test to another. These difficulties are not to be confused with disorders of semantic (meaning) memory access that are characterized by difficulties in producing the right word, but without loss of concept. Here, it is a case of language disorders, which are very frequent in AD and are revealed by means of image and vocabulary tests. The written language is also impacted in AD, the most telltale symptom being a tendency to even out the writing of irregular words.

Working memory, by which small quantities of information are stored and manipulated for a limited time, is also disturbed at a very early stage of AD. The "central executive" (the most important yet least understood component of the working memory model), which is responsible for the allocation of attention resources and the coordination of the other working memory subsystems, is particularly sensitive to the disease. Working memory is commonly evaluated through attention span tasks (repetition of series of figures, in the right order and back to front) or dual-task paradigms. The impairment of the central executive should be considered as one of the fundamental cognitive disturbances of AD with repercussions on multiple tasks.

To sum up, AD affects first of all episodic memory, semantic (meaning) memory, and working memory—the three most elaborate memory systems. On the other hand, lower-level systems, such as the perceptual system and procedural memory (habits), show more staying power, at least during the early stages of the disease.

Other cognitive functions, in addition to memory and language, are impaired in AD. Of particular note are the executive functions, or high-level mental processes implicated in the accomplishment of a purposeful activity. Disruption often occurs early on and can appear at a predementia stage. However, they may not occur with great frequency. This was the case of my father when he decided that letting out water by drilling holes in the floor rather than fixing the leaking roof was an appropriate decision. Keep in mind that this event occurred at least 5 years prior to us realizing he had any type of memory issue. The identification of executive disorders is an important challenge given their repercussions for the patient not only in their everyday life but also in becoming aware of the mounting cognitive deficit.

COGNITIVE FUNCTIONING TESTS EXPLORED

There are several standard tests used by neurologists to characterize and classify a person with failing memory. Besides episodic, semantic, and working memory, other disorders associated with brain disease appear at a fairly early stage of AD: apraxia (difficulty in executing coordinated movements), Agnosia (difficulty in identifying objects), or visuospatial disorders expressed by difficulties in producing (spontaneously or by copying) geometrical or figurative drawings. There are cognitive tests that are designed to differentiate between these various dysfunctions.

The key tests are summarized as follows:

- **Wisconsin Card Sorting Test:** Initially, a number of stimulus cards are presented to the participant. The shapes on the cards are different in color, quantity, and design. The person administering the test decides whether the cards are to be matched by color, design, or quantity. The participant is then given a stack of additional cards and asked to match each one to one of the stimulus cards, thereby forming separate piles of cards for each. The participant is not told how to match the cards; however, he or she is told whether a particular match is right or wrong. During the course of the test the matching rules are changed and the time taken for the participant to learn the new rules and the mistakes made during this learning process are analyzed to arrive at a score.

- **The Rey–Osterrieth Test: The Rey–Osterrieth Complex Figure Test (ROCF)**, which was developed by Rey in 1941 and standardized by Osterrieth in 1944, is a widely used neuropsychological test for the evaluation of visuospatial constructional ability and visual memory. Recently, the ROCF has been a useful tool for measuring executive function that is mediated by the prefrontal lobe. The ROCF consists of three test conditions: Copy, Immediate Recall, and Delayed Recall. At the first step, subjects are given the ROCF stimulus card and then asked to draw the same figure. Subsequently, they are instructed to draw what they remembered. Then, after a delay of 30 minutes, they are required to draw the same figure once again. The anticipated results vary according to the scoring system used, but commonly include scores related to location, accuracy, and organization. Each condition of the ROCF takes 10 minutes to complete, and the overall time of completion is about 30 minutes.

The Mini-Mental State Examination

The Mini-Mental State Examination (MMSE) has been the most common method for diagnosing AD and other neurodegenerative diseases affecting the brain. It was devised in 1975 by Folstein et al. as a simple standardized test for evaluating the cognitive performance of subjects, and where appropriate to qualify and quantify their deficit [7]. It is now the standard bearer for the neuropsychological evaluation of dementia, mild cognitive impairment, and AD.

The MMSE was designed to give a practical clinical assessment of change in cognitive status in geriatric patients. It covers the person's orientation to time and place, recall ability, short-term memory, and arithmetic ability. It may be used as a screening test for cognitive loss or as a brief bedside cognitive assessment. By definition, it cannot be used to diagnose dementia, yet this has turned into its main purpose.

The MMSE was termed "mini" because it concentrates only on the cognitive aspects of mental function and excludes mood and abnormal mental functions that are covered, for example, in Blessed Dementia Scale. It is administered by clinical or lay personnel after brief training and requires 5–10 minutes for completion.

It is a brief and practical scale, so it cannot be expected to perform perfectly in every situation. Various limitations have been identified. It may miss impairments resulting from right hemisphere lesions and may miss mild impairments. Instructions for administration and scoring lack detail. Many users have reported that people with low education tend to give false-positive responses.

The MMSE includes 11 items, divided into 2 sections. The first requires verbal responses to orientation, memory, and attention questions. The second section requires reading and writing and covers ability to name, follow verbal and written commands, write a sentence, and copy a polygon. All questions are asked in a specific order and can be scored immediately by summing the points assigned to each successfully completed task; the maximum score is 30.

Details of scoring have led to considerable discussion. For example, it was originally proposed that counting backwards by sevens could be replaced by spelling "world" backwards. Folstein has clarified that he uses the serial sevens if at all possible; it is more difficult than the spelling alternative. The challenge of scoring the overlapping pentagon diagram (Fig. 3.1; Box 3.1) has

BOX 3.1

STANDARD MINI-MENTAL STATE EXAMINATION

Orientation:

1. What is the?	Year?	1
	Season?	1
	Date?	1
	Month?	1
2. Where are we?	State?	1
	County?	1
	Town or city?	1
	Hospital or clinic?	1
	Floor?	1

Registration:

3. Name three objects, taking one second to say each. Then ask the patient all three after you have said them. Give one point for each correct answer. Repeat answers until patient learns all three   3

Attention and calculation:

4. Serial sevens (count backwards from 100 by sevens). Give one point for each correct answer. Stop after five answers. Alternatively: Spell WORLD backwards   5

Recall:

5. Ask for the names of the three objects learned in Question 3. Give one point for each correct answer   3

Language:

6. Point to a pencil and watch. Have the patient name them as you point   2

7. Have the patient repeat "No ifs, ands, or buts"   1

8. Have the patient follow a three-stage command: Take the paper in your right hand. Fold the paper in half. Put the paper on the floor   3

9. Have the patient read and obey the following: "CLOSE YOUR EYES." (Write it in large letters)   1

10. Have the patient write a sentence of his or her own choice. (The sentence should contain a subject and an object and should make sense. Ignore spelling errors when scoring)   1

11. Enlarge the design printed below to 3–5 cm per side and have the patient copy it. (Give one point if all sides and angles are preserved and if the intersecting sides for a quadrilateral)   1

Total out of 30

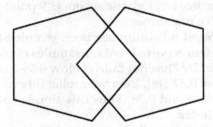

FIGURE 3.1 Mini-Mental State Examination intersecting pentagons.

even been addressed by computer digitizing and analysis. Treating questions not answered as errors is recommended. The issue of how to handle nonresponses due to illiteracy or blindness has been handled either by treating these as errors or by prorating the overall score. Folstein has commented that he administers the items without regard to the reason for failure (deafness, etc.), and then, after scoring, comments on possible reasons for failure. "A basic rule of clinical medicine is to collect the facts or observations before making interpretations."

The break point most commonly used to indicate cognitive impairment deserving further investigation is 23/24 while some recommend 24/25 to enhance sensitivity for mild dementia. The break point is commonly modulated according to educational level because a single point may miss cases among more educated people and generate false-positives among those with less education.

Cognitive decline measured by the MMSE is not linear over time. It seems to advance more slowly during the mild and severe stages of the disease and faster at the moderate stages. The rate of the initial decline predicts the subsequent (fast or slow) decline. The severity of the initial cognitive decline (at the time of the first visit) is linked to a prognosis. On the other hand, a measurement remaining stable for at least 2 years ("plateau") is a good prognosis and patients often have reasonable function over a 7-year period of follow-up. However, cognitive decline varies from one individual to another. A patient losing three or more points on the MMSE per year is considered a "rapid decliner" while a "slow decliner" is one who loses less than two points per year on the same examination. What these variable rates of decline tell us is that there is individuality to the progression of Alzheimer's. It thus infers that the rate of decline may be modulated by factors within your control including proper diagnosis and treatment.

The score of 16 on the MMSE seems to mark a transition point below which disruptions of basic everyday activities begin to emerge within 12 months. At this level, the patient requires partial or total assistance with everyday activities. Many studies show, however, that the decrease in daily activities in AD is present from the very early stages, affecting social life and leisure activities early on. Often, the reduction in social activities is one of the first signs arousing the attention of the patient's family in the same way as memory difficulties. It is closely associated with apathy, diminished motivation, and difficulties in planning ahead. Other studies show that even at the stage of MCI, some people may already suffer from an inability to perform certain tasks of everyday life. When followed up, these subjects showed a significantly higher risk of an evolution toward dementia compared to those free from such disabilities.

In summary, any score greater than or equal to 25 points (out of 30) is effectively normal (intact). But any score of less than 30 should result in the patient being kept under a watchful eye and retested frequently. Multiple scores averaged are far more telling compared to a single data point. Below 25, scores can indicate severe (≤ 9 points), moderate (10–20 points), or mild (21–24 points) cognitive impairment.

For the MMSE, test and retest reliability has been examined in many studies. In a review of his own studies, Folstein reported that for samples of psychiatric and neurological patients, the test–retest reliability "has not fallen below 0.89 (on a scale of 0–1); inter-rater reliability has not fallen below 0.82" [8]. Interrater reliability has also been widely studied with interrater reliability of 0.69 and 0.78. Thus this simple test is reproducible from patient to patient and tester to tester.

The test has predictive validity as all five respondents in one study, whose score decreased by more than seven points in 3 years, were diagnosed with neurological deficits [9].

Conflicting with this, another report, however, found that the MMSE had limited utility in predicting the psychological functioning of 90 psychiatric inpatients [10]. Generally, most studies showed that the MMSE performed very well in identifying moderate and severe cases of dementia, but less well in identifying mild cases. The MMSE also suffers from variations due to educational levels of the patient.

The MMSE forms the leading screening instrument in North America but is somewhat less popular in Europe. While it has known weaknesses, it has the great virtue of being well understood. The diversity of efforts to improve it illustrates the difficulty of developing the ideal dementia-screening instrument.

In summary, this simple test, the MMSE, has become widespread as indicated by the speed with which large numbers of papers incorporating it were published. The MMSE is brief enough for routine clinical use, and nonprofessionals can administer it in survey settings. Validity results appear as good as, or slightly better than, those of other scales that have subsequently been developed with the intent of replacing it with a presumably better test. But it is an aid to the clinician, and too much should not be expected of it. A comprehensive diagnosis requires a full mental status examination, history, physical examination, and other supporting tests to understand the "why" of the disease and not just the "what." In his 1998 retrospective, Folstein wrote, "The MMSE is now 22 years old and can speak for itself. It travels around the world" [11].

Neurological Tests Beyond the MMSE

The MMSE is not the only verbal/visual/written test. Several other tests have emerged but have not replaced the MMSE. However, there is no need to choose one in favor of the other as they each add their own nuance and provide information about the patient's health.

- **ADAS-Cog (Alzheimer's Disease Assessment Scale-Cognitive):** For the cognitive criterion, another scale commonly used by neurology is the ADAS-Cog (Alzheimer's Disease Assessment Scale-Cognitive). ADAS-Cog was designed to measure the severity of the most important symptoms of AD. Its subscale, ADAS-cog, is the most popular cognitive testing instrument used in clinical trials of memory-enhancing drugs. It consists of 11 tasks that measure the disturbances of memory, language, verbalizing ideas, attention, and other cognitive abilities, which are often referred to as the core symptoms of AD. This is a 70-point scale. The importance of this test is its apparent ability to differentiate between AD patients and those with depression disorders. In comparison with the MMSE, ADAS-cog seems to be more helpful in early diagnostics of AD. Patients may decline by an average of four points in 6 months, six to eight points in 1 year, and in nonlinear fashion according to the stages of neurodegenerative disease. An average improvement of at least 2.5 points on this scale has been considered relevant in trials designed to reveal a symptomatic gain.
- **Wechsler Test of Adult Reading (WTAR):** This test is used to estimate an individual's level of intellectual functioning before the onset of injury or illness. This assessment tool is for older adolescents and adults and takes less than 10 minutes to complete. The WTAR is reported to be an effective method for predicting a person's preinjury IQ and memory abilities. The WTAR allows measurement of premorbid (preinjury) level of

intellectual functioning for individuals aged 16–89 years. This reading test is composed of a list of 50 words that have atypical grapheme to phoneme translations. The intent in using words with irregular pronunciations is to minimize the current ability of the client to apply standard pronunciation rules and assess previous learning of the word. Unlike many intellectual and memory abilities, reading recognition is relatively stable in the presence of cognitive declines associated with normal aging or brain injury. The purpose of the WTAR is not for the assessment and diagnosis of developmental reading disorders, but rather for an initial estimation of premorbid intellectual and memory abilities (assuming a normal development of reading skills prior to injury or cognitive decline). The WTAR is advertised to be reliable and valid because of the following: there is a large national norming sample carefully matched to the United States population; demographic data exist to accurately predict premorbid IQ in neuropsychological cases; and there is extensive clinical validity with group studies including AD, Huntington's disease, PD, Korsakoff's syndrome, and Traumatic Brain Injury.

- **Dementia Rating Scale—Second Edition (DRS-2):** DRS-2 is an enhanced version of the original DRS designed to provide standardized, quantitative cognitive functioning assessment in neurologically impaired populations. The 36-task and 32-stimulus card instrument is individually administered and designed to assess levels of cognitive functioning for individuals between 55 and 89 years of age with brain dysfunction. The DRS-2 is sensitive at the lower ends of functioning and differentiating levels of severity deficits. Conversely, the instrument generally will not discriminate individual functioning in the average or higher range of intelligence due to the design to minimize floor effects of clinically impaired individuals. The initial pool of items from DRS was revised for comprehensive and brief administration allowing for a low floor so that even severely impaired individuals could be evaluated. The DRS-2 has a wider age range than the original DRS, and the age-corrected scaled and percentile ranks are more sensitive to change in cognitive status. The task and stimulus card have not been changed from the original. In the hands of an experienced neuropsychologist or clinical psychologist, this is an adequate instrument but is highly dependent upon the skills of each individual test administrator. The DRS-2 assesses cognitive functioning on the following 5 subscales: Attention (8 items), Initiation/Perseveration (11 items), Construction (6 items), Conceptualization (6 items), and Memory (5 items). Stimulus items contain material familiar to the majority of individuals. Time to Administer is as follows: 10–15 minutes for high-functioning-dementia patients and 30–40 minutes for low-functioning-dementia patients.

Limits of Cognitive Tests

A group in Australia reviewed the state of early diagnosis of AD and dementias, with particular focus on psychological testing. The establishment of a diagnosis is critical in Australia for a patient to be enrolled in the Pharmaceutical Benefits Scheme. They have recently concluded that, "despite the many advances in our understanding of AD and the technology to measure changes analytically, primary diagnosis still relies on the identification of cognitive decline."

Any brief screen or assessment of a complex behavior such as cognition has limitations. Despite its widespread clinical use, and like all brief dementia-screening tests, the MMSE has been criticized for the following: lacking sensitivity to patients with mild cognitive impairment, lacking diagnostic specificity, and not taking into account levels of education, premorbid ability, and other patient variables such as visual problems or poor command of English. Dementia may be missed in some patients, and other patients without dementia may be misclassified. A normal score on the MMSE does not necessarily exclude a brain abnormality or dementia.

The ADAS-Cog shares many of the limitations reported for the MMSE. Scores on the ADAS-Cog are also variable. For example, in the original clinical study of the scale, 27 patients with AD and 28 normal elderly people were rated and then retested 12 months later. The range of scores corresponding to 1 standard deviation from the mean in the AD group was 0–31 at baseline and 0–38 at 12 months, demonstrating wide variability in scores. Perhaps not surprisingly, given this variability, only eight of the patients with AD showed a significant increase in the severity of their dysfunction after 12 months.

The limitations of the tests in indexing change highlight the importance of referring patients with suspected AD for a broader range of both psychological and analytical tests. The benefits of broader and deeper testing are potentially lifesaving. For example, additional psychological testing may provide important information about other confounding cognitive, mood, or personality changes. Testing for markers of inflammation may reveal other diseases that are comorbid with cognitive decline that, when left untreated, may lead to early mortality.

ADVANCED DIAGNOSIS WITHIN THE STANDARD-OF-CARE

At the present time, AD is clinically defined as a dementia whose diagnosis is founded on the presence of a cognitive decline with repercussions on everyday life. Thus, the diagnosis is based on a two-stage approach: first, the demonstration of a dementia syndrome and then, the identification of elements suggesting AD (slow and insidious encroachment of cognitive disorders).

The disease was long considered a degenerative disorder of the period preceding old age (before the age of 65). The cognitive and behavioral disorders observed in the elderly were then grouped under the term "senile dementia." It was not until the 1960s that the uniqueness of AD, the most frequent cause of dementia, was recognized irrespective of the age at which it began.

Many criteria for the diagnosis of AD have been put forward, including ICD-10 (World Health Organization, 1993), DSM-IV (American Psychiatric Association, 1994), and NINCDS–ADRDA (National Institute of Neurological and Communicative Diseases and Stroke/Alzheimer's Disease and Related Disorders Association, 1984). All refer to a gradual impairment of memory and other cognitive functions in the absence of any other disease that could account for the emergence of a dementia syndrome. The sensitivity of these criteria is globally satisfactory (an average of 80% over all the studies), but there is a lesser specificity (around 70%) for the diagnosis of probable AD with postmortem confirmation.

Generally speaking, the diagnosis of AD is particularly difficult at the beginning and end of the development of the disease. At the outset, the symptoms are discreet and may be masked or confused with difficulties related to the normal aging process. At the end of the evolution, at the final stages of cognitive and behavioral degeneration, it is difficult to find, from examination, specific marks of a disease. This being the case, it is all the more important to question the patient's family circle about the manner in which the disorders emerged.

Advanced Brain Imaging

Cognitive tests document various aspects of brain dysfunction, and neurology often knows what part of the brain is likely compromised based on measurable deficits. The next stage of diagnosis looks into the brain to correlate the conjecture of the cognition tests with actual physical alterations in the brain. The intent is multifold. Clearly, it is important to correlate the cognitive decline with brain structures. This correlation should also help better characterize the disease as AD as opposed to another type of dementia. Finally, as we learn more about brain physiology, knowing the impacted part of the brain may help direct treatments in the future.

Today, medicine has a variety of very powerful imaging tools that map the brain quite accurately and are able to paint a picture of the deteriorating brain. The primary tests are MRI, CT Scans, and PET scans.

Magnetic Resonance Imaging

A magnetic resonance imaging (MRI) uses computer-generated radio waves and a strong magnetic field to produce a detailed image of the brain. MRIs are helpful in the diagnosis of tumors, eye diseases, infections, inflammation, atrophy caused by neurodegenerative diseases such as AD, damage to vessels and microvessels, and damage due to head injury. Similar to the CT scan, an MRI requires the patient to lie on a table that slides into a tube that contains the imaging equipment. Also, because of the strong magnetic field involved in the procedure, those with medical implants like pacemakers should avoid this test. MRIs take up to an hour to complete and produce both two-dimensional and three-dimensional images. MRI tests are very expensive.

MRI is used to investigate for AD, mainly to rule out other possible causes for cognitive impairment, such as a brain tumor or blood clot. Recent research suggests that MRI could become a key diagnostic tool by revealing changes in the brain even before Alzheimer's symptoms appear. We do know that even early AD patients show significant brain atrophy and small vessel disease. Thus it is logical that MRI screening may be able to show early signs of these two manifestations in asymptomatic people and those with early, mild cognitive impairment.

Thus MRI is not used to predict who will develop AD. However, it does assist in the diagnosis of Alzheimer's by evaluating for particular patterns of brain atrophy that occur in patients with the disease. AD affects the brain in many ways, but one of the most apparent involves an area called the hippocampus. This part of the brain is responsible for memory and processing emotion; it also plays a role in an individual's motor skills. In a study conducted in 2008, researchers using MRI to evaluate people with AD found that the hippocampus in those already diagnosed was nearly one-third smaller than average [12]. The hippocampus was 19% smaller in people who had not been diagnosed but were experiencing mental impairment.

Researchers who are studying MRI as a diagnostic tool for AD say the technique is promising, although there are no guidelines or recommendations yet for its use. In the above-mentioned study, which involved 74 subjects, physicians reported being able to classify those with AD and those without symptoms with 84% accuracy based on measurement of the hippocampus. The researchers were accurate 73% of the time when distinguishing between patients without symptoms and those with mild cognitive impairment. However, it's important to note that this was a small study.

In another study focused on the diagnostic ability of MRI for early AD, researchers looked at MRI results for 119 patients with varying degrees of cognitive impairment [13]. Some patients were normal, some had cognitive impairment at the time of the MRI, and others were already diagnosed with AD. The researchers were 100% accurate when determining which patients had been diagnosed with AD and which had no symptoms. The study reported a 93% accuracy rate when researchers were asked to distinguish between patients with no symptoms and patients who had only mild cognitive impairment, but were not yet diagnosed with AD.

Specifically, magnetic resonance imaging studies have shown cerebral morphological alterations associated with AD and concerning first of all the hippocampal region, in line with the distribution of neurofibrillary tangles. Numerous authors have demonstrated a marked atrophy of the medial region of the temporal lobe compared to healthy old subjects, even at a predementia stage of the disease. The atrophy then spreads to other areas (external temporal cortex, posterior cingulate gyrus, temporoparietal cortex), in line with the expansion of the neurofibrillary degeneration.

MRI provides an elegant, yet expensive way to contribute to a diagnosis of AD early and also to track its progression. It fails to explain why the disease occurs except superficially. For example, MRI is able to show the region of the brain impacted and if there are signs of vascular damage, mini-strokes, or structures like Lewy bodies or neurofibrillary tangles.

New research is focusing on using MRI in combination with a drug that highlights beta-amyloid deposits. The early results indicate that this method will be even more expensive because the staining agent is as (or more) expensive as the MRI. Knowing the presence and location is helpful but does not really add to the development of new treatments. Medical science has known about the connection between beta-amyloid and Alzheimer's-type dementia for decades. With AD becoming epidemic or even pandemic, promoting a roughly $2000–8000 per patient diagnostic does not appear to be a fruitful direction for the medical clinic of the future. However, Chapter 6 covers a diagnostic method that provides results similar to MRI, costs about one-hundredth of MRI, is less invasive compared to MRI, and is much more patient and doctor friendly compared to MRI.

Positron Emission Tomography

A positron emission tomography (PET) scan provides both two- and three-dimensional pictures of brain activity by measuring radioactive isotopes (elements that attach to chemicals that flow to the brain) injected into the bloodstream. PET scans are used to detect tumors and damaged tissue, measure metabolism, and view how blood flows in the brain. They are often ordered as follow-ups to CT scans or MRIs. PET scans are performed in a hospital or outpatient imaging clinic. After the isotope is injected into the patient's bloodstream, overhead sensors detect the isotope's activity. The information is processed by a computer and displayed on a monitor or film. The length of time to complete a PET scan varies depending on the reason for the test.

PET scans can help detect plaques in the brain (amyloid lesions), which are associated with AD, reported by researchers in the journal *Archives of Neurology*. The authors explain as background information that researchers are trying to understand AD more deeply, as well as other forms of dementia. In doing so, the use of PET scans has been explored. PET scans use nuclear medicine imaging (radiation) to create three-dimensional color images of how things function inside the human body. The device detects pairs of gamma rays, which are emitted indirectly by a positron-emitting radionuclide (tracer). This is placed in the body on a biologically active molecule. Computers reconstruct the images.

Before performing a PET scan, a radioactive substance is produced in a machine (cyclotron); it is then tagged to a natural chemical, which is known as a radiotracer, or simply a tracer. The radiotracer is then inserted into the body. Various teams of scientists are examining how effective various types of tracers are in identifying brain findings linked to Alzheimer's.

In a key PET scan study, patients with probable AD, people with mild cognitive impairment, and healthy volunteers (controls) were assessed [14]. Variations in the brain uptake of a tracer between the three groups were noted, and differences were considered wide enough to help tell the difference between the conditions, particularly in amyloid burden. "With the potential emergence of disease-specific interventions for AD," the authors noted, "biomarkers that provide molecular specificity will likely become of greater importance in the differential diagnosis of cognitive impairment in older adults. Amyloid imaging through PET scanning offers great promise to facilitate the evaluation of patients in a clinical setting."

Additionally, Alzheimer's PET scans use FDG, a derivative of glucose. The use of FDG, which shares a similar structure to glucose, is important, as the absorption of glucose is effective in determining the metabolic activity of the brain. In AD and other forms of dementia, the brain produces a metabolic pattern that is significantly different from the metabolic pattern of healthy brain cells. As PET imaging examines the metabolic activity of brain cells by tracing how FDG is absorbed, it is able to detect AD and other forms of dementia.

Recent studies have confirmed the effectiveness of PET in distinguishing AD from other forms of dementia. This is because AD has a metabolic abnormality (bilateral temporoparietal hypometabolism) that is significantly different from metabolic abnormalities found in other forms of dementia. PET scan of Alzheimer's has increased in recent years as PET imaging provides a noninvasive, painless way for physicians to tentatively confirm the presence of Alzheimer's in patients. Traditionally, autopsy or biopsy was considered the only methods to absolutely confirm the presence of AD. With PET technology, it may now be possible to identify Alzheimer's in a fairly early phase.

More studies have pointed to the possible effectiveness of using PET scanning of the hippocampus as a way to detect AD while in its early stages. It is a well-known medical fact that the hippocampus, a region of the brain that is instrumental in learning and short-term memory, is affected in the early stages of AD. It is believed that through a PET scan of hippocampus it will be possible to see the first signs of AD long before it has spread to the cerebral cortex, which damages cognitive function and impairs the memory. Future studies on the viability of a PET scan of hippocampus have been undertaken to further the use of PET scanning for detecting AD.

In one of the largest studies ever to compare tests for AD, researchers from the University of California, Berkeley found that word-recollection memory testing, like the MMSE combined with PET scans of the brain, was best able to predict who would develop AD in those with early cognitive impairment [15]. Study participants whose PET scans and memory tests were abnormal were nearly 12 times more likely to develop AD than people whose scores on both tests were normal. This study shows the clear benefit of using multiple tests in a diagnostic program to best characterize the patient's risk and the potential course of the disease. Clearly, adding even more types of tests will add even more predictive value. Importantly, however, PET scans provide very little, if any, information as to the cause of AD.

Many studies, within the standard-of-care, have concluded that PET scanning may be the best single test for predicting progression to Alzheimer's. But the cost of the imaging test will limit its use. The cost of a PET scan is anywhere from $3000 to $6000—about twice the cost of normal MRI brain scans (those without special amyloid-sensitive dyes). Is it realistic to think PET can be used for general disease screening or clinical care? Can our society absorb the tremendous costs associated with this type of testing? A goal must be to develop less costly predictors of progression to Alzheimer's that are as good as or better than PET. The main value of PET may be as a reference to compare to new strategies. Read Chapter 6 for a thorough understanding of tests that have the diagnostic value of PET and can be used in an inexpensive screening mode.

X-Ray Computer Tomography Scan

An x-ray computer tomography (CT) scan produces a clear, two-dimensional image of the brain that shows abnormalities such as brain tumors, blood clots, strokes, or damage due to head injury. CT scans are painless and noninvasive (nothing is inserted into the head in order to get the image), but they might be difficult for someone claustrophobic. The 20-minute procedure is usually done at a hospital or clinic that specializes in imaging. It involves lying on a table that's inserted into a chamber where the pictures are taken.

A CT scan provides a picture of the anatomy of the brain by taking multiple X-rays and reconstructing the image of the brain with a computer. CT scans are used in patients with dementia to rule out stroke, tumor, or hydrocephalus. In the early stages of AD, a CT scan may look normal because the changes in the brain occur at a microscopic level. However, in later stages, one of the memory centers of the brain, the hippocampus, may be smaller. Sometimes a contrast agent or dye is injected into a vein in the arm before the CT scan to obtain a more detailed picture of the brain's anatomy.

Although CT scans do not usually contribute to the recognition of AD, the presence of ventricular enlargement may help distinguish AD from other dementias.

In summary, the three imaging techniques, MRI, PET, and CT, are best used to rule out causes other than dementia for cognitive impairment. They each contribute structural information that is useful for a diagnosis. MRI is very expensive but is usually preferred as it provides more detailed information compared to the other tests. None of these tests is considered useful in providing an early diagnosis, but methods are being developed to enhance their sensitivity. These tests should be used to support a diagnosis and help track the path of the disease over time. These tests do not provide information as to the root cause of the disease.

WHO CREATES DIAGNOSTIC CRITERIA?

Where does diagnostic information come from? Like most scientific findings, it comes from the research literature. We caught a glimpse into the future early in this chapter through the workgroups established by governmental and academic experts. Our existing criteria came from work published in 1985 in *Archives of Neurology* titled "Diagnosis of Alzheimer's disease" [16].

Early and accurate diagnosis of Alzheimer's disease (AD) has a major impact on the progress of research on dementia. To address the problems involved in diagnosing AD in its earliest stages, the National Institute on Aging, the American Association of Retired Persons, the National Institute of Neurological and Communicative Disorders and Stroke, and the National Institute of Mental Health jointly sponsored a workshop for planning research.

The purpose of the meeting was to identify the most important scientific research opportunities and the crucial clinical and technical issues that influence the progress of research on the diagnosis of AD. The 37 participants included some of the most knowledgeable and eminent scientists and physicians actively involved in the study of AD. The participants were divided among six panels representing the disciplines of neurochemistry, neuropathology, neuroradiology, neurology, neuropsychology, and psychiatry. Within each of the panels, participants discussed specific areas of research requiring further investigation.

Hopefully this excerpt assists in understanding the issues patients face in overcoming AD. The list of participants is all from the field of neurology. And the representation has not changed since these early days, as evidenced by the list of neurologists and pathologists involved in the next generation of standards. That is what is meant by "neurology owns Alzheimer's disease." Neurology has established the standards of care for disease management of Alzheimer's from diagnosis to treatment, to palliative and compassionate care. As Einstein said, "Insanity is doing the same thing over and over again and expecting a different result."

The *Journal of the American Medical Association* published a so-called "Consensus Statement" titled "Diagnosis and treatment of Alzheimer's disease and related disorders, consensus statement of the American Association for Geriatric Psychiatry, the Alzheimer's Association, and the American Geriatrics Society" [17]. The abstract of that paper is presented as follows:

Objective. A consensus conference on the diagnosis and treatment of Alzheimer's disease (AD) and related disorders was organized by the American Association for Geriatric Psychiatry, the Alzheimer's Association, and the American Geriatrics Society. The target audience was primary care physicians, and the following questions were addressed: (1) How prevalent is AD and what are its risk factors? What is its impact on society? (2) What are the different forms of dementia and how can they be recognized? (3) What constitutes safe and effective treatment for AD? What are the indications and contraindications for specific treatments? (4) What management strategies are available to the primary care practitioner? (5) What are the available medical specialty and community resources? (6) What are the important policy issues and how can policy makers improve access to care for dementia patients? (7) What are the most promising questions for future research?

Participants. Consensus panel members and expert presenters were drawn from psychiatry, neurology, geriatrics, primary care, psychology, nursing, social work, occupational therapy, epidemiology, and public health and policy.

Evidence. The expert presenters summarized data from the world scientific literature on the questions posed to the panel.

Consensus Process. The panelists listened to the experts' presentations, reviewed their background papers, and then provided responses to the questions based on these materials. The panel chairs prepared the initial drafts of the consensus statement, and these drafts were read by all panelists and edited until consensus was reached.

Conclusions. Alzheimer's disease is the most common disorder causing cognitive decline in old age and exacts a substantial cost on society. Although the diagnosis of AD is often missed or delayed, it is primarily one of inclusion, not exclusion, and usually can be made using standardized clinical criteria. Most cases can be diagnosed and managed in primary care settings, yet some patients with atypical presentations, severe impairment, or complex co-morbidity benefit from specialist referral. Alzheimer's disease is progressive and irreversible, but pharmacologic therapies for cognitive impairment and nonpharmacologic and pharmacologic treatments for the behavioral problems associated with dementia can enhance quality of life. Psychotherapeutic intervention with family members is often indicated, as nearly half of all caregivers become depressed. Health care delivery to these patients is fragmented and inadequate, and changes in disease management models are adding stresses to the system. New approaches are needed to ensure patients' access to essential resources, and future research should aim to improve diagnostic and therapeutic effectiveness.

Work like this leads to the establishment of published and mandated policies and procedures in medicine for the management of disease. We see from the participants that a multimodal group of researchers and physicians was not assembled. There was hardly a "consensus" because those who really have the potential to understand this disease were not invited to the discussion. Interestingly, the consensus suggested that primary care could handle AD. What do you believe? Your PCP could coordinate diagnosis and care but by coordinating a multidisciplinary medical team to first perform a differential diagnosis.

References

[1] Andreassi K. New clues to a long life. National Geographic Magazine, May 2013.

[2] Jack CR Jr, et al. Introduction to the recommendations from the National Institute on Aging–Alzheimer's Association workgroups on diagnostic guidelines for Alzheimer's disease. Alzheimers Dement 2011;7(3):257–62.

[3] McKhann GM, et al. The diagnosis of dementia due to Alzheimer's disease: recommendations from the National Institute on Aging–Alzheimer's Association workgroups on diagnostic guidelines for Alzheimer's disease. Alzheimers Dement 2011;7(3):263–9.

[4] Albert MS, et al. The diagnosis of mild cognitive impairment due to Alzheimer's disease: recommendations from the National Institute on Aging–Alzheimer's Association workgroups on diagnostic guidelines for Alzheimer's disease. Alzheimers Dement 2011;7(3):270–9.

[5] Sperling RA, et al. Toward defining the preclinical stages of Alzheimer's disease: recommendations from the National Institute on Aging–Alzheimer's Association workgroups on diagnostic guidelines for Alzheimer's disease. Alzheimers Dement 2011;7(3):280–92.

[6] Montine TJ, et al. National Institute on Aging–Alzheimer's Association guidelines for the neuropathologic assessment of Alzheimer's disease: a practical approach. Acta Neuropathol 2012;123(1):1–11.

[7] Folstein MF, Folstein SE, McHugh PR. "Mini-mental state": a practical method for grading the cognitive state of patients for the clinician. J Psychiatr Res 1975;12(3):189–98.

[8] Cockrell JR, Folstein MF. Mini-mental state examination. In: Copeland JRM, Abou-Saleh MT, Blazer DG, editors. Principles and practice of geriatric psychiatry. Chichester, United Kingdom: John Wiley & Sons Ltd; 2002. p. 140.

[9] Mitrushina M, Satz P. Reliability and validity of the Mini-Mental State Exam in neurologically intact elderly. J Clin Psychol 1991;47:537–43.

[10] Faustman WO, Moses JA Jr, Csernansky JG. Limitations of the Mini-Mental State Examination in predicting neuropsychological functioning in a psychiatric sample. Acta Psychiatr Scand 1990;81:126–31.

[11] Folstein M. Mini-mental and son. Int J Geriatr Psychiatry 1998;13:290–4.

[12] Colliot O, Chételat G, Chupin M, Desgranges B, Magnin B, Benali H, et al. Discrimination between Alzheimer disease, mild cognitive impairment, and normal aging by using automated segmentation of the hippocampus. Neuroradiology 2008;248:194–201.

[13] MRI may prove powerful tool in predicting development of Alzheimer's disease. National Institute on Aging; March 29, 2000. National Institutes of Health; July 1, 2008.

[14] Fleisher AS, Chen K, Liu X, et al. Using positron emission tomography and florbetapir F18 to image cortical amyloid in patients with mild cognitive impairment or dementia due to Alzheimer disease. Arch Neurol 2011;68(11):1404–11.

[15] Landau SM, et al. Comparing predictors of conversion and decline in mild cognitive impairment. Neurology 2010;75(3):230–8.

[16] Khachaturian ZS. Diagnosis of Alzheimer's disease. Arch Neurol 1985;42(11):1097.

[17] Small GW, et al. Diagnosis and treatment of Alzheimer disease and related disorders. Consensus statement of the American Association for Geriatric Psychiatry, the Alzheimer's Association, and the American Geriatrics Society. JAMA 1997;278(16):1363–71.

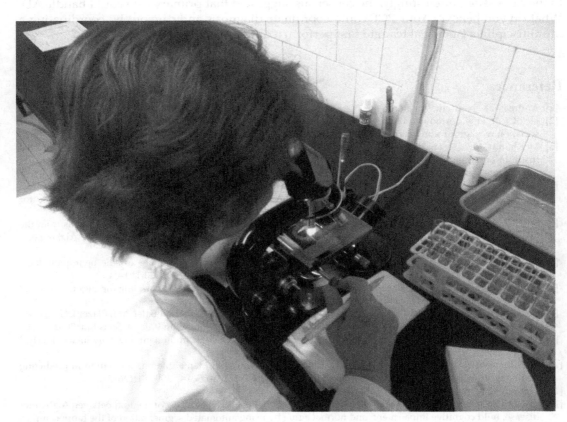

Image Credit: Microscopic observation in biochemical laboratory, by Делфина (Own work). Available at Wikimedia Commons: https://commons.wikimedia.org/wiki/File%3AMicroscopic_observation%2C_%D0%9C%D0%B8%D0%BA%D1%80%D0%BE%D1%81%D0%BA%D0%BE%D0%BF%D0%B8%D1%80%D0%B0%D1%9A%D0%B5.jpg

Diagnostic Accuracy

We automatically have faith in our healthcare system. A doctor is one of the most well-respected professions, and doctors are highly educated people who go through more education prior to entry into the workforce compared to any other. And they take and must master tough courses like Chemistry (yuck). Doctors deserve a great deal of respect, but we must recognize that they are not God; rather, they are human. Doctors need not shoulder blame for the woes of modern medicine. This chapter points to a myriad of causes that impact the quality of care delivered to you, the patient.

The human body is so complex that medicine teeters between being a science and an art. Of course the key to crossing the meridian from art to science is all in the diagnosis. A prescriptive, subjective, or narrow approach to diagnosis will frequently yield an incorrect diagnosis for our most complex diseases. The Society for Functional Medicine states:

> If you are not testing, you are guessing.

Many of the tools used by neurology, the main medical discipline for diagnosing Alzheimer's disease (AD), put this branch of medicine clearly in the "art" category. They do have sophisticated tools like PET and MRI (and other alphabet soup). But they are only for measuring the brain. That's only 2% of the body! What about the other 98%? Does it count? And what does an MRI tell you about root cause? For example, the MRI might say you have atrophy and mini-strokes (microinfarcts). Great, my brain is shrinking, and I've had some mini-strokes. So what does this diagnosis tell the doctor about treatment strategies? The answer is little to nothing as it is just used to confirm a diagnosis of, for example, AD. But Alzheimer's should not be the end point of a diagnosis. Some things are causing and contributing to the progression of AD. Medicine must dig deeper.

There is a significant body of literature that explores true medical misdiagnoses. This chapter reviews some of that work. Those studying the accuracy of medicine largely ignore certain types of misdiagnoses. Some of the diseases we suffer today have fancy names and describe a range of symptoms. A diagnosis of one of these "names" is considered correct within the standard-of-care but discourages deeper and broader diagnosis that might reveal root causes that could lead to better treatment. For example, a diagnosis of "Alzheimer's" is considered correct.

This chapter explores diagnoses and misdiagnoses in medicine in general. Neurology is not singled out. The reason this topic is important is that we, the consumer, need to understand

The End of Alzheimer's. http://dx.doi.org/10.1016/B978-0-12-812112-2.00004-5

the credibility of the diagnosis we are given. Doctors, if they want to be paid for their work, must arrive at a diagnosis that is found in a book of codes. What if your ailment doesn't fit? You still will leave the doctor's office with a diagnosis, period. What do you do? Appreciate that medicine today has a severe limitation; thus it is incumbent upon you to take a proactive approach to your own well-being.

In the remainder of this book, the concept of a true differential diagnosis as it relates to Alzheimer's and dementias is described. Hopefully this chapter and the remainder of the book will empower you with knowledge to ask questions of your physicians about diagnoses that are truly important to your health. Do not allow your doctors to "wash their hands" because you meet the minimal descriptive requirement of medicine that simply triggers medical reimbursements. You deserve better, and they are capable of better when time and money allows.

The Motley Fool published an article titled "Make money from Alzheimer's whether drugs succeed or not." We spent Chapter 2 evaluating the shortcomings of the Amyloid Cascade Hypothesis. Yet this approach continues to be pursued since funds are flowing into this approach. In the Motley Fool article, they discuss such topics as not having the right target and not treating the right patients. In their summary they state, "It used to be that the only way to diagnose an Alzheimer's patient was postmortem; an autopsy would reveal the plaques in the brain. Cognitive tests while the patient was alive could show that the patient's mind was failing, but it's hard to distinguish between Alzheimer's and dementia." But a misdiagnosis between Alzheimer's and dementia is a relatively minor mistake. All these diseases, as you will see, are associated with chronic systemic inflammation; thus they are very much connected.

Do you want to be diagnosed? As it turns out, most people want to know the risk of getting a disease. Tufts School of Medicine came to that conclusion when they evaluated the results of a recent study they conducted [1]. They presented hypothetical scenarios to 1463 people that they generated randomly, including the scenario of being diagnosed with AD. As part of their matrix, they presented the possibility of getting a disease, based on the diagnosis, as 10% and 25%. As another variable, they indicated that test accuracy might be perfectly accurate or somewhat accurate.

The majority of people said they would be tested, and almost regardless of the scenarios. Most people wanted to know, even if there wasn't a cure based on the diagnosis. "The results suggest that most people prefer to take predictive tests even in the absence of direct treatment consequences, and they are willing to pay reasonably large amounts for the opportunity. Our study thus adds to research indicating that people desire information for its own sake." Positive responses to diagnosis and testing ranged from a low of 70.4% for an imperfect test for Alzheimer's and a 10% average risk of disease to 88% for a perfectly accurate prostate-cancer test and a 25% average risk of the disease.

The result for AD is quite surprising. Historically, there have been ongoing discussions in the medical community about communicating dreadful diagnoses. A frequent topic is whether patients in the early stages of AD should be told their diagnosis. This has become of interest in the course of evaluating and caring for patients with AD and dealing with their caregivers and families. Unlike patients with other chronic illnesses that have signs and symptoms that prompt the patients to seek consultation on their own, family members who suspect the diagnosis of AD usually bring patients with memory problems. Often any evaluation is preceded by impassioned pleas from the family not to disclose findings to the patient should this diagnosis be confirmed. Thus apparently a patient wants to know if the disease is brewing while loved ones chose to protect the would-be sufferer from the truth.

DIAGNOSIS EXPLORED

When someone in your family has been given an Alzheimer's diagnosis, you may be severely distressed and wonder just how impacted your future will be. Some may have purposely put off seeing their doctor when their family member has shown symptoms of the disease simply because they don't want to hear the bad news of an Alzheimer's diagnosis. There are many sites devoted to successful diagnoses and treatment of a myriad of diseases but Alzheimer's is a clear exception. An up and down search of the web to find diagnosis success stories about Alzheimer's yields nothing. This isn't surprising because the diagnosis, although comforting by providing an understanding of your loved one's behavior, is still considered a tortuous death sentence.

Medicine consists of two distinct parts: (1) Diagnosis—figuring out what is wrong with the patient; and (2) Treatment—deciding what to do for the patient, and then carrying out the plan. Far too many healthcare professionals and laypeople seem to feel that medical diagnosis isn't really all that complicated. We want the treatment—that is, we want our pills! You can rely on a doctor's hunch based on a set of criteria or plug symptoms and exam findings into an appropriately sophisticated algorithm and out pops the answer. Hey, patients can even do it themselves on *WebMD*, right? Humans are too complex to relegate diagnosis and treatment solely to "processes." Even auto repair requires input from an experienced repairman. Great physicians always apply an artist's sensibility. Diagnosis requires an artist's eye, a writer's ear for dialogue, and a sense of composition. Medical professionals apply scientific knowledge but that alone is not enough.

Medical diagnosis is truly an art that takes years to fully master eliciting nuances of the patient medical history and a broad and deep set of diagnostic information. Appreciating a subtle physical finding or findings with painstakingly honed physical examination techniques can be accomplished only with time. These are skills attained only with copious hands-on experience backed up by years of study and updating of old and emerging medical information. Great physicians are great diagnosticians first and foremost. And it is all those background years of education and training that prepares the healthcare practitioner to master this critical skill. Thus you can see the value of having several doctors presiding over your diagnosis.

Treatment is arguably far more straightforward than diagnosis. "Cookbook" medicine (i.e., providing a standard treatment based on a diagnosis) often works well (we hope, because that is what we have), but only to the extent that the patient's condition has been correctly diagnosed. Indeed treatment often needs tweaking for individual patients, but this is seldom as complex an endeavor as diagnosis. And this is where medicine could fail the patient. There are great treatments available, but that becomes mute in the face of a poor diagnosis. The 80/20 rule adequately describes the weight of effort health care should apply to diagnosis versus treatment. Diagnosis should be a minimum of 80% of the clinical process and treatment only 20%. In today's 10-minute doctor visit, that gives diagnosis a paltry 9 minutes. That is not enough to go beyond a diagnosis of symptoms. We all need and deserve a diagnosis of the root causes of our affliction.

MISDIAGNOSIS

Surveys and other studies make a strong case that many of us are touched by misdiagnoses. Case in point, my daughter, many years ago when she was 8, became quite ill. Her symptoms included vomiting, a tummy ache, and a general feeling of sickness and malaise. She saw her local pediatrician on a number of occasions over a week and tests were extensive, including an MRI. Yet she went undiagnosed and her condition worsened, as she was bedridden. I called a friend whom I respected, an emergency room doctor, and explained the situation. He without hesitation said, "She has FOS syndrome." She was constipated! Yes, constipated. And our local doctors with all their tools couldn't even see it on the MRI. We gave her an enema and problem solved.

If we cannot diagnose constipation, how can we expect a diagnosis of AD? The answer is fourfold:

- There are lower expectations.
- Test, test, do more tests, and ask for tests that are not normally performed, particularly those associated with system-wide inflammation in the blood.
- Get multiple opinions from different areas of medical specialization.
- Pay for your own tests because medical insurance will limit you otherwise. Do you want your care limited by your insurance coverage?

Misdiagnoses are rampant and on the rise. "With all the tools available to modern medicine you might think that misdiagnosis has become a rare thing. But you would be wrong. Studies of autopsies have shown that doctors misdiagnose fatal illnesses about 20 percent of the time. Thus, millions of patients are being treated for the wrong disease or not being treated at all." This was excerpted from a *New York Times* article, from February 22, 2006. The rate of serious misdiagnosis has not really changed since the 1930s. This information comes from the prestigious *Journal of the American Medical Association*. This is the richest country in the world, where one-seventh of the economy is devoted to health care, and yet misdiagnosis is killing tens of thousands of Americans every year.

The issue holding diagnosis back is that financial incentives are not aligned with the rigor of the diagnosis. Under the current medical system, doctors, nurses, lab technicians, and hospital executives are not actually paid to come up with the right diagnosis. They are paid to perform tests, to do surgery, and to dispense drugs. There is no bonus for curing someone and no penalty for failing, except when the mistakes rise to the level of malpractice. But that rarely happens when the medical team stays within the confines of the standard-of-care.

Misdiagnosis can and does occur and is reasonably common with error rates ranging from 1.4% in cancer biopsies to a high 20%–40% misdiagnosis rate in emergency or ICU care. Surveys of patients also indicate the chance of experiencing a misdiagnosis to range from 8% to 40%. This makes misdiagnosis one of the most common types of medical mistakes. There are various reasons as to why a misdiagnosis can occur including errors by doctors, specialists, and laboratory tests. The patient can also contribute to an error in various ways. But arguably the biggest reason for misdiagnosis is the prescriptive nature of medicine. Doctors are compelled, for reimbursement reasons, to assign a diagnosis based on a book of diagnostic codes, and these codes may not address the underlying cause of disease.

Misdiagnosis need not be a feared outcome. There are various ways to prevent a misdiagnosis such as seeking a second opinion or a specialist referral. Getting educated about the possible alternative or underlying diagnoses for a condition is useful information to discuss with your doctor. And don't submit to the diagnosis. Dig as deep as you can by seeking medical professionals who provide a supplemental or alternative view. If you rely solely on the healthcare system as it is, you get a prescription.

Lawrence Weed, MD, is a physician who spent the later years of his career investigating the pitfalls of medicine. His findings are embodied in a book titled *Medicine in denial."* A couple of excerpts from his book further the case of a system likely to continue to have misdiagnoses:

> Contrary to what the public is asked to believe, physicians are not trained to connect patient data with medical knowledge safely and effectively. Rather than building that foundation for decisions, autonomous physicians traditionally rely on personal knowledge and judgment, in denial of the need for external standards and tools. Medical decision-making thus lacks the order, transparency and power that external standards and tools would bring to it. Physicians are left to carry a prohibitive burden. Acting under severe time constraints, they must connect intricate patient data with crucial details from vast and growing medical knowledge. The outcome is that the entire health care enterprise lacks a secure foundation.
>
> In short, essential standards of care, information tools, and feedback mechanisms are missing from the medical marketplace. And the underlying medical culture does not even recognize their absence. This does not prevent some caregivers from becoming virtuoso performers in narrow specialties. But that virtuosity is personal, not systemic, and limited, not comprehensive. Missing is a secure system for enforcing care of high quality by all caregivers for all patients.

PREVALENCE OF MISDIAGNOSIS

An accurate initial medical diagnosis is the foundation upon which all subsequent healthcare decisions are based. An error in diagnosis can cause a cascade of negative events to occur, affecting the individual patient and their families as well as the healthcare system and our society as a whole. This is quite true for AD; after all, "Alzheimer's" is just a label for a constellation of symptoms relative to cognitive deficits. Alzheimer's is a diagnosis of a syndrome but it is not a diagnosis of a disease or a set of diseases. In that sense, "Alzheimer's disease" is a misdiagnosis.

Medical misdiagnosis has three accepted major categories:

- False-positive: misdiagnosis of a disease that is not actually present
- False-negative: failure to diagnose a disease that is present
- Equivocal results: inconclusive interpretation without a definite diagnosis

Consider a fourth category: insufficient breadth and depth of diagnosis. This happens because of assumptions about the causes of disease. AD, for example, is looked at narrowly as "brain only." This approach lacks adequate depth and breadth to arrive at a complete and accurate diagnosis. We are not looking far enough with available tools based on available knowledge (the Trillion Dollar Conundrum).

There have been multiple autopsy studies that have uncovered frequent clinical errors and misdiagnoses, with some rates as high as 47% [2]. A study of autopsies published in the *Mayo Clinic Proceedings* comparing clinical diagnoses with postmortem diagnoses for medical intensive care unit patients revealed that in 26% of cases, a diagnosis was missed clinically [3]. If the true diagnosis were known prior to death, it might have resulted in a change in treatment and prolonged survival in most of these misdiagnosed cases. The study's researchers concluded, "Despite the introduction of more modern diagnostic techniques and of intensive and invasive monitoring, the number of missed major diagnoses has not essentially changed over the past 20 to 30 years."

It is difficult to really measure the frequency of misdiagnoses, except in the relatively rare studies of living diagnoses versus actual autopsy findings. There is a general feeling that misdiagnosis is quite common, with many people giving anecdotal accounts of their own experiences. Whereas there are many studies of adverse drug events and nosocomial infections (infections obtained during a hospital visit), there is a relative lack of misdiagnosis studies. A study of Patient Safety Incidents by HealthGrades found that "Failure to Rescue" (meaning failure to diagnose and treat in time) was the most common cause of a patient safety incident, with a rate of 155 per 1000 hospitalized patients [4]. Unfortunately, the study did not further break down statistics into the types of misdiagnosis, delayed diagnosis, or other factors.

The National Patient Safety Foundation (NPSF) commissioned a phone survey in 1997 to review patient opinions about medical mistakes [5]. Of the people reporting a medical mistake (42%), 40% reported a "misdiagnosis or treatment error," but did not separate misdiagnosis from treatment errors. Respondents also reported that their doctor failed to make an adequate diagnosis in 9% of cases, and 8% of people cited misdiagnosis as a primary causal factor in the medical mistake. Loosely interpreting these facts gives a range of 8%–42% rate for misdiagnoses.

There is clearly a cost for misdiagnoses. The national costs of medical errors resulting in injury are estimated to be between $17 and 29 billion annually. The costs to the United States healthcare system represent over 50% of these additional expenses. These expenditures burden not only health plans and insurers but also employers who are already reeling with escalating premiums as well as individuals who must dig deeper to cover copays.

The failure to diagnose a condition is one of the most common types of misdiagnosis. Malpractice lawsuits from failure or delayed diagnosis occur mostly, in terms of dollar value, from conditions such as heart attack, breast cancer, appendicitis, lung cancer, and colon cancer. However, these are not the most common undiagnosed conditions, but are simply the ones that lead to the most rapid damages. Serious conditions such as diabetes and hypertension are very commonly undiagnosed, but do not lead as rapidly to severe injury.

To examine how commonly failure to diagnose or delayed diagnosis occurs, in Table 4.1 is given a list of conditions according to the number of people undiagnosed. This is an estimate of how many people unknowingly currently have the condition. In most cases, the rates refer to the United States or other industrialized nations.

Have you ever heard of Toxoplasmosis? Yet it is at the top of the list of most misdiagnosed disease. A search of the medical literature yields 5600 articles that include both AD and Toxoplasmosis.

Consider this personal story. The son of a friend's wife was diagnosed with a lymphoma-based cancer by a major prestigious university medical center in New England. His case was desperate, and he was descending rapidly. My friend called me (T.J.L.) to discuss the case, and I contacted my coauthor, Dr. Trempe. Upon describing the symptoms, Dr. Trempe said, "He could have Toxoplasmosis infection. Toxoplasmosis can get into the lymph system and

TABLE 4.1 List of Conditions Commonly Undiagnosed or Diagnosed Late

Condition	Percent (%)	United States People
Toxoplasmosis[a]	22.06	60 million
Sleep disorders	14.71	40 million
Otosclerosis	10.00	27.2 million
Extra Nipples	10.00	27.2 million
Osteoporosis	6.62	18 million
Hypertension[a]	5.51	15 million
COPD	5.51	15 million
Chronic lower respiratory diseases	5.51	15 million
Migraine	5.15	14 million
Thyroid disorders[a]	4.78	13 million
Latent tuberculosis[a]	3.68	10 million
Obstructive sleep apnea	3.68	10 million
Diabetes[a]	2.10	5.7 million
Sleep apnea[a]	2.00	5.4 million
Chlamydia[a]	1.25	3.4 million
Parkinson's Disease[a]	1.10	3 million
Age-related macular degeneration[a]	0.83	2.3 million
Sjögren's Syndrome	0.74	2 million
Aneurysm	0.74	2 million
Hemochromatosis	0.55	1.5 million
Salmonella food poisoning	0.51	1.4 million
Breast Cancer	0.37	1 million
Glaucoma[a]	0.37	1 million
Celiac Disease	0.37	1 million
von Willebrand disease	0.37	1 million
Open-angle glaucoma[a]	0.37	1 million
Cryptosporiosis	0.18	500,000
Gonorrhea	0.15	400,000
HIV/AIDS	0.08	225,000
Narcolepsy	0.06	150,000

COPD, chronic obstructive pulmonary disease.
[a] Conditions that may have a relationship with Alzheimer's disease.

cause swelling of the glands that look like lymphoma tumors. Did the medical center do an actual biopsy or just use imaging?" My friend confirmed that the biopsy was equivocal. He also confided that the young man (in his thirties) had HIV.

Within short order, the university medical center began a regiment of chemotherapy. In 3 weeks, the young man was dead. Now we can only speculate about what went wrong but certainly no one really probed deeply into his history to determine his HIV status. HIV and chemo is well known as a deadly combination. The family was hesitant to come to Boston for a supplemental opinion because whoever heard of Toxoplasmosis, and this doctor colleague of mine is not even an oncologist. This story illustrates the many important reasons to get a supplemental opinion. Oncologists are going to find a cancer (hammer and nail philosophy), whereas doctors outside of that discipline are likely to look elsewhere. And do not be turned off by the unknown compared to the "fashionable" diagnosis.

There are other diseases like Toxoplasmosis that do not, but probably should, be on the CDC list. Have you ever heard of Q-fever? This is of the same class of "insult" as Toxoplasmosis, yet few know about it. There are many more factors that can contribute to Alzheimer's and other diseases (such as cardiovascular disease) that are not on the radar of modern medicine. If you have some amorphous symptoms associated with your memory, what is the chance you will be ordered a Toxoplasmosis test? Answer: 0. Why? No healthcare plan will cover the cost of the test based on that diagnosis. Getting a Toxoplasmosis test and having it covered is a very challenging proposition in the descriptive standard-of-care to which we are subject.

WHY DOES MISDIAGNOSIS OCCUR?

There are many ways that a diagnosis can go wrong. There can be contributing factors from any of the participants: Patient, Doctor, Specialist, Other healthcare professional, and Tests (laboratory or pathology tests).

Patient

It seems poor form to blame a patient for a wrong diagnosis, and indeed blame is probably the wrong word. Nevertheless, the patient is involved and can contribute to a wrong diagnosis. The most likely way for a patient to contribute to a misdiagnosis is attempting to do so themselves without professional medical advice. Sometimes patients don't tell the doctors everything. Other times a patient might feel symptoms are not worth mentioning. Some people won't mention a symptom unless the doctor asks, and they assume it must be irrelevant if the doctor doesn't ask about it directly. Embarrassment is another reason a patient interferes with a thorough diagnosis. Many patients are known to neglect to mention poor oral hygiene, sexually transmitted diseases, and other aches and pains.

Compliance is a major problem in medicine. Patients just do not follow "doctor's orders." In some cases, patients don't get diagnostic tests done, even when a doctor has ordered the tests. This can occur due to oversight, complacency, denial, or other reasons. For example, one contributing factor to delayed diagnosis of colon cancer is patients' perceived embarrassment over tests such as colonoscopy and sigmoidoscopy. For the diagnosis of a complex disease, it is critical for the healthcare professional to have all medical information at their disposal, including results of blood tests done over the past several years.

Doctor

Since the doctor has to make a diagnosis, it is certainly possible to make the wrong diagnosis. There are many ways that this can occur. There are more than 20,000 human diseases, and doctors know only the most common. The doctor needs to be aware of all medical information so that a referral to a specialist can be made. Any diseases that get a lot of attention tend to get somewhat overdiagnosed. This means that less common diseases that might have similar symptoms are sometimes overlooked. Not all doctors are alike. A general practitioner is well versed in common diseases but not in more rare disease areas that a specialist would know better.

All doctors are human and have biases. They can have the "hammer-and-nail" bias: if you have a hammer, everything looks like a nail. If they see a certain disease frequently, they will diagnose it frequently and might make an error if it is not that disease, but something with similar symptoms. This tendency to go with what is familiar is also seen in treatments, where a surgeon will recommend surgery more often, but an endocrinologist will recommend pills more often.

Some doctors will avoid tests, assuming that you don't want to pay extra costs. For example, if there is a very rare condition, say 1-in-200, should your doctor get you to test for it? Some doctors won't even tell you about this type of test. This works fine for the majority, but fails for the very small percentage that might have the rare disease. To exacerbate this issue, many tests are not covered by insurance when based on a doctor's hunch. Often times a doctor may have a very clear belief but cannot justify the advanced testing needed to confirm the belief because the system does not accommodate tests that are not of "medical necessity," a favorite term of denial by insurance companies to doctors working on behalf of their patients.

The reality that medicine is a big business is interfering with the process of a full, deep, and broad diagnosis. Receding reimbursement rates for a diagnostic office visit, by both Medicare and private insurance, is limiting the time a doctor can spend with a patient. Typical reimbursement for an office visit for a patient over 65 and on Medicare is as low as $65. Consider that overhead rates for most medical practices are 75%. That means the doctor gets about $20 for a visit. The doctor must then see several patients each hour to make a living and pay for all associated debts of being a doctor, including their having attended 12 years of expensive postgraduate education.

The 1-hour appointment of old is now 10 minutes or less. That doesn't give the doctor much time to ask a few questions, make a tentative diagnosis, order some blood tests to confirm it, and then answer some questions from the patient. We'd all like to think that the doctor went and double-checked our disease in their books, discussed it with other specialists, and consulted the latest research about how to diagnose and treat it correctly, but all that requires time the doctor doesn't have, and these efforts do not receive financial compensation. In reality, doctors apply educated guesses, particularly with more complicated and rare conditions. No matter your skill level, if you are not testing, you are guessing.

The various medical tests that are used to confirm or rule out diagnoses can also sometimes fail. They are useful diagnostic tools, but are not perfect. Of course, a simple human error can occur in any of the various tests. For example, samples could get contaminated or mixed up, or the test procedure might get done improperly. Some tests require visual inspection, such as cell tests for cancer (e.g., Pap smears), and rely on the human judgment of the person inspecting them. Natural errors are rare, but they can occur. Recall that my daughter's constipated bowel didn't show well on MRI.

All laboratory tests have known conditions under which they fail. They can fail with either a false-positive, wrongly indicating that you have a condition when you don't, or a false-negative, wrongly indicating you don't when you actually do. Either way will get you the wrong diagnosis. Most tests fail very infrequently, but if you read the documentation about each test, you'll see that each have known limitations. Some tests fail on some people because of special features about a person. Some tests for one disease will fail if you have some other rare diseases.

Doctors Are Not the Problem

We know doctors are highly intelligent, highly educated, and highly trained individuals. The following is excerpted from *Occupational outlook handbook*, 2010–11 edition, from the Bureau of Labor Statistics:

- Many physicians and surgeons work long, irregular hours.
- Acceptance to medical school is highly competitive.
- Formal education and training requirements, typically four years of undergraduate school, four years of medical school, and three to eight years of internship and residency are among the most demanding of any occupation.

So what has gone wrong? Much modern research points to the system rather than the doctor. Simply consider how the complexity of technology has advanced in the past couple of decades. Indeed, the technology is there at our disposal and, in some ways, should make our jobs easier and more productive. However, technology can be overwhelming, too, even to doctors, especially when they are pressed for time by the business of medicine.

Causes of Misdiagnosis

In 1999, The Institute of Medicine released "To err is human," which asserts that the problem in medical errors and misdiagnoses is not bad people in health care; it is that good people are working in bad systems that need to be made safer [6]. Their major finding was that communications within the system were the single biggest cause of mistakes and errors in treatment and assessment. The problems in the communication system include the following elements:

- poor communication, unclear lines of authority of physicians, nurses, and other care providers;
- disconnected reporting systems within a hospital: fragmented systems in which numerous handoffs of a patient result in lack of coordination;
- the impression that action is being taken by other groups within the institution;
- reliance on automated systems to prevent error and assist with diagnosis;

- inadequate systems to share information about errors hampering analysis of contributory causes and improvement strategies;
- cost-cutting measures by hospitals in response to reimbursement cutbacks.

The Joint Commission's annual report on quality and safety in 2007 found that inadequate communication between healthcare providers, or between providers and the patient and family members, was the root cause of over half of the serious adverse events in accredited hospitals [7]. Other leading causes included inadequate assessment of the patient's condition and poor leadership or training.

PREVENTING MISDIAGNOSIS

A billionaire was highlighted on Fox Business News as having apparently been misdiagnosed. He was initially diagnosed with meningitis but over a year later he was rediagnosed and determined to have a serious form of brain cancer. How could such a misfortune befall a billionaire, who is able to afford the best doctors, the best technology, and the most sophisticated diagnostic tests? Sure, a billionaire potentially has better access. However, the playing field has leveled in terms of equipment and methods. And a busy doctor is a busy doctor even for a billionaire who pays by the same method as you and me, some type of insurance. Thus, even for a billionaire, getting the right care is "still a bit of a crap shoot." This further illustrates the complexity of medicine, the body, and the art of diagnoses.

A presumed "Alzheimer's" patient of substantial means along with his family flew to Boston in his private jet for an evaluation and hopefully a treatment. They were a delightful group, and the father, with apparent dementia, was a robust gentleman of about 65 years of age. He was diagnosed with AD by the Head of Neurology and Dean at a well-known medical school. This former prominent businessman raised millions in funds for that medical school. He was also obtaining treatment from a Florida clinic that claimed success in treating AD. A colleague examined him and determined he had certain pathologies that did not suggest Alzheimer's per se, but appeared to be contributing to his condition. The family flew back to their home, and a family member went to see the head of Neurology and Dean of the University her father had assisted to validate the diagnosis. The Dean said, "Everyone is positive for these things." He basically dismissed this alternative diagnosis that was not, by the way, a "brain-only" diagnosis. My colleague asked if the Dean would treat based on the diagnosis, as the target of the diagnosis, similar to Toxoplasmosis, causes a variety of diseases. He received no reply from the Dean. The neurology team continued to test and treat with standard Alzheimer's medications, and the patient continued to decline.

So how can you improve your odds? Be prepared, be educated, and be your own or your loved one's advocate. The medical profession is too busy to advocate for you so take charge and quarterback your wellness. There are a lot of things that you can do to avoid or reduce the risk of a wrong diagnosis or incomplete diagnosis, but you will have to push the system as it exists today.

Consider the following excerpt from the biography of Steve Jobs by Walter Isaacson:

> Jobs allowed his wife to convene a meeting of his doctors. He realized that he was facing the type of problem that he never permitted at Apple. His treatment was fragmented rather than integrated. Each of his

myriad maladies was being treated by different specialists—oncologists, pain specialists, nutritionists, hepatologists, and hematologists—but they were not being coordinated in a cohesive approach, the way James Eason had done in Memphis. 'One of the big issues in the health care industry is the lack of caseworkers or advocates that are the quarterback of each team,' Powell (Jobs's wife) said. This was particularly true at Stanford, where nobody seemed in charge of figuring out how nutrition was related to pain care and to oncology.

Who figured this out and then quarterbacked Mr. Jobs's care? A family member became his advocate, specifically his wife. For complicated diseases like Alzheimer's, especially as this disease becomes recognized and managed as a whole-body disease, having a family advocate will become even more important because more and more specialists will become involved. The multifactorial nature of Alzheimer's will exacerbate the fragmentation that is systemic to the healthcare system already.

Always consider getting a second opinion or what is called a supplementary diagnosis. Getting the opinion of two or more doctors makes the chances of a wrong diagnosis even lower. And it is better to get a second opinion from different disciplines. For example, two surgeons will both recommend surgery while an internist may have an opposing or even complementary but different view. Have the second healthcare professional look at your case from scratch; hear you talk about your symptoms in your own words, and think about your case without being influenced by the conclusions of your original doctor. Don't say, "I was seen by Dr. X and he/she tells me I have X and need treatment Y, what do you think?" Instead, describe your symptoms, tell him or her about your family history and the tests you've had done, and help this doctor come to his or her own conclusion. If the two diagnoses match, then the chances of a wrong diagnosis are much lower. And if they don't match, then there is a puzzle to solve. You have only one life and maybe only one chance to get this right.

Know your family history and remind your doctor of it. Don't assume your doctor remembers that time you told him that two of your aunts died of breast cancer, or that your grandfather and father have a history of malformed blood vessels in their brains. Research studies have shown that a family history may be a better predictor of disease than even genetic testing. Family history is not solely a description of family genes. History gives much more broad information about environment, too. Find out about your family's medical history, write it down (the Surgeon General has a good online tool to help you do this), and make sure your doctor knows about it, especially if you're sick and they're trying to find out what's wrong.

Ask questions. The typical doctor sees as a many as 40 patients a day. It's all too easy to be referred to a specialist and start treatment without having all of your questions answered. But asking questions won't just make you feel more comfortable; it can disrupt your doctor's thought process and make him or her think about your case in a way that may save your life. Dr. Jerome Groopman, one of the world's foremost researchers on how doctors think (he's written the definitive book on it), agrees [8]:

> Doctors desperately need patients and their families and friends to help them think. Without their help, physicians are denied key clues to what is really wrong. I learned this not as a doctor but when I was sick, when I was the patient.

You can find some useful tips on what questions to ask at the United States government's web site called "Questions are the answer." Also, advocate for yourself. Seek the latest information of which your doctor may not be aware. Also, some tests are not covered by insurance so your doctor will be hesitant to order these tests. They are tests either outside the standard-of-care or not covered based on the initial diagnosis. Many of these tests, appropriate to AD, are covered in subsequent chapters.

Do not assume technology will save you. The best medical technology is available today. Still, studies show it is no more effective at getting the right diagnosis or keeping you alive than a doctor piecing together your family history along with more traditional, low-tech tests. If you had to pick between getting a high-tech test and a doctor who will spend an hour talking to you, thinking about your case, and putting all of the pieces together, research says you should pick the doctor.

You are not alone if you agree with "picking the doctor." Guess who agrees with you? The doctor! Doctors are frustrated by the "10-minute visits" as much as you. But that doesn't change the reality of medicine today. Imagine you are a patient waiting to see your physician, who arrives, greets you, sits down, and performs some perfunctory small talk. The doctor inquires about your ongoing medical problems, listens intently, maintains eye contact, and elicits pertinent information with skillful questioning. The doc gives you the impression that he or she is empathetic. It's hard to believe, but doctors now undergo training on how to make a short office visit seem longer, and they don't like it, but it has become a requirement for compensation purposes in some instances.

Suddenly the doctor checks the time, stands up, and says, "I'm sorry, but the 10 minutes allotted for your appointment are up. I'm afraid I don't have time to review your current medications, examine you, recommend appropriate treatment and preventive measures, or answer any questions you may have. You'll have to schedule two more visits."

Would you accept a partial appointment from your auto repair guy or your attorney? Would either shortchange you? They might if they are paid a fixed fee for the appointment, as is the case in medicine, rather than being paid by an hourly rate or an appropriate fee based on the service delivered. At that point, you would likely express significant chagrin and disgust. Although this scenario is contrived, it is becoming a more normal and real-world dilemma that raises a question for physicians: Can doctors provide quality care (defined by the Institute of Medicine as safe, timely, efficient, effective, equitable, and patient-centered) in a system that, for financial reasons, essentially mandates no more than 10 minutes of doctor/patient interaction? Clearly the answer is a decided no.

In the recent *New England Journal of Medicine* article, "The value of DNKs," author Susan Mackie, MD, a one-time internal medicine resident at Beth Israel Deaconess Medical Center in Boston, described her struggle to provide patient-centered care during 10-minute clinic appointments [9]. She explained how she survived the grueling pace of her clinic only because of "DNKs," an acronym for the patients who "did not keep" their appointments. These no-shows allowed her and her fellow residents to spend more time with their other patients.

Relying on DNKs to solve the time crunch is hardly a solution to the problem. What happens on days when there are no DNKs? How can any physician provide patient-centered care in 10 or even 15 minutes, especially to patients with complex medical problems who are taking a long list of medications? Entering this information into electronic medical record systems takes 10 minutes. Of course, there are patients for whom brief appointments may be appropriate, such as those with minor trauma, sore throat, and earache. They often can be evaluated and appropriately treated in a short amount of time. But even for those patients, such brief appointments may not be sufficient. For example, a clinician may need to explain to a parent why she would not immediately recommend antibiotics for a child with an ear infection and why it's important to see if symptoms are resolved within 48 hours before prescribing them.

In her article, Mackie describes what she's learned from her preceptor about how to make the most out of a short appointment: "My preceptor, a seasoned primary care physician, has been teaching me how to 'make a 10-minute visit feel like a 60-minute visit.' I've learned to incorporate some of her tricks—constructive listening to demonstrate empathy, adept questioning to elicit pertinent information, and good doses of eye contact thrown in at every step." But after having the luxury of an extra 20 minutes with a patient, thanks to the day's DNKs, Mackie questions whether these techniques can really make up for the lack of time. "My impression," she wrote, "is that there is no substitute for time. Either I am not skilled enough to make 10 minutes be 60 minutes, or there is something real about clock time. I suspect it's the latter … Yet I firmly believe that adequate time—not simply perceived time, but real time—is an indispensable component of our encounters with patients if we are to be good doctors."

Do not have sympathy for the apparent frenetic state of your doctor. Calmly ask your doctor to explicitly name the condition that is their diagnosis. You have to do this because sometimes the doctor won't actually tell you the name of the condition they suspect. The reasons for a doctor doing this are not usually intentionally misleading, but are more like not wanting to scare you with a serious-sounding name, thinking you won't understand it anyway, or that you don't want to know. In fact, many patients don't want to know, but just want the doctor to tell them what to do (i.e., what treatments to use). However, if you want to be clear on the diagnosis, you need to have it clearly called out.

It is hard to assess the accuracy of your diagnosis unless you understand what it is and why it is given. Exactly what is the diagnosis? How sure is your doctor? What tests has your doctor done? Are any other diagnoses possible? What other diagnoses has your doctor ruled out? Are any extra complications possible? What other related diseases are possible? Which ones have been tested for or ruled out? These are some of the many questions to ask your doctor about your diagnosis.

As Dr. Lisa Sanders, who writes the *New York Times*' Diagnosis column, puts it: "There are lots of diseases that can look like something else. And that's where clinical judgment and experience are essential. Doctors see results as coming straight from God. But just because a test gives you a yes or no answer doesn't mean its right." This happens rarely, but can be very severe, particularly if your diagnosis is strongly based on one test result. Doctors tend to trust laboratory tests because they are rarely wrong. There are several possible methods to reduce your risk of a wrong diagnosis based on a wrong test.

Don't fear requesting a repeat of the same test: This reduces the likelihood of a simple laboratory error or administrative mix-up, since it shouldn't happen twice. Scientists never rely on a single data point to draw a conclusion, and neither should medicine, particularly if that lab result is critical to the overall diagnosis. Also, if there are multiple diagnostic tests for your disease, consider having another type of test done, even if you incur the cost for the test. Lyme disease is a good example where there are at least 13 different tests for the Lyme bug. Interestingly, a recent article published in the *Journal of the American Medical Association* shows that these tests, all approved by the FDA for the same indication, often give different results [10].

Scientists never rely on a single data point to draw a conclusion. If medicine is a science, should doctors rely on one piece of information or a single test result?

A $300-BILLION DIAGNOSTIC ERROR

The diagnosis and treatment of AD is at its infancy. We know very little about the disease, and diagnoses are woefully inadequate, while treatments are nonexistent. We need to be prepared and educated to evaluate the merit of emerging diagnoses and treatment so that viable solutions can be made available in a timely manner. The best way to efficiently drive the future is to understand the strengths and weaknesses of the past.

Cardiovascular disease is presumably a well-known disease, certainly compared to AD. There exist a myriad of diagnostic tests, treatment procedures, and pharmaceutical interventions for the various stages and types of cardiovascular disease. In 2007, nearly 1 million Americans died of cardiovascular disease, accounting for 34% of all deaths, and it continues to be either the largest or the second largest medical cause of death, competing with all types of cancer. There has been a decrease in deaths since a peak in 1968, and the decrease is attributable to many factors but is closely tied to the Surgeon General report on smoking in 1964. One could infer that all medical advancement, procedures, and treatments have had a small impact on cardiovascular mortality.

In the 1980s a miracle drug was introduced that was designed to attack the root cause of cardiovascular disease—cholesterol. Cholesterol offered a great culprit for disease and provided an easy, one-step and one-parameter diagnostic for cardiovascular risk. Since then this class of drug is second only to penicillin with regard to its pervasiveness into medicine. Some might use the term "impact" over "pervasiveness," but that might mislead you into believing this class of drug is effective. Over the past 25 years, $300 billion was spent on statin prescriptions to treat those at "high risk" for cardiovascular disease. Now anyone with the slightest elevation in cholesterol gets the prescription sheet handed to him or her. Also, statins are starting to be recommended for noncardiovascular diseases. Statins are clearly a wonder drug, right? You should wonder.

You might be wondering why there is such an obsession on statins and cholesterol in a book on Alzheimer's: Chapters 7 and 8 show that there is a profound connection between cardiovascular disease and AD. Thus both statin and cholesterol are important considerations for anyone suffering or concerned about AD.

A picture is worth a thousand words. Let's examine how well modern medicine, and their use of statins, has performed to protect us from cardiovascular diseases. First, let's establish a baseline for what a cure looks like.

Baseline Disease 1—Pellagra

It is a vitamin deficiency disease most commonly caused by a chronic lack of niacin (vitamin B3) in the diet. Thus the cause is well known and so is the treatment. Pellagra is a deadly disease, and in Fig. 4.1 is given a death rate trend curve that shows what a cure can do to rates.

Baseline Disease 2—Tuberculosis

Tuberculosis is a common, and in many cases lethal, infectious disease caused by various strains of mycobacteria, usually *Mycobacterium tuberculosis*. Tuberculosis typically attacks the

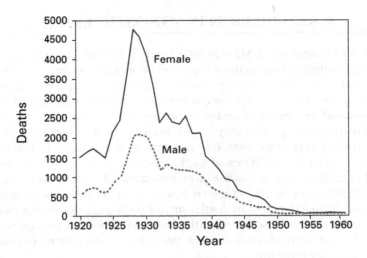

FIGURE 4.1 Number of reported pellagra deaths in the United States, 1920–60. *From CDC. Achievements in public health, 1900–1999: safer and healthier foods. MMWR 1999;48(40):905–13. Available at: https://www.cdc.gov/mmwr/preview/mmwrhtml/mm4840a1.htm*

lungs, but can also affect other parts of the body. Not so incidentally, tuberculosis appears to be connected with AD, at least in some cases.

The tuberculosis death rate curve given in Fig. 4.2 shows the power of knowing the cause of the disease and treating it. However, there is apparently some complacency or lack of understanding about tuberculosis, as it is making a slight comeback in the 21st century (Fig. 4.2).

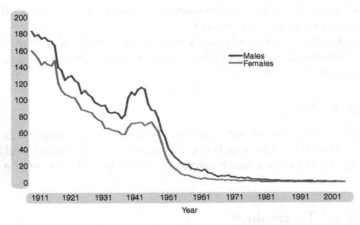

FIGURE 4.2 Tuberculosis death rate per 100,000 population, 1900–60. *From Scotland's Population 2005: The Registrar General's Annual Review of Demographic Trends: 151st Edition, Chapter 2, Causes of Death. Available at: https://www.nrscotland.gov.uk/files/statistics/scotlands-population-2005-the-register-generals-annual-review-151stedition/j9085e05.htm*

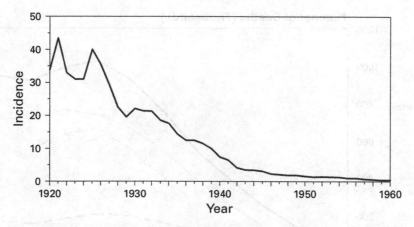

FIGURE 4.3 **Incidence of typhoid fever per 100,000 population, United States, 1920–60.** *From CDC. Achievements in public health, 1900–1999: safer and healthier foods. MMWR 1999;48(40):905–13. Available at: https://www.cdc.gov/mmwr/preview/mmwrhtml/mm4840a1.htm*

Baseline Disease 3—Typhoid Fever

Typhoid fever, also known simply as typhoid, is a common worldwide bacterial disease transmitted by the ingestion of food or water contaminated with the feces of an infected person, which contain the bacterium. It is treated (cured) with antibiotics (Fig. 4.3).

Statins and Cardiovascular Disease

A key question is as follows: are statins a cure for cardiovascular diseases or do they just address symptoms? Recent data show that over the 25 years of their use, overall mortality was not reduced. Not one net life appears to be spared. These drugs do reduce cardiovascular mortality, their intended target, but only slightly. However, all-cause mortality was not reduced. Did statins therefore address the root cause of disease, symptoms, or either for that matter? Let's look at the cardiovascular disease mortality curve and compare it to our three baseline diseases. Keep in mind that statin therapy to "cure" cardiovascular disease began in the 1980s and over 30 million Americans are on statin drugs today.

Let's look at the cardiovascular disease mortality curve (Fig. 4.4) and compare it to our three baseline diseases. Also, compare the cardiovascular disease mortality curve to the smoking trend curve (Fig. 4.5 [11]). Keep in mind that statin therapy to "cure" cardiovascular disease began in the 1980s, as did beta-blockers, ACEs, and ARBs. Over 30 million Americans are on statin drugs today.

Focus on the slope of the curves in Figs. 4.4 and 4.5 starting at about 1980. Notice that the curve in Fig. 4.4 actually shows a slight "uptick" when statins came into vogue. It is clear that statin does not provide a cure. A comparison between 4.4 and 4.5 clearly indicates that reduction in cardiovascular deaths is due to smoking reduction. Any changes in death rates due to statin use are in the "noise" of the curve (their impact is trivial). Do a Google search on "statin" and "drinking water." Some doctors and other health professionals were

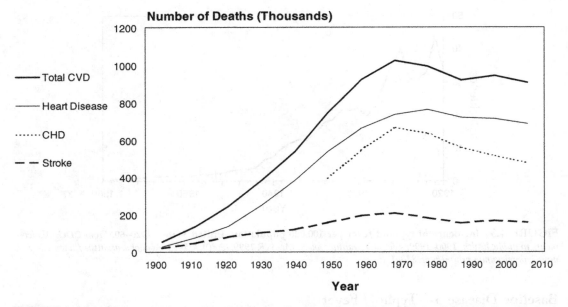

FIGURE 4.4　Number of deaths from cardiovascular diseases by year, United States, 1900–2003. *From National Heart, Lung and Blood Institute. Disease statistics. In: NHLBI fact book fiscal year 2005. NIH, US Department of Health & human Services; 2014. Available at: https://www.nhlbi.nih.gov/about/documents/factbook/2005/chapter4.*

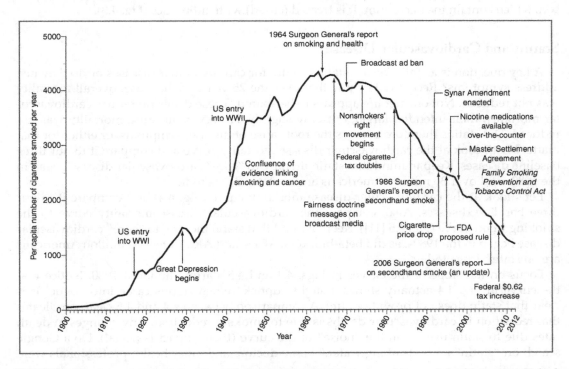

FIGURE 4.5　Annual per capita cigarette consumption and major smoking and health events, United States, 1900–98. *From CDC. Achievements in public health, 1900–1999: tobacco use—United States, 1900–1999. MMWR 1999;48(43):986–93. Available at: http://www.cdc.gov/mmwr/preview/mmwrhtml/mm4843a2.htm.*

so convinced about the value of statin that there were actually proposals to put this drug in our drinking water. So how does statin compare as a curative treatment to the methods used for tuberculosis, Typhoid fever, and Pellagra? Not so well!

The figures illustrate the shortcomings of statin drugs. What are some experts saying 30 years into the cholesterol diagnosis and statin treatment? *Proto* (*Protomag*) is a publication from Massachusetts General Hospital and is considered their "Dispatches from the frontiers of medicine." In 2011, a feature article was published called "Questioning statins" [12]. The byline was "WHAT STATINS MIGHT DO FOR YOU: lower cholesterol; reduce risk of cardiovascular disease; cause muscle pain and fatigue; fail to significantly prolong your life." Other gems in this publication include "Statins don't seem to confer the ultimate health benefit— longer life. So is lowering cholesterol as important as everyone has been led to believe?" And then they dropped the bomb, "Why did statins appear to protect the hearts of people who didn't have high cholesterol? It could be that they not only lower cholesterol but also reduce inflammation." If you read between the lines, it appears that Harvard Medical School is saying that cholesterol is not the cause of cardiovascular disease but inflammation is. We can also infer that statins are antiinflammatory, but based on their results against cardiovascular disease, they are very poor at the job and have too many side effects. Yet doctors prescribe these like crazy (read Chapter 5 for a detailed explanation). Could it be that Harvard is suddenly taking this position because most statins are coming "off patent," and thus the exorbitant profits that come with prescribing statins are disappearing?

Based on the *Protomag* article, where do medical centers like Mass General stand on statin prescriptions? Let's fast-forward 2 years from the *Protomag* publication to a story about patient "Bob" (not his real name but his story is true and recent). Bob had a bout of ischemia (loss of blood flow to the heart). He has been complaining about shortness of breath and general fatigue. He went to a local but well-known clinic where they almost put a stent in him, but he knew, prior to his visit, that stents do not change mortality. How can a tube inserted in your vessels treat the cause of the disease? A week or two later, Bob was unhappy with the follow-up and went to Mass General. The appointment appeared thorough, and Bob was impressed with MGH but said, "It was a bit of a zoo." Finally his doctor prescribed a "new" cholesterol/heart medication, according to Bob. Bob is a very "up-front" person who is aware that statins do not work for him as he has a long history of elevated cholesterol, most likely genetic in origin.

What did Bob get prescribed? Lipitor, the biggest selling drug of all time and a statin! Not only that, the doctor gave the impression that this is a new medication. Let's explore that for a moment. Did you know that drug companies give "unrestricted" funds to major teaching hospitals like Mass General? That particular hospital gets more than nine figures every year with which to do what they want (private and unsubstantiated communication). Yes, that is (well) over $100,000,000 every year—no strings attached, and that is just one of the partner's hospitals. Now you know why they are pushing a "new" medicine made by the drug companies.

This major medical institution is "raising the white flag" over cardiovascular disease. If the internal staff knows that statins are relatively useless based on their own publication and that inflammation could play a part, then why prescribe statin? The answer is it is the standard-of-care. Right or wrong, it is the accepted practice of medicine. It is a "no-risk" prescription even if the patient dies from cardiovascular disease. Can a patient's family sue the entire medical community? The hospital gets paid, and the doctor is a hero to the patient and the hospital, because it is possible to perform the diagnosis and prescribe the treatment in a 10-minute visit. Even sadder, although cardiovascular disease is recognized as a disease of inflammation, medicine currently does not know how to treat this cause, at least within the standard-of-care.

The American College of Cardiology and the American Heart Association have also surrendered to heart disease. They have given up on you, the patient. Thus the "religion" of statin pervades the highest level of cardiovascular thought leaders. These groups give statins "high marks" in the new cardiac prevention guidelines. These guidelines were published in the *Journal of the American College of Cardiology* and *Circulation: Journal of the American Heart Association* [13]. The American College of Cardiology president, John Harold, said:

> The overarching goal of both the ACC and the AHA is to prevent cardiovascular diseases and improve the care of people living with or at risk of these diseases.

They even recommend statin therapy for diabetes, but these drugs cause that disease. Based on this latest guideline, here is a list of those who have given up on finding a solution to cardiovascular disease (based on an article titled "Statins get high marks in new cardiac prevention guidelines" by Todd Neale, Senior Staff Writer of MedPage Today):

- The American Heart Association
- The American College of Cardiology
- Cedars-Sinai Heart Institute in Los Angeles
- Northwestern University

Others who have thrown in the towel and have no solution to your cardiovascular disease are those who endorsed the guidelines. These include:

- The American Association of Cardiovascular and Pulmonary Rehabilitation
- American Pharmacists Association
- American Society for Preventive Cardiology
- Association of Black Cardiologists
- Preventive Cardiovascular Nurses Association, and WomenHeart
- The National Coalition for Women With Heart Disease

Breaking news: In one of the largest studies ever done on the subject, researchers have found that taking statins, the widely used cholesterol-lowering drugs, is associated with an increased risk for cataracts. "Okay," you say to yourself, "no big deal. My doctor has convinced me with nice glossy drug company literature that the statin will prevent me from dying tomorrow, so what is the big deal about cataracts? It is a small price to pay for staying alive, right?" First, look at the chart on death rates one more time as it is worth a double take. Second, when you read the chapter on the eye (Chapter 6), you will find out that a cataract is actually a biomarker for high cardiovascular death risk. Isn't there just too much unknown about this "miracle" drug to be worth the risk based on its lack of benefit?

Lesson learned? The statement about inflammation in the *Proto* publication basically says that $300 billion was spent on the diagnosis of cholesterol followed by statin therapy, but for the wrong reason. Cholesterol is the wrong diagnosis! We have spent 30 years chasing one cause, one root cause, of cardiovascular disease at the expense of launching a major effort to fully understand, diagnose, and treat this disease. Do you see parallels with Alzheimer's? Let's not make the same mistake, although, to some degree, it is happening before our eyes as almost all research is focused on the presumed evil "amyloid protein." We have seen this movie already. Research must diverge from the Amyloid Cascade Hypothesis.

The statin experiment is a solid illustration of the complexity of disease. Medicine believes it knows a lot about cardiovascular disease. Cholesterol, particularly LDL, has been characterized (and marketed) as the "bad" cholesterol. Medicine thinks it is the cause of cardiovascular disease. When statins are used to dramatically reduce human cholesterol levels, cardiovascular disease diminishes only slightly. Clearly the root cause of this disease is much more complex than just LDL, if LDL is really even a root cause. To give statin therapy its due, it is marginally curative for cardiovascular disease, but how it is curative is yet to be well understood and has little to do with LDL, at least as it was characterized and as it is still explained in your doctor's office.

The point of this discussion on cardiovascular disease is to demonstrate that the simple terms "diagnosis" and "misdiagnosis" are much more complicated than they appear. This is especially true with regard to complex diseases that slowly develop over time as we age. Alzheimer's and dementia are arguably the most complex of these diseases, thus warrant deeper and broader diagnostic analysis compared to that provided in the standard-of-care.

Statins Trigger Memory Problems

Cholesterol, once thought to be the simple culprit for cardiovascular disease, is now part of a complex picture that possibly overlaps into AD, but again, not as a cause and certainly not as a root cause. As long-term data on the use of cholesterol-lowering medications are rolling in, there are more and more reported cases with memory and cognition loss. A 2012 *New York Times* article had the title "A heart helper may come at a price for the brain: statins use causes problems with the brain."

The article outlined cases in which people with heart attacks and other cardiovascular problems were prescribed statins soon began experiencing memory problems. One statin patient reported that thinking and remembering became so laborious that he could not even recall his three-digit telephone extension or computer password at work and that his brain felt like mush. His doctor suggested a "drug vacation," and when he stopped taking the statin for 6 weeks, the problems disappeared. Then he tried a different statin at a high dose, but the cognitive difficulties returned. His doctor has since lowered his dose by more than half, and while the memory lapses have not disappeared, he has learned to cope.

Statin therapy has become a religion of modern medicine, and patients don't often leave the medication for fear of sudden and early death. The patient highlighted in the *Times* wanted to keep his numbers in "an acceptable range." An acceptable range is determined by your physiology, not by a drug company. Your brain tells your liver to produce cholesterol. Who is right—the statin industry or your brain?

Statins are the most prescribed drugs in the world, and there is evidence that for people at high risk of cardiovascular problems, the drugs lower cholesterol and also the risk of heart attack and stroke, but only marginally at best. But for years doctors have been fielding reports from patients that the drugs leave them feeling "fuzzy" and unable to remember small and big things, like where they left the car, a favorite poem, or a recently memorized presentation. The FDA finally acknowledged what many patients and doctors have believed for a long time: statin drugs carry a risk of cognitive side effects. The agency also warned users about diabetes risk and muscle pain.

Nearly 21 million patients in the United States were prescribed statins at its peak, but nobody knows how many of them have experienced cognitive side effects. A doctor at the University of California, San Diego, has collected more than 3000 reports of side effects related to statin use. She said doctors have too often dismissed the complaints, writing off the memory lapses and muscle pain in particular as a normal sign of getting older.

The evidence against cholesterol-lowering therapy is now mounting, just like the evidence in favor of cholesterol-lowering therapy mounted 30 years ago as statins were developed. Consider the following recently published report titled "New cholesterol guidelines for longevity (2010)" [14]:

> The 'cholesterol hypothesis' was established several decades ago and has been accepted in medical fields worldwide. Recommendations based on this hypothesis have been repeatedly issued from authoritative organizations such as the WHO and NIH (ATPIII), and most medical societies in Japan have accepted the recommendations. However, the guidelines based on the cholesterol hypothesis included some misinterpretations of the reported data and were found to be even risky for the majority of people (general populations of over 40–50 years of age).
>
> A law came into effect in the EU to secure the clarity and fairness of clinical trials in 2004. Since then, clinical trials on cholesterol-lowering medications performed by scientists with no conflicts of interest with industry have revealed that 'these (statins) are effective in lowering LDL levels but are essentially ineffective in preventing atherosclerotic diseases.'

Statins were found to be essentially ineffective for the prevention of coronary heart disease. Among the data evaluated, only the JUPITER study claimed a reduction in coronary heart disease complication among participants with normal or low total cholesterol levels, which was severely criticized. Not surprising, one of the authors of the JUPITER study is on the Crestor (statin) patent.

The authors of the 2010 cholesterol guideline article conclude:

- Statins lower LDL cholesterol levels, but exhibit no beneficial effects on coronary heart disease.
- It was found that the higher the LDL cholesterol level, the lower is the all-cause mortality among the general populations.
- Statins increase HDL level and decrease LDL/HDL ratio, but are ineffective in preventing coronary heart disease; instead, they increase all-cause mortality.
- The lower the LDL cholesterol level, the higher are the cancer and all-cause mortality rates.

> In view of the serious and irreversible side effects reported for statins, e.g. cancer and the effects on the central nervous system as well as the widely recognized rhabdomyolysis, the number of cases for which statins are applicable, if any, seems to be very limited.

Definition of Rhabdomyolysis The breakdown of muscle fibers that leads to the release of muscle fiber contents (myoglobin) into the bloodstream. Myoglobin is harmful to the kidney and often causes kidney damage.

The conventional guidelines were primarily based on medical papers before 2004 that claimed statins were evidence-based medicines. In light of these new reports on the ineffectiveness on coronary heart disease of statins and other lipid-modifying drugs, the conventional cholesterol guidelines should be revised substantially. Thus, the so-called 'good and bad cholesterol' theory over-simplifies the roles of HDL and LDL and should not be used to convince people to take the cholesterol-lowering medications.

Link Between Low Cholesterol and Mortality

You are probably not surprised there is a powerful connection between cholesterol levels and premature death, but the actual trend is not likely what you expect.

The *Journal of Acquired Immune Deficiency Syndromes and Human Retrovirology* published a paper in 1998 titled "Association between serum total cholesterol and HIV infection in a high-risk cohort of young men" [15]. The authors stated, "Low serum total cholesterol (TC) is associated with a variety of nonatherosclerotic diseases, but the association of TC with infectious disease has been little studied. In this study, we examined the relationship between serum TC and HIV infection in members of a large health maintenance organization in Northern California." They found that the men in the study under the age of 50 had a significantly higher rate of infection if their total cholesterol was less than 160. They also found a similar excess risk of AIDS and AIDS-related death. These findings suggest that low serum total cholesterol levels should be considered a marker of increased risk of HIV infection in men already at a heightened risk of HIV infection. Thus cholesterol is apparently protective to our bodies against viruses and other infectious species. Chapter 9 delves deep into the connection between Alzheimer's and infection. After reading that chapter, you may decide to quit your statin and raise your cholesterol.

A study titled "Lowering cholesterol concentrations and mortality: a quantitative review of primary prevention trials" also shows the connection between a low cholesterol and a higher likelihood of early death [16]. The authors concluded, "The association between reduction of cholesterol concentrations and deaths not related to illness warrants further investigation. Additionally, the failure of cholesterol-lowering to affect overall survival justifies a more cautious appraisal of the probable benefits of reducing cholesterol concentrations in the general population." It's important to note that this paper was published more than 20 years ago, yet we have been infiltrated with statin therapy. It is simple, convenient, and backed by billions in marketing dollars.

In 2012, a paper was published regarding the connection between cholesterol and early mortality titled "Is the use of cholesterol in mortality risk algorithms in clinical guidelines valid? Ten years prospective data from the Norwegian HUNT 2 study" [17]. The research sought to "document the strength and validity of total cholesterol as a risk factor for mortality in a well-defined, general Norwegian population without known CVD at baseline." The authors concluded, "Our study provides an updated epidemiological indication of possible errors in the CVD risk algorithms of many clinical guidelines. If our findings are generalizable, clinical and public health recommendations regarding the 'dangers' of cholesterol should be revised. This is especially true for women, for whom moderately elevated cholesterol (by current standards) may prove to be not only harmless but even beneficial."

If you are a woman on statin, you should consider obtaining and reading this paper. Over 15 studies have suggested an inverse relationship between total cholesterol and mortality. That is, people with "higher" cholesterol have lower overall mortality. Some studies have shown an inverse or a U-shaped association between cholesterol and death from causes other than cardiovascular diseases, such as cancer and AD. The phrase "U-shaped association" indicates that higher mortality (or incidences) can be observed in individuals with both low and high levels of cholesterol compared with individuals with levels in between. It is important to note that a high cholesterol level, at the far end of the U shape, is 350 or more. A level of approximately 250 is ideal in this context, whereas the standard-of-care wants all of us to be less than 190.

> Our results contradict the guidelines' well-established demarcation line (190) between 'good' and 'too high' levels of cholesterol. They also contradict the popularized idea of a positive, linear relationship between cholesterol and fatal disease. Guideline-based advice regarding CVD prevention may thus be outdated and misleading, particularly regarding many women who have cholesterol levels in the range of 190–270 and are currently encouraged to 'take better care of their health' through the use of statins.

"Know your numbers" (a concept pertaining to medical risk factor levels, including cholesterol) is currently considered part of responsible citizenship, as well as an essential element of preventive medical care. Many individuals who otherwise consider themselves healthy struggle conscientiously to push their cholesterol under the presumed "danger" limit coached by health personnel, personal trainers, and caring family members. Massive commercial interests are linked to drugs and other remedies marketed for this purpose.

Cholesterol and Alzheimer's Disease

There appears to be an emerging and logical connection between low lipid levels (cholesterols) and Alzheimer's. The paper "Decreased serum lipids in patients with probable Alzheimer's disease" and several others indicate that low cholesterol levels and Alzheimer's may go hand-in-hand [18]. They investigated the cholesterol/Alzheimer's link in 30 probable Alzheimer's patients and 30 matched controls. "Subjects with probable AD had significantly lower serum triglycerides compared to the control group." The researchers reported a negative correlation between triglycerides and MMSE (cognitive impairment) values in both the Alzheimer's group and the control group.

To understand the flaw associated with statin use for the lowering of cholesterol and its connection with AD, basic brain physiology must be understood. Cholesterol plays a central role in the brain's metabolism; the fact that the brain accounts for only 2% of the body mass and brain cholesterol represents 25% of the total body cholesterol speaks for itself. Overall, the brain is the organ with the highest content of cholesterol in the body. It is clear that fats are important for brain health, and cholesterol is a (healthy) fat.

> This is not conjecture. This is fact.

MISDIAGNOSIS: CASE STUDY

By way of example, here is a story that clarifies an "obligatory" of misdiagnosis from a true diagnosis.

The granddaughter of a colleague had bouts of constipation, stomach upset, and gastrointestinal issues, diagnosed as "Dyspepsia." My colleague asked where his daughter could go to get help real help. They being in Australia, I recommended one of the top gastroenterologists in the world, Thomas Brody, 1-hour flight from their home in Melbourne. Before embarking on the flight, Phillip, the grandfather, asked me if I knew of any "fixes" for the problem. Admittedly without testing I gave a "guess" and offered a harmless remedy, magnesium supplementation (Chapter 10). Magnesium has many health benefits and the vast majority of our children are magnesium deficient. One of the benefits of magnesium is, once in the bowel, it draws water like some of the synthetic chemical prescription medications for constipation.

Phillip's granddaughter went on a product called Slow-Mag, a slow-release magnesium chloride, at about half of the recommended adult dose. About a week later I checked in with Phillip and he said his granddaughter was getting along fine and all issues were resolved. A month later I had a conference call with Phillip and his daughter, the mother of the patient, to discuss the case with focus on prognosis. The daughter asked me a very pointed question, "What was the underlying cause of the disease that the magnesium cured?" I bit my tongue and told her the simple truth. "The disease was a magnesium deficiency." By good luck I happened to be correct. Six months later Phillip's granddaughter is still "spot on."

You see we have been brainwashed into believing that diseases are some esoteric aspect of one gene gone aberrant or astray and that we have to have $1 trillion worth of research to discover the one gene in our particular body that's making us sick. It's actually not that complicated. Look at Clayton's work on the mid-Victorian diet. How do you cut chronic disease by a factor of 10? Overcome specific micronutrient and phytonutrient deficiencies that are causing these diseases. Our children, for example, are not eating enough vegetables and nuts rich in magnesium and, as a result, all types of fancy ailments are cropping up.

The original, natural, God-given antibiotic is vitamin D, yet 95% of us are vitamin D deficient or insufficient. During the tuberculosis epidemic, did people have tuberculosis or a vitamin D deficiency, as those with vitamin D at sufficient levels hardly got the disease? The root cause of the disease, thus the deserving diagnosis, should have been a vitamin D deficiency. Are children who are obese have "metabolic syndrome" or is it because most of them are magnesium deficient and PUFA imbalanced as these factors are critical to metabolism and inflammation control? Based on our diet choices, our bodies are "inflamed" and insulin resistant. Do we have type II diabetes, or are we deficient in Omega 3s, which reduces inflammation, and have excess Omega 6s, which promotes inflammation? What is the name of that disease, "omega-3 and magnesium deficiency" or "diabetes?"

What is AD then? Only a differential diagnosis will reveal the truth.

References

[1] Neumann PJ, et al. Willingness-to-pay for predictive tests with no immediate treatment implications: a survey of US residents. Health Econ 2012;21(3):238–51.

[2] Scarborough N. Medical misdiagnosis in America 2008: a persistent problem with a promising solution. Health-Leaders Media white paper. HealthGrades; 2008.

[3] Roosen J, Frans E, Wilmer A, et al. Comparison of premortem clinical diagnoses in critically ill patients and subsequent autopsy findings. Mayo Clinic Proc 2000;75:562–7.

[4] Patient safety in American hospitals, HealthGrades Quality Study. HealthGrades; July 2004.

[5] Harris L. Public opinion of patient safety issues research findings. Chicago, IL: National Patient Safety Foundation; 1997.

[6] In: Kohn LT, Corrigan JM, Donaldson MS, editors. To err is human: building a safer health system, vol. 627. Washington, DC: National Academies Press; 2000.

[7] Joint Commission. Improving America's hospitals: the Joint Commission's annual report on quality and safety, 2007. Retrieved February 25, 2008; 2009.

[8] Groopman JE, Prichard M. How doctors think. Boston: Houghton Mifflin; 2007.

[9] Mackie S. The value of DNKs. N Engl J Med 2010;362(17):1561.

[10] Ang CW, et al. Large differences between test strategies for the detection of anti-*Borrelia* antibodies are revealed by comparing eight ELISAs and five immunoblots. Eur J Clin Microbiol Infect Dis 2011;30(8):1027–32.

[11] CDC. Achievements in public health, 1900–1999: tobacco use—United States, 1900–1999. MMWR 1999;48(43):986–93.

[12] Questioning statins. Proto Magazine; January 15, 2011. Available from: http://protomag.com/articles/statins-change-of-heart

[13] Stone NJ, Robinson J, Lichtenstein AH, et al. 2013 ACC/AHA guideline on the treatment of blood cholesterol to reduce atherosclerotic cardiovascular risk in adults: a report of the American College of Cardiology/American Heart Association Task Force on Practice Guidelines. J Am Coll Cardiol 2014;63:2889–934.

[14] Okuyama H, Hamazaki T, Ogushi Y. Committee on Cholesterol Guidelines for Longevity, Japan Society for Lipid Nutrition. New cholesterol guidelines for longevity (2010). World Rev Nutr Diet 2011;102:124–36.

[15] Claxton AJ, et al. Association between serum total cholesterol and HIV infection in a high-risk cohort of young men. J Acquir Immune Defic Syndr 1998;17(1):51–7.

[16] Muldoon MF, Manuck SB, Matthews KA. Lowering cholesterol concentrations and mortality: a quantitative review of primary prevention trials. BMJ 1990;301(6747):309.

[17] Thelle DS, Tverdal A, Selmer R. Is the use of cholesterol in mortality risk algorithms in clinical guidelines valid? Ten years prospective data from the Norwegian HUNT 2 study. J Eval Clin Pract 2012;18(6):1219.

[18] Lepara O, et al. Decreased serum lipids in patients with probable Alzheimer's disease. Bosnian J Basic Med Sci 2009;9(3):215–20.

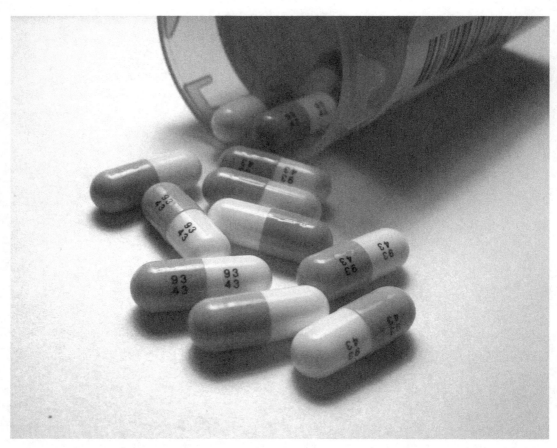

Image Credit: Fluoxetine HCl 20mg Capsules (Prozac), by Tom Varco (Own work). Available at Wikimedia Commons: https://commons.wikimedia.org/wiki/File%3AProzac_pills_cropped.jpg

5

Can Medicine Save You?

The research presented throughout this book shows that there are many solutions to Alzheimer's prevention, diagnosis, and treatment. However, due to the Trillion Dollar Conundrum, this knowledge, or methods associated with this knowledge, is not translating into clinical delivery. We (the patient) must take some blame as we rely on medicine to compensate for our behaviors. There is a middle ground where we take charge of our health, and medicine assists with wellness and intervenes under more extreme cases.

A short review of history provides insights into three key turning points in medicine that led to modern medicine becoming "sick care" as opposed to "health care," such that diseases like Alzheimer's disease (AD) go untreated:

1. the presumption that longevity accompanied by good health could come about only through pharmaceutical intervention
2. the commercialization of health care; examples include:
   a. allowing drug companies to invade the most basic levels of health care such as participating in medical school education
   b. permitting "pull-through" marketing of prescription medication by means of direct-to-consumer advertisements
3. the evolution of medical research as a business, rather than a partner, subordinate to clinical medicine

DO WE NEED DRUGS TO BE HEALTHY?

This sounds like a really good book title but, in some respects, Paul Clayton and Claude Bernard have already properly written it. Claude Bernard of 19th-century France defined the concept of internal balance. Specifically, he coined the term "milieu interieur," or homeostasis. Examples of homeostasis are the constancy of our internal temperature and the regulation of glucose levels on a daily basis. He informed Louis Pasteur, an author of the "Germ Theory" of disease, that Pasteur's diseases were far less likely to occur in a body with a balanced "milieu interieur." Proof positive of this concept was the eradication of certain infectious diseases, long before antibiotics, by simply "balancing" vitamin D and other micronutrient and macronutrient levels. Clayton wrote a series of articles titled "The mid-Victorian diet," where he constructs an elegant theory explaining how presumed poor health of 1870s Brits ushered in the "drug" culture [1–4].

The End of Alzheimer's. http://dx.doi.org/10.1016/B978-0-12-812112-2.00005-7

THE COMMERCIALIZATION OF HEALTH CARE

Dr. Alzheimer had it right a century ago; observations in the clinic should drive drug and treatment discovery. Today new drugs are developed from "discoveries" in test tubes. Somehow we have the cart before the horse and in our modern focus, the lab and drug development supersedes good medicine and translating the knowledge that we gain from direct experience with patients. Medical research is a big and independent business. A recent article in *The Economist* highlights the ugly side of the well-known mantra, "publish or perish" [5].

The Trillion Dollar Conundrum exists in both fact and fiction. At the two extremes we have the movie *The Fugitive* with Harrison Ford. In this fictitious account, Dr. Richard Kimble (Ford) is wrongfully convicted for the murder of his wife, and when he escapes from federal custody he is declared a fugitive. He sets out to prove his innocence and bring those who were responsible to justice. What was the motive for the crime? A cover-up of falsified medical research that led to the approval of a dangerous drug with immense financial upside. At the other extreme is a true story of a major medical school doctor who ignored his own research to promote an unproven therapeutic for Alzheimer's. His motive was the approval of a useless drug with immense financial upside. "Publish or perish" disregards any consideration of the Hippocratic oath or patient well-being. It is purely an issue of gain.

Consider that there are approximately 20,000 journals in the general medical arena (estimated by searching the National Library of Medicine medical and biochemical database). Then assume that journals are published monthly, on average with 10 technical articles per journal. That amounts to an estimated total of over 2 million papers germane to medicine and science published each year. Consider that each publication reflects years of work and required funding of about $500,000 (considering professor salaries, staff time, laboratory equipment, space rental, and more). Then the annual cost of all these publications is (drum roll) $1,000,000,000,000. One trillion dollars! Are you getting $1 trillion' worth of health care? And this is every year. This is the basis of the Trillion Dollar Conundrum.

Most of us find it difficult to read the daily paper, an occasional book, or another favorite magazine. Most magazines pile high and find their way into recycling with pages still crisp and clean. How can a doctor with a very busy practice and a family life keep up with emerging information on diseases such as Alzheimer's? Google and scholar.google provide a quick means of sourcing information, but many articles are available only as abstracts, and very often the abstract does not paint an adequate picture for action.

The "publish or perish" doctrine is not aligned with quality of content. *The Economist* article makes that case well. Indeed, publications beget publications beget project funding whether from the National Institutes of Health, National Science Foundation, or private foundations, or through internal sources. The more a researcher publishes, the better ticket he or she can write to a more prestigious university where that person is more likely to publish more, bring in more grants, and get more pay. It is a vicious cycle, and no one in research or academia is exempt from the process.

Erlend Hem published an editorial in the *Norwegian Medical Journal*, declaring, "Too many journals—too little good research" [6]. He made the point that it may pay to publish many mediocre articles rather than a few good ones. He states, "I sent my first manuscript to an international scientific journal in 1998. In those days it was normal for manuscripts to 'go wandering.'" This meant that when a manuscript was rejected by one journal, it was sent to another, normally less prestigious journal than the first you had tried your luck with. When the manuscript was rejected by the second journal, it was sent on to a third. In the meantime, you waited. The wait for a response from the editors of journals was well known by researchers. It caused a lot of frustration and cost time, but the process had a purpose. External peer reviewers often evaluated the manuscript, and as a result authors received sound advice on how they could improve it. Peer reviewers were also able to check for errors and fraud.

In recent years, this practice has changed. Your manuscripts may still be rejected if you send them to the most prestigious journals, but manuscripts do not necessarily go wandering the way they did in the past. One reason for this is the emergence of a large number of new electronic periodicals. Hern states,

> If I had done a mediocre study, I would send the manuscript to one of these journals. I haven't been rejected yet. I have the impression that if a manuscript possesses a modicum of quality, it will be accepted for publication. In the beginning, I was delighted not to have to send manuscripts out wandering. It was wonderful to be able to submit a manuscript and assume it would be accepted. I have mentioned this to colleagues in different fields. Many of them have now reflected over it. 'Yes, now that you mention it, I've actually never had a manuscript rejected by one of those journals either,' tends to be the response. They are not necessarily inferior publications, they have the normal external peer review, and they are indexed in the central databases and are freely available on the Internet.

So is there a problem?

One reason that these journals accept many manuscripts is the manner of publication. The articles are published only on the Internet, and as we all know, cyberspace has virtually no space limitations, and the costs of publication are close to zero. Neither the number of articles nor the number of pages concerns editors or publishers. There is room for everything. The problem is that the number of good studies and manuscripts has not increased at the same pace. The result is that a lot of research that is not particularly good, particularly important, or particularly valid is published.

A primary responsibility of the editors of the foremost scientific journals is still to publish the best and most groundbreaking studies. These are few and far between. Editors compete for the really good articles. They contact researchers at professional congresses and scientific meetings to try to get them to send the very best for their own publications. The motivation is naturally that if the journal publishes these articles, they will attract more readers, have more citations, create a higher impact factor, and earn more money. What editor can resist the temptation to be a little less exacting under the circumstances? If the study is big and important, fast track publication may also be offered. The journal will then guarantee publication within, for example, 4 weeks of the manuscript being received by the editorial staff. There is clearly a risk that the professional quality assurance will then be poorer than it would otherwise have been. And who might read this questionable work?

These two phenomena, that it is simpler to publish than it used to be and that quality control is under pressure, create new challenges. It has become easier for those who want to publish, but more difficult for those who are going to use the research. How do we know whether a study is both qualitatively sound and important? PubMed, the most important biomedical database, has indexed over 5500 periodicals and 21 million articles, and a new article is indexed every minute on average. The struggle to keep up professionally, whether one is a researcher or a clinician, is generally said to be formidable. In the view of Dr. Hern, this is not so.

The bulk of what is published can be safely ignored. The researchers ought perhaps to have dropped the project, and rather used the sorely needed funding on studies that are really needed. When I read information summaries or review articles, I am often struck more by the lack of knowledge than the reverse.

In the case of journals that earn their money by demanding payment from authors, as many of the new electronic periodicals do, the more articles they publish, the more they earn. The researchers will earn more in the form of promotions, research funding, and impact on the publication indicator, if they publish extensively. Thus it may pay both researchers and editors to publish many mediocre studies rather than a few good ones. This may be advantageous for the economy of the journals and the careers of the researchers, but whether it benefits many others is open to question.

The main target audience of general medical journals is practitioners, and the ultimate aim is to improve patient care. In a paper titled "Medical journals and dissemination of health research: have they fulfilled their role," the authors argue that the information needs of practitioners are poorly met by these journals and propose various schemes to improve the value of journals to practitioners [7]. They further suggest that journals need to move from a passive dissemination mode to focus on active dissemination of clinically relevant information and suggest various strategies for improvement. The following is excerpted from their detailed but very clear evaluation of the problems of current medical publications:

The ultimate purpose of research in healthcare is to improve patient care. However, lack of effective strategies to disseminate the findings of such research has been an age-old problem. Lemon juice was shown to be effective in preventing scurvy in 1601 but it was almost 200 years later that British Navy took this intervention on board. Such delays in implementation of research findings have serious and often deleterious effects on patient care.

The first medical journal, *Medicina Curiosa*, was published in 1684 in England. Although it ceased to exist after two issues in the same year, medical journals have proliferated since then. In 1996, over 30,000 clinical journals were in publication with nearly 3,500 journals cited in the Medline. One might imagine that these journals would provide an effective medium of communication between scientists who do research and clinicians who use their results in practice, hopefully leading to timely implementation of effective interventions. Unfortunately, this does not seem to have been the case.

Case studies of research, including brilliant research that has gone largely unnoticed or purposefully ignored with significant clinical implications, are abundant and can fill an entire book. Consider Warren and Marshall who elucidated the link between stomach ulcers and bacteria beginning in 1979. Antibiotic treatment of stomach ulcers and the investigation of the cause of stomach cancer, as a result of their work, are still not practiced in all corners of medicine even though they were awarded the Nobel Prize in Medicine in 2005. This case has strong relevance to AD, as you will see by reading further.

In some instances, there is simply a refusal to accept information that is published, even when the scientific proof is overwhelming. The famous Framingham Heart Study, for example, which has been testing 5209 people in the town of Framingham, MA, since 1948 (now looking at their kids and grandkids), is arguably the largest continuous study in medicine and has spawned a slew of publications. Its intent was to assess the cause and effect of cardiovascular diseases. However, from the very outset of the study, since they didn't know the outcome in advance, a broad and deep range of tests were administered and critically reviewed.

An important finding of the Framingham study was that certain lifestyle habits reduce the risk of Alzheimer's. People who had high levels of DHA and EPA fatty acids (an antiinflammatory) in their blood due to the habit of regularly eating fatty fish had fewer cases of Alzheimer's, for example. But "that bit of data needs to be studied with higher scientific standards," says Dr. Martha L. Daviglus, Professor of Preventive Medicine at Northwestern University who chaired the NIH panel to review such findings. Here is a case where there are clear evidence-based data found throughout medical literature, but influential pundits do not accept their validity at the expense of discouraging doctors from recommending such a simple measure of prevention and treatment. The justification? "The Framingham Study was not intended to evaluate Alzheimer's risk factors." Is this reasonable or is there a deeper motivation for statements like this?

The authors of "Medical journals and dissemination of health research: have they fulfilled their role" continue with more gems that help us further understand the Trillion Dollar Conundrum [7].

> Health research often begins in the laboratory but for it to lead to clinically important changes in practice; it would have to be evaluated in patient centered research first. Scientists test most new ideas in laboratories on animals (bench studies) or on small human population (field studies). Many successful bench studies and field studies fail at the next stage of rigorous 'clinical trials.' It is, therefore, unwise to base a change in practice on bench or field studies except in a few circumstances such as when a study result is exceptionally impressive in a cohort of patients who would have otherwise had a uniformly bad outcome. Such studies are rare and overall the value of bench and field studies to clinicians is minimal.

Getting findings of research into practice often needs to involve a range of techniques to increase awareness, develop and disseminate guidance, promote them, and then maintain the adoption of guidance at the local level. This goes beyond a single publication or a lecture at a single medical conference. It requires a champion who can run with the concept from initial research to clinical practice. That doesn't happen often due to the highly "vertical" nature of medicine. The key stumbling block is that researchers are researchers, clinicians are clinicians, and they don't work together. They, as a group, do not have the same financial incentives. The researchers are funded through research grants and propagate their career through publishing and obtaining more grants to further their studies. Clinicians clearly have a different reward system. Thus the research is "lost in the translation."

The major shortcoming defined by the Trillion Dollar Conundrum is that little money or effort is put forth to translate the research into clinical practice. A few deep-pocketed companies are responsible for almost all new developments in medicine as they are the only ones with the financial means to translate research to the market. Thus translations occur in the current system only if there is a strong, novel intellectual property (patents) that provides the company with exclusivity and profit. Brilliant ideas that show how to use old technologies to solve new problems do not have such intellectual property control, which limits the financial reward associated with the development effort; thus these activities often do not occur. We buy what we are sold. The Warren and Marshall case (the Nobel Prize Winners of 2005) clarifies the issue of research "translating" to medical practice. They discovered that antibiotics attack the root cause of stomach ulcers, a bacterium named *H. pylori*. Did any large company profit from having doctors prescribe more antibiotics? Hardly. Thus more palliative drugs like Tagamet and others, with strong new patent portfolios, found their way to the market with the help of huge pharmaceutical sponsors that led to multibillion dollar profits.

Supposing the busy practitioner has gone through the laborious task of gathering the scattered articles and selected the clinical studies, discarding the bench and field studies, can he or she now be sure of the methodological quality of these papers? Most people rely on peer review to ensure methodological quality and may not even read the methods section. However, a study by Haynes showed over 90% of articles to be of poor methodological quality and thus should not be the basis of clinical practice [8].

A significant issue is that authors speculate and exaggerate their results. Many studies are methodologically flawed. Journals have a duty to educate their readers in being able to select the good ones from the bad ones, yet paper journals have traditionally neglected the role of educating the readers in critical appraisal of literature so that effectiveness and safety in considering studies could be assessed. It is rare for a product to be sold without some sort of user's guide, but this is what journals tend to have done in the past. Recently the *Journal of the American Medical Association* has been publishing user guides, and *British Medical Journal* (*BMJ*) has focused on the subject of getting evidence into practice.

Peer reviewing is intended to act as an intellectual quality control to reject methodologically unsound studies, avoid publication bias, curb the exaggeration of results, and improve presentation. There is published evidence for publication bias, nationality bias, language bias, and for the inability to detect defective studies. Peer reviewing is widely accepted to be ineffective and even corrupt. Another disadvantage is the delay it incurs from submission to publication of about 12–16 months, which can unnecessarily delay clinically useful information reaching clinicians. Some journals have now started using fast track procedures [7].

MEDICAL JOURNALS RETRACTING MORE RESEARCH

Medical research errors in publications are growing rapidly. We get a small snapshot into the extent of this problem by measuring the number of articles and stories that are retracted due to some inaccuracy. The number of retractions is going up according to a *Wall Street Journal* investigation conducted by Thomson Reuters. It says there were only 22 retractions in 2001 but 339 in 2011, a 15-fold increase. John Budd, Professor at the University of Missouri who spent years studying why publications are retracted, found that between 1997 and 2008, 47% of the articles were pulled because of "misconduct or presumed misconduct." Errors accounted for 25%; 21% were taken down because the authors could not get the same results consistently. The remaining 7% were unclassified.

Budd says errors such as an accidentally contaminated tissue sample can be understandable—"it is just the way human beings are"—but the misconduct and fraud are "harder to understand." Budd's research suggests it's "almost certain that some people are motivated by the need or desire to advance." Publication in a major medical journal can help a researcher's career and lead to promotions or funding for additional research.

"A single paper in *Lancet* and you get your chair and you get your money. It's your passport to success," Richard Horton, editor of that journal, told the *Wall Street Journal*. The number of retractions is small compared with the overall amount of research published, but it can have a big impact and is likely just the tip of the iceberg. For example, a British study by Dr. Andrew Wakefield in 1998 reported that autism was linked to childhood vaccines [9]. The paper led to some parents' refusal to vaccinate their children for measles, mumps, and rubella. In January 2011, the journal *BMJ* retracted Wakefield's study, calling it an "elaborate fraud."

Fiona Godlee, *BMJ*'s editor-in-chief, reported that it was a "deliberate attempt to create an impression that there was a link by falsifying the data." Wakefield has defended the research.

In 2005, the journal *Science*, one of the top three scientific journals in the world, published an article by South Korean scientist Hwang Woo-Suk who claimed to have cloned human embryonic stem cells [10]. A year later the journal retracted it, saying, "Data presented in the Woo-suk papers are fabricated." Woo later admitted, "It is true that the research papers had fabricated data, and I will take full responsibility. I acknowledge this and apologize."

When retractions happen, journals publish notices, and on their websites many indicate in red type that the published report has been retracted. Since the initial publication, however, other authors may have based their research or cited parts of their studies on the now retracted study. Budd finds that most troubling. His research shows that only 5% of the citations for works retracted in 1999 acknowledged the cited work had been retracted. The Woo-suk article is clearly retracted but that is not indicated based on initial search results displayed in scholar.google, whereas the Wakefield article clearly indicates "RETRACTED" in the title.

VALIDITY OF CONTENT IN THE MEDICAL LITERATURE

As if the proliferation of medical publications and papers isn't enough for the bedraggled doctors, the validity of material is strongly coming into question. Dr. John Ioannidis, Professor and chairman at the Department of Hygiene and Epidemiology, University of Ioannina School of Medicine, as well as tenured Adjunct Professor at Tufts University School of Medicine and Professor of Medicine and Director of the Stanford Prevention Research Center at Stanford University School of Medicine, makes this case better than anyone else. He has made a career of doing what much of medical research has not—that is, unequivocally validating his work. And that work is the review of that literature.

Dr. Ioannidis is known as a metaresearcher, who, not so simply put, has become one of the world's foremost experts on the credibility of medical research. He and his team have shown, again and again, and in many different ways, that much of what medical and biomedical researchers conclude in published studies, conclusions that doctors keep in mind when they perform a diagnosis or prescribe medications or treatment, is misleading, exaggerated, and often flat-out wrong.

He charges that as much as 90% of the published medical information upon which doctors rely is flawed to an extent. His work has been widely accepted by the medical community; it has been published in the field's top journals where it is heavily cited, and he is a big draw at conferences. And his work broadly targets everyone else's work in medicine, as well as everything that physicians do and all the health advice they give. You would think that would put a large target on his back, yet his work seems to be universally accepted as at least somewhat on point. His major worry is that the field of medical research is so pervasively flawed and so riddled with conflicts of interest that it might be chronically resistant to change or even to publicly admitting that there's a problem. Change will really occur only when financial incentives are changed and aligned in favor of rewards for excellence of content.

One of his major theses is that an obsession with winning funding has gone a long way toward weakening the reliability of medical research.

He first stumbled on the sorts of problems plaguing the field in the early 1990s at Harvard Medical School. In poring over medical journals, he was struck by how many findings of all types were refuted by later research. Examples of medical misinterpretations, often in the news, include mammograms, colonoscopies, and PSA tests that are far less useful cancer-detection tools than we had been told previously. Other examples include widely prescribed antidepressants such as Prozac, Zoloft, and Paxil that were revealed to be no more effective than a placebo for most cases of depression, or when we learned that staying out of the sun entirely could actually increase cancer risks. Ioannidis makes a point that medical headlines through regular news outlets are most often in error as colleagues and competitors announce them before rigorous review.

But beyond the headlines, Ioannidis was shocked at the range and reach of the reversals he was seeing in everyday medical research. "Randomized controlled trials," which compare how one group responds to a treatment against how an identical group fares without the treatment, had long been considered nearly unshakable evidence, but they, too, ended up being wrong some of the time. "I realized even our gold standard research had a lot of problems," he says. Baffled, he started looking for the specific ways in which studies were going wrong. And before long he discovered that the range of errors being committed was astonishing, including:

- what questions researchers posed;
- how they set up the studies;
- which patients they recruited for the studies;
- which measurements they took;
- how they analyzed the data;
- how they presented their results;
- how particular studies came to be published in medical journals.

Large pharmaceutical company research wasn't measuring critically important, "hard" outcomes for patients, such as survival versus death, and instead tended to measure "softer" outcomes, such as self-reported symptoms ("my chest doesn't hurt as much today"). Also, often when drug-company data seemed to show patients' health improving, the data often failed to show that the drug was responsible or that the improvement was more than marginal.

Ioannidis said, "There is an intellectual conflict of interest that pressures researchers to find whatever it is that are most likely to get them funded." Perhaps only minorities of researchers were succumbing to this bias, but their distorted findings were having a significant effect on published research. To get funding and tenured positions, and often merely to stay afloat, researchers have to get their work published in well-regarded journals. The studies that do often make the grade are those with eye-catching findings. Revolutionary theories are important, but getting reality to bear them out is another matter. The great majority collapse under the weight of contradictory data when studied rigorously. Further he said, "If you're attracted to ideas that have a good chance of being wrong, and if you're motivated to prove them right, and if you have a little wiggle room in how you assemble the evidence, you'll probably succeed in proving wrong theories right." After all, as medicine in the human body is so complex, nothing is ever 100% correct. So if some researchers find that something is 80% wrong, others might focus on the 20% to prove it right. Does the Amyloid Cascade Hypothesis fit into this model?

Rates of "wrongness" in different areas of medical research reached levels as high as 80% in nonrandomized studies (by far the most common type) and 25% for supposedly gold-standard randomized trials. The Ioannidis article spelled out his belief that researchers were frequently manipulating data analyses, chasing career-advancing findings rather than good science, and even using the peer-review process, in which journals ask researchers to help decide which studies to publish, to suppress opposing views. This is evident in Alzheimer's where Dr. Campbell communicated that she left the anti–beta-amyloid research area due to lack of funding. Her opposing view was suppressed.

Dr. Ioannidis considered potential claims that researchers and doctors know how to scan the literature and discern the credible from the incredible. He carefully studied 49 of the most highly regarded research findings in medicine over the previous dozen or so years as judged by the science community's 2 standard measures: the papers had appeared in the journals most widely cited in research articles, and the 49 articles themselves were the most widely cited articles in these journals. These were articles that helped lead to the widespread popularity of treatments such as the use of hormone replacement therapy for menopausal women, vitamin E to reduce the risk of heart disease, coronary stents to ward off heart attacks, and daily low-dose aspirin to control blood pressure and prevent heart attacks and strokes.

Ioannidis was testing the pinnacle of scientific and medical literature. Of the 49 articles, 45 claimed to have uncovered effective interventions. Thirty-four of these claims had been retested, and 14 of these, or 41%, had been convincingly shown to be wrong or significantly exaggerated. If between a third and a half of the most acclaimed research in medicine was proving untrustworthy, the scope and the impact of the problem were undeniable. Ioannidis published many of his findings in the *Journal of the American Medical Association*, one of the highest regarded journals among medical professionals [11–19].

Findings of incorrect results are not just in the purview of treatment, as research exploring diagnostic testing also showed significant flaws. How was the problem of errors and fraud revealed in medicine? More testing (specifically, more robust and critical testing) revealed it. The same principles apply to you when considering an approach or a doctor's approach to your diagnosis that, of course, dictates treatment. Don't accept one test regardless of how it is acclaimed. Endeavor to find an array of tests as broad and deep as possible. You can feel comfortable with a final diagnosis when multiple tests from multiple sources add up to essentially the same conclusion.

ERRORS IN MEDICAL JOURNALS

The concept of accuracy in scientific research became more mainstream through an article in the *Wall Street Journal* titled "Scientists' elusive goal: reproducing study results" [20]. This is one of medicine's best-kept secrets. Most results, including those that appear in top-flight peer-reviewed journals, can't be reproduced as we learned from the work of Dr. Ioannidis.

A few years ago, several groups of scientists began to seek out new cancer drugs by targeting a protein called KRAS. The mutated form of KRAS is believed to be responsible for more than 60% of pancreatic cancers and half of colorectal cancers. It has also been implicated in the growth of tumors in many other organs, such as the lung. So scientists have been especially keen to impede KRAS and thus stop the constant signaling that leads to tumor growth.

In 2008, researchers at Harvard Medical School used cell culture experiments to show that by inhibiting another protein, they could prevent the growth of tumor cell lines driven by the malfunctioning KRAS. The finding caught the interest of Amgen, the world's largest biotechnology company, who first heard about the experiments at a scientific conference. When the Harvard researchers published their results in the prestigious journal *Cell*, in May 2009, Amgen quickly set out to assess the opportunity.

Amgen took to the task in two ways: (1) to find and test molecules that might inhibit STK33 and (2) reproduce the Harvard data. Since drug development can cost up to $1 billion, their approach made sense, especially in the wake of such recently failed science like that of Sirtris (Glaxo bought it for $728 million only to shut the project down due to lack of reproducibility). Amgen scientists, it turned out, could not reproduce any of the key findings published in the scientific journal. What could account for the irreproducibility of the results? The Harvard team offered a cornucopia of reasons. Interestingly, the Harvard team (at least in the accessible public domain) never offered to, and then never did, reproduce the results.

Amgen smartly killed its STK33 program. They published a paper in the journal *Cancer Research* describing their failure to reproduce the main *Cell* findings. Amgen suggests that academic scientists, like drug companies, should perform more experiments in a "blinded" manner to reduce any bias toward positive findings. Otherwise, there is a human desire to get the results your boss wants you to get. Amgen indicated that they were not surprised by the lack of reproducibility, and that is why they tested the results first. More often than not, Amgen states they are unable to reproduce findings published by researchers in journals.

Unlike pharmaceutical companies, academic researchers rarely conduct experiments in a "blinded" manner. This makes it easier to pick statistical findings that support a positive result. In the quest for jobs and funding, especially in an era of economic malaise, scientists need more successful experiments to their names, not failed ones. An explosion of scientific and academic journals has added to the pressure. Yes, this is a consequence of the Trillion Dollar Conundrum.

Reproducibility is the foundation of all modern research, the standard by which scientific claims are evaluated. In the United States alone, biomedical research is a $100 billion per year enterprise. Research is big business. According to a report published by the United Kingdom's Royal Society, there were 7.1 million researchers working globally across all scientific fields—academic and corporate—in 2007, a 25% increase from 5 years earlier.

Why are researchers tempted to err on the side of positive results? It is far more than just a salary, a tenured position, and a pension. Due to the FDA requirement to study animals first, many candidate drugs perform as anticipated, with minor investment. The next phase is to study humans. Phase 1 does not involve tests for effectiveness of the drug; rather it focuses on safety as evaluated in healthy volunteers. Thus, to this point, the "funnel" is wide, yet the value of the company has increased dramatically. A fledgling company that has obtained some initial venture financing may seek to be purchased by a larger drug development company at the Phase 1 or Phase 2 stage. The "exit" cost is peanuts for a large drug company (most are swimming in cash; check the Fortune 500 list). A $50–500 million buyout is typical, and the founders, including the professors and sometimes researchers who own shares, stand to make high returns. So when published medical findings can't be validated and developed by others, there are major repercussions. Researchers get rich (anyway) but, for example, bona fide development projects that are "less sexy" may get scuttled. Who suffers? Patients in need of new medicines do.

Pharmaceutical companies contribute to the problems associated with valid science. Many of the research dollars funding academic research come from the pharmaceutical industry that often collaborates with the academic teams. The key question: Is the problem getting worse? It is hard to drill down into the bowels of a scientific laboratory to evaluate methods. However, medicine is an "outcome-driven" science (and art). The success rate of Phase 2 human trials, where a drug's efficacy is measured, fell to 18% in 2010 from 28% in 2006, according to a global analysis published in the journal *Nature Reviews* [21]. At the Phase 2 level of drug development, spending may reach $100 million on the testing of the drug.

According to the *Wall Street Journal* article, Bayer published a study describing how it had halted nearly two-thirds of its early drug target projects because in-house experiments failed to match claims made in the literature [22]. The German pharmaceutical company says that none of the claims it attempted to validate were in papers that had been retracted or were suspected of being flawed. Most astounding, even the data in the most prestigious journals could not be confirmed, Bayer said.

ALZHEIMER'S DISEASE AND MEDICAL RESEARCH ERRORS

The failed Alzheimer's drug called Dimebon started out as a project with high promise based on research. Pfizer gambled that this 25-year-old Russian cold medicine could be an effective drug for AD based upon published data from researchers at Baylor College of Medicine and elsewhere [23,24]. These studies suggested that the drug, an antihistamine, could improve symptoms in Alzheimer's patients. According to Medivation, Dimebon's developer and a partner with Pfizer, "Statistically, the studies were very robust."

In 2010, Medivation along with Pfizer released data from their own clinical trial for Dimebon, involving nearly 600 patients with mild to moderate Alzheimer's symptoms. The companies said they were unable to reproduce the published results. They also indicated they had found no statistically significant difference between patients on the drug and those on the inactive placebo. Again, it is not the failure of Dimebon, but the "opportunity cost" of the lost time and effort by big pharma to pursue a meaningful target for AD. Someone profited along the way at the expense of millions of Alzheimer's sufferers.

The Dimebon case was discussed in Chapter 2. This is another in the failed group of anti-amyloid therapies. In this case, Dimebon did not outperform a sugar pill. Interestingly, there are no requirements for Drug Company reporting so the term "outperform" must be called into question. It is highly likely in light of Chapter 2 on the Amyloid Cascade Hypothesis that the sugar pill worked better because amyloid protein appears to be important to brain health.

A major problem in the drug development process is that the overseeing agency, the FDA, does not require a drug company to submit all their data for review. The FDA acknowledges, "The agency considers all data it is given when reviewing a drug but does not have the authority to control what a company chooses to publish." Did you get the meaning of that? The FDA considers all the data it is given. How does the agency know what was withheld? The prescribing doctors have access to less data (presumably not the bad stuff) because they do not delve into FDA submissions and cannot read what is not published. At best, they rely on the scientific publications, or worse, the pharmaceutical sales literature. The article "Bias, spin, and misreporting: time for full access to trial protocols and results" adds perspective to the state of the drugs patients are getting today [25,26]. The author states:

> Although randomized trials provide key guidance for how we practice medicine, trust in their published results has been eroded in recent years due to several high-profile cases of alleged data suppression, misrepresentation, and manipulation. While most publicized cases have involved pharmaceutical industry trials, accumulating empiric evidence has shown that selective reporting of results is a systemic problem afflicting all types of trials, including those with no commercial input. These examples highlight the harmful potential impact of biased reporting on patient care, and the violation of ethical responsibilities of researchers and sponsors to disseminate results accurately and comprehensively.
>
> Biased reporting arises when two main decisions are made based on the direction and statistical significance of the data—whether to publish the trial at all, and if so, which analyses and results to report in the publication. Strong evidence for the selective publication of positive trials has been available for decades. More recent cohort studies have focused on the misreporting of trials within publications by comparing journal articles either with documents from regulatory agencies or with trial protocols from research ethics committees, funding agencies, research groups, and journals. These cohort studies identified major discrepancies—favorable results were often highlighted while unfavorable data were suppressed; definitions of primary outcomes were changed; and methods of statistical analysis were modified without explanation in the journal article.

You, the patient, may have great trust in your doctor, but what about his or her suppliers? According to Ioannidis and his colleagues, "We have to take it [on faith] that the findings are OK." Basically, if you are on a drug, especially a new drug without a long history, you are a guinea pig for the drug companies. It makes sense to have a healthy suspicion of the medications you take. Just watch TV and you are barraged with drug ads that have a laundry list of side effects. How is it that our society has become so dependent upon these "miracle" medications? It started in the early 1900s as described by Paul Clayton [1]. The next phase of the proliferations of drugs for "health" started within the hallowed halls of our medical schools.

DRUG COMPANY'S INFLUENCE ON MEDICINE

The pharmaceutical industry supports the medical community in a variety of ways. Their main purpose, as a business, is to produce medications and make them available to patients through medical doctors. This industry is competitive, complicated, costly, and yet extremely lucrative. The top 10 pharmaceutical companies all occupy comfortable locations on the Fortune 500 listing of the top worldwide companies, based on revenues and profits.

Drug companies have a profound influence on doctors who prescribe their drugs. The influence of drug companies extends further to sponsorship of opinion leaders promoting their drugs and groups producing clinical guidelines. Legislation does attempt to control the influence of pharmaceutical companies on the medical community. However, many loopholes exist that allow the drug companies to exert undue influence on medical doctors and the medical profession. Some of many areas that drug companies exercise influence on medicine include:

- contributing monies for the salaries of FDA employees;
- contributions to political campaigns, PACs, and lobbyists;

- sponsoring continuing education credit seminars for medical doctors (there are many perks associated with these conferences and seminars);
- providing monies (sometimes multiple millions) as unstructured research grants to medical schools and hospitals;
- publishing research papers on their new drugs, many of which are written by ghostwriters and often without including unpublished and unfavorable findings;
- unleashing multimillion dollar marketing campaigns spearheaded by sales representative whose mission is to "educate" doctors on the benefits of their new drugs;
- influencing medical school curriculum (since about 1980).

How do these practices influence diagnosis and treatment of AD? They have a profound impact on patient care, regardless of the disease or condition. There are currently no new, novel, or on-patent drugs that change the course of AD. However, there are off-patent drugs that are strongly suggested to impact the direction of the disease. Since these drugs do not have a financial sponsor (a drug company), they are almost never used for the purpose of treating AD. When a drug company seeks drug approval, it is usually for one (or maybe a couple) of diseases only. These drugs are not "designated" for Alzheimer's treatment, but the FDA does permit so-called "off-label" use. That is, the drug "label," or more importantly, the drug master file maintained by the FDA, does not explicitly allow or prohibit the use of the drug for the treatment of Alzheimer's or other diseases.

A few years ago, Michael J. Fox presented the keynote address at the annual biotech meeting and convention in Boston. He informed the audience that a generic drug called Minocin (minocycline) was the only drug that improved his Parkinson's conditions. Vast medical literature suggests this drug also has a very positive effect on Alzheimer's sufferers. However, the only drugs that are presented to doctors on a daily basis are essentially ineffective drugs such as Aricept that have a financial incentive behind their sale.

You do not have to venture far into Google to verify the influence of drug companies on Alzheimer's treatment. The Business Day section of the *New York Times* published an article titled "Drug dosage was approved despite warning." This was sleuthed by Katie Thomas and printed on March 22, 2012. Here is a case of a drug company doing everything it can to extend the patent life of an ineffective Alzheimer's drug, strictly for profit reasons. The report states that 4 months before Aricept, the multibillion-dollar Alzheimer's drug, was set to lose its patent protection, its makers received approval for a higher dosage that extended their exclusive right to sell the drug. But the higher dosage caused potentially dangerous side effects and worked only slightly better than the existing drugs, according to an article published in the *British Medical Journal* [27].

"How the FDA forgot the evidence: the case of donepezil 23 mg" is a sad but true story by Dartmouth Institute for Health Policy and Clinical Practice Professors, Lisa M. Schwartz and Steven Woloshin [27]. They write, "What is the difference between 20 and 23? If you said three, you are off by millions—of dollars in sales, that is—at least from the perspective of Eisai, the manufacturer of donepezil (marketed as Aricept by Pfizer)" for Alzheimer's sufferers.

A clinical trial with 1400 patients found that the larger dosage led to substantially more nausea and vomiting and potentially dangerous side effects for elderly patients struggling with advanced AD. The drug called Aricept 23 was approved in July 2010 against the advice of reviewers at the FDA. They noted that the clinical trial had failed to show that the higher dosage of 23 mg versus the previous dosages of 5 and 10 mg met its goals of improving both cognitive and overall functioning in people with moderate to severe AD. Interestingly, the 5- and 10-mg doses of Aricept went "off-patent" in November 2010, a scant 4 months after the approval of Aricept 23. Also, the dose, 23 mg, could not be matched through any combination of the 5- and 10-mg versions. This stinks for patients.

Aricept is now a $2 billion-a-year blockbuster in large part because people caring for elderly patients with dementia are desperate for something, anything to slow their loved ones' inexorable decline. The original dose for the drug, which was approved in 1996, provided a short-term improvement in memory that faded to insignificance within 6 months. While the clinical trial at the higher dose showed that patients did slightly better in cognition (like recognizing numbers), the drug had no impact whatsoever on their actual functioning in day-to-day life, at least none that their caregivers could notice. Yet the major side effects of the drug, namely nausea and vomiting, increased significantly. The article claimed that the FDA had specifically said to the trial sponsors that the higher dose had to have an impact that caregivers could notice to win approval. Schwartz and Woloshin charged the FDA with violating its own standards.

With approval in hand, the drug's sponsors launched a major new advertising campaign featuring emotional scenes of people caring for spouses or parents with Alzheimer's. The ads implied the drug-improved cognition, which it did on tests, but didn't mention anything about overall functioning, which did not improve. Like all drug ads, they warned about side effects, but gave no sense of their seriousness. The drug companies also made erroneous claims about the benefits of Aricept 23 in advertisements to doctors and on the label. They claimed that the drug had improved both clinical and overall functioning when that was not the case. The FDA, when alerted to the error, said it "was an oversight," and the label has since been corrected.

Before continuing to bash doctors, researchers, the FDA, and drug companies, let's review my motivations. Simply stated, it is to protect you, the patient, from succumbing to the philosophy that modern medicine, equipped mainly with a prescription pad, can protect and maintain your health. Are modern drugs helping us? Let's return to the writings of Paul Clayton through his series of articles, for perspective [1].

> Analysis of the mid-Victorian period in the United Kingdom reveals that life expectancy at age five (in 1870) was as good or better than exists today, and the incidence of degenerative disease was 10% of ours. Their levels of physical activity and hence calorific intakes were approximately twice ours. They had relatively little access to alcohol and tobacco, and due to their correspondingly high intake of fruits, whole grains, oily fish, and vegetables, they consumed levels of micro- and phytonutrients at approximately 10 times the levels considered normal today. This paper relates the nutritional status of the mid-Victorians to their freedom from degenerative disease and extrapolates recommendations for the cost-effective improvement of public health today.

Do you need pills with the myriad of side effects we hear about on TV daily, or do we need knowledge and personal responsibility? It is clear that the degenerative diseases we suffer today are:

- epidemic
- largely self-induced

- modern (hardly present 100 years ago)
- not being solved by the pills and procedures of modern medicine

In light of the unknowns associated with new drugs and adverse side effects, why would you take prescription medications? Before taking any medication, ask your doctor a couple of simple, open-ended questions:

- How does this drug work? You are looking to determine if it is for symptoms or root causes.
- How does this drug improve my overall mortality? Emphasize overall, as opposed to a specific type of mortality. For example, statin drugs slightly improve cardiovascular mortality but patients on statins die just as fast from "all-cause" mortality due to increased cancer, diabetes, and other afflictions.
- How long has this drug been on the market, how many prescriptions have you written, and for how long, and have any of your patients complained of side effects?
- Please suggest a couple of nonpharmaceutical alternatives for the drug you are prescribing.

A patient came to a clinic, apparently suffering from Parkinson's disease. She was on 30 medications. At the level of medication, adding a 31st was hopeless and dangerous. A statistician could do factorial analysis on the list of 30 and not be able to predict the drug–drug interactions of the 31st without a very powerful computer. Could she reduce her medication burden? No. Several of the drugs were addictive, and in her state of confusion, attempting to wean her proved to be a challenge. The addictive drugs were antipsychotic and designed to calm her. The husband became perturbed by the doctors' reluctance to move forward until the doctor suggested that he take the 30 meds his wife was on and then try the 31st on her behalf (he was not serious but the husband got the point).

Many of the top-selling drugs are addictive. Most of the drugs prescribed by psychiatry create dependency while doing nothing to treat the root cause. Even the stomach acid–relieving drugs create a type of dependence because, when the drugs are cut back or stopped, the "acid" problem becomes exacerbated. So what do you do? Stay on the drug.

We have seen that researchers are tempted to err on the side of positive results. That is a tendency of human nature, but in medicine, it has potentially dire consequences. At least these researchers have autonomy in their actions, or do they?

GHOSTWRITING OF MEDICAL ARTICLES

Are you familiar with the term "ghostwriting?" Before becoming educated on medical ghostwriting, I viewed this as a harmless exercise in which a seasoned "writer-for-hire" wrote on behalf of a busy, and possibly inarticulate, celebrity. But ghostwriting occurs for scientific and medical papers, too. This is an acceptable method of creating an article; presuming the author conducted the research and guided the writer. Sadly, this is not always the case.

Consider reading the following articles:

2009: Ghostwriting: The dirty little secret of medical publishing that just got bigger [28]
2010: Ghostwriting at elite academic medical centers in the United States [29]

2010: How ghost-writing threatens the credibility of medical knowledge and medical journals [30]

According to Galvin Yamey, MD:

> If you are an editor, author, reviewer, or reader of medical journals, or if you depend on your doctor or health care provider for getting unbiased information from medical journals, then run for cover and bow your head in disgust. What the authors reveal amounts to one of the most compelling expositions ever seen of the systematic manipulation and abuse of scholarly publishing by the pharmaceutical industry and its commercial partners in their attempt to influence the health care decisions of physicians and the general public.

These articles educate us on how drug and medical writing companies have invented a new way to create a research paper for publication in medical journals. Exposed in the articles is a writing company that was commissioned to produce a manuscript on a piece of research to fit the drug company's needs, and then a person was identified to be the "author."
Yamey continues,

> An email from a writer employed by the medical writing company, DesignWrite, to employees of Wyeth, the company that performed the study, and Parthenon (another medical writing company) on November 10, 2003 concerning manuscripts on Totelle (a brand of hormone replacement therapy manufactured by Wyeth) tells the story concisely. 'Thanks to all who have reviewed and approved the manuscripts … I have received no word on authors for the Totelle 2 mg bone manuscript P3, and need input on this matter before this manuscript can move forwards.'
> The story came to light due to an ongoing court case in which women were suing Wyeth, the manufacturers of Prempro, which is a hormone replacement therapy. During the discovery process for this case, one of the lawyers representing injured women in the litigation, Jim Szaller of Cleveland, Ohio, became aware of many documents that laid out in detail the company's (mostly successful) attempts to publish papers written by unacknowledged professional medical writers in which the message, tone, and content had been determined by the company but the paper was subsequently nominally 'authored' by respected academics—in sum a coordinated and carefully monitored campaign of ghostwriting.

The details of the Prempro case include Wyeth paying ghostwriters to produce medical journal articles favorable to its hormone replacement therapy, according to congressional letters seeking more information about the company's involvement in medical ghostwriting. At least one article was published even after a federal study found the drug raised the risk of breast cancer. Senator Charles E. Grassley, an Iowa Republican, asked Wyeth and Design-Write to provide internal documents about the process of creating medical publications about their hormone therapy drugs. The documents show that company executives came up with ideas for medical journal articles, titled them, drafted outlines, paid writers to draft the manuscripts, recruited academic authors, and identified publications to run the articles—all without disclosing the companies' roles to journal editors or readers.
"This is not the place to review everything written on this topic. Others have written about ghostwriting campaigns concerning single drugs that have led to catastrophic health effects, and how even research papers and clinical trials are affected by ghost authors. What's clear is that ghostwriting can no longer be considered one of the 'dirty little secrets' of medical publishing that nothing can be done about. While editors, medical schools, and universities have turned a blind eye to, or at the least failed to tackle head-on the pervasive presence of

ghostwriting, drug companies and medical education and communication companies have built a vast and profitable ghostwriting industry," said Yamey.

How pervasive is the ghostwriting problem? In some academic circles, it has come to be considered acceptable, and marketing campaigns center on "evidence" provided by seemingly respectable academic review articles, original research articles, and even reports of clinical trials that have the drug company as the author. What research is unbiased? We just don't know. Maybe a rule of thumb is to only take drugs that have a >20-year track record and avoid the new "blockbuster" drugs until they prove safe in the real world rather than in some laboratory.

> It's time to get serious about tackling ghostwriting. As has been shown in the documents released after the Vioxx scandal, this practice can result in lasting injury and even deaths as a result of prescribers and patients being misinformed about risks [31]. Without action, the practice will undoubtedly continue. How did we get to the point that falsifying the medical literature is acceptable? Whatever the reasons, as the pipeline for new drugs dries up and companies increasingly scramble for an ever-diminishing proportion of the market in 'me-too' drugs, the medical publishing and pharmaceutical industries and the medical academic community have become locked into a cycle of mutual dependency, in which truth and a lack of bias have come to be seen as optional extras. Medical journal editors need to decide whether they want to roll over and just join the marketing departments of pharmaceutical companies. Politicians need to consider the harm done by an environment that incites companies into insane races for profit rather than for medical need. After all, even drug company employees get sick; do they trust ghost authors?
>
> —*Galvin Yamey* [28]

Guest authorship, a polite way to infer ghostwriting, is a disturbing violation of academic integrity standards, which form the basis of scientific reliability. The scientific base guiding clinical practice and decision making is to a large degree formed by the peer-reviewed medical literature. Indeed, pharmaceutical sponsors borrow the names of academic experts precisely because of the value and prestige attached to the presumed integrity and independence of academic researchers. In turn, academics receive considerable credit for publication, thus providing an incentive for their willingness to act as "guests."

In the United States, cases relating to gabapentin, rofecoxib, paroxetine, sertraline, fenfluramine/phentermine (fen-phen), and Prempro are well documented, while many others, relating to rosiglitazone, olanzapine, quetiapine, valdecoxib, and celecoxib, remain under seal by the courts. These cases demonstrate the dangers inherent in permitting pharmaceutical companies to have too much influence in the published literature.

In 2008, the overall prevalence of articles with honorary authorship, ghost authorship, or both was 21.0%. The study that determined this statistic concluded that inappropriate authorship remains a significant problem in high-impact biomedical publications.

The Food, Drug, and Cosmetic Act (FDCA) governs drug safety; under it, manufacturers are forbidden from directly marketing a drug for a use other than the FDA-approved indication. Under the FCA, lawsuits have been brought for FDCA violations against drug companies, based in part upon the company's utilization of ghostwritten articles to support off-label use, through illegal means, that induces physicians to prescribe medication for unapproved uses. In 2004, Pfizer pleaded guilty to charges that its Warner-Lambert unit flouted federal laws (FDCA and FCA) by promoting nonapproved uses for a drug, alleging it used an illegal marketing strategy to drive up sales. Pfizer paid $430 million in settlement. The lawsuit alleged that the drug marketing campaign included compensating doctors for putting their names on ghostwritten articles, paying them hefty speakers' fees, and covering the costs of "educational" trips at lavish resorts.

To safeguard yourself against the "unknown," ask your doctor "how long has this drug been on the market, how many prescriptions have you written, and for how long, and have any of your patients complained of side effects?" Also, avoid drugs that haven't been in public use for >20 years. But what about AD and the sufferers who are waiting for that one miracle pill that the drug companies will produce to save us? There are many available products and drugs with long track records that substantially change the course of Alzheimer's, are available now, and have been around for quite some time. The issue, if you have read closely, is that the lack of financial incentive keeps these from you. But take heart (and brain) because there are viable options. Start by taking charge of your own health and not relying upon the prescription pad.

Now the statistics on inaccurate medical publications is more understandable. But the drug company process to infiltrate medicine and promote their products does not stop at presenting their data through journals and conferences.

DRUG COMPANIES PENETRATE CURRICULUM AT MAJOR MEDICAL SCHOOLS

Interestingly, the interactions between drug companies and medicine appear most involved at the highest levels of academia and medical education. An article titled "Harvard Medical School in ethics quandary" in the *New York Times* by Duff Wilson in 2009 sheds light on this conundrum [32].

According to the article, at a first-year pharmacology class at Harvard Medical School, a student grew wary as the professor promoted the benefits of cholesterol drugs and seemed to belittle a student who asked about side effects. A little research by the student revealed the professor was not only a full-time member of the Harvard Medical faculty but also a paid consultant to 10 drug companies, including 5 makers of cholesterol treatments. This finding led the student to question the integrity of his education from Harvard, often thought of as the premier medical school in the world.

"I felt really violated," a student said, as quoted in the article. "Here we have 160 open minds trying to learn the basics in a protected space, and the information he was giving wasn't as pure as I think it should be."

Some Harvard Medical School students and a few faculty members started a campaign to stop outside influences in their classrooms and laboratories, as well as in Harvard's 17 affiliated teaching hospitals and institutes. They say they are concerned that the same money that helped build the school's world-class status may in fact be hurting its reputation and affecting its teaching.

The American Medical Student Association gave Harvard a grade of F based on how well medical schools monitor and control drug industry money. Harvard Medical School's peers received much higher grades, ranging from the A for the University of Pennsylvania to B's received by Stanford, Columbia, and New York University, to a C for Yale. An unofficial excuse came from Harvard claiming the problem occurred because its teaching hospitals are not owned by the university, complicating reform because the dean is fairly new and his predecessor was such an industry booster that he served on a pharmaceutical company board, and because a crackdown, simply put, could cost it money or faculty.

The Harvard students have already secured a requirement that all professors and lecturers disclose their industry ties in class, a blanket policy that has been adopted by no other leading medical school. One Harvard professor's disclosure in class listed 47 company affiliations.

The students at Harvard leading the charge against drug company indoctrination say they worry that pharmaceutical industry scandals, including some criminal convictions, billions of dollars in fines, proof of bias in research, and publishing and false marketing claims, impact their future ability to serve patients. These types of activities have cast a bad light on the medical profession.

The school said it was unable to provide annual measures of the money flow to its faculty beyond the $8.6 million that pharmaceutical companies contributed last year for basic science research and the $3 million for continuing education classes on campus. Most of the money goes to professors at the Harvard-affiliated teaching hospitals, and the dean's office does not keep track of the total.

Harvard Medical faculty members receive tens or even hundreds of thousands of dollars a year through industry consulting and speaking fees. Under the school's disclosure rules, about 1600 of 8900 professors and lecturers have reported to the dean that they or a family member had a financial interest in a business related to their teaching, research, or clinical care. The reports show 149 with financial ties to Pfizer and 130 with Merck. The rules, though, do not require them to report specific amounts received for speaking or consulting, other than broad indications like "more than $30,000."

Dr. Jean Haddad wrote an interesting paper that was published in the San Francisco Medical Society website. It was titled "The pharmaceutical industry's influence on physician behavior and health care costs."

The development of new drugs and therapies is responsible for improving health and longevity. Yet, these improvements in health care have been accompanied by a dramatic increase in cost. The National Institute for Healthcare Management found that US spending on prescription drugs went from $111.1 billion to $131.9 billion in one year, an increase of $20.8 billion (18.8 percent). The bulk of the increase was due to spending on a relatively small group of drugs. Increases in the sales of 23 drugs accounted for 50.7 percent of the 20.8 billion. The NIHCM concluded that the overall increase in prescriptions and especially the shift toward use of costlier and newer drugs. These are the drugs, statistically, that are causing the most harm.

PHARMACEUTICAL MARKETING

The pharmaceutical industry is the most profitable of any major industry. It also has the most sophisticated and effective marketing techniques of any industry in the world. Among their myriad of marketing techniques, four major initiatives are described here:

1. Supporting required Continual Medical Education (CME)
2. Direct marketing to physicians
3. Direct-to-consumer advertising
4. Free drug samples

Supporting Required Continual Medical Education (CME): The pharmaceutical industry currently provides a substantial proportion of the cost of CME in this country and uses that support as a marketing tool. This conclusion was drawn in a *Journal of the American Medical Society* article and others [33–37]. They organize events, prepare slides and curriculum materials, pay speakers, display and promote products, subsidize attendance at meeting, and provide free meals, transportation, and lodging.

The conflicts of interest and likelihood of biased presentations are inherent in such practices. There is evidence that physicians attending such conferences later prescribe these products more often than competing drugs. New industry-medical education and communication companies, mainly funded by the pharmaceutical industry, prepare CME courses. Dr. Trempe presented a forward-thinking way to treat macular degeneration at a conference, primarily sponsored by Roche, the drug maker for the "standard-of-care" drug. The talk did not recommend against the Roche product, but it did argue for, what could be considered, a more "root-cause" therapeutic approach. He was never invited back to that, or any other conference.

The use of two drugs was reviewed at a hospital before and after all-expenses-paid symposia at a "luxurious resort" on the West Coast, the other in the Caribbean. Usage of both drugs increased following the symposia, in contrast to national usage patterns at the time. This occurred despite the stated belief of the participating physicians that these enticements would not alter their prescribing patterns.

Direct Marketing to Physicians: It has been estimated that pharmaceutical companies spend over $8 billion a year on marketing to physicians, 80% of their marketing budget. This averages to over $9000 annually per practicing physician in this country. Drug sales representatives come to the office, and leave samples, lunch, pens, pads, and selected articles in order to "educate" the doctor about their new products. There is 1 of them for every 10 practicing physicians in this country. Do you think that sales literature is accurate and virtuous? Is any sales literature? However, these products influence life and death.

Physicians think they are above being influenced. Studies have shown, however, that changes in physician prescribing patterns occur following symposia, expense-paid trips, honoraria, speaking engagements, and drug rep visits. Of course, this is why this practice continues. Pharmaceutical companies have compiled extensive data on doctor prescribing patterns. They buy this information, as well as prescribing data from pharmacies, allowing them to "profile" doctors and tailor their marketing to each physician.

A recent study in the *Journal of the American Medical Association* looked at authors of Clinical Practice Guidelines [38]. Eighty-seven percent of them had prior financial arrangements with the pharmaceutical industry and 59% had a relationship with manufacturers of the drugs they recommended in the guidelines. In practically all instances there was no disclosure of these relationships in the published guidelines.

Direct-to-Consumer Advertising: Direct-to-consumer (DTC) advertising has been legal since 1985. The industry spent $55 million in 1991, $80 million in 1996, and $2.5 billion in 2000 for DTC, 16% of the industry's marketing budget. Claritin has more advertising dollars spent on it than does either Coca Cola or Budweiser. The company estimates it generates $3.50 in extra sales for every dollar spent on advertising.

All Western nations, with the exception of New Zealand and the United States, have historically banned direct advertising of pharmaceuticals to consumers.

The FDA position on direct-to-consumer marketing of drugs is as follows:

In sum, prescription drug advertising can provide consumers with important information about new prescriptions and new indications for existing prescription drugs, as well as information about symptoms of treatable illnesses and other conditions. Done properly, prescription drug advertising can assist consumers in taking a pro-active role in improving their health. However, to be of value, these advertisements must not be false and misleading. As a result, FDA continues to closely monitor DTC advertising to help ensure that this promotional activity is accurate and balanced. FDA will complete evaluation of its own research and that of other groups to help ensure that FDA's policies in regulating DTC advertising are optimal.

DTC is effective. A study in 1999 showed that 80% of patients who asked for an advertised drug were prescribed it. DTC can have a deleterious effect on the doctor–patient relationship, putting physicians in the uncomfortable and possibly adversarial position of discouraging patients from using new expensive therapies that may not be in their best interest. DTC is wrong for the doctor and patient but is right for drug company profits.

Free Drug Samples: These "premiums" influence prescription practices based on a survey of 154 general medicine and family physicians at an academic medical center. Nearly all physicians surveyed said that they would ideally choose a certain type of drug, a diuretic or β-blocker, as initial therapy for hypertension. However, of the physicians who said they would use a sample for an uninsured patient with hypertension, more than 90% chose a sample that differed from their preferred choice. The existence of samples influenced them to use medication they would not have otherwise prescribed. Drug companies provided several billion dollars' worth of pharmaceutical samples, most of these the newest, most expensive products. Very often, the drug rep does not reveal the actual cost of the new drug claiming it is covered by insurance or that its cost varies among plans.

Are things changing today, because much of the information presented in this chapter covers a decade or more? Surely policies and procedures are being put in place to correct the inappropriate influence of drug companies on medicine. But not so fast! While editing this chapter, the following article popped up on an alert: "Guideline$: following the money in acne treatment" [39].

When the American Academy of Pediatrics endorsed guidelines recommending expensive prescription drugs to treat childhood acne, it didn't tell doctors this: 13 of the 15 experts who drafted the guidelines were paid consultants or speakers for companies that market the drugs recommended in the guidelines. Or this: the organization that developed the guidelines—paid the academy to publish them—received 98% of its 2011 revenue from companies that make acne drugs.

The guidelines recommend prescription drugs that cost as much as $1700 for a year's treatment. By contrast, benzoyl peroxide, an effective over-the-counter product that is a primary component in some of the prescription drugs, costs less than $80 a year.

Buyers beware! (And get a second or third job to cover your "scripts" cost.)

"The medical profession is being bought by the pharmaceutical industry, not only in terms of the practice of medicine, but also in terms of teaching and research," says

Arnold Relman, a Harvard Professor and former editor of *The New England Journal of Medicine*, whose recent critique of the industry's influence in health care, published in the *New Republic*, won him and his coauthor one of the top awards for magazine journalism in the United States. "The academic institutions of this country are allowing themselves to be the paid agents of the pharmaceutical industry. I think it's disgraceful" [40].

A pharmaceutical executive states, "A physician's prescribing value is a function of the opportunity to prescribe, plus his or her attitude toward prescribing, along with outside influences. By building these multiple dimensions into physicians' profiles, it is possible to understand the 'why' behind the 'what' and 'how' of their behavior."

Too bad health care does not use sophisticated methodology like this to diagnose patients.

References

[1] Clayton P, Rowbotham J. How the mid-Victorians worked, ate and died. Int J Environ Res Public Health 2009;6(3):1235–53.
[2] Clayton P, Rowbotham J. An unsuitable and degraded diet? Part one: public health lessons from the mid-Victorian working class diet. J R Soc Med 2008;101(6):282–9.
[3] Rowbotham J, Clayton P. An unsuitable and degraded diet? Part three: Victorian consumption patterns and their health benefits. J R Soc Med 2008;101(9):454–62.
[4] Clayton P, Rowbotham J. An unsuitable and degraded diet? Part two: realities of the mid-Victorian diet. J R Soc Med 2008;101(7):350–7.
[5] How science goes wrong. The Economist, October 19, 2013. Available from: http://www.economist.com/news/leaders/21588069-scientific-research-has-changed-world-now-it-needs-change-itself-how-science-goes-wrong.
[6] Hem E. Too many journals—too little good research. Tidsskrift for den Norske lægeforening: tidsskrift for praktisk medicin, ny række 2011;131(19): 1871.
[7] Coomarasamy A, et al. Medical journals and dissemination of health research: have they fulfilled their role? Clin Med NetPrints 2000;1.
[8] Haynes RB. Where is the meat in clinical journals? ACP J Club 1993;119:A-22–3.
[9] Wakefield AJ, et al. RETRACTED: ileal-lymphoid-nodular hyperplasia, non-specific colitis, and pervasive developmental disorder in children. Lancet 1998;351(9103):637–41.
[10] Hwang WS, et al. Patient-specific embryonic stem cells derived from human SCNT blastocysts. Science 2005;308(5729):1777–83.
[11] Ioannidis JPA. Contradicted and initially stronger effects in highly cited clinical research. JAMA 2005;294(2):218–28.
[12] Ioannidis JPA, et al. Comparison of evidence of treatment effects in randomized and nonrandomized studies. JAMA 2001;286(7):821–30.
[13] Ioannidis JPA, Lau J. Completeness of safety reporting in randomized trials. JAMA 2001;285(4):437–43.
[14] Balk EM, et al. Correlation of quality measures with estimates of treatment effect in meta-analyses of randomized controlled trials. JAMA 2002;287(22):2973–82.
[15] Ioannidis JPA. The importance of potential studies that have not existed and registration of observational data sets registration of observational data sets. JAMA 2012;308(6):575–6.
[16] Patsopoulos NA, Analatos AA, Ioannidis JPA. Relative citation impact of various study designs in the health sciences. JAMA 2005;293(19):2362–6.
[17] Ioannidis JPA, Cappelleri JC, Lau J. Issues in comparisons between meta-analyses and large trials. JAMA 1998;279(14):1089–93.
[18] Tatsioni A, Bonitsis NG, Ioannidis JPA. Persistence of contradicted claims in the literature. JAMA 2007;298(21):2517–26.
[19] Ioannidis J, Panagiotou OA. Comparison of effect sizes associated with biomarkers reported in highly cited individual articles and in subsequent meta-analyses. JAMA 2011;305(21):2200.

[20] Naik G. Scientists' elusive goal: reproducing study results Wall Street Journal, vol. 258, issue 130; 2011. p. A1.

[21] Arrowsmith J. Trial watch: phase III and submission failures: 2007–2010. Nat Rev Drug Discov 2011;10(2):87.

[22] Mullard A. Reliability of 'new drug target' claims called into question. Nat Rev Drug Discov 2011;10(9):643–4.

[23] Ustyugov A, et al. Dimebon slows the progression of neurodegeneration in transgenic model of synucleinopathy. Lancet 2008;372:207–15.

[24] Doody RS, et al. Effect of dimebon on cognition, activities of daily living, behaviour, and global function in patients with mild-to-moderate Alzheimer's disease: a randomised, double-blind, placebo-controlled study. Lancet 2008;372(9634):207–15.

[25] Chan A-W. Bias, spin, and misreporting: time for full access to trial protocols and results. PLoS Med 2008;5(11):e230.

[26] Rising K, Bacchetti P, Bero L. Reporting bias in drug trials submitted to the Food and Drug Administration: a review of publication and presentation. PLoS Med 2008;5:e217.

[27] Schwartz LM, Woloshin S. How the FDA forgot the evidence: the case of donepezil 23 mg. BMJ 2012;344:e1086.

[28] Yamey G. Ghostwriting: the dirty little secret of medical publishing that just got bigger. PLoS Med 2009;6:e1000156.

[29] Lacasse JR, Leo J. Ghostwriting at elite academic medical centers in the United States. PLoS Med 2010;7(2):e1000230.

[30] Barbour V. How ghost-writing threatens the credibility of medical knowledge and medical journals. Haematologica 2010;95(1):1–2.

[31] Ross JS, Hill KP, Egilman DS, Krumholz HM. Guest authorship and ghostwriting in publications related to rofecoxib. A case study of industry documents from rofecoxib litigation. JAMA 2008;299:1800–12.

[32] Wilson D. Harvard Medical School in ethics quandary. The New York Times, vol. 2; 2009.

[33] Wazana A. Physicians and the pharmaceutical industry. JAMA 2000;283(3):373–80.

[34] Relman AS. Separating continuing medical education from pharmaceutical marketing. JAMA 2001;285(15):2009–12.

[35] Blumenthal D. Doctors and drug companies. N Engl J Med 2004;351(18):1885–90.

[36] Lexchin J. Interactions between physicians and the pharmaceutical industry: what does the literature say? CMAJ 1993;149(10):1401.

[37] Dorman T, Silver IL. Continuing medical education: comment on clinician attitudes about commercial support of continuing medical education. Arch Intern Med 2011;171(9):847.

[38] Choudhry NK, Stelfox HT, Detsky AS. Relationships between authors of clinical practice guidelines and the pharmaceutical industry. JAMA 2002;287(5):612–7.

[39] Guideline$: following the money in acne treatment. MedPage Today; September 15, 2013. Available from: http://www.medpagetoday.com/Pediatrics/GeneralPediatrics/41618.

[40] Moynihan R. Who pays for the pizza? Redefining the relationships between doctors and drug companies. 1: entanglement. BMJ 2003;326(7400):1189–92.

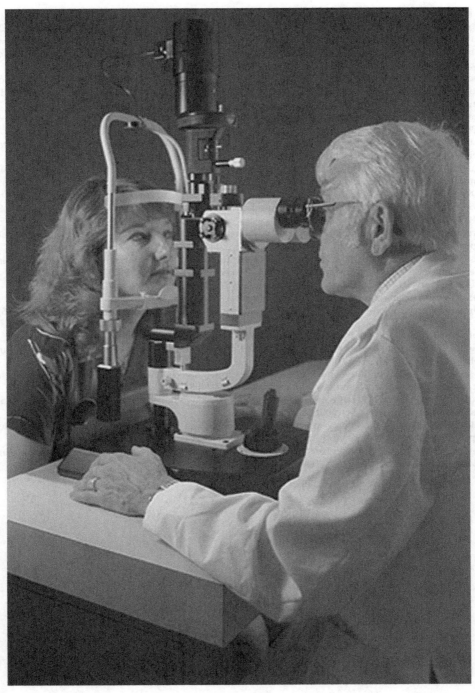

Image Credit: *An optometrist examining a patient with a biomicroscope and a slit lamp, by US Government (http://www.bls.gov/oco/ocos073.htm[dead link]). Available at Wikimedia Commons: https://commons.wikimedia.org/wiki/File%3AOptometrista.jpeg*

CHAPTER

6

A New Diagnostic Paradigm

Ever-increasing specialization is made necessary and inevitable by the information explosion of our times. It is, under these circumstances, easy to lose sight of the underlying interconnectedness of things. This same information explosion has, somewhat paradoxically, also enabled us to see a more fundamental unity within the diversity. We find that medical problems that may seem different or independent when viewed at a superficial level are actually manifestations of a common underlying pathophysiologic mechanism acting simultaneously at different sites throughout the body [1].

—*Daniel H. Gold in "The eye in systemic disease"*

Clayton Christensen is the author of "The innovators dilemma" [2]. Over a decade ago he wrote an article in the *Harvard Business Review* titled "Will disruptive innovations cure health care?" [3]. He clearly possessed a crystal ball, as most of what was wrong in medicine then is now even more against the patient and good health today. His fundamental lesson is that disruptive innovation takes complex ideas and makes them available to regular folks. Examples he cites are as follows: the computer mainframes and punch card evolving to the laptop (or iPad) or George Eastman's inventions making amateur photography widespread.

In the context of medicine, disruptive innovation brings technology to low-level healthcare works that affords an accurate and complete differential diagnosis. Thus nurses, medical assistants, technicians, and physicians' assistants meet this requirement. Technology should be bringing elegant yet simple methods to the clinic. What has happened instead is that specialization has grown stronger in medicine, and the diagnosis is layers upon layers away from the patient and the treating doctor and his or her staff. This model does not work; it hampers innovation and allows specialists to sit in an "ivory tower" without adequate accountability to patient outcomes.

Let's single out neurology as an example. This specialty in medicine is primarily responsible for the diagnosis and treatment of Alzheimer's disease (AD). In their "standard-of-care," the diagnosis of Alzheimer's is largely a guess. Can you see a gathering of neurologists debating the merits of this statement? No one reading this should care what neurology decides because a "definitive" Alzheimer's diagnosis by neurology is still irrelevant to the patient. It does not provide the medical profession with any tools with which to make you better. Let them debate mild cognitive impairment versus Alzheimer's versus dementia, or any other neurodegenerative label. It is of little consequence to the sufferer and the family.

If you are the patient, ask what the diagnosis means in terms of what medicine can do to improve your health. Here is where neurology must say, "Sorry, there is nothing we

The End of Alzheimer's. http://dx.doi.org/10.1016/B978-0-12-812112-2.00006-9

can do in any case." They might include a missive like "it is important to characterize your neurological condition in case treatments are developed for your specific condition sometime in the future." This is a plausible reason for their diagnostic process but it provides no immediate help to patients. Let them evaluate you, but keep your expectations very low. Appendix 2 contains a lengthy diagnosis for a patient by a neurologist. Read the entire text so you can appreciate the inability of this profession to move the needle on neurodegenerative diseases.

Where can you go for a diagnosis of Alzheimer's that is meaningful for designing treatments that work? Sadly just a few places offer a truly differential diagnosis of Alzheimer's at the moment, and these few isolated islands of doctors eschew the standard-of-care. This is due to the lack of innovation or the lack of any systems that allow for truly innovative change in medicine. However, if you are willing to quarterback your own diagnosis and care (or that of your loved ones), then read what follows carefully. There is emerging innovation, and the story unfolds in a blink of an eye. However, since what is described is mostly outside the standard-of-care, you will have to search hard for cooperative doctors, most of whom will require cash payment because insurance does not cover these tests that search for the root cause(s) of AD. Yes, this will be challenging. The good and hopeful news is that tests are available that unlock the Alzheimer's mystery and can lead to treatment that helps you now.

Whence does this new innovation come? It comes through the eye.

"There are many systemic diseases we see in the eye," said Dr. Roy Chuck, chair of the Department of Ophthalmology and Visual Sciences at Albert Einstein College of Medicine and Montefiore Medical Center in New York City.

"The eye is quite literally a 'real window' to the rest of the body," according to Dr. Noel Bairey Merz, director of the Women's Heart Center at Cedars Sinai Heart Institute in Los Angeles. "The vitreous fluid is clear, and we can look through the opening in the iris and see the blood vessels quite easily," she said. "They taught us in medical school to look with the ophthalmoscope as part of the general exam. Sadly, it's not done by most practitioners, and they have lost the skill set."

"Diagnosing illness through the eye is nothing new," according to Dr. Marco Zarbin, Chief of Ophthalmology at the University of Medicine and Dentistry, New Jersey. "It happens all the time," he said, "from rare conditions to diseases like multiple sclerosis, leukemia, and brain tumors. If you look at your brain, two-thirds of it is dedicated to some aspect of vision. It's a big deal."

Researchers at the University of Pennsylvania School of Medicine estimate that the human retina can transmit visual input at about the same rate as an Ethernet connection. Much research on the basic science of vision asks what types of information the brain receives; the U Penn researchers instead asked how much. Using an intact retina from a guinea pig, the researchers recorded spikes of electrical impulses from ganglion cells using a miniature multielectrode array. The investigators calculate that the human retina can transmit data at roughly 10 million bits/second. By comparison, an Ethernet can transmit information between computers at speeds of 10–100 million bits/second.

The retina is actually a piece of the brain that has grown into the eye and processes neural signals when it detects light, say the U Penn researchers. Ganglion cells carry information from the retina to the higher brain centers. Other nerve cells within the retina perform the first stages of analysis of the visual world. The axons of the retinal ganglion cells, with the support of other types of cells, form the optic nerve and carry these signals to the brain.

The eye is a window to the soul. But are you aware that the eye is also a window to the brain and the cardiovascular system? This fact is slowly emerging as important in the diagnosis of AD. Many compelling and important connections between the eye and the brain in AD have been made but are largely ignored. We need go no further than the *Harvard University Press* that lists a book for sale titled *"The retina, an approachable part of the brain,"* written by John E. Dowling [4]. According to the Harvard University Press overview:

> John Dowling's *The Retina*, published in 1987 (and revised in 2012), is the most widely recognized introduction to the structure and function of retinal cells. Dowling draws on 25 years of new research to produce an interdisciplinary synthesis focused on how retinal function contributes to our understanding of brain mechanisms.
>
> The retina is a part of the brain pushed out into the eye during development. It retains many characteristics of other brain regions and hence has yielded significant insights on brain mechanisms. Visual processing begins there as a result of neuronal interactions in two synaptic layers that initiate an analysis of space, color, and movement. In humans, visual signals from 126 million photoreceptors funnel down to one million ganglion cells that convey at least a dozen representations of a visual scene to higher brain regions.

Dr. Dowling is no rookie trying to make a name for himself. He is a Gordon and Llura Gund Professor of Neurosciences at Harvard University and a Professor of Ophthalmology (Neuroscience) at Harvard Medical School. He is a member of the National Academy of Sciences, the American Philosophical Society, and the American Academy of Arts and Sciences; he also has won the Helen Keller Prize for Vision Research, the Paul Kayser International Eye Research Award of the International Society for Eye Research, and the Glenn A. Fry Medal in Physiological Optics.

EYE/ALZHEIMER'S LINK

The eye and AD is a natural link based on Dowling's work. A simple Google search for "Alzheimer's" and the "eye" yields 21 million web hits. Here is a list of some of the top results from that search along with the part of the eye that is connected to AD:

- Retina: "Can we predict Alzheimer's a decade before symptoms?" The amount of beta-amyloid protein in the brain corresponded closely to the amount of that same protein in the retina, in the very back of the eye.
- Retina: "Age-related macular degeneration (AMD): Alzheimer's disease in the eye?" AMD is a late-onset, neurodegenerative retinal disease that shares several clinical and pathological features with AD.
- Eye General: "Ocular biomarkers for early detection of Alzheimer's disease." The eye is the only place in the body where vasculature or neural tissue is available for noninvasive optical imaging.

- Eye General: "Alzheimer's disease and the eye." A variety of visual problems have been reported in patients with AD including loss of visual acuity (VA), color vision and visual fields, changes in pupillary response to mydriatics, defects in fixation and in smooth and saccadic eye movements, changes in contrast sensitivity and in visual evoked potentials (VEPs), and disturbances of complex visual functions such as reading and visuospatial function.
- Lens: "Eye test identifies Alzheimer's disease in patients." Detecting a specific fluorescent signature of beta-amyloid in the human lens.
- Blood vessels in the Retina: "Eye tests may predict Alzheimer's risk." Abnormalities in retinal vascular parameters (simply put, blood vessels) may indicate increased amyloid plaque in the brain and can serve as biomarkers for preclinical AD.
- Visual Cortex: "Improving Alzheimer's and dementia care: the eyes have it." For AD and other dementias, loss of function in the visual cortex of the brain helps us better understand a person's condition.
- Visual Field: "Visual field loss and Alzheimer's disease." AD may occur with visual field loss (loss of peripheral vision).

"Alzheimer's disease and the eye" in the previous list is from an article published in the *Journal of Optometry* in 2009 by Richard A. Armstrong of the United Kingdom [5]. Armstrong performed an exhaustive review of the impact of AD on the eye. He divided the review into two sections:

1. functional changes in vision;
2. changes to eye pathology.

If seen in a patient, any one of these changes could suggest (at a minimum) that further tests should be conducted. Table 6.1 shows an abbreviated listing of functional eye changes that sometimes occur in concert with AD. Many of these may be detected before the clinical aspects of cognitive deficit from Alzheimer's occur.

This is a mature area of study as many of these articles extend back into the 1980s and 1990s. Armstrong himself first published a paper titled "Alzheimer's disease and the eye," in 1996 [37].

Some of the data presented about functional changes in vision and AD are controversial because of a limited amount of data and the lack of consistency in results from study to study. Also, function variations in the eye do not give particularly useful information about the cause of the disease. However, it is important to note the vast number of ways the neurodegenerative brain process of Alzheimer's can impact the eye. Consider any type of eye change in yourself or a loved one cause to "look further." Take eye changes as a clue as they are more obvious than, say, your father drilling holes in your porch floor to expel rainwater (read "my story" in Chapter 12).

Sure, functional eye changes are not definitive for AD. We should not be fast to conclude that these eye tests are not useful because of false-negatives and maybe false-positives. However, isn't it presumptuous to assume that AD for each patient is the same considering the vagaries of current standard-of-care diagnosis? Clearly the disease is multifactorial; thus the brain of each patient is impacted differently, depending upon the contribution of each of the multifactorial aspects of the disease. So what is the best conclusion about eye tests and

TABLE 6.1 Functional Vision Changes in Alzheimer's Disease

Visual Function	Change in Alzheimer's	References
Visual acuity	Changes in some patients	[6–11]
Color vision	50% of patients impacted	[5,7,9]
Visual fields	Inferior visual field affected	[12]
Depth perception	Reduced in some patients	[13]
Fixation	Reduced in some patients	[6,7]
Saccadic (fast movement)	Delayed. 50% of patients show abnormalities	[6,7,14,15]
Slow pursuit movements	Impaired	[13,16]
Contrast sensitivity	Impacted	[8,10,17–19]
Visual masking	Significantly affected	[19,20]
Pattern ERG	Reduction in wave amplitude	[21–26]
Cortical VEP	Delayed in some studies	[27–31]
Visuospatial	Deficits in 50% of patients	[7,32,33]
Reading	Problems understanding words	[34]
Object recognition	Problems in 50% of patients	[5,19]
Eye–head coordination	Impaired	[35]
Visual hallucinations	20% of patients experience visual hallucinations	[36]

ERG, electroretinogram; *VEP*, visual evoked potential.

Alzheimer's? Several of the eye symptoms that are listed earlier point to a correlation with Alzheimer's, providing a good baseline for an inference of the disease. And these eye problems usually occur before obvious clinical symptoms of AD emerge.

Medicine continues to seek the one-test "holy grail." With a disease this complex, it is an unrealistic pursuit. However, the differential diagnosis process amalgamates all potentially relevant data to create a portfolio for disease. Have you watched CSI (Crime Scene Investigation)? The case isn't solved based on one piece of evidence. To give neurology credit, they do perform multiple tests; however, they do not characterize the disease beyond the brain. In CSI, it is like trying to solve a mystery by just using the sense of touch. It is inadequate.

EYE PATHOLOGY CHANGES AND ALZHEIMER'S DISEASE

The response and function of the eye is the least of the eye changes that occur in association with AD. Eye pathologies, the physiological changes in the eye, also occur in concert with AD. These changes are not just coincidence or "comorbid" occurrences. It turns out that many of these eye changes are due to the same disease processes that is AD. More importantly, since the eye is so easily accessed compared to any other complex body tissue,

TABLE 6.2   Pathological Changes in the Visual System in Alzheimer's Disease

Region	Change in Alzheimer's Disease	References
Lens	Deposition of β-amyloid	[39]
Retina	Reduction in retinal ganglion cells, thinning of the nerve cell layer	[19,26,40,41]
Optic disk	Disk pallor, optic atrophy, disk cupping in the absence of open-angle glaucoma in some patients	[6,12]
Optic nerve	Decline in nerve axons, preferentially affecting the large-diameter axons	[14,26,42]
Suprachiasmatic nucleus	Degeneration reported in some patients	[43]
Visual cortex	Rarely atrophic, myelin reduced in outer laminae, loss of pyramidal cells	[44–47]
Color vision	Numerous cored senile plaques but few tangles	[42]

the eye provides the means to diagnose and study AD long before it shows clinical symptoms in a person. The eye is also useful for measuring and monitoring someone who does have AD.

> "The eyes truly are unique real estate," says Andrew Iwach, MD, Associate Clinical Professor of Ophthalmology at the University of California San Francisco and executive director of the Glaucoma Center of San Francisco. "They're the only place in the body where you can see a bare nerve, a bare artery, and a bare vein without doing any cutting. And the disease processes we see occurring in the eye are probably occurring in the rest of the body [38].

In Armstrong's review of the eye and Alzheimer's, he explored the connection between eye pathology and AD. These manifestations are provided in Table 6.2.

THE RETINA AND ALZHEIMER'S DISEASE

The Alzheimer's Association International Conference (AAIC) is the world's largest conference of its kind, bringing together researchers from around the world to report and discuss groundbreaking research and information on the cause, diagnosis, treatment, and prevention of AD and related disorders. As a part of the Alzheimer's Association's research program, AAIC serves as a catalyst for generating new knowledge about dementia and fostering a vital, collegial research community.

At the Alzheimer's Association International Conference in Paris in 2011, the connection between the eye and the brain finally and inevitably took center stage, mainly due to the lack of any breakthroughs in the traditional realms of research on AD, specifically neurology.

At the Paris meeting in 2011 there were several thousand research ideas and findings presented from keynote speeches, workshops, lectures, poster sessions, and conversations over a libation. Of the 5000 or so presentations, the news media picked up on 2 significant findings. One "revelation" may change the path of Alzheimer's diagnosis in the future.

Researchers from Australia showed that measuring blood flow in the eye, specifically the retina, might predict or portend the development of AD. The study's leader, Shaun Frost of Australia's national science agency, CSIRO, said more study is planned on larger groups to see how accurate the test might be. Dr. Lee Goldstein, who is an Alzheimer's researcher at Boston University and formerly at Harvard Medical School, captures the potential merit of this result.

It's a small study but suggestive and encouraging, Goldstein said. My hat's off to them for looking outside the brain for other areas where we might see other evidence of this disease.

This is a powerful statement in many respects. Here, an AD "insider" suggests that it is a breakthrough that other researchers are beginning to look outside the brain for signs and causes of the disease. Actually, nonneurology researchers have been looking beyond the brain for decades. Goldstein himself, along with a slew of Harvard Medical School colleagues, published a paper in 2003 that did just that—looking outside the brain for signs and symptoms of AD. Astonishingly, they found clear indication of AD beyond the brain. Just like the Australian group, they found their evidence of AD in the eye [39].

The Australian eye study involved photographing blood vessels in the retina, the nerve layer lining the back of the eyes. "Most eye doctors have the cameras used for this, but it takes a special computer program to measure blood vessels for the experimental test doctors are using in the Alzheimer's research," said Dr. Frost.

"Eye doctors often are the first to see patients with signs of Alzheimer's, which can start with vision changes, not just the memory problems the disease is most known for," said Dr. Ronald Petersen, a Mayo Clinic dementia expert. "Brain scans can find evidence of Alzheimer's before it causes memory and thinking problems, but their high cost makes them impractical for routine use. That's why a simple eye test for the disease could be very helpful to families who need to arrange for appropriate care."

Here is a summary of the Australians' work connecting blood flow in the retina with AD. In a small pilot study, Frost and colleagues examined retinal photographs of people with Alzheimer's, people with mild cognitive impairment, and healthy participants from the larger Australian Imaging, Biomarkers and Lifestyle (AIBL) Flagship Study of Ageing. They examined a variety of parameters, including the width of retinal blood vessels. They found that the widths of certain blood vessels in the back of the eye were significantly different for people with Alzheimer's versus those for healthy controls, and that this correlated with a brain imaging benchmark indicative of AD, the deposition of amyloid plaque in the brain as measured by expensive PET imaging.

How significant is this finding by the Australians who captured the headlines at the Alzheimer's big meeting? First, let's consider the number of research dollars represented at the meeting. Using a formula of $500,000/research paper published, and considering there were 5000 or so presentations, that 1 conference presented data costing:

$2,500,000,000 ($2.5 billion)

Does this test (that by the way won't be validated for clinical use for a decade or more) tell us anything about the cause of the disease? No. Is it useful? Yes. It does help further the case that the disease extends beyond the brain and that the eye is sensitive to processes in the brain. Keep in mind that the retina is part of the brain so this is a bit of an erroneous statement, but most people are not aware of this truth. However, very little, if anything, was learned about the root causes except that we need to consider looking beyond the brain cavity.

Let's look deeper into the most publicized findings from the 2011 annual AAIC:

- The first was that AD could be detected early by measuring blood flow changes in the retina of the eye.
- The second was that people in the earliest stages of the disease are subject to more falls.

Are these research headlines that will fast disappear into anonymity and never reach the clinic where patients benefit? We now know to question everything because researchers tend to overdramatize the importance of research. But is there a golden nugget in this finding?

The discovery of a connection between retinal blood flow and AD was reported as new at the AAIC meeting. However, a team at Harvard Medical School published the same findings 4 years before, in 2007 [48]. Why did one team grab the headlines even though they were 4 years late to the party? Let's not forget that there are 2,000,000 medical and scientific publications each year. The 2007 publication regarding retinal blood flow and AD was a needle in the haystack. Praise goes to the Australian team who studied this area and advanced the finding. There is no foul here. They also attended the meeting where it grabbed the right attention. Importantly, the Australians essentially confirmed the Harvard finding, thus adding credibility to the result of both teams.

The abstract from the Harvard team who performed the original work in 2007 reads:

PURPOSE. There is evidence suggesting that visual disturbances in patients with Alzheimer's Disease (AD) are due to pathologic changes in the retina and optic nerve, as well as to higher cortical impairment. The purpose of this study was to evaluate retinal hemodynamic parameters and to characterize patterns of retinal nerve fiber layer (RNFL) loss in patients with early AD.

METHODS. Nine patients with mild to moderate probable AD (mean Mini Mental State Examination score 24 of a possible 30 (age 74.3 ± 3.3 years; mean ± SD) and eight age-matched control subjects (age, 74.3 ± 5.8 years) were included in this prospective cross-sectional study. Blood column diameter, blood velocity, and blood flow rate were measured in the major superior temporal retinal vein in each subject by using a laser Doppler instrument. Peripapillary RNFL was measured by optical coherence tomography.

RESULTS. Patients with AD showed a significant narrowing of the venous blood column diameter (131.7 ± 10.8 μm) compared with control subjects (148.3 ± 12.7 μm, P = 0.01), and a significantly reduced venous blood flow rate (9.7 ± 3.1 μL/min) compared with the control subjects (15.9 ± 3.7 μL/min, P = 0.002). A significant thinning of the RNFL was found in the superior quadrant in patients with AD (92.2 ± 21.6 μm) compared with control subjects (113.6 ± 10.7 μm, P = 0.02). There were no significant differences in the inferior, temporal, or nasal RNFL thicknesses between the groups.

CONCLUSIONS. Retinal abnormalities in early AD include a specific pattern of RNFL loss, narrow veins, and decreased retinal blood flow in these veins. The results show that AD produces quantifiable abnormalities in the retina.

Thus the Harvard group reported a clear relationship between blood flows in the back of the eye, very close to the brain (specifically in the retina). Also they showed that the nerve connecting the retina and the brain, the retinal nerve fiber layer (RNFL), showed atrophy. This is an interesting part of the result because the nerve fiber layer is much more like the white matter of the brain compared to the blood vessels. Maybe atrophy of the nerve fiber layer is more telling compared to the blood flow data? Blood flow can change from a wide variety of health factors including blood pressure medication, stress, diabetes, and a host of other causes. And in some cases, blood flows can rebound, whereas atrophy (loss of tissue) in the nerve fiber layer is much more permanent, like loss of brain tissue. The connection between atrophy of the retinal nerve fiber layer and AD will be discussed in detail later in this chapter.

The group from Australia did acknowledge that "another group previously presented similar data and this is beneficial to both groups as it lends great credibility to the result."

No one can argue that the Harvard team, headed by Drs. Trempe and Feke, are nothing short of experts on the subject of retinal blood flow and their measurements. Dr. Feke has published 45 research papers and patent filing on this very topic, dating back over 35 years, starting in 1976. A list of those papers is provided later. There are some very important educated guesses about retinal blood flow and its connection to AD that can be gleaned from this body of work. They are given at the end of this listing:

1976: Laser Doppler measurement of blood flow in the fundus of the human eye
1977: Flow and diffusion of indocyanine green and fluorescein dyes in the fovea centralis
1978: Laser Doppler measurements of blood velocity in human retinal vessels
1978: Fluorescein dye-dilution technique and retinal circulation
1979: Bidirectional LDV system for absolute measurement of blood speed in retinal vessels
1979: Mean circulation time of fluorescein in retinal vascular segments
1982: Prolongation of the retinal mean circulation time in diabetes
1983: Retinal blood flow alterations during progression of diabetic retinopathy
1983: Clinical application of the laser Doppler technique for retinal blood flow studies
1983: Response of human retinal blood flow to light and dark
1985: Anterior optic nerve blood flow in experimental optic atrophy
1985: Retinal circulatory changes related to retinopathy progression in insulin-dependent diabetes mellitus
1987: Laser Doppler technique for absolute measurement of blood speed in retinal vessels
1988: Evaluation of micrometric and microdensitometric methods for measuring the width of retinal vessel images on fundus photographs
1989: Blood flow in the normal human retina
1989: Effects of optic atrophy on retinal blood flow and oxygen saturation in humans
1991: Optic nerve head blood speed as a function of age in normal human subjects
1992: Retinal blood flow alterations associated with scleral buckling and encircling procedures
1992: Retinal blood flow changes in eyes with rhegmatogenous retinal detachment and scleral buckling procedures
1994: Retinal circulatory abnormalities in type 1 diabetes
1994: Regional retinal blood flow reduction following half fundus photocoagulation treatment
1994: Quantitative circulatory measurements in branch retinal vessel occlusion
1995: Optic nerve head circulation in untreated ocular hypertension
1996: Retinal blood flow changes in type I diabetes. A long-term follow-up study
1997: Retinal laser doppler apparatus
1998: Method and apparatus for examining optic nerve head circulation

2002: Effect of nocturnal blood pressure reduction on retrobulbar hemodynamics in glaucoma

2002: Retinal hemodynamic autoregulation during postural change

2003: Reproducibility and clinical application of a newly developed stabilized retinal laser Doppler instrument

2004: Effect of brimonidine versus latanoprost on the maintenance of retinal blood flow homeostasis during postural change in normal tension glaucoma

2004: Optic nerve head circulation in nonarteritic anterior ischemic optic neuropathy and optic neuritis

2005: Retinal haemodynamics in patients with age-related macular degeneration

2005: Effect of plasmapheresis on retinal hemodynamics in patients with Waldenstrom's macroglobulinemia

2005: Association between systemic arterial stiffness and age-related macular degeneration

2006: Laser Doppler instrumentation for the measurement of retinal blood flow: theory and practice

2006: Non-invasive methods for evaluating retinal affecting neurodegenerative diseases

2006: Association between systemic arterial stiffness and age-related macular degeneration

2006: Hyperviscosity-related retinopathy in Waldenstrom macroglobulinemia

2007: Retinal abnormalities in early Alzheimer's disease

2008: Retinal blood flow response to posture change in glaucoma patients compared with healthy subjects

2008: Retinal blood flow and nerve fiber layer measurements in early-stage open-angle glaucoma

2008: Effect of plasmapheresis on hyperviscosity-related retinopathy and retinal hemodynamics in patients with Waldenström's macroglobulinemia

2008: Retinal haemodynamics in individuals with well-controlled type 1 diabetes

2009: Ophthalmologic techniques to assess the severity of hyperviscosity syndrome and the effect of plasmapheresis in patients with Waldenström's macroglobulinemia

2011: Effect of brimonidine on retinal blood flow autoregulation in primary open-angle glaucoma

2013: Effects of dorzolamide–timolol and brimonidine–timolol on retinal vascular autoregulation and ocular perfusion pressure in primary open angle glaucoma

2014: Effect of brimonidine on retinal vascular autoregulation and short-term visual function in normal tension glaucoma

What did we learn from these titles? First, Dr. Feke et al. published their work on the association between AD and retinal blood flow in 2007. Second, Feke is clearly the world expert on interpreting retinal blood flow. Third, there are several diseases that also have retinal blood flow changes like Alzheimer's, including type 1 diabetes, glaucoma, macular degeneration, aging, and diabetic retinopathy. From these data, two conclusions can be drawn:

1. Because other diseases impact retinal blood flow, the test is confounded by too many factors and is not useful for Alzheimer's.

2. All these diseases are interrelated and lead to similar responses in the body. Thus they have an overlapping root cause(s), and the test results are important as an indicator for all the diseases, including Alzheimer's.

The latter makes sense based on the body of literature, including plenty that is presented in this and subsequent chapters.

Ideas that came out of the Paris Alzheimer's meeting didn't just pop out of the sky. Feke has put 35 years into understanding a concept that now appears to have gained acceptance relevant to Alzheimer's and may add to the growing body of knowledge that explains the link between Alzheimer's, the eye, and other chronic diseases. The Feke team did show us that 35 years of diligent research makes a significant contribution to medical science, in ways that were probably not appreciated at the outset of a career.

The Frost group from Australia published their findings in 2012 [49]. One of the many news outlets that grabbed the story from the Alzheimer's conference presented the following summary of their work in 2011:

> Another study featured at AAIC 2011 explored whether characteristics of blood vessels in the retina (the light-sensitive layer at the back of the eye) might serve as possible biomarkers for Alzheimer's disease. While most Alzheimer's-related pathology occurs in the brain (author's note—this statement is subject to debate as we learn more), the disease has also been reported to create changes in the eye, which is closely connected to the brain and more easily accessible for examination in a doctor's office.

What will happen with the austere work of Feke, Trempe, and Frost? Frost knows what is in his plan. "Our studies are very preliminary, but encouraging," said Frost. "Since amyloid plaque build-up in the brain occurs years before cognitive symptoms of Alzheimer's are evident, a noninvasive and cost-effective retinal test may hold promise as an early detection tool for the disease. We hope that, in the future, our measure could be used with blood-based tests to help doctors identify who needs further assessment with PET imaging and MRI for Alzheimer's, but more research is needed."

You will find that researchers frequently add the caveat, "but more research is needed." Why? Again, researchers are paid to do research and not to advance their findings into the clinic. Again we see the effects of the previously described Trillion Dollar Conundrum because we spend $1 trillion on medical and scientific research each year, and little of that research advances into the clinic to help you, the patient.

To give you an idea how thirsty news media is for stories on AD successes, here is a list of 23 (yes, 23) of the news outlets that broadcasted these very initial findings from the Paris meeting. If you read each of the articles, you will note that not one refers to the previous (2007) study on this very topic. The take-home lesson here is that most (if not all) scientific findings published in the general news media are superficial at best and lack detailed analysis. Included is a link to a YouTube video of Shaun Frost discussing his research. He did acknowledge that there has been "one other study on this topic but we've taken it a bit further."

1. Eye test for Alzheimer's disease in early stages. www.aarp.org
2. An eye test could diagnose Alzheimer's disease in its early stages. www.medicalnewstoday.com *Medical News*
3. ICAD: Eyes spot early Alzheimer's. www.medpagetoday.com
4. Eye test may give clues to Alzheimer's. www.smh.com.au

5. Falls, eye tests may hint at early Alzheimer's. www.reuters.com
6. Eye test may detect Alzheimer's, Australian scientists find. www.christianpost.com
7. Eye test may help diagnose Alzheimer's. www.mnn.com
8. Researchers in Australia look at changes in the eye to determine what's going on in the brain. AAIC 2011. *Alzheimer's Association*
9. Possible new "eye test" for Alzheimer's. www.actionalz.org
10. Alzheimer's early detection possible with eye test and falls. www.ibtimes.com
11. New-eye-exam-could-identify-signs-of-Alzheimer's. www.newsy.com
12. Can Alzheimer's be tracked via the eye. www.reviewofophthalmology.com
13. Eye test 'may help detect Alzheimer's'. www.thehindu.com
14. Eye cells could help diagnose Alzheimer's disease. www.bbc.co.uk
15. Eye examination may lead to early diagnosis for Alzheimer's. www.catholic.org
16. ADHD: Do the eyes have it? www.medscape.com
17. Simple eye test could spot Alzheimer's early on. kmtv.com
18. Eye test to aid Alzheimer's diagnosis. www.abc.net.au
19. Eye test could be used to detect Alzheimer's disease. https://bluesci.wordpress.com/
20. Study: early detection promising for Alzheimer's treatment. http://wtnh.com/
21. Eye test could predict Alzheimer's early. http://thirdage.com/
22. New eye exam could identify signs of Alzheimer's. www.dailymotion.com
23. Retinal eye test catches dementia inexpensively. www.alzheimersweekly.com
24. https://www.youtube.com/watch?v=tq31o8WXyO8

Shaun Frost published a review article in 2010, and the abstract is included in the subsequent text. He has put together a nice review titled "Ocular biomarkers for early detection of Alzheimer's disease" [50].

Alzheimer's disease (AD) is the most common form of dementia and is clinically characterized by a progressive decline in memory, learning, and executive functions, and neuropathologically characterized by the presence of cerebral amyloid deposits. Despite a century of research, there is still no cure or conclusive premortem diagnosis for the disease. A number of symptom-modifying drugs for AD have been developed, but their efficacy is minimal and short-lived. AD cognitive symptoms arise only after significant, irreversible neural deterioration has occurred; hence there is an urgent need to detect AD early, before the onset of cognitive symptoms.

An accurate, early diagnostic test for AD would enable current and future treatments to be more effective, as well as contribute to the development of new treatments. While most AD related pathology occurs in the brain, the disease has also been reported to affect the eye, which is more accessible for imaging than the brain. AD-related proteins exist in the normal human eye and may produce ocular pathology in AD. There is some homology between the retinal and cerebral vasculatures, and the retina also contains nerve cells and fibers that form a sensory extension of the brain. The eye is the only place in the body where vasculature or neural tissue is available for non-invasive optical imaging. This article presents a review of current literature on ocular morphology in AD and discusses the potential for an ocular-based screening test for AD.

The blood vessels of the retina display diagnostic versatility as we saw from the work of Trempe and Feke. Its depth and breadth is growing daily. Schizophrenia, a neuropsychological disease, also shows signs in the retinal vessels. A paper titled "Microvascular abnormality in schizophrenia as shown by retinal imaging" [51] is enlightening. "Retinal and cerebral microvessels are structurally and functionally homologous, but unlike cerebral microvessels, retinal microvessels can be noninvasively measured in vivo by retinal imaging," state the authors. "Retinal imaging is a simple, noninvasive technology for assessing microvascular abnormalities in living individuals diagnosed with schizophrenia. Cerebrovascular abnormalities have been discussed as a pathological feature in schizophrenia." The authors conclude:

> The findings provide initial support for the hypothesis that individuals with schizophrenia show microvascular abnormality. Moreover, the results suggest that the same vascular mechanisms underlie subthreshold symptoms and clinical disorder and that these associations may begin early in life. These findings highlight the promise of retinal imaging as a tool for understanding the pathogenesis of schizophrenia.

THE LENS OF THE EYE AND ALZHEIMER'S DISEASE

The lens of the eye became front-and-center in AD based on a Harvard Medical School study published in 2003 [39]. This finding is far more diagnostic of AD compared to the retinal blood flow work described previously. The finding of the Frost and Feke groups on retinal blood flow is important, but it is clear from the many publications of Feke that retinal blood flow is not particularly specific for the diagnosis of Alzheimer's disease, as many conditions and medications contribute to changes in retinal blood flow.

This Harvard Medical School group performed detailed research to determine the connection between the lens of the eye and the brain, and the presence of beta-amyloid in tissues of both. The paper is entitled "Cytosolic β-amyloid deposition and supranuclear cataracts in lenses from people with Alzheimer's disease" [39], and parts of the abstract are provided here:

> Pathological hallmarks of Alzheimer's disease include cerebral β-amyloid (beta-amyloid) deposition, amyloid accumulation, and neuritic plaque formation. We aimed to investigate the hypothesis that molecular pathological findings associated with Alzheimer's disease overlap in the lens and brain.
>
> Methods: We obtained postmortem specimens of eyes and brain from nine individuals with Alzheimer's disease and eight controls without the disorder, and samples of primary aqueous humour from three people without the disorder who were undergoing cataract surgery. Dissected lenses were analyzed by slit-lamp stereophotomicroscopy, western blot, tryptic-digest/mass spectrometry electrospray ionization, and anti-beta-amyloid surface-enhanced laser desorption ionization (SELDI) mass spectrometry, immunohistochemistry, and immunogold electron microscopy. Aqueous humour was analyzed by anti-beta-amyloid SELDI mass spectrometry. We did binding and aggregation studies to investigate beta-amyloid-lens protein interactions.
>
> Findings: We identified beta-amyloid1-40 and beta-amyloid1-42 (beta-amyloid) in lenses from people with and without Alzheimer's disease at concentrations comparable with the brain, and beta-amyloid1-40 in primary aqueous humour at concentrations comparable with cerebrospinal fluid. Beta-amyloid accumulated in lenses from individuals with Alzheimer's disease as electron-dense deposits located exclusively in the cytoplasm of supranuclear/deep cortical lens fibre cells (n=4). We consistently saw equatorial supranuclear cataracts in lenses from people with Alzheimer's disease (n=9) but not in controls (n=8). These supranuclear cataracts colocalized with enhanced beta-amyloid immunoreactivity and birefringent Congo Red staining. Synthetic beta-amyloid bound βB-crystallin, an abundant cytosolic lens protein. Beta-amyloid promoted lens protein aggregation that showed protofibrils, birefringent Congo Red staining, and beta-amyloid/αB-crystallin coimmunoreactivity.

Interpretation: Beta-amyloid is present in the cytosol of lens fiber cells of people with AD. Lens beta-amyloid might promote regionally specific lens protein aggregation, extracerebral amyloid formation, and supranuclear cataracts.

Here is a layperson's interpretation of this very fine and important work:

1. From the study: "We identified beta-amyloid1-40 and beta-amyloid1-42 (beta-amyloid) in lenses from people with and without Alzheimer's disease at concentrations comparable with the brain …."

   Interpretation: Beta-amyloid appears in the brain of both Alzheimer's and non-Alzheimer's patients. Thus either beta-amyloid can appear in the brain independent of AD or beta-amyloid appears before a diagnosis of Alzheimer's is possible. Most important, the amount of beta-amyloid found in the brain matched the amount found in the lens of the eye. The lens of eye, then, is likely a very powerful diagnostic for the presence of beta-amyloid in the brain. This appears to be the Holy Grail of diagnosis and detection. But since the connection between Alzheimer's and beta-amyloid is still suspect, this test is most important for detecting brain beta-amyloid first, and Alzheimer's second. As you recall from Chapter 2, it is evident that beta-amyloid is not a root cause of AD and thus is not a therapeutic target. However, beta-amyloid is a very good (but not perfect) diagnostic predictor of Alzheimer's.

2. From the study: "and beta-amyloid1—40 in primary aqueous humour at concentrations comparable with cerebrospinal fluid."

   Interpretation: Beta-amyloid is found in the fluid in the eye at the same level as in the fluid in the brain. It may be that the disease propagates through the fluid. It is known that the fluid in the eye is the same fluid as in the brain. It circulates because the eye cavity and the brain cavity are intimately connected. This is well known in the field of ophthalmology because an air bubble placed in the eye, as a surgical aid, sometimes migrates into the brain. Some researchers suggest that Alzheimer's beta-amyloid should be detected by drawing cerebrospinal fluid, by needle, from the spine. There were few takers for that clinical trial because few want an experimental spinal tap. A better way, clearly, is to evaluate beta-amyloid in the eye since the correlation is so strong. Also, the test is inexpensive and noninvasive.

3. From the study: "We consistently saw equatorial supranuclear cataracts in lenses from people with Alzheimer's disease (n = 9) but not in controls (n = 8)."

   Interpretation: Supranuclear cataracts are formed by beta-amyloid. If you have supranuclear cataracts, we now know that it is highly likely that you have beta-amyloid plaques in your brain and thus are more likely to have Alzheimer's compared to people without these cataracts. The common name for these cataracts is cortical cataracts. These cortical cataracts start in the periphery of the lens and do not produce visual impairment during their early stage. Over many years they slowly progress toward the center of the lens and may eventually have to be removed surgically if they start to cause visual disturbance. It is nuclear cataracts, however, that are the subject of most lens surgeries. Cortical cataracts progress over many years, giving ocular professionals plenty of time to observe and monitor them during a routine dilated eye examination. Also, there are no codes for treatment of cortical-type cataracts within the standard-of-care, so there is no financial reimbursement for their management. Thus eye doctors simply observe these cataracts. The definition of these two types of cataracts is provided here:

Cataract /cat·a·ract/ (-rakt) An opacity of the crystalline lens of the eye or its capsule.
Nuclear Cataract Opacification involving the nucleus (center) of the lens; the common form of age-related (senile) cataracts.
Cortical Cataract A cataract in which the opacity affects the cortex (periphery) of the lens.

The Harvard Medical School authors make some valuable points about their research.

A limitation of this study is the small sample sizes. Nevertheless, our findings do provide evidence for extracerebral Alzheimer's disease-associated amyloid pathology. In particular, we have seen apparent beta-amyloid pathological findings in the equatorial supranuclear and deep cortical subregions of lenses from people with Alzheimer's disease. It is noteworthy that equatorial supranuclear cataracts in this peripheral lens subregion are rare compared with common age-related (nuclear) cataracts. By contrast with age-related cataracts, equatorial supranuclear cataracts (cortical cataracts) are anatomically obscured from inspection by the iris, but are visible on routine medical examination in most cases or are easily exposed with dilation. These cataracts are not associated with visual impairment.

I've (T.J.L.) met one of the authors of the 2003 *Lancet* paper over lunch in 2011, and he made his point perfectly clear. "We have to have this (eye/AD case) absolutely right, iron clad, before we advance to a clinic stage. We will only get one chance at this, and it has to be done with the utmost of integrity." I completely disagree with that researcher. There is more than enough evidence to support incorporating an advanced ocular assessment as part of an examination and diagnosis of Alzheimer's. However, the author freely admitted to being a "pure researcher," even though he holds the title "MD." His world revolves around publishing more papers. At the time, he was seeking $1 million of additional research funding.

The best way to gather data is to get these methods out of the theoretical—that is, out of the lab—and apply them to patients. These tests are noninvasive, quick, and inexpensive. For example, every patient who fails the neurology standard-of-care diagnosis should have an advanced eye diagnosis as well, so that we can quickly obtain a large "cohort" of data to show its value. But medicine just doesn't work cooperatively or progressively anymore. This is the very issue Clayton Christensen discussed in "Will disruptive innovations cure healthcare?" [3].

There was a time when the concept of "Grand Rounds" was in vogue. None other than Charles Mayo, founder of the Mayo Clinic, established this concept. Dr. Mayo would gather doctors from all disciplines to discuss various cases. Today we shuttle patients, not the doctors. As a result, there is no deliberate collaboration between medical disciplines. Thus valuable assessments, like cortical (Alzheimer's) cataracts, do not become part of the Alzheimer's assessment. An advanced eye exam is noninvasive and very inexpensive. As you now may realize, "inexpensive" is the kiss of death for a medical technique because no one in medicine makes any money with such tests. MRI, for example, is preferred to a simple eye test because reimbursement is so much richer. It creates a vicious cycle in which doctors are not informed about the importance of cortical cataracts. Without this knowledge, your doctor will tell you that it is not important.

The work at Harvard did spawn a company called Neuroptix that has received an estimated $10 million in funding over the past 10 years. The goal of Neuroptix (now called Cognipix) was to develop a special machine that could quantitatively measure the beta-amyloid (AD protein) in the lens of the human eye. Over a decade has passed and still there is no real commercial product from Neuroptix/Cognipix. The supposition is that, even with fancy tools, it is quite difficult to quantify the beta-amyloid in the lens of the eye (and brain), possibly because the lens of the eye is quite fluid; thus the protein moves around, and it is difficult to measure the same protein over time. Imagine trying to quantify seaweed floating in the ocean. That, however, is the goal of the instrument—to measure changes (increases or decreases) of beta-amyloid in the lens of the eye, in the same person, over time. This would allow researchers and clinicians to evaluate the progression or regression of AD over time and as a function of external influences such as treatment. Neuroptix may have sold some of their instruments, but strictly to pharmaceutical companies who will use the tool as a way to measure the efficacy of new drugs being developed to combat AD. This is not confirmed, however, as the company is private.

Clearly the Neuroptix model is one that could gain traction assuming they can get the instrument to work as theorized. The reason for the projected success is that they can sell the instruments for a profit, and the test requires a dye or other such consumable that enhances the imaging. The consumable also adds to the profits associated with the sale of the device and strengthens the patent and proprietary nature of their technology, the razor blade model, if you will. Ultimately, the goal of Neuroptix is to be purchased by a large diagnostics company such as Boston Scientific, which will make the company's founders a tidy sum of money. Do I blame this approach? Certainly not. It is the reality of business, and they have potentially created a valuable mouse trap that, if successful, can be made widely available and help people significantly by diagnosing AD at its earliest stages of development. However, there are standard instruments used by optometrists and ophthalmologists that achieve most of the objectives of the Neuroptix device without the fancy instrument and costs. They are discussed later in this chapter. How can you be sure these tools and methods already exist? These are the tools the Harvard Medical School team used to create their initial findings.

NEUROLOGY AND OCULAR PROFESSIONALS

Why hasn't the eye/Alzheimer's connection progressed into the clinic? One can only make an educated guess. Lack of collaboration is a big part of the issue. AD is "owned" by neurology and these doctors simply do not use the eye as part of their medical protocol. It's not part of their core training, and they look down upon the eye as a surrogate for the brain. The eye, and its simplicity for diagnosis and access, is a threat to the standard-of-care used by neurology.

Why haven't eye doctors picked up on these tests? There are two types of eye doctors: ophthalmologists and optometrists. Ophthalmologists hold the title Medical Doctor (MD), and they are primarily surgeons. They derive a significant portion of their income from performing surgeries on the eye. They do equally well applying monthly needle injections into the back of the eye that is the current standard-of-care for many of the major eye diseases. These doctors practice "eye-only" disease management and are rarely interested in investigating the systemic causes of eye diseases.

Performing simple eye diagnoses for detection of AD pales, financially, in comparison to surgical procedures and injections. And, if a patient is diagnosed with AD through eye tests, what do the ophthalmologists do next? To whom do they refer the patient? You guessed it— neurology. Neurology then returns the patient to the vicious cycle of their testing and treatment protocols. Patients with cortical cataract but no memory impairment are considered without disease and healthy by neurology.

Optometrists are doctors of optometry. They do not go through the rigorous medical training that includes the general medical curriculum but have the same eye training as ophthalmologists. They are primarily trained to manage vision problems. They are very low on the totem pole of the medical hierarchy. Optometrists, however, are perfectly suited to use these emerging techniques for diagnosis of major diseases like Alzheimer's because what they see, as part of a normal or advanced eye exam, includes markers for Alzheimer's. Hopefully there will be a way to build partnerships between optometry and receptive disciplines in medicine to provide a broad and deep diagnosis and treatment protocol. The most likely partnerships are between optometrists and functional medical doctors who believe in looking for root causes of disease through whole-body evaluation.

The connection between the beta-amyloid in the eye lens and the Alzheimer's brain has more proof than the one Harvard study. According to the story behind the headline "Cataract lens study may lead to early Alzheimer's detection and diagnosis," there is more evidence supporting the eye lens/Alzheimer's connection. The article states that the results of a study by a team of United States researchers suggest that zinc found in lenses in individuals with AD as well as those with Down syndrome strongly supports the hypothesis that Alzheimer's is a systemic disorder. Researchers presented their findings at the Annual Meeting of the Association for Research in Vision and Ophthalmology, and subsequently it was published [52].

Accumulation of amyloid-β (A β) in the brain is a principal feature of Alzheimer's disease and Down syndrome. The researchers, from Boston University, Lawrence Berkeley National Laboratory, and UC Davis, analyzed human Alzheimer's and Down syndrome lenses and identified co-localized zinc and beta-amyloid in the same cytosolic compartments of lens fiber cells. According to the scientists, these results demonstrate that zinc contributes mechanistically to the distinctive age-dependent supranuclear lens phenotype associated with Alzheimer's and Down syndrome.

The news outlet headlines indicate that these data are the first to establish an Alzheimer's-linked amyloid pathology outside the brain and may ultimately "pave the way for development of novel ophthalmic technology for early Alzheimer's detection and diagnosis." However, the actual abstract by the researchers does not make that claim. This "new" finding of an association with levels of zinc in the lens is somewhat novel and may bear some significance. However, zinc is not a likely therapeutic candidate because a very large study, called the AREDS, looked at the impact of antioxidant supplementation, including minerals like zinc, on macular degeneration. Macular degeneration and Alzheimer's, as you will see, are very closely related diseases. There were both favorable and unfavorable outcomes associated with this supplementation. A 1992 correspondence by Dr. Trempe in the *Archives of Ophthalmology* points out the issues associated with large doses of zinc used in the AREDS [53]:

During the past year, ophthalmologists have been subjected to an extensive publicity campaign regarding the role of micronutrients, especially zinc supplementation in the treatment of macular degeneration. The article by Newsome et al. [54] in the February 1988 issue of the *Archives* is often quoted to substantiate this advertising. In the study done by the authors, 200 mg of zinc was given daily to patients with macular degeneration. This is a toxic amount of zinc that can produce serious complications. The recommended daily allowance is 15 mg. When more than 150 mg of zinc is taken, it can produce serious copper deficiency, sideroblastic anemia, and bone marrow depression [55]. Such patients must often undergo costly and unpleasant tests, such as bone marrow biopsy, to determine the cause of the anemia. It is important for clinicians to be aware of this possible complication when prescribing large amounts of zinc to patients.

The first paper that described beta-amyloid in the eye was published in 1996 in *The Journal of Biological Chemistry* and was titled "Oxidative stress increases production of beta-amyloid precursor protein and beta-amyloid (Aβ) in mammalian lenses, and Aβ has toxic effects on lens epithelial cells," by researchers at the Laboratory of Molecular and Developmental Biology, National Eye Institute, National Institutes of Health, Bethesda, MD [56]. The abstract is reproduced here:

> Many amyloid diseases are characterized by protein aggregations linked to oxidative stress. Such diseases including those of the brain, muscle, and blood vessels exhibit plaques containing beta-amyloid (Aβ). Here we demonstrate that the Alzheimer's precursor protein (betaAPP) and A beta are present at low levels in normal lenses and increase in intact cultured monkey lenses. Rat lenses exposed to oxidative stress showed increased betaAPP in the anterior epithelium and cortex. … Aβ cross-reacting protein was readily detected in the cortex of a cataractous human lens. Our data show that betaAPP and Aβ increase in mammalian lenses as part of a response to hydrogen peroxide or UV radiation and suggest that they may contribute to the mechanism by which oxidative damage leads to lens opacification.

This initial finding of beta-amyloid in a lens was not on humans, and animal models have limited value for complex diseases such as Alzheimer's, but this paper shows a clear demonstration of the AD protein occurring outside the brain. More importantly, it demonstrates that the beta-amyloid appears in tissue that is easily seen with low-cost, noninvasive, and readily available instruments. Also, this research does not make a clear connection between the lens of the eye and Alzheimer's, but it set the stage for the Harvard work that did show such a definitive connection in 2003. Finally, this paper by the National Institutes of Health reinforces that potential overlap between Alzheimer's and other diseases as they state that the beta-amyloid of Alzheimer's is also found in brain, muscle, and blood vessels.

It is important to understand that oxidative damage to the lens protein (misfolding) is reversible in a normal healthy lens (and, by extension, tissue in general). Aging and the acceleration of aging coincide with buildup of misfolded protein like beta-amyloid. Excessive production of this protein is not compensated by an equal amount of removal from tissue in aging and accelerated aging.

THE EYE AND WHOLE-BODY DISEASES

We have learned that there is a strong and interesting connection between the eye and AD. Retinal blood flow changes appear to be a simple and easy way to assess the Alzheimer's process early on. Cortical cataracts that appear in the lens of the eye are beta-amyloid and match the beta-amyloid of Alzheimer's in the brain. Does the eye have more to tell about Alzheimer's and disease in general? Can it give us clues as to root causes?

Research on the association between eye diseases and system-wide (systemic) diseases has a long but somewhat disjointed history. In the past two decades, however, a compelling case connecting the eye with whole-body diseases emerged. Eyes can reveal an existing health problem or an impending one. Retina images, for example, reveal the health condition of a patient. Subtle changes in the retina vessels can give warnings of a stroke and other latent cardiovascular and neurodegenerative diseases.

The blood vessels in the eyes are part of the circulatory system in the body. What is happening in the vessels of the eye may be/is probably/are occurring elsewhere in the circulatory system, including the heart and the peripheral circulatory system. The circulatory system is highly connected and the eye provides an easy window into its observation. Leukemia can be noticed by a hemorrhage of the eye. Certain types of brain tumors, even early in their development, lead to visual disturbances that are perceptible to a trained ocular professional. Brain tumors may affect vision by causing swelling of the optic nerve.

In addition to a vascular system, the eye contains a well-developed nervous system that is readily observable simply by looking into the eye under magnification or using various commonplace instruments that are able to map the nervous system components. The optic disk, optic nerve, and retinal nerve fiber layer in the posterior chamber (back) of the eye provide a great deal of information about nervous system health.

The fluid in the eye, which nourishes its various tissues, is also present in the nervous system and is shared with the brain. The health of this fluid tells about the health or pending health of the eye tissues and tissues connected to the eye cavity including the brain. Fluid health may be measured directly or by evaluating structures that result from the deteriorating health of the fluid. For example, cataracts of various types are the result of "sick" fluid in the eye. There is an almost never-ending supply of whole-body health information supplied by the eye, particularly with regard to the brain. Consider that 60% of brain function is involved in eyesight. Thus a sick eye frequently implies a sick brain and vice versa. Sadly, the valuable information provided by the eye is largely lost or ignored in medicine.

What we once thought were markers of "eye-only" diseases are turning out to tell much more about our systemic or whole-body health. Cataracts, glaucoma, macular degeneration, and many of the associated formations are very much signals of diseases that go beyond the eye. Beta-amyloid, which is the hallmark of AD, appears in the eye, in the front (anterior) chamber as cortical cataracts, and in the back (posterior) chamber as drusen. Amyloid diseases are now widely recognized to occur in various parts of the body either coincidentally or simultaneously with their formation in the eye. Thus, the eye is emerging as a powerful diagnostic tool for a wide range of both acute (cardiovascular) and chronic (aging) diseases.

AGE-RELATED EYE DISEASE STUDY (AREDS)

The eye tells us about aging and, most importantly, about the rate of relative aging according to a study sponsored by the National Institutes of Health (NIH) and other organizations from around the world. The NIH sponsored a formal trial on eye diseases in the 1990s. That trial was called the AREDS, short for the Age-Related Eye Disease Study. The goal of this study was to learn about macular degeneration and cataract, two leading causes of vision loss in older adults. The study looked at how these two diseases progress and what their causes may be. In addition, the study tested certain vitamins and minerals to find out if they can help to prevent or slow these diseases. There is an emerging connection between macular degeneration and AD so the results of this study help untangle the Alzheimer's story, too.

Eleven medical centers in the United States took part in the study, and more than 4700 people across the country enrolled in AREDS. The study was supported by the National Eye Institute, part of the federal government's National Institutes of Health. The clinical trial portion of the study also received support from Bausch & Lomb Pharmaceuticals and was completed in October 2001. This study eventually led to a very different scope and very profound findings compared to its initial goal. It turns out that certain eye diseases are predictors of premature or early death (mortality). In other words, what this study revealed is that a rapidly aging eye occurs in a rapidly aging body. Since AD is primarily a disease of rapid or accelerated aging (advancing age is the biggest risk factor for AD), the eye disease and AD are likely linked.

Based on the data coming from the study, the Age-Related Eye Disease Study (AREDS) Research Group evaluated whether visual impairment, type of lens opacity, cataract surgery, and advanced AMD are associated with overall or cause-specific premature mortality. During a median follow-up of 6.5 years, 524 of 4753 participants (11%) died. Note that this rate of death is more than two times higher compared to breast cancer. (Has anyone sponsored a cataract or macular degeneration walk for this deadly disease?) Participants who had advanced AMD compared with those who had few, if any, drusen (a precursor to AMD), had increased mortality, and advanced AMD was associated with excessive cardiovascular deaths compared to people without advanced AMD (Fig. 6.1).

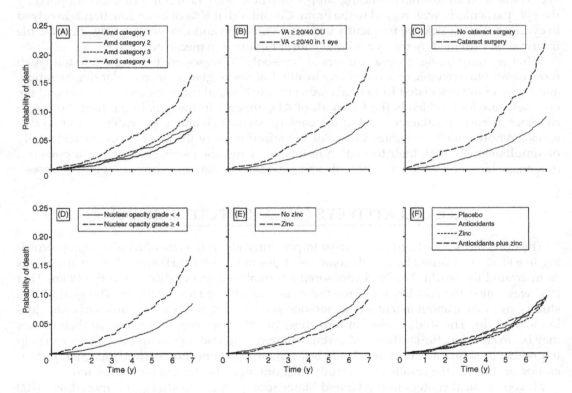

FIGURE 6.1 Increased probability of death with eye diseases. Covariate-adjusted estimates of the probability of all-cause mortality by category: age-related macular degeneration (*Amd*) (A), visual acuity (*VA*) (B), cataract surgery (C), nuclear opacity grade (D), zinc treatment (E), and Age-Related Eye Disease Study treatment (F). *From AREDS Research Group. Associations of mortality with ocular disorders and an intervention of high-dose antioxidants and zinc in the Age-Related Eye Disease Study. AREDS report no. 13. Arch Ophthalmol 2004;122(5):716–26.*

Those with visual acuity worse than 20/40 in one eye had increased mortality, when the cause of the visual impairment was an eye disease (early AMD, e.g.). That is, people with deteriorating vision died sooner compared to those people with perfect vision. And, generally, the cause of death was cardiovascular in nature. The eye tells us we have cardiovascular disease. Authors of one part of the AREDS stated, "Nuclear opacity and cataract surgery were associated with increased all-cause mortality and cancer deaths." The authors concluded, "The decreased survival of AREDS participants with AMD and cataract suggests these conditions may reflect systemic processes rather than only localized disease."

According to Frederick L. Ferris III, MD, from the National Eye Research Institute and the National Institutes of Health (NIH), part of the focus of AREDS shifted to a focus on natural history and mortality data. "With this study, we have the ability to look at a large population with AMD and cataract, and examine such things as mortality and ocular disorders. The findings suggest some interesting things," Dr. Ferris said. "One is that perhaps cataract and macular degeneration—both of which are associated with age—may actually serve as a marker for one's physiologic age in contrast to one's chronological age. We all know there are plenty of patients who are chronologically 70 years old but look 50—and vice versa."

Dr. Ferris of the NIH stated that the AREDS findings have relevance in clinical practice. "If patients are experiencing AMD and cataract, it may be a wake-up call for them to review their habits and see if there are any modifiable factors that might be increasing their risk of mortality," Dr. Ferris said. "Lifestyle changes such as stopping cigarette smoking and changing dietary behaviors may increase their chance of survival if they are diagnosed with these two diseases (cataract and/or macular degeneration)."

The other issue explored in this study was the relationship between high doses of antioxidants and zinc on mortality. "There is extensive information in the literature about how antioxidants are good for you and prolong life, but we didn't find any evidence of that," said Dr. Ferris. He noted this study was not population-based. Instead, it used volunteers. "We know that in general, people who agree to participate in clinical trials have a different outlook on their well-being than people who don't," he said. "For example, they tend to be more health conscious, which means that they may be at lower cardiovascular risk than the general population. However, if we look at this study together with other population-based studies, we can be reasonably confident that these results are real." We now know that dosing with antioxidants is a misguided approach and actually leads to higher morbidity and mortality. The immune system is generally an oxidative process. Too much antioxidant intake may compromise the efficacy of the immune system.

Joan W. Miller, MD, from Harvard Medical School noted that the association of advanced AMD and mortality was "made well." She said, "Many of these epidemiologic studies focus on persons with early or intermediate AMD, but this study was actually able to look at people in the advanced stages, which has provided key data." She said, "This study strengthens the message that AMD is not just an ocular problem but that it involves systemic factors, including a vascular component. Cardiovascular problems are an important cause of mortality, and the eye can provide vital information about what is going on systemically."

These findings also indicate that there is an opportunity for ophthalmologists and optometrists to educate primary care doctors about the relationship between advanced AMD and mortality. "A paper such as this one (AREDS #13) is important to distribute to primary care doctors [57]. Those are the people who need to hear about the association," she said. "Indeed, if a patient has worsening AMD and cardiovascular illness, this should be a sign to the physician that this individual may be at an increased risk of mortality."

The earlier given quote is by Miller in a study conducted in 2004. Today few, if any, eye care professionals work with other doctors on the eye/systemic disease connection. Miller is no exception and continues to promote "eye-only" treatments including laser therapy and eye injections. How does a laser used to correct an eye problem protect a patient from cardiovascular disease risk? How does a drug that stops new vessel growth, when injected into the back of the eye, stop people from dying of cardiovascular disease?

Dr. Trempe of the New England Eye Institute responded to a *New England Journal of Medicine* article promoting the use of Lucentis for macular degeneration [58]. The technical name for Lucentis is Ranibizumab. This drug slows or stops the growth of new vessels in the back of the eye, leading to slight improvements in vision. The downside is it stops the growth of new vessels in the rest of the body, some of which save lives when an existing vessel closes from disease. Lucentis patients spend more time yo-yoing into emergency rooms, and some die sooner compared to those not treated. Here is Dr. Trempe's response:

> To the Editor:
> In their Clinical Therapeutics article on the use of ranibizumab for neovascular age-related macular degeneration (AMD), Folk and Stone (Oct. 21 issue) [59] do not mention the significant risk of death from cardiovascular disease among such patients. In the Age-Related Eye Disease Study (ClinicalTrials.gov number, NCT00000145), during a median follow-up of 6.5 years, 534 of 4753 participants (11.2%) died [57]. Furthermore, development of disease in the other eye is common. In the Minimally Classic/Occult Trial of the Anti-VEGF Antibody Ranibizumab in the Treatment of Neovascular Age-Related Macular Degeneration (NCT00056836) and Anti-VEGF Antibody for the Treatment of Predominantly Classic Choroidal Neovascularization in Age-Related Macular Degeneration (NCT00061594) trials, in the entire treatment group, the same destructive wet type of AMD developed in the other eye on average within 1 year in 22% of the patients and within 2 years in 33% of the patients [60]. The risk factors associated with AMD and cardiovascular disease are the same [61,62]. Treatment to control those risk factors should start early, when drusen are first detected. The goals of treatment should be the following: first, to decrease the rate of death from cardiovascular disease; second, to prevent the disease from affecting the good eye; and, finally, to treat the eye involved with advanced disease. Giving repeated intraocular injections to control the disease when it is far advanced is only part of the treatment.

The real question is the following: should intraocular injections be any part of the treatment? The answer to this question is possibly no, but only after the patient is informed of risk and benefit. The impression is that anti-VEGF treatment greatly improves vision, when in fact vision improves only marginally. And risks of anti-VEGF treatments are seldom discussed with the patient. In Fig. 6.2 is a chart showing significant risk of "adverse events" associated with continued and prolonged use of the anti-VEGF drugs Avastin and Lucentis. The eye treatment is injuring and killing patients. This is taken from an article titled "Treatment of exudative AMD: data from the CATT and IVAN trials" [63]. In the context of Fig. 6.2, serious adverse events are "mostly hospitalizations." The text of the article indicates that the types of "events" landing you in the hospital are heart attack and stroke. Is the eye then isolated from the rest of the body? Maybe those blood vessels that the artificial anti-VEGF treatment blocks are actually there to help and protect you.

In addition to the AREDS, there are many other studies that connect the eye to whole-body health.

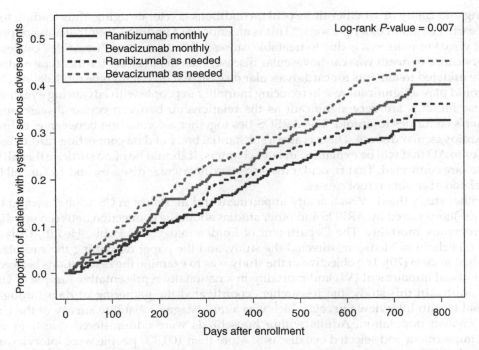

FIGURE 6.2 **Heart attacks and strokes causes by Lucentis and Avastin.** *From Martin DF, Maguire MG, Fine SL, et al. Ranibizumab and bevacizumab for treatment of neovascular age-related macular degeneration: two-year results. Ophthalmology. 2012;119(7):1388–1398.*

THE BLUE MOUNTAIN EYE STUDY

In the Blue Mountain Eye Study, the researchers aimed to assess the long-term (11-year) mortality risk associated with visual impairment and its two principal causes: AMD and cataract. The study revealed that visual impairment significantly predicted higher all-cause mortality and that signs of AMD significantly predicted higher all-cause and vascular mortality. This was especially true of individuals aged 75 years or older. And the AMD–mortality associations in persons younger than 75 years remained significant after further adjustment for visual impairment, cataract, and other potential economic and biological confounders [64–69].

The researchers found that any cataract and cortical, nuclear, and posterior subcapsular cataracts considered separately were significant predictors of all-cause mortality in the overall population. Any cataract significantly predicted vascular mortality in all persons 49 years and older even after accounting for visual impairment and other potential economic and biological confounders. This association of visual impairment and mortality has been consistently reported from different populations and different studies. Recall the connection between AD and cortical cataracts. The connection between the eye, Alzheimer's, and other diseases is building.

In their conclusions they were uncertain whether an association between eye-specific eyesight decline indicates that visual impairment, age-related eye disease, or both are markers of aging and frailty or whether these ocular conditions accelerate aging, thus leading to relatively earlier death in older persons. "This is an important finding given that a major proportion of visual impairment is due to treatable causes." What are these "treatable" causes? The primary cause of death was cardiovascular disease; thus treatment is for this type of disease. Clearly updated treatments for cardiovascular disease that do not involve modulating cholesterol could play a significant role in reducing mortality in people with advancing eye diseases.

These findings are very significant as the relationship between ocular disease and Alzheimer's continues to unfold. The AREDS ties together a connection between eye diseases and cardiovascular disease. There is also a substantial body of data connecting cardiovascular diseases to AD that will be explained in later chapters. It should be of no surprise that all these diseases are connected. That is, ocular diseases, cardiovascular diseases, and AD are all likely interrelated at common root causes.

Another study titled "Visual acuity impairment and mortality in US adults" sought to answer questions raised by AREDS and other studies about the connection between eye disease and premature mortality. The Department of Epidemiology and Public Health, University of Miami School of Medicine, directed the study, and the paper describing their results was published in 2003 [70]. The objective of the study was to examine the associations between reported visual impairment (VI) and mortality in a nationally representative sample of United States adults. In this study, the researchers coordinated the gathering of data through the National Health Interview Survey, which was a multistage probability survey of the United States civilian population. Adults within households were administered questions about visual impairment and selected eye diseases. More than 100,000 people were interviewed.

A total of 327 participants (0.3%) had severe visual impairment in both eyes; an additional 4754 (4%) had some visual impairment and/or severe visual impairment in at least 1 eye. Through the survey 8949 deaths were identified. After controlling for survey design, age, race, marital status, educational level, reported health status, glaucoma, cataract, and retinopathy, women, but not men, with reported severe bilateral (both eyes) visual impairment were at a significantly increased risk of death relative to their counterparts without visual impairment. Risk of mortality was also slightly but significantly elevated in women and men with some reported visual impairment compared with those reporting no visual impairment. Similar patterns of associations were found for cardiovascular disease–related mortality. Risk of cancer-related mortality was not associated with visual impairment according to this study.

The reports concluded that severe bilateral visual impairment and, to a smaller extent, less severe visual impairment were associated with an increased risk of all-cause mortality and cardiovascular disease–related mortality in United States women.

THE BEAVER DAM STUDY

The National Eye Institute funded the Beaver Dam Eye Study. The purpose of the study was to collect information on the prevalence and incidence of age-related cataract, macular degeneration, and diabetic retinopathy, which are all common eye diseases causing loss of vision in an aging population. The study was designed to discover (or detect) causes of these conditions. It also examined other aging problems, such as decline in overall health and quality of life and development of kidney and heart disease.

The study was initially funded in 1987. A private census was conducted in the city and township of Beaver Dam, WI, and found that there were approximately 6000 people aged 43–84 years. Approximately 5000 of them participated in a baseline examination between 1988 and 1990. Five-, 10-, 15-, and 20-year follow-up examinations have taken place and 3700, 2800, 2100, 2000, and more than 1900 people participated in each of the respective examination phases.

Several papers presenting results from the Beaver Dam Study were published in peer-reviewed medical journals [71–75]. A key paper is referenced and analyzed here: "Age-related eye disease, visual impairment, and survival: the Beaver Dam Eye Study," by M.D. Knudtson, B.E. Klein, and R. Klein from the Department of Ophthalmology and Visual Sciences, University of Wisconsin School of Medicine and Public Health [73]. The objective of the study was to investigate the relationship of AMD, cataract, glaucoma, visual impairment, and diabetic retinopathy to survival during a 14-year period. Persons ranging in age from 43 to 84 years in the period from September 15, 1987, to May 4, 1988, participated in the baseline examination. Standardized methods, including photography, were used to determine the presence of ocular disease. Survival was followed after baselines were recorded.

RESULTS: As of December 31, 2002, 32% of the baseline population had died (median follow-up, 13.2 years). After adjusting for age, sex, and systemic and lifestyle factors, poorer survival was associated with cortical cataract (the Alzheimer's beta-amyloid containing cataract), any cataract, diabetic retinopathy, and visual impairment, and marginally associated with increasing severity of nuclear sclerosis.

CONCLUSIONS: Cataract, diabetic retinopathy, and visual impairment were associated with poorer survival and not explained by traditional risk factors for mortality. These ocular conditions may serve as markers for mortality in the general population.

MORE STUDIES ON EYE DISEASE AND EARLY DEATH

There are more studies on eye and whole-body health, and they all reach the same conclusions:

The Priverno Eye Study: This was a population-based cohort study of the incidence of blindness, low vision, and survival. Lens opacities are associated with a higher risk of death. The purpose of the present study was to further investigate the relationships between different types of lens opacity and patient survival. The analysis of the Priverno data confirms an association between lower survival and cataracts, particularly those confined to the lens nucleus and those that had already prompted surgery. An example research article is titled "Association between lens opacities and mortality in the Priverno Eye Study" [64].

The Barbados Eye Study: The purpose of this study was to determine incidence and risk factors for each main cause of visual loss in an African-Caribbean population. Incidence of visual impairment was high and significantly affected quality of life. Age-related cataract and open-angle glaucoma caused ~75% of blindness, indicating the need for early detection and treatment. The connection between metabolic and cardiovascular disease and ocular indications and diseases is strong. A research paper that resulted from the Barbados Eye Study is "Lens opacities and mortality: the Barbados Eye Studies" [76].

The Blue Mountain Eye Study (BMES): It was the first large, population-based assessment of visual impairment and common eye diseases of a representative, older

Australian community sample. The findings demonstrate the connection between eye and systemic diseases. In particular, cardiovascular risk factors were prominent for eye diseases including cataract, macular degeneration, glaucoma, and retinopathy. An example of a research paper that resulted from the Blue Mountain Eye Study is "Open-angle glaucoma and cardiovascular mortality: the Blue Mountain Eye Study" [64].

The Beijing Eye Study: The Beijing Eye Study is a population-based study that included 4439 subjects who were initially examined in 2001 through blood tests and ocular assessment. The data suggest that glaucoma, particularly angle-closure glaucoma, may be associated with an increased rate of mortality in adult Chinese in Greater Beijing. A research paper that resulted from the Beijing Eye Study is "Mortality and ocular diseases: the Beijing Eye Study" [77].

The Rotterdam Eye study: This study started in 1990 in a suburb of Rotterdam, among 10,994 men and women aged 55 and over. Major risk factors that were found for macular degeneration included atherosclerosis (cardiovascular disease). Retinal venular (microvessel) diameters play a role in predicting cardiovascular disorders. Dilated retinal venules at baseline were predictive for stroke, cerebral infarction, dementia, white brain matter lesions, impaired glucose tolerance, diabetes mellitus, and mortality. Inflammation is part of these diseases. A research paper that resulted from the Rotterdam Eye Study is "Is there a direct association between age-related eye diseases and mortality? The Rotterdam Study" [78]. This paper concluded, "Both ARM and cataract are predictors of shorter survival because they have risk factors that also affect mortality."

The Andhra Pradesh Eye Disease Study: A large-scale prevalence survey of blindness and visual impairment (The Andhra Pradesh Eye Diseases Study [APEDS1]) was conducted between 1996–2000 on 10,293 individuals of all ages in three rural and one urban clusters in Andhra Pradesh, Southern India. More than a decade later (June 2009–March 2010), APEDS1 participants in rural clusters were traced (termed APEDS2) to determine ocular risk factors for mortality in this longitudinal cohort [78a]. This study, along with the other studies highlighted in this section simply show the profound connection between visual acuity loss through cataract formation and increasing mortality risk (Fig. 6.3).

These authoritative eye and whole-body health studies have led to research publications from all over the world that show a link between eye diseases and premature mortality. Selected titles are provided in the subsequent text. In no way is this list comprehensive, and note that this work extends back to 1992. The point of this exercise is that there is enough research. We need to get this information into the clinic to help real people like you.

Minassian DC. Mortality and cataract: findings from a population-based longitudinal study. Bull World Health Organ 1992;70:219–23.

Thompson JR. Cataract and survival in an elderly nondiabetic population. Arch Ophthalmol 1993;111:675–9.

Klein R, Klein BE, Moss SE. Age-related eye disease and survival: the Beaver Dam Eye Study. Arch Ophthalmol 1995;113:333–9.

West SK. Mixed lens opacities and subsequent mortality. Arch Ophthalmol 2000; 118:393–7.

Hu FB. Prospective study of cataract extraction and risk of coronary heart disease in women. Am J Epidemiol 2001;153:875–81.

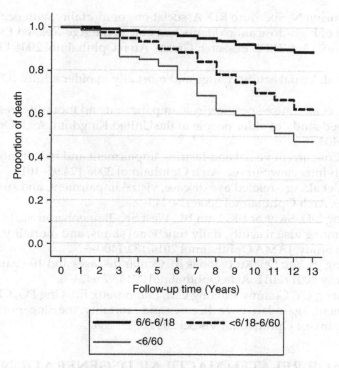

FIGURE 6.3   Increased probability of death with visual impairment. *From Khanna RC, Murthy GVS, Giridhar P, et al. Cataract, Visual Impairment and Long-Term Mortality in a Rural Cohort in India: The Andhra Pradesh Eye Disease Study. PLOS One 2013;8(10):e78002.*

McKibbin M. Short-term mortality among middle-aged cataract patients. Eye 2001;15:209–12.

Hennis A. Lens opacities and mortality: the Barbados eye studies. Ophthalmology 2001;108:498–504.

McCarty CA, Nanjan MB, Taylor HR. Vision impairment predicts five-year mortality. Br J Ophthalmol 2001;85:322–6.

Wang JJ. Visual impairment, age-related cataract, and mortality. Arch Ophthalmol 2001;119:1186–90.

Reidy A. Increased mortality in women with cataract: a population-based follow up of the North London Eye Study. Br J Ophthalmol 2002;86:424–8.

Williams SL, et al. Baseline cataract type and 10-year mortality in the Italian–American case–control study of age-related cataract. Am J Epidemiol 2002;156:127–31.

Lee DJ, et al. Visual acuity impairment and mortality in US adults. Arch Ophthalmol 2002;120:1544–50.

Panchapakesan J, et al. Five-year incidence of cataract surgery: the Blue Mountains Eye Study. Br J Ophthalmol. 2003;87:168–72.

Thompson JR, Gibson JM, Jagger C. The association between visual impairment and mortality in elderly people. Age Ageing 1989;18:83–8.

Clemons TE, Kurinij N, Sperduto RD. Associations of mortality with ocular disorders and an intervention of high-dose antioxidants and zinc in the Age-Related Eye Disease Study: AREDS report no. 13. AREDS Research Group. Arch Ophthalmol 2004;122: 716–26.

Freeman EE, et al. Visual acuity change and mortality in older adults. IOVS 2005;46:4040–5.

Thiagarajan M, et al. Cause-specific visual impairment and mortality: results from a population-based study of older people in the United Kingdom. Arch Ophthalmol 2005;123:1397–403.

Lam BL, et al. Concurrent visual and hearing impairment and risk of mortality: the National Health Interview Survey. Arch Ophthalmol 2006;124:95–101.

Knudtson MD, et al. Age-related eye disease, visual impairment, and survival: the Beaver Dam Eye Study. Arch Ophthalmol 2006;124:243–9.

Christ SL, Zheng DD, Swenor BK, Lam BL, West SK, Tannenbaum SL, et al. Longitudinal relationships among visual acuity, daily functional status, and mortality: the Salisbury Eye Evaluation Study. JAMA Ophthalmol 2014;132:1400–6.

Wang YX, Zhang JS, You QS, Xu L, Jonas JB. Ocular diseases and 10-year mortality: the Beijing Eye Study 2001/2011. Acta Ophthalmol 2014;92:e424–8.

Siantar RG, Cheng CY, Gemmy Cheung CM, Lamoureux EL, Ong PG, Chow KY, et al. Visual impairment, age-related eye diseases and mortality: the Singapore Malay Eye Study (SiMES). Invest Ophthalmol Vis Sci 2014;55(5):2682.

AGE-RELATED MACULAR DEGENERATION, A MARKER FOR ALZHEIMER'S DISEASE

A large body of medical research suggests that AMD and AD are highly connected. This section explores that possibility.

AMD is the major cause of central blindness in the elderly. It is characterized by a progressive loss of color and fine vision attributable to degeneration of the macula that is a highly specialized region of the central retina unique to humans and other primates. In addition to degeneration, new vessels (neovascularization) occur in response to deterioration of the blood vessels supporting the health of the back of the eye. Why does this occur? This is the subject of subsequent chapters.

Definition of Macula A small, sensitive area of the retina responsible for central vision. This region provides the most distinct vision in the retina and is only 1.5 mm (0.06 inches) in diameter.

The global eye disease survey conservatively indicates that 50 million persons worldwide suffer from AMD symptoms and one-third of them are blind or severely visually impaired because of AMD. The disease has a tremendous impact on the physical and mental health of the geriatric population and their families, and it is becoming a major public health and financial burden. In the absence of an effective treatment for AMD, the number of patients severely disabled by it is expected to double in the next 20 years. These statistics are curiously similar to that of AD. The prevalence of early AMD in the age category of 65–74 years is 15%, in the age category of 75–84 years 25%, and in persons 85 years and older 30%. The trend in people with

AMD and even the numbers are very similar to AD. Sadly, the only stage of AMD that even has a treatment is the latest stage when patients are severely afflicted. And, as you will see, the treatment is marginal at best and comes with high-priced side effects.

The AMD cause is known to be multifactorial, that is, in addition to a genetic component, environmental risk factors such as smoking, obesity, arteriosclerosis, hypertension, and infection may predispose a patient to AMD. At present, chronic oxidative stress like free radicals and inflammation are strongly linked to development of the disease, just like with many diseases of aging, including AD. The presence of extracellular drusen deposit (small globules of unspecified material including beta-amyloid—the hallmark of Alzheimer's) is common in AMD. Prior to the onset of AMD, visual defects such as reduced contrast sensitivity, central visual field loss, and space/time sensitivity are experienced by many patients that lead to difficulties in coping with routine daily tasks. These are signs and symptoms that often go undiagnosed but could be used as part of an early diagnosis for both AMD and AD.

"Wet" AMD is an advanced form of AMD that causes vision loss due to abnormal blood vessel growth (choroidal neovascularization), ultimately leading to blood and protein leakage behind the macula. In other words, the blood vessels in the retina (nourishing the optic nerve) are so weak that the blood leaks out of the vessels. Technically, abnormal neovascular lesions leak and eventually rupture, causing hemorrhage. Do you think only these specific vessels in the body are diseased? Unlikely. Bleeding, leaking, and scarring from these blood vessels eventually cause irreversible damage to the overlying photoreceptors (the nervous tissue that is similar to the nervous tissue in the brain) and rapid vision loss if left untreated. Only about 10% of patients suffering from macular degeneration have the wet type. Macular degeneration is not painful, and this may allow it to go unnoticed for some time. And, within the standard-of-care, only the wet form is treated. The so-called "dry" form that constitutes 90% is simply observed for progression. That is our standard-of-care. We wait for the disease to be destructive and then we reactively take action.

The present standard-of-care pays little attention to drusen and early AMD except to watch the disease (yes, this is a disease) progress until the drusen have significantly enlarged. The National Eye Institute and the American Academy of Ophthalmology support this "observational" approach. At later stages of the disease, the vitamin and antioxidant treatment is recommended, and even this simple approach is not without complications as discussed earlier regarding zinc supplementation.

What defines a patient as high risk? This is an often-asked question in medicine. Dr. Fred Pelzman posed this question in the January 17, 2014, issue of *MedPage Today*. He asks, "What's a high-risk patient?" and then answers, "Interesting question, and, as you would expect, one without an easy answer. We believe the AREDS study clearly defines a high-risk patient. Those with eye diseases are the patients with high future risk of disease and death. When drusen is first detected during a routine eye exam, it is of utmost importance that the patient be informed about the high (cardiovascular) mortality associated with AMD and about the association with AD down the road."

Similarly with AMD, the principal risk factor for AD is aging. Cardiovascular dysfunction, such as the underlying cause of hypertension, obesity, and diabetes in midlife, contributes to the development of dementia, just like for AMD. The major pathological hallmarks of AD are extracellular beta-amyloid plaque deposition and neurofibrillary tangle (NFT) formation. Emerging evidence suggests that progressive inflammation and increased oxidative stress play a key role in the early development of AD features, and this is common to AMD. These processes are known to play a central role in the loss of brain neurons and optic nerve tissue. This may precede the appearance of the AD hallmark amyloid plaques and NFTs in the brains of affected individuals.

Both Alzheimer's and macular degeneration have an early phase that often leads to the full disease. In Alzheimer's, it is referred to as Mild Cognitive Impairment (MCI). In macular degeneration, it's when accumulations of beta-amyloid deposits occur under the retina, called drusen. Drusen in the retina are an early sign of macular degeneration. The beta-amyloid protein, we well know, occurs in the senile plaques of the brains of Alzheimer's suffers.

In this section, we review findings that show a very clear link between AD and macular degeneration. This is extremely important from the point of view of Alzheimer's diagnosis and treatment. From a diagnosis perspective, AMD may (but not always) precede AD by 5–10 years. That is, on average, the age group afflicted with AMD is 5–10 years younger compared to those afflicted with AD. Thus, AMD may be useful as an early diagnostic tool for AD. Little attention is paid to early AMD because there are no interventions or treatments for it under the standard-of-care until it progresses to the wet form. If more doctors become aware of the important link between AD and AMD, the earliest form of macular degeneration will get more diagnostic attention. It is probably underdiagnosed currently since it is a condition that is simply observed. And, even though you are told there is no treatment for AD, be sure that healthy people with just the earliest beginnings of the disease are much more likely to respond to therapy than those fully engaged in the disease. In fact, the drug industry is now looking to test their (ineffectual) antiamyloid therapies on people without apparent Alzheimer's to see if these drugs will lead to a reduction in Alzheimer's incidences some years down the road. It won't work, obviously, but do you think they will recruit patients based upon the occurrence of macular degeneration?

ALZHEIMER'S DISEASE AND MACULAR DEGENERATION COMORBIDITIES

With all the discussion about an association between AD and AMD, a logical question is the following: has anyone investigated if patients with one of these diseases also have the other simultaneously? In medical jargon, this is referred to as comorbidities, or diseases occurring at the same time in the same patient. Papers on this topic have been published dating back decades. The Rotterdam Study led to a paper titled "Is age-related maculopathy associated with Alzheimer's disease? The Rotterdam Study," which concluded, "These findings suggest that the neuronal degeneration occurring in age-related maculopathy and Alzheimer's disease may, to some extent, have a common pathogenesis" [79].

"Prevalence and patterns of co-morbid cognitive impairment in low vision rehabilitation (LVR) for macular disease" is a study from 2010 [80]. In this study, researchers found that almost 20% of older adults with macular degeneration screened positive for cognitive impairment. They didn't define the extent of cognitive impairment, but we can assume that it is MCI rather than full dementia, as dementia sufferers were unlikely to be referred to them for their low vision work. In a 2007 study titled "The combined effect of visual impairment and cognitive impairment on disability in older people," the results were essentially equivalent, making these studies very believable [81].

These studies connecting AD and AMD indicate that there is evidence that these may share common causes, and macular disease and cognitive impairment may develop through common underlying conditions, such as atherosclerosis. The authors of the 2007 study conclude:

By revealing such patterns of cognitive impairment in this population, the study raises intriguing questions about the etiology (causes, origins) of the supposed link between visual and cognitive functioning. To the authors' knowledge, this is the first study to demonstrate markedly poor performance on the FAS* test among older adults with macular disease. One possible explanation for this observation is that the brain structures required for verbal fluency are at risk for similar pathophysiological insults (disease causes) as the macula, such that these specific impairments in cognition and vision are likely to arise concurrently.

Summary: AD and AMD often occur at the same time in the same patient, and these diseases likely stem from the same root cause.

*The FAS test is a controlled oral word association test in which the participant is given 60 seconds to name words beginning with a particular letter (F, A, or S), excluding certain types of words such as proper names.

There is plenty of research espousing the connection between AMD and AD. A scholar. google search on the terms "Alzheimer's" and "macular degeneration" leads to 15,900 hits. That is, there are 15,900 research publications that contain both AD and AMD somewhere in the paper. Using the formula that each publication costs roughly $500,000, the total research dollars spent to arrive at these 15,900 publications is a staggering $8 billion. Seventeen of those papers have the two terms in the title of the paper. This does not include a search of synonyms or related terms like dementia or the abbreviations such as AD for Alzheimer's or ARMD, MD, or AMD for macular degeneration.

Macular degeneration is considered a neurodegenerative disease, as is AD. In fact, neurodegenerative disorders include multiple sclerosis (MS), AD, Parkinson's disease (PD), Huntington's disease, amyotrophic lateral sclerosis (ALS), tauopathies, and AMD. In all the neurodegenerative diseases, nerve cell loss occurs. Although loss of nerve cells in AD is classically said to occur in specific regions of the brain, as the disease progresses cells all over the central nervous system are affected. Of great interest to eye physicians is the strong association of the loss of nerve cells in the eye with this disease. Moreover, there is increasing evidence that similar mechanisms in the eye, the brain, and the body cause the development of this nerve cell loss. This suggests that the processes of cell death occurring in the eye may be an indicator of, or window on, cell death occurring in other parts of the brain.

One of the most common early signs of dry AMD is drusen, the yellow deposits under the retina. New evidence indicates that substructural elements within drusen contain beta-amyloid (Aβ), a major proinflammatory component of AD plaques. Thus macular degeneration presents with the same beta-amyloid protein known to be the hallmark of AD. A startling statistic that comes out of the AREDS we already discussed is that 10% of 54-year-old males have drusen in their eye. Clearly these are the people to watch, evaluate, measure, diagnose, and treat if we want to eradicate AD.

Here are the stages of the development of macular degeneration, which is quite similar to stages of development of Alzheimer's:

- Early AMD: People with early AMD have either several small drusen or a few medium-sized drusen. At this stage, there are no symptoms and no vision loss. Up to 10% of 54-year-old males have this early AMD yet most are unaware.
- Intermediate AMD: People with intermediate AMD have either many medium-sized drusen or one or more large drusen. Some people see a blurred spot in the center of their vision. More light may be needed for reading and other tasks.
- Advanced Dry AMD: In addition to drusen, people with advanced dry AMD have a breakdown of light-sensitive cells and supporting tissue in the central retinal area. This

breakdown can cause a blind spot (scotoma) in the center of their vision. Over time, the scotoma may get bigger and darker, taking away more of their central vision. Reading or recognizing faces will become more difficult unless the objects are very close.

- Wet AMD: This occurs when abnormal blood vessels start to grow under the macula. These abnormal new blood vessels are fragile and often leak blood and fluid. The blood and fluid raise the macula from its normal place at the back of the eye. Damage to the macula occurs rapidly. The accumulation of blood and fluid under the retina causes rapid deterioration of visual acuity.

In both these diseases (AD and AMD), we don't understand exactly why some patients progress rapidly while others do not. But there are many illnesses that "self-rectify" or go into a remission, from headache to cancer. AMD is no exception. In the "Avastin" study, 50% of the placebo-treated group improved spontaneously, at least to some degree, versus 90% for the treated group. Alzheimer's does not appear to have similar outcomes. Our assumption is that these patients are truly very sick and the body has become so weak that the immune system and other reparation processes are overwhelmed by the depth and breadth of the disease. However, for those patients in early stages of disease, the conditions can improve as the immune system responds to the disease and wages a successful affront to it. Also, those with early signs of the disease may become more self-aware and take "environmental" (lifestyle) measures to help themselves. Clearly, those with the earliest signs of it are younger and generally healthier; thus their bodies' ability to fight the affliction is enhanced compared to an older and much more sick individual.

In 2011 a team from Finland published a review article in the *Journal of Alzheimer's Disease* with the title "Age-related macular degeneration (AMD): Alzheimer's disease in the eye?" [82]. In this paper, the authors reviewed and cited 197 separate scientific publications. They combined the conclusions in all the research to arrive at a fundamental conclusion, which is that AD and macular degeneration have, for the most part, a common disease mechanism. Interestingly, they showed that there are different genetic factors between the two diseases that could explain why they do not develop simultaneously. That is, some patients have macular degeneration with no clinical signs of AD. Also, AD tends to develop later in life compared to AMD, probably because the brain affords extra protection through the blood–brain barrier.

A comprehensive study of macular degeneration patients who have preclinical (asymptomatic) Alzheimer's has not been conducted. There have been studies where the two diseases are present in the same patient (comorbidity). It may be that patients with macular degeneration are suffering from Alzheimer's at the same time but, due to "cognitive reserve," the patient does not show signs or symptoms of the disease. Only a detailed differential diagnosis will reveal the truth. What we do know is that patients with diagnosed AD have extensive brain damage and atrophy even before they are aware of any problems. Very small lesions in the macula of the eye cause severe symptoms while fairly large lesions in the brain do not. This is the power of the eye in early screening and diagnosis.

Let's explore the Finnish study that emphasizes the major similarities between AD and age-related macular diseases. In some respects, AMD is better studied and has more treatment options compared to AD. Thus, an important extension of their work is learning about the potentially treatable risk factors and causes of macular degeneration. If Alzheimer's and macular degeneration are the same or similar diseases, then early diagnosis of Alzheimer's is likely possible through a diagnosis of macular degeneration. Effective treatments for macular degeneration should also have some efficacy toward AD, especially if a diagnosis can be made early into the disease process when the patient is most healthy and thus most receptive to treatment interventions. The following is the abstract from the Finnish review:

Age-related macular degeneration (AMD) is a late-onset, neurodegenerative retinal disease that shares several clinical and pathological features with Alzheimer's disease (AD), including stress stimuli such as oxidative stress and inflammation. In both diseases, the detrimental intra- and extracellular deposits have many similarities. Aging, hypercholesterolemia, hypertension, obesity, arteriosclerosis, and smoking are risk factors to develop AMD and AD. Cellular aging processes have similar organelle and signaling association in the retina and brain tissues. However, it seems that these diseases have a different genetic background. In this review, differences and similarities of AMD and AD are thoroughly discussed.

Note that the authors refer to AMD as a neurodegenerative disease of the retina. Indeed, the retina contains both nervous system and vascular system components just like the brain. As mentioned before, the eye is substantially part of the brain, including such organ systems as the retina (the photographic plate of the eye), the optic nerve, the lens, and the retinal nerve fiber layer that is a lengthy nerve that connects the optic disk to the visual cortex of the brain. They stress that both diseases, AMD and AD, have a significant inflammatory component. In addition, in both diseases, "deposits" are found. In fact, these deposits are chemically and biochemically similar and both contain the beta-amyloid that is the hallmark of AD. That is, the AD protein is also found in the retina of the eye in AMD patients. Importantly, the authors point out that the risk factors for AMD and AD are the same, or at least overlap substantially. Those risk factors are related to general cardiovascular diseases and diabetes that have well-established treatments.

Interestingly, even though cardiovascular disease has many treatment options, it remains the number one killer of Americans and is either first or second, globally, as a deadly disease. Clearly medicine is not detecting the disease at its earliest stages when lives can be saved by proper treatments. If we can reach back early in the diagnostic spectrum with methods to assess AMD and other cardiovascular diseases, it might provide the benefit of allowing treatment of both cardiovascular disease and AD. Using the eye as an early diagnostic predictor of future disease could help earlier and more effective interventions.

AMD and AD share several physiological features that are present before the disease is clinically detectable and continue to manifest with the disease, including oxidative stress and inflammation. These are well-known inducers of protein aggregation, the so-called misfolded protein response or amyloidosis, which results in the appearance of the beta-amyloid material. These accumulated deposits have many molecular similarities in AMD and AD. A decreased capacity to remove damaged cellular proteins during aging has been strongly implicated in both diseases. In other words, as we age, these misfolded proteins both form more often and overwhelm the cellular removal mechanism, or the mechanism to remove these errant proteins is not as efficient, so they build up in tissue and, in some cases, cause harm. These proteins may also be a marker for something more fundamental that causes harm.

In healthy people, macrophage cells and other immune components diligently sweep away waste products, errant materials, and other debris. They even recycle these damaged proteins for fuel. Through a process with the expressive name of autophagy, or "self-eating," cells create specialized membranes that engulf junk in the cell's cytoplasm and carry it, by way of a phagosome, to a part of the cell known as the lysosome, where the trash is broken apart and then burned by the cell for energy. Apparently, as we age this system for waste removal either falters or is overwhelmed by waste. In patients who are ill, thus aging more quickly than the norm, this material builds more quickly. Finding and treating the cause will stop or delay the disease process.

In the prospective population-based Rotterdam Study, the presence of AMD and AD was screened, and follow-up examinations on the surviving participants were conducted from mid-1993 to the end of 1994, an average of 25 months after the initial exam [79]. Subjects with advanced AMD at baseline showed an increased risk for AD. These findings suggested that the neuronal degeneration occurring in AMD and AD might, to some extent, have a common mechanism. This study showed a possible health-determinant connection between AMD and AD. Interestingly, none of the AD-related genes are known to overlap with those related to AMD. Therefore, it appears that the two diseases are distinct from a genetic perspective, but not from an environmental or mechanistic perspective.

PROTEIN AGGREGATION PROCESS IN AGE-RELATED MACULAR DEGENERATION AND ALZHEIMER'S DISEASE

Oxidation and other physiological processes ultimately lead to unfolding or conformational changes in proteins. These altered proteins do not function as intended and often, as is the case with beta-amyloid, have very unexpected effects. In the case of beta-amyloid and Alzheimer's, the effects are assumed deleterious, but that remains to be definitely proven in actual patients. In some cases, the unfolding may lead to a loss of structural or functional activity of proteins and in other cases may lead to toxicity or, as emerging research by Harvard Medical School and others suggests, actually provide protection [83]. Accumulation of proteins is a recurring event in many age-related diseases, including AMD, Alzheimer's, Inclusion Body Myositis, and AD of the heart.

The accepted dogma for the formation and accumulation of these unfolded proteins is that during aging (more appropriately, accelerated aging), the capacity to repair damaged, nonfunctional proteins is decreased or overwhelmed. This may trigger increased oligomeric (short-chain molecules) proteins to aggregate into plaque and, at high enough levels, become detrimental. The AD brain is also characterized by the accumulation of oxidized proteins that may further reduce the body's ability to remove these materials. Alternatively, if the buildup of these proteins was due to defensive measures, then they logically would not be cleared from tissue until their purpose was served. Little work has been devoted to this hypothesis but it is emerging as evidenced by the Harvard research.

Again, from the standard view, the presumed detrimentally altered structure or functional activity of proteins and a compromised clearance system are observed in both AMD and AD, which results in excess accumulation of potentially toxic material, beta-amyloid. The accumulation of this material, if not toxic, may simply interfere with the removal of waste essential for cellular health. Regardless of the reason for the buildup of these materials, one thing is certain—the same processes occur for both Alzheimer's and macular degeneration. At a minimum, macular degeneration serves as a diagnostic asset for Alzheimer's. However, by extension, treatment for macular degeneration could serve as a model for Alzheimer's treatments. Evaluating the impact of treatment on macular degeneration is substantially easier compared to that on AD due to the accessibility of the eye, and we have two eyes that do not develop disease at the same time.

The Finnish group concludes with the following remarks:

> AMD and AD are both age-related neurodegenerative diseases that share similar environmental risk factors comprising of smoking, hypertension, hypercholesterolemia, atherosclerosis, obesity, and unhealthy diet. Cellular pathology associates with increased oxidative stress, inflammation, and impaired proteasomal and lysosomal function that evoke formation of intra- and extracellular deposits. The detrimental deposits consist of largely similar aggregated proteins in both diseases. However, the genetic background seems to be different between AMD and AD. The bright ocular tissues and advanced imaging technology to study the posterior pole of the eye provide interesting opportunities to understand the early signs of AMD that might be associated with AD pathology as well.

What they appear to be saying in their last concluding line is that the eye, with its ready accessibility and advanced probing instruments available for its study, is perfectly positioned to look for the early signs of macular degeneration and AD.

DIABETIC RETINOPATHY AND THE BRAIN

Another eye condition that portends Alzheimer's or dementia is Diabetic vasculopathy or Diabetic retinopathy that occurs in the retina. Simply put, if you have diabetic retinopathy long enough, you will eventually have dementia. More accurately, if you have diabetic retinopathy, you have the early development of dementia, but you are just not aware of this ongoing process. As with all the major eye diseases that are tied to systemic disease, here are three scenarios that describe the potential progression into or regression from systemic disease:

1. The eye disease, or more accurately, the underlying systemic cause of the eye disease, rectifies through some mechanism, and the brain disease also rectifies before the individual experiences tangible symptoms.
2. The eye disease and the underlying systemic disease progress rapidly and the person dies before the brain becomes clinically sick.
3. The eye disease and the underlying systemic disease progress and the brain disease eventually manifests clinically.

Strong evidence of the systemic and inflammatory nature of diabetic retinopathy is provided by many research articles including one titled "Bone marrow–CNS connections: implications in the pathogenesis of diabetic retinopathy." In this paper, the authors state:

> This review provides an overview of a novel, innovative approach to viewing diabetic retinopathy as the result of an inflammatory cycle that affects the bone marrow (BM) and the central and sympathetic nervous systems. Diabetes associated inflammation may be the result of BM neuropathy which skews hematopoiesis (formation of blood cells) towards generation of increased inflammatory cells but also reduced production of endothelial progenitor cells responsible for maintaining healthy endothelial function and renewal. The resulting systemic inflammation further impacts the hypothalamus, promoting insulin resistance and diabetes, and initiates an inflammatory cascade that adversely impacts both macrovascular and microvascular complications, including diabetic retinopathy (DR). This review examines the idea of using anti-inflammatory agents that cross not only the blood–retinal barrier to enter the retina but also have the capability to target the central

nervous system and cross the blood–brain barrier to reduce neuroinflammation. This neuroinflammation in key sympathetic centers serves to not only perpetuate BM pathology but promote insulin resistance which is characteristic of type 2 diabetic patients (T2D) but is also seen in T1D.

Not one part of our body functions in isolation of any other. For example, it is apparent that there is a two-way communication between the brain and the bone marrow. The sympathetic centers of the brain regulate the release of bone marrow cells into the circulation, and the stimulation of connective tissue cells also regulates hematopoiesis. In certain diabetes, the neural pathways involved are adversely affected and excessive numbers of white blood cells get released, infiltrate the brain, and directly affect brain activity. Neuroinflammation results in impacting the retina and, eventually, the brain.

THE EYE IN SYSTEMIC AND ALZHEIMER'S DISEASE

The eye in systemic disease by Daniel H. Gold, MD (Author), and Thomas A. Weingeist (Editor) illustrates the profound connection between eye and system-wide disease [1]. The author showed a clear ocular–systemic interrelationship. They stated that ocular manifestations play a part in many systemic diseases. Indeed, the eye is often the first clue that a more serious disease process may be at work. Theirs is the first comprehensive reference to examine the connection between the eye and the systemic disease. This book offered the viewpoint of over 318 clinicians who specialize in specific diseases. Together, they presented an overview that showed why ocular manifestations occur and illustrated the underlying patterns that produce the recognizable signs and symptoms of an individual clinical disease.

An ocular manifestation of a systemic disease is an eye condition that directly or indirectly results from a disease process in another part of the body. There are many diseases known to cause ocular or visual changes. Diabetes, for example, is the leading cause of new cases of blindness in those aged 20–74, with ocular manifestations such as diabetic retinopathy and macular edema affecting up to 80% of those who have had the disease for 15 years or more. Other diseases such as acquired immunodeficiency syndrome (AIDS) and hypertension are commonly found to have associated ocular symptoms. For a list of body-wide diseases with an ocular component, please see Appendix 5.

SICK EYE IN A SICK BODY—GLAUCOMA— OCULAR ALZHEIMER'S DISEASE?

This title is an amalgamation of two important research papers. If you split the title given earlier, you get "Sick eye in a sick body" and "Glaucoma—ocular Alzheimer's disease?" These two articles weave a story to help us understand how a simple and common eye disease, glaucoma, is essentially the same as AD. Thus glaucoma is a potential diagnosis for present or future AD.

In 2006, European researchers published a review article titled "A sick eye in a sick body? Systemic findings in patients with primary open-angle glaucoma" [84]. Prior to this publication, few in the medical community recognized glaucoma, a significant neurological disease, to be a systemic disease. The assumption was that the disease was isolated to the eye. Now there is recognition that glaucoma is actually a precursor to AD. It is the same type of disease, but it shows up in the nervous system in the eye often before symptoms of brain degradation appear. This time lag may be due to the extraordinary "reserve" of the brain compared to the nervous system of the eye. Despite our recognition of the association between glaucoma and a sick body, no measures have been made to utilize this information for the benefit of patients. If anything, due to the 10-minute office visit and the marketing behind the drug Lucentis, eye-only philosophies and treatments are expanding. The abstract of the "A sick eye in a sick body" is replicated here:

> Despite intense research, the pathogenesis of primary open-angle glaucoma (POAG) is still not completely understood. There is ample evidence for a pathophysiological role of elevated intraocular pressure; however, several systemic factors may influence onset and progression of the disease. Systemic peculiarities found in POAG include alterations of the cardiovascular system, autonomic nervous system, immune system, as well as endocrinological, psychological, and sleep disturbances. An association between POAG and other neuro-degenerative diseases, such as Alzheimer's disease and Parkinson's disease, has also been described. Furthermore, the diagnosis of glaucoma can affect the patient's quality of life. By highlighting the systemic alterations found in POAG, this review attempts to bring glaucoma into a broader medical context.

The authors go on to say that their findings suggest that glaucoma is not just a process involving the visual system, but more likely the manifestation of a more generalized systemic dysfunction. "Glaucoma is a multifactorial disease, and a complex cascade of events and interactions between interocular pressure, vascular, immunological, and various other systemic factors that must be postulated to explain the development of glaucomatous damage." Interestingly, AD is also viewed as a multifactorial disease. Is it possible that some of these "factors" overlap?

Rewrite the above paragraph and replace the word "glaucoma" with "Alzheimer's disease" and "visual" with "brain" (or some other term like memory or cerebral). It still is perfectly consistent with what we now know about AD, at least in informed segments of the research community.

Definition of Glaucoma Glaucoma refers to a group of eye conditions that lead to damage to the optic nerve. This nerve carries visual information from the eye to the brain and has the highest concentration of the amyloid precursor protein. Intraocular pressure (IOP), commonly associated with glaucoma, is not part of the American Academy of Ophthalmology definition. Visual field loss is a part of glaucoma pathogenesis. Corneal thinning may be added to the definition because it is more associated with visual field loss compared to IOP.

Amyloid precursor protein (APP) is an integral membrane protein found in many tissues and concentrated in the synapses of neurons. Its primary function is not known, although it has been implicated as a regulator of synapse formation, neural plasticity, and iron export. APP is best known as the precursor molecule whose breakdown (cleavage) generates beta-amyloid. Beta-amyloid refers to a group of similar molecules whose length varies from 37 to 49 amino acids in length. It is the beta-amyloid 42 that is most implicated in AD.

The APP gene provides instructions for making amyloid precursor protein. Even though little is known about the function of amyloid precursor protein, researchers speculate that it may bind to other proteins on the surface of cells or help cells attach to one another. Studies suggest that in the brain, it helps direct the movement (migration) of nerve cells (neurons) during early development. Mutations in the APP gene can lead to an increased amount of the amyloid β peptide or to the production of a slightly longer and stickier form of the peptide. When these protein fragments are released from the cell, they can accumulate in the brain and form clumps called amyloid plaques. These plaques are characteristic of AD. A buildup of amyloid β peptide and amyloid plaques may lead or be spectators to the death of neurons and the progressive signs and symptoms of this disorder. But note, from the definition of glaucoma, that this process happens to the nervous tissue of the eye as well as the brain.

There are many descriptions for glaucoma: open-angle glaucoma, chronic glaucoma, chronic open-angle glaucoma, primary open-angle glaucoma, closed-angle glaucoma, narrow-angle glaucoma, angle-closure glaucoma, acute glaucoma, secondary glaucoma, and congenital glaucoma. It's the second most common cause of blindness in the United States. There are four major types of glaucoma:

- open-angle (chronic) glaucoma
- angle-closure (acute) glaucoma
- congenital glaucoma
- secondary glaucoma

Here is a simple explanation of the disease process for glaucoma. The front part of the eye is filled with a clear fluid called aqueous humour. This fluid is always being made behind the colored part of the eye (the iris). It leaves the eye through channels in the front of the eye in an area called the anterior chamber angle, or simply the angle. Anything that slows or blocks the flow of this fluid out of the eye will cause pressure to build up. One important cause is inflammation and associated tissue damage that causes disruption in the flow of fluids. In some instances, the result is increased pressure, but this is not always the case. However, in the standard-of-care, treatment is based on this pressure being high, and it is presumed that the pressure damages the optic nerve. It is the pressure, not the inflammation, which is treated in the standard-of-care. New approaches that target the inflammation are emerging as more effective compared to pressure-reducing methods, but these are considered experimental and have not made it into patient care yet.

Open-angle (chronic) glaucoma is the most common type of glaucoma. The cause is unknown, at least within the standard-of-care. An increase in eye pressure occurs slowly over time. The pressure pushes on the optic nerve. Open-angle glaucoma tends to run in families. Your risk is higher if you have a parent or grandparent with open-angle glaucoma, showing that genetics and environment play a role. People of African descent are at particularly high risk for this disease. Just like AD and macular degeneration, there is a genetic component, and like the other diseases, glaucoma is not a purely genetic disorder. Environmental factors are more important compared to genetics, so you have control over the occurrence of this disease.

Angle-closure (acute) glaucoma occurs when the exit of the aqueous humour fluid is suddenly blocked. This causes a quick, severe, and painful rise in the pressure in the eye. Angle-closure glaucoma is an emergency. This is very different from open-angle glaucoma, which painlessly and slowly damages vision. If you have had acute glaucoma in one eye, you are at risk for an attack in the second eye, and your doctor is likely to recommend preventive treatment. Dilating eye drops and certain medications may trigger an acute glaucoma attack.

The fact that the disease can migrate from one eye to the other illustrates the same anatomical change is occurring in both eyes. The disease isn't isolated; it's systemic.

Secondary glaucoma is the type of glaucoma most associated with AD. It is caused by drugs such as corticosteroids, eye diseases such as uveitis, systemic diseases, and trauma.

GLAUCOMA AND ALZHEIMER'S DISEASE

There has been an absolute flurry of activity by researchers studying the connections between glaucoma and AD. But like most discoveries in medicine, the piqued activity occurs long after the initial connection. In this case, a paper in 1986 presented the possibility that glaucoma and AD are connected. A listing of journal articles with AD and glaucoma in the title is presented here, including the 1986 paper that does not specifically have the word "glaucoma" in the title:

1986: Optic-nerve degeneration in Alzheimer's disease. *New England Journal of Medicine*

1989: Retinal ganglion cell degeneration in Alzheimer's disease. *Brain Research*

1990: Optic nerve damage in Alzheimer's disease. *Ophthalmology*

1994: Intracranial pressure and Alzheimer's disease: a hypothesis. *Medical Hypotheses*

1997: The cellular mechanism underlying neuronal degeneration in glaucoma: parallels with Alzheimer's disease. *Australian and New Zealand Journal of Ophthalmology*

2001: Alzheimer's peptide: a possible link between glaucoma, exfoliation syndrome and Alzheimer's disease. *Acta Ophthalmologica*

2002: Association of glaucoma with neurodegenerative diseases with apoptotic cell death: Alzheimer's disease and Parkinson's disease. *American Journal of Ophthalmology*

2002: High occurrence rate of glaucoma among patients with Alzheimer's disease. *European Neurology*

2003: Alzheimer's peptide and serine proteinase inhibitors in glaucoma and exfoliation syndrome. *Documenta Ophthalmologica*

2003: Glaucoma: ocular Alzheimer's disease. *Frontiers in Bioscience*

2003: Hypothesis for a common basis for neuroprotection in glaucoma and Alzheimer's disease: anti-apoptosis by alpha-2-adrenergic receptor activation. *Survey of Ophthalmology*

2004: Molecular aspects of glaucoma related to Alzheimer's disease. Thesis (OD), Inter American University of Puerto Rico, School of Optometry.

2004: Progressive glaucoma in patients with Alzheimer's disease. *Eye*

2006: High frequency of open-angle glaucoma in Japanese patients with Alzheimer's disease. *Journal of the Neurological Sciences*

2006: Normal-tension glaucoma and Alzheimer's disease: hypothesis of a possible common underlying risk factor. *Medical Hypotheses*

2007: Expression of protein markers of Alzheimer's disease in human glaucoma eyes. *Investigative Ophthalmology & Visual Science*

2007: Nerve link: Alzheimer's suspect shows up in glaucoma. *Science News*

2007: Normal-tension glaucoma and Alzheimer's disease: *Helicobacter pylori* as a possible common underlying risk factor. *Medical Hypothesis*

2008: An abnormal high trans-lamina cribrosa pressure: a missing link between Alzheimer's disease and glaucoma? *Clinical Neurology and Neurosurgery*

2008: An abnormal high trans-lamina cribrosa pressure difference: a missing link between Alzheimer's disease and glaucoma? *Clinical Neurology and Neurosurgery*

2008: Molecular mechanism of neuroprotection in glaucoma and Alzheimer's disease using Gouqizi. The 6th international symposium of Chinese Medicinal Chemists

2009: Alzheimer's disease and glaucoma: is there a causal relationship? *British Journal of Ophthalmology*

2009: Evaluation of cerebrospinal fluid pressure in patients with Alzheimer's disease as a possible cause of glaucoma. *Acta Ophthalmologica*

2009: The role of cerebrospinal fluid pressure in the development of glaucoma in patients with Alzheimer's disease. *Bulletin de la Société Belge*

2010: Alzheimer's disease and glaucoma: imaging the biomarkers of neurodegenerative disease. *International Journal of Alzheimer's Disease*

2010: Alzheimer's disease: cerebral glaucoma? *Medical Hypotheses*

2010: Role of synuclein-beta (SNCB) and synuclein gamma (SNCG) in glaucoma and Alzheimer's diseases—a bioinformatic approach. *International Journal on Computer Science and Engineering*

2010: The role of cerebrospinal fluid pressure in the development of glaucoma in patients with Alzheimer's disease. *Bulletin de la Societe belge d'ophtalmologie*

2011: Alzheimer's disease and glaucoma. *British Journal of Ophthalmology*

2011: Alzheimer's disease and primary open-angle glaucoma: is there a connection? *Clinical Ophthalmology*

2011: Egr1 expression is induced following glatiramer acetate immunotherapy in rodent models of glaucoma and Alzheimer's disease. *Investigative Ophthalmology & Visual Science*

2011: Glaucoma and Alzheimer's disease in the elderly. *Aging Health*

2011: Hypothesis of optineurin as a new common risk factor in normal-tension glaucoma and Alzheimer's disease. *Medical Hypotheses*

2011: Normal-tension glaucoma and Alzheimer's disease: retinal vessel signs as a possible common underlying risk factor. *Medical Hypotheses*

2011: The dark side of the nerve—what's the link between glaucoma and Alzheimer's or Parkinson's? Can we use one to predict the other? *Review of Optometry*

2013: Alzheimer's disease and glaucoma: mechanistic similarities and differences. *Journal of Glaucoma*

2013: Progressive neurodegeneration of retina in Alzheimer's disease—are β-amyloid peptide and tau new pathological factors in glaucoma? cdn.intechopen.com

2013: Elevated levels of multiple biomarkers of Alzheimer's disease in the aqueous humour of eyes with open-angle glaucoma. *Investigative Ophthalmology & Visual Science*

One of the articles referring to the connection between glaucoma and AD was titled "The dark side of the nerve—what's the link between glaucoma and Alzheimer's or Parkinson's? Can we use one to predict the other?" by James L. Fanelli, OD [85]. He presented a case of a woman with multiple comorbidities that is very instructive in considering the overlap of Alzheimer's with the many diseases discussed so far. The patient description was as follows:

- The patient was a 76-year-old female with decreasing vision.
- The medications were as follows: simvastatin, Cymbalta (duloxetine, Lilly), Synthroid (levothyroxine, Abbott), 81-mg aspirin, and vitamins.
- Husband reported that she also had the early stages of dementia. Also noted were mild tremors in the extremities and in her head and neck, and involuntary quivering of the eyelids.
- Interocular pressure was normal (no high-tension glaucoma).
- Both nuclear and cortical cataracts were noted (cardiovascular and Alzheimer's cataracts).
- There was some indication of optic disk neuritis.
- Somewhat unhealthy vessels were present in the back of the eye, "mild arteriolar narrowing."
- Retinal topography demonstrated slightly thinned retinal nerve fiber layers.
- Early glaucomatous field loss was present.

A variety of "eye-only" treatments were applied to this patient, and she showed improvement in eye-only symptoms. However, approximately 3 months later, the patient returned with recurrence. Again, an eye treatment was carried out. "Interestingly, her overall motor skills and fine motions had deteriorated significantly to the point where she was continuously moving her arms, legs, trunk, neck, and head in a pronounced manner. Her husband relayed that she had been to a neurologist for her PD and similar processes."

This patient received a very thorough ocular diagnosis that provided a host of information about the patient's ocular, brain, and systemic health. Clues about the patient's systemic health were provided by the nuclear cataracts, which are tied to cardiovascular disease mortality (AREDS). The connection to Alzheimer's was provided by the optic disk, cortical cataracts (the Alzheimer's cataract), thinned retinal nerve fiber layer, and testimony from her husband. Later she was diagnosed with low-tension glaucoma that further supported an Alzheimer's diagnosis. Based on the diagnoses by her optometrist, neurologist, and (maybe) her cardiologist, she was being treated for high cholesterol and management of certain symptoms of her disease. Lowering cholesterol and managing cardiovascular risk with aspirin in the proper approach, at least based on new research, actually exacerbates her neurodegenerative disease conditions. Aspirin causes weak and leaky vessels to leak even more. Lowering cholesterol is wrong, period, as the previous and the next chapter reveal.

Sadly, this woman continued to deteriorate and descend into both PD and AD. Eye markers are superb screening tools for disease, but the diagnostic process must continue to evaluate "why" through a series of blood tests that could reveal a range of potentially treatable root cause targets. This was not done because the standard-of-care does not provide a mechanism for a doctor to dig deeper into root causes. Symptomatic treatment may work for a headache but not for the myriad of "diseases" this poor woman had.

After reviewing this patient case, Fanelli went on to review the connection between the eye and AD with particular emphasis on glaucoma. His review is presented here:

Is there possibly a link between this patient's early dementia, loss of motor skills, and optic nerve appearance? Several years ago, widespread scientific news reported that 'having an eye examination' could provide an early diagnosis of Alzheimer's disease (AD). At that time, several novel studies looked at optic nerve characteristics in patients who were diagnosed with middle- to late-stage AD. Researchers were looking at various models that might link the diagnosis of AD to something else that would facilitate a much earlier diagnosis. The norm at the time was to wait until a constellation of symptoms developed and then make the diagnosis. As with any disease, early diagnosis would lead to early intervention.

The optic nerve became the focus of research because of its unique ability to be visualized (compared to other, less accessible central nervous system structures like the brain). A 1986 study compared the optic nerves and retina of postmortem patients diagnosed with AD to those without AD [86]. The researchers found that there was a significant decrease in the number of retinal ganglion cells in the optic nerves and retinal nerve fiber layers of patients with AD. This work occurred a generation ago, and patients await a translation into early diagnosis or patient care.

In 1989, researchers looked at the physical characteristics of the retinal nerve fiber layer (RNFL), and ganglion cells in humans with AD [87]. Although this study also looked at postmortem tissues, fundamental differences were seen histologically in the RNFL and ganglion cells of patients with AD, namely, a loss of the glial cell support structure, as well as cytoplasmic changes in the retinal ganglion cells. The authors then suggested that optic nerve evaluation should become a standard part of the analysis of patients with AD. Again, this work is a generation old. Ask your healthcare team for an eye evaluation as part of your health review.

According to Fanelli, what should clinicians have an "eye" out for? One 2003 study from Italy showed that patients with pressure-dependent glaucoma, ocular hypertension, and AD all show similar optic nerve characteristics, namely, a thinning of the RNFL (to be discussed later in this chapter), increased cupping, and abnormal pattern electroretinogram (ERG) recordings (also to be discussed later in this chapter) [88]. More recently, the use of optical coherence tomography (OCT) in the field of neurology has been gaining wider interest. Investigators have found that RNFL thickness, as measured by OCT, is diminished in patients with MS, AD, and Parkinson's disease. So far, an ironclad cause-and-effect relationship between ocular hypertension, glaucoma, and neurodegenerative diseases is not established; however, there is an overlap in the clinical findings in the eye.

Regarding his patient case, Fanelli concluded, "In our patient's case, she exhibited progressive difficulty with motor skills, was in the early stages of Alzheimer's, and was a glaucoma suspect based on the appearance of her optic nerves. Is there a link? Quite possibly."

Just how connected is glaucoma with systemic diseases? Table 6.3 shows associations and the relative weight of these associations. The key question to consider is the following: are these associations mere overlaps or do the diseases stem from the same set of root causes?

MORE ON GLAUCOMA AND ALZHEIMER'S DISEASE

The strength of the association between AD and glaucoma is illustrated by the 17,200 separate articles published that include the terms "Alzheimer's" and "glaucoma" dating back 30 years. One of the critical findings was published in 2007, in which British scientists reported a major link between Alzheimer's and glaucoma [89]. They state that the discovery could lead to the eye disease being regarded as an early warning for dementia.

TABLE 6.3 Relationship to Glaucoma (Either Likely or Possible)

Arteriosclerosis	Blood pressure	Arterial hypertension
Arterial hypotension	Vasospasm	Electrocardiographic changes
Headache and Migraine	Autonomic nervous system	Immune system
Autoimmunity	Leukocyte activation	Platelet aggregation
Blood viscosity	Diabetes mellitus	Thyroid disease
Graves disease	Hypothyroidism	Pituitary system
Neurodegenerative diseases	Alzheimer's disease	Parkinson's disease
Sleep disturbances	Sleep apnea syndrome	Psychological alterations
Ischemic brain lesions	Hearing loss	*Helicobacter pylori* infection

The belief of the Brits is that the research could speed up the development of new treatments for Alzheimer's and revolutionize the treatment of glaucoma, the second most common cause of blindness. If glaucoma is confirmed as a major risk factor for Alzheimer's, then the early warning signs it gives could help ensure that patients have more opportunities to delay the onset of dementia, they argued. The researchers discovered that the same "plaque" proteins are a key process in the development of both diseases. Clumps or plaques of beta-amyloid proteins, which kill brain cells in Alzheimer's patients, also kill optic nerve cells in the eyes of glaucoma sufferers. Be careful not to become confused by claims that the beta-amyloid kills brain cells. They are just repeating the accepted belief but do not provide proof of this cause/effect. The important finding is that the processes for both diseases are the same. Of course, the NIH and Harvard Medical School showed the association between the lens, the brain, and the amyloid plaques by way of cortical cataracts. Because glaucoma tends to develop years earlier than Alzheimer's, it may provide a useful warning signal. The same holds true for the cortical cataracts of the lens of the eye. These two pieces of information, taken together, amplify the diagnostic potency of the eye for AD.

Research carried out by Dr. Cordeiro and colleagues suggested that the retina could provide a window into the brain, allowing doctors to diagnose AD by looking for evidence of nerve cell death [90]. She said, "Main Street opticians have been routinely looking at the brain in a more direct way than has been possible by high tech brain scanners." In the article published in 2011, the British researcher states the connection between Alzheimer's and glaucoma.

Primary open angle glaucoma and Alzheimer's disease have long been established as two separate pathological entities, primarily affecting the elderly. The progressive, irreversible course of both diseases has significant implications on an aging population. As the complex nature of the two diseases has progressively unraveled over the past two decades, common changes have also been elucidated. The mutual neurological changes in primary open angle glaucoma and Alzheimer's disease have facilitated the development of neuroprotective strategies. Further understanding of the common physiology of primary open angle glaucoma and Alzheimer's disease and their timeline may have great implications on early diagnosis and effective therapeutic targeting.

The timeline issue is of critical importance to diagnosis. The assertion thus far is that macular degeneration, cortical cataract formation, and glaucoma all (statistically) precede the clinical signs of Alzheimer's. However, this may be purely from a diagnostic perspective and not from a disease progression perspective. For example, the brain may be more resistant to the impacts of disease due to a variety of factors compared to the eye. Or the eye simply shows disease early due to its accessibility. The most likely situation is that these diseases start their development at close to the same time, as the individual becomes more susceptible to disease with age. Consider macular degeneration for perspective. This disease develops somewhat independently in the left eye compared to in the right eye, but as the disease progresses, it eventually impacts both eyes. Thus it is logical that the (same) disease clinically develops at slightly different times for Alzheimer's, glaucoma, and macular degeneration.

"Alzheimer's disease: cerebral glaucoma?" was published in 2010 [91]. Here the authors took an approach that is the reverse of most others. Rather than viewing glaucoma as a risk factor for AD, they looked at it the other way around. Maybe AD is really a cerebral form of another disease—glaucoma. First, they paid tribute to work from 15 years before where the authors speculated that the high pressure of glaucoma might increase the odds of a patient coming down with Alzheimer's [92]. Furthermore, the authors from 1994 suggested that, in more advanced stages of AD, such pressure factors could already be missing due to the disease process. This is essentially what is seen in late-stage glaucoma that is comorbid with AD. Increased pressure is not always up even when nervous tissue damage is evident.

The 2010 authors support the initial findings of their predecessors. They also show how emerging research has revealed similarities in the process, leading to retinal ganglion cell death in glaucoma and neuronal cell death in AD. These authors are arguably the first to raise the question of whether AD could be a cerebral form of glaucoma. The linking of glaucoma to mechanisms of AD could reflect the anatomical and functional similarities between the intraocular space and the intracranial space.

It is quite clear that the eye provides a beautiful and elegant window into the diagnosis of AD through glaucoma and other processes and structures. As with most chronic eye diseases, they are known early on but treated only in later stages of disease development. That means that the earliest stages are often paid little heed other than to document the stage of the disease and to measure future deterioration. It is at this early stage that all pertinent diagnostic strategies must be used to determine if a complete case for disease can be made, root causes elucidated, and then treated. The patient is going to be more receptive to treatment at the earliest signs of disease. The eye is that early warning system.

New definition of Glaucoma Glaucoma is Alzheimer's disease of the eye (and vice versa).

RETINAL NERVE FIBER LAYER AND ALZHEIMER'S DISEASE

The retina is part of the brain and the retinal nerve fiber layer, connecting the retina and the (rest of the) brain, and providing additional information about the health and disease of the nervous tissue of the brain. The eye, due to its transparency, affords examination of the retina by means of direct or indirect methods. Direct methods are as simple as using a microscope to look deep into the eye at the finest visible structures, whereas indirect methods involve more sophisticated instrumentation akin to ultrasound and other imaging strategies.

The retinal nerve fiber layer (RNFL) can also be seen with microscopy and can be very accurately measured using imaging. The thickness of the RNFL, which contains the axons of the retinal ganglion cells, can be objectively measured with imaging techniques. This tissue represents the brain as different "fibers" connect with different areas of the brain and allows doctors to understand both positive and negative forces affecting it. We are just learning how these easily quantifiable properties of the retina provide insight into the concealed parts of the brain.

Both the retina and the brain areas that are responsible for cognitive functioning originate from the embryonic forward part of the brain (the prosencephalon). The premise of retinal involvement in cognitive functioning is supported by studies describing an increased prevalence of glaucoma in patients with AD. Other supportive evidence comes from postmortem tissue studies demonstrating retinal nerve cell loss in patients with AD and from studies in living patients who have a reduced number of retinal ganglion cells and associated thinner RNFL thickness when they have AD. These studies make a compelling case for the connection between the processes occurring in the retinal nerve fiber layer and the brain of people with neurodegenerative conditions like AD.

This idea is not exactly new. The first report of the association between the atrophy in the retina of the eye and AD was in 1986 in a paper titled "Optic nerve degeneration in Alzheimer's disease." This was published in the most prestigious *New England Journal of Medicine* [86]. Their very telling abstract is provided as follows:

> Alzheimer's disease is a dementing disorder of unknown cause in which there is degeneration of neuronal subpopulations in the central nervous system. In postmortem studies, we found widespread axonal degeneration in the optic nerves of 8 of 10 patients with Alzheimer's disease. The retinas of four of the patients were also examined histologically, and three had a reduction in the number of ganglion cells and in the thickness of the nerve-fiber layer. There was no retinal neurofibrillary degeneration or amyloid angiopathy, which are typically seen in the brains of patients with Alzheimer's disease. The changes we observed in the patients with Alzheimer's disease were clearly distinguishable from the findings in 10 age-matched controls and represent a sensory-system degeneration that occurs in Alzheimer's disease. Study of the retina in patients with this disease may be helpful diagnostically, and isolation of the affected ganglion cells may facilitate molecular analysis of the disorder.

With new technology for measuring the eye, it is now possible to measure the thickness of the ganglion cells in the macular area with a noninvasive test that is done in less than 5 minutes on an undilated eye.

In any such study, the first question to ask is the following: how was the diagnosis of AD made? Note that the authors suggested that the study of the retina, and particularly the study of the nerve that connects the retina to the brain, might be helpful diagnostically. An important question from a diagnostic perspective is the following: does the change in the RNFL track changes in the brain in a predictable way from the beginning of a disease process to the end? Another study hinted at a possible correlation between the amount of retinal ganglion cell loss and the severity of cognitive impairment in a group of 14 patients with AD.

A Rotterdam research team looked at the association between cognitive functioning and RNFL thickness in a large, population-based sample of healthy subjects [93]. They assessed a broad range of cognitive functions by means of an extensive neuropsychological examination and measured RNFL thickness with advanced instrumentation. They found a very clear connection between cognitive function and RNFL thickness especially in younger people. Clarity was lost with older people in the study group who did not suffer from disease such as dementia or Alzheimer's. Regardless, the authors of that study concluded, "We also speculate that any damage to any of these tissues, including the RNFL, would be unlikely to run an equal course. Generalized loss might, in principle, differentially affect the various neuronal tissues of prosencephalic (forebrain) origin."

Scientists from Rome assessed the retinal thickness in patients with mild cognitive impairment and AD [22]. They noted that previous studies have shown that degenerative changes occur in optic nerve fibers and manifest as thinning of RNFL in patients with AD. The objective of the study was to assess the relationship between mild cognitive impairment (MCI), AD, and loss of RNFL. They found a significant decrease in RNFL thickness in both study groups (AD and MCI) compared to in the control group, particularly in the inferior quadrants of the optic nerve head, while the superior quadrants were significantly thinner only in AD. This, for the first time, showed how the RNFL changes over time as cognitive impairment progresses into full AD.

A team from the United Kingdom summed up using RNFL in the analysis of AD, in a major review [94]:

> There is increasing evidence of RNFL thinning or retinal ganglion cell loss in patients with AD, but the relationship between the degree of cognitive impairment and the degree of RNFL loss has not been established yet. There are a few possibilities that could explain the findings. These include AD change in the retina, abnormal trans-lamina cribrosa pressure, and retrograde trans-synaptic degeneration. The degenerative changes in the brain and retina vary among AD patients because of different AD subtype, severity, and duration. It seems that the RNFL measurement has a good potential to be a monitoring tool in AD patients in the near future. Further investigations are required to understand more about AD pathology in these areas.

Note the response from research. "Further investigations are required …." It is hard to argue that we don't need more data. However, the time is now to include methods like this in the diagnosis of AD.

When you perform a google.scholar search on retinal nerve fiber layer and Alzheimer's, the search returns over 8000 records ($4 billion dollars' worth or research). Some titles are presented chronologically as follows:

1986: Optic nerve degeneration in Alzheimer's disease. *New England Journal of Medicine*

1990: Optic nerve damage in Alzheimer's disease. *Ophthalmology*

1996: Retinal nerve fiber layer abnormalities in Alzheimer's disease. *Acta Ophthalmologica Scandinavica*

2001: An evaluation of the retinal nerve fiber layer thickness by scanning laser polarimetry in individuals with dementia of the Alzheimer type. *Acta Ophthalmologica Scandinavica*

2006: Reduction of optic nerve fibers in patients with Alzheimer disease identified by laser imaging. *Neurology*

2007: Retinal abnormalities in early Alzheimer's disease. *Investigative Ophthalmology and Visual Science*

2010: Retinal nerve fiber layer structure abnormalities in early Alzheimer's disease: evidence in optical coherence tomography. *Neuroscience Letters*

NUCLEAR CATARACT AND ALZHEIMER'S DISEASE

Uncomplicated (nuclear) cataracts are visual biomarkers of the aging process. Nuclear cataracts have a strong connection to cardiovascular diseases, and cortical cataracts are associated with AD.

Nuclear cataracts are the most common type of cataract and are recognized as the result of the aging process. They involve the center of the lens and cause visual problems earlier in their development compared to cortical cataracts. Cortical cataracts involve the periphery of the lens and are caused by accumulation of beta-amyloid protein 1-42, the same type of protein that accumulates in the brain of AD patients. The opacity in nuclear cataract, although a "misfolded" protein, is not beta-amyloid 1-42.

To illustrate the complexity of cause and effect in AD, consider nuclear cataracts. Many studies, including the AREDS, show a clear correlation between cataracts, cardiovascular diseases, and increased mortality. The main cause of early death in cataract patients tends to be from a cardiovascular disease. AD has a profound connection with inflammation and cardiovascular diseases, as you will see in the next two chapters. One might surmise that there is a connection between nuclear cataract and AD. If there is, the case has yet to be made.

A British and Swiss group published a review paper in 2011 titled "Cataract and cognitive impairment: a review of the literature" [95]. Their abstract and introduction (excerpts as follows) provide a great introduction to other eye/Alzheimer's diagnostic indicators that will be examined later:

> Acquired cataract and cognitive impairment are both common age-related problems, and ophthalmologists are increasingly likely to encounter patients who have both. Dementia types that display early visuoperceptual impairment may present first to ophthalmology services. When these patients have coexisting cataract it may be difficult to distinguish visual complaints due to cataract from those due to dementia.
>
> It has been postulated that assessing the eyes and vision could help with the diagnosis of dementia and the monitoring of dementia treatment. For example, glaucoma and dementia have been proposed to have a causal relationship; early age-related macular degeneration and cognitive function seem to be related; visual field defects are recognized as a presenting feature in some dementias, and eye movements can distinguish between different causes of dementia. However, the interaction between age-related cataract and neurodegenerative disorders affecting the central visual pathway is a neglected research area, despite its implications for clinical practice. Many of the issues raised here in regard to the interaction or coexistence of cataract and neurodegeneration will also be relevant to other age-related eye diseases.

Cataract and dementia are degenerative processes and have been proposed to share common mechanisms. Risk factors for cataract include advancing age, female gender, smoking, and lower socioeconomic class (and associated lower educational attainment). These are also risk factors for dementia including AD. Vascular risk factors contribute to both vascular dementia and AD. These risk factors are also important for the generation of cataract. Diabetes, smoking, and obesity are risk factors for age-related cataract as well as dementia. Antihypertensives appear to be protective in dementia, and there is some association between hypertension and cataract. The ApoE4 allele, which is known to be a genetic risk factor for AD, is not associated with increased cataract risk. Given many common risk factors, dementia patients may be expected to have a higher incidence of cataract. However, a single study of self-reported previous cataract surgery found no association between cataract and cognitive impairment.

A scholar.google search confirms the lack of literature on the connection between normal age-related cataracts (nuclear cataracts) and AD. A title search yielded a meager four articles, and a couple of those were not about nuclear cataracts. J.J. Harding from Oxford University, in a paper titled "Cataract, Alzheimer's disease, and other conformational diseases," did tie the two diseases together [96]. The abstract does imply a strong connection between Alzheimer's and cataract, but a paucity of subsequent publications is confounding. The abstract reads, "Unfolding of proteins was shown to occur in human cataract more than 25 years ago. Recently, the term 'conformational diseases' was applied to a whole group of diseases in which unfolded or misfolded proteins accumulate. In this article, common features in the biochemistry and epidemiology of cataract, Alzheimer's disease, and other conformational diseases are explored."

ELECTRICAL IMPULSES IN THE BRAIN AND THE EYE

Electroretinography (ERG)

The retina is the photographic plate of the body and operates similarly to a digital camera by making an instantaneous electrical imprint on various nerve cells of the retina that is then transmitted to the brain through the retinal nerve fiber layer. In optometry and ophthalmology, methods of detecting, measuring, and evaluating the resulting electrical signals are available for the evaluation of the health of the nervous system of the eye. These same tests are emerging as ways to measure the electrophysiological impacts of general neurodegenerative disease processes including AD. However, these tests are more geared toward research and are seldom used in a clinical setting.

Electroretinography is the testing method used to measure the electrical responses of various cell types in the retina, including the photoreceptors (rods and cones), inner retinal cells (bipolar and amacrine cells), and the ganglion cells. For AD, the focus is on ganglion cells. A complimentary test to electroretinography is optical coherence tomography that directly measures retinal ganglion cells.

Definition of Retinal Ganglion Cell (RGC) A type of neuron located near the inner surface (the ganglion cell layer) of the retina of the eye.

Electrodes are usually placed on the cornea and the skin near the eye, although it is possible to record the ERG from skin electrodes. During a recording, the patient's eyes are exposed to stimuli, and the resulting signal is displayed showing the time course of the signal's amplitude (voltage). Signals are very small and are typically measured in microvolts or nanovolts.

The ERG is composed of electrical potentials contributed by different cell types within the retina and the type of stimulus applied. For measuring the response of retinal ganglion cells, the stimulus is created by alternating a checkerboard in front of the subject. Clinically, ophthalmologists and optometrists use ERG for the diagnosis of various retinal diseases. Recent literature shows that the results of these tests contribute to a "differential" diagnosis of Alzheimer's and may provide data when other eye markers appear normal.

A Polish team shed new light on the value of electroretinography as a contributor to the overall diagnosis of dementia and AD. The paper was titled "Pattern electroretinogram (PERG) and pattern visual evoked potential (PVEP) in the early stages of Alzheimer's disease" [97]. Eastern Bloc medicine has more of a focus on physics compared to the Western emphasis on biology. The authors pointed out that patients with AD frequently complain of vision disturbances that are not often observed in routine ophthalmological examination. For example, early, undiagnosed, and thus "preclinical" Alzheimer's patients may have minor visual disturbances, including problems with reading, blurred vision, and vague complaints of poor vision that is difficult to assess but may be associated with disease upon more sophisticated evaluation. In these patients, optic nerve degeneration and loss of retinal cells, specifically the disappearance of ganglion cells and their axons, may be evident. Some believe that these changes most likely emanate from the brain (visual cortex) with less likelihood that the retina and the optic nerve are involved.

The aim of the Polish study was to determine whether retinal and optic nerve function, measured by PERG and PVEP tests, is changed in individuals in the early stages of AD with normal routine ophthalmological examination results. Standard PERG and PVEP tests were performed in 30 eyes of 30 patients with the early stages of AD. The results were compared to 30 eyes of 30 normal healthy controls. The data from their study allowed the authors to conclude that dysfunction of the retinal ganglion cells as well as of the optic nerve is present in the AD patients. They concluded that the dysfunctions were at least partially caused by visual disturbances observed in patients with the early stages of AD. Their concluding remarks are provided here:

> The results of our study strongly suggest, for the first time, that in patients with the early stages of AD, bioelectrical dysfunction of retinal ganglion cells in the optic nerve is present, and this is registered by PERG and PVEP tests.
>
> It is worthwhile to note that recording of PERG and PVEP tests in patients with AD is not easy because of the possible secondary effects of defocus or other behavioral problems. To minimize these problems in our study, appropriate selection of patients was made. We performed electrophysiological tests only in early stages of AD when patients had minimal disturbances of cognitive function and their cooperation was very good.
>
> In our group of patients with the early stages of AD, the observed reduced amplitude of N95-wave confirms the dysfunction of ganglion cells, whereas reduced amplitude and implicit time increase in P50-wave may be related to dysfunction, not only of the ganglion cells, but also of the outer layers of the retina in relation to the ganglion cells.
>
> Most authors suggest that in patients with AD, visual disturbances are caused by neuropathological changes in the visual cortex. The results obtained in the current study show that visual disturbances in AD may also be related to the retinal ganglion cells and optic nerve dysfunction.

This idea that the visual disturbances in AD may also be related to retinal ganglion cells and optic nerve dysfunction is an important conclusion. This ties together the concept that glaucoma and other diseases of the retina are occurring at the same time and mostly likely by the same pathway, as is deterioration of neurons in AD. Thus we have the same conclusion about the connection between glaucoma and AD from two distinctly different research disciplines.

Berisha et al. from the Harvard Medical School observed that patients with early AD showed a significant thinning of retinal nerve fiber layer (RNFL), as measured by Optical Coherence Tomography (OCT). OCT is analogous to ultrasound but more precise [48]. This may be connected with beta-amyloid aggregation and neuronal degeneration in the retina. Another important conclusion was the difference between structural and electrical measurements. OCT is a formidable way to measure the structural integrity of the retina, especially the retinal nerve fiber layers.

In the Polish study, OCT was performed on a few patients in addition to the PERG and PVEP tests. The few patients tested in the earliest stages of presumed AD had relatively normal OCT results, but abnormal PERG and PVEP examinations. Therefore, these examples suggest that dysfunction of the ganglion cells measured by electrophysiological tests precedes the structural changes in the ganglion cell layer as measured by OCT. Both tests may have merit for identifying preclinical Alzheimer's but it is possible that the electrophysiological tests may preempt OCT as being able to catch the disease process the earliest.

It might be somewhat obvious that deteriorating neurons contribute to a delayed electrical signal and measurements of this type should be considered in a differential diagnosis of dementia or AD. This dysfunction associated with these measurements might be a characteristic feature of AD and is certainly useful in the differential diagnosis.

Visual Evoked Potential (Response)

The visual evoked potential (VEP) tests the function of the visual pathway from the retina to the visual cortex. The visual cortex is at the far end of the brain from the retina, and electrical signals must pass through multiple synapses to get from the retina to the visual cortex. This makes the measurement of electrical impulse between the two areas very diagnostic for potential deficiencies in signal pathways. It measures the conduction of the visual pathways from the optic nerve, optic chiasm, and optic radiations to the occipital cortex.

Definition of Synapse In the nervous system, a synapse is a structure that permits a neuron to pass an electrical or chemical signal to another cell.

In the VEP test, the output is the measurement of the electrical response received at the visual cortex upon stimulation of the retina, for example, with changes in light or other visual alterations. The shape of the "wave" that the electric response creates looks a bit like an electrocardiogram signal, with some very characteristic forms (amplitude and wavelength). The VEP is useful for detecting a visual conduction disturbance and possibly conduction through the various synapses; however, it is not specific with regard to cause. A tumor compressing the optic nerve, an ischemic disturbance, or a demyelinating disease, may cause changes; only additional clinical history and often MRI are needed to uncover the area of brain impacted. However, it is a particularly sensitive test.

Definition of Ischemia A decrease in the blood supply to a bodily organ, tissue, or part caused by constriction or obstruction of the blood vessels. In the brain, a stroke is an ischemic disturbance.

With an abnormal VEP, the differential diagnostic may lead to the conclusion that one or more diseases are present:

- Alzheimer's disease (AD)
- Parkinson's disease (PD)
- optic neuropathy
- optic neuritis
- ocular hypertension
- glaucoma
- diabetes
- toxic amblyopia
- Leber's hereditary optic neuropathy
- aluminum neurotoxicity
- manganese intoxication
- retrobulbar neuritis
- ischemic optic neuropathy
- multiple sclerosis (MS)
- tumors compressing the optic nerve—optic nerve gliomas, meningiomas, craniopharyngiomas, giant aneurysms, and pituitary tumors

VEP has significant clinical usefulness. For example, it is more sensitive than MRI or physical examination, it is an objective and reproducible test for optic nerve function, the abnormality observed persists over long periods, and it is inexpensive compared to MRI. Under certain circumstances, it may be helpful for positively establishing optic nerve function in patients with a subjective complaint of visual loss; a normal VEP virtually excludes an optic nerve problem. VEP provides significant value in the detection and evaluation of MS. It can indicate problems along the pathways of certain nerves that are too subtle to be noticed or found on an exam using standardized tests and procedures including MRI. Problems along the nerve pathways are a direct result of the disease. The demyelination in MS causes the nerve impulses to be slowed, garbled, or halted altogether.

In AD and dementia, VEP has an important role. Researchers continue to look for the Holy Grail, that one test that definitively characterizes Alzheimer's and effectively excludes all other possible neurodegenerative processes or other confounding factors. VEP is not that test, and no such tests exist. That's why differential diagnosis is important for a complex disease like Alzheimer's. What VEP does provide is yet another useful and inexpensive way to measure and characterize brain function and behavior. It is useful for establishing a baseline of health in both healthy people and those already diseased. The problem with VEP is that it cannot be performed unless a patient is willing to pay cash for the test, as it is seldom covered by health insurance. What patient knows to ask for a VEP test and what clinic still has the equipment? Sadly few, if any, clinics or labs conduct this test. This simple, cost-effective test is strictly a research tool used in the bowels of some university laboratory.

A substantial literature documents delays in the VEP response in groups of patients with probable Alzheimer's dementia compared with in groups of age-matched normal control subjects. These studies looked at certain attributes of the VEP output for unique signatures applicable to Alzheimer's only. But because many visual system pathologies can produce VEP component delays, a selective delay (the "P2" delay) is often measured relative to other responses in Alzheimer's. The P2 may be selectively delayed while other features are often normal in probable Alzheimer's patients. The P2 delay was studied because it is not found in patient groups suffering from nondementing psychiatric disorders or other forms of dementia, but in AD groups it increases over time, paralleling dementia severity.

A Mercer University Medical School study, published in 2003 titled "Diagnostic utility of visual evoked potential changes in Alzheimer's disease," explored the value of VEP [98]. They looked at 45 AD patients and 60 age-equivalent, healthy control subjects. Although significant between-group differences were found, classification accuracies for individual patients and controls were too low for the P2 delay to contribute meaningfully to clinical diagnosis. A more recent study, using different experimental conditions, did show strong correlations between AD and the P2 delay [99]. A recent article by an Italian group appears to substantiate the value of new methods for the VEP test [99]. They conclude, "The presence of concurrent changes of independent parameters suggests that the neurodegenerative process can impair a control system active in Eye-closed Condition which the electrophysiological parameters depend upon. F-VEP can be viewed as a reliable marker of such impairment."

The jury is still out regarding VEP being the definitive early diagnostic for AD but, since it is one of the few diagnostic methods that measures electrical signal across the brain, it clearly needs to be considered as part of the thorough diagnostic protocol.

EYE FUNCTION AND ALZHEIMER'S DISEASE

Visual Field Alterations (Peripheral Vision)

Symptoms referable to the visual system may be the earliest signs of Alzheimer's and dementia. Visual field alterations are one such symptom. The visual field refers to the total area in which objects can be seen in the side (peripheral) vision when the patient focuses their eyes on a central point. Visual fields tend to contract with accelerated aging; however, the contraction of field is known statistically for healthy people as they age, too. Visual field contraction is more impacted in Alzheimer's patients compared to healthy age mates. This test is interesting and inexpensive. It is difficult to perform in those patients with advanced neurodegenerative disease symptoms, however.

Research into visual performance of Alzheimer's patients, including visual fields, dates back decades. A team from the University of Montreal wrote a paper in 1995 titled "Visual field loss in senile dementia of the Alzheimer's type," which sheds some insight on this topic [12]. The authors recognized, from previous works, that visual performance is impaired in patients with AD. They investigated the visual field in patients with and without disease totaling 122 people. They noted that fewer "reliable" visual fields were obtained for the Alzheimer's subjects compared to those for the controls. In the AD group they noted reduced visual fields including differential luminance sensitivity that was significantly reduced relative to in the control group. Visual sensitivity was reduced throughout the visual field. Patients with more severe dementia exhibited greater reductions in visual sensitivity. On follow-up, the majority of the diseased patients exhibited progression of visual field loss. They noted that, although challenging, Alzheimer's patients are able to participate in a visual field assessment.

Some of the key research in this area includes:

1987: Visual field loss in Alzheimer's disease. *Journal of the American Geriatrics*

1987: Visual field limitation in the patient with dementia of the Alzheimer's type. *Journal of the American Geriatrics Society*

1993: Delayed late component of visual global field power in probable Alzheimer's disease. *Journal of Geriatric Psychiatry and Neurology*

1995: Visual field loss in senile dementia of the Alzheimer's type. *Neurology*

1996: Visual field defects in Alzheimer's disease patients may reflect differential pathology in the primary visual cortex. *Optometry & Vision Science*

2002: Visual field loss and Alzheimer's disease. *Eye*

2010: White and gray matter of visual cortex in Alzheimer's disease: visual field maps, population receptive fields, and diffusion tensor imaging. *Alzheimer's and Dementia*

2011: Aging and dementia in human visual cortex: visual field map organization and population receptive fields. *Journal of Vision*

A University of California group wrote a paper in 2011 titled "Aging and dementia in human visual cortex: visual field map organization and population receptive fields" [100]. The authors suggest, "It is possible that measurements of changes in visual cortex in these patients (those with the earliest signs of visual complaints) could aid early detection, accurate diagnosis and timely treatment of dementia." They conclude that normally aging subjects do not show major visual field map organizational deficits, whereas AD patients do show visual field map organizational deficits.

Fixation (Visual Concentration) and Alzheimer's Disease

Fixation stability, the ability to maintain an image on the fovea, is a quantitative way to measure the ability of a person to concentrate. Concentration is one of the first mental capacities to deteriorate in dementia. Look upon this test as being akin to a target shooter at a rifle range. The marksman focuses on the target and attempts to coordinate all of his or her muscles to "lock" in one position that sets the gun barrel aimed at the center of the target. In the fixation test, the patient uses all their muscles, including the six muscles that regulate eye movement, to focus on a spot in the back of an instrument. Of course, the brain controls our ability to control our muscles.

Definition of Fovea In the eye, a tiny pit located in the macula of the retina that provides the clearest vision of all. Only in the fovea are the layers of the retina spread aside to let light fall directly on the cones, the cells that give the sharpest image.

Fixation measurement instruments are extremely precise and accurate. As the person "fixates" on the dot at the back of the instrument, lasers measure the movement of the pupil. Well-defined ranges have been established for the level of normal fixation with age. For example, significant ocular motor control is considered lost if the target image moves 1 degree from foveal center, or if random movement of the image on the fovea exceeds 2 degrees/second. Either of these conditions may occur if deficiencies in oculomotor control compromise the ability to maintain target alignment within these limits. Controlling the coordination of eye movement change is compromised with normal aging, but only slightly. For example, smooth, sweeping "pursuit" movements slow with age. Also, the range of voluntary eye movements becomes restricted, especially for upward gaze. The duration and peak velocity of rapid eye movements (saccades) are reduced in older people, especially when eye movements are large.

The overall measure of fixation stability does not change significantly with age as reported by researchers at Northwestern University in a 1986 study titled "Visual fixation in older adults" [101]. Older observers were often as stable in their fixations as the young observers. Moreover, the older observers were no more variable in their trial-to-trial fixations than their younger counterparts. Third, the older observers maintained the same degree of stability over the course of the test session. This last finding indicates that older adults' oculomotor mechanisms controlling fixation did not show signs of fatigue over the hour or so of testing. Thus as we age, we continue to be able to fixate our eyes on a given object that, to some degree, measures our ability to concentrate.

How do neurodegenerative diseases impact our ability to fixate a "stare" on a fixed point, and is this meaningful as a diagnostic measure? A group from Bristol University in the United Kingdom wrote "Visual attention-related processing in Alzheimer's disease" [102]. The abstract, presented in the subsequent text, sends a strong message that attention-based studies, which provide measurable values, are important to support a diagnosis. These tests complement and augment existing tests like the Mini-Mental State Examination.

> The clinical diagnosis of Alzheimer's disease (AD) involves neuropsychological testing to assess the integrity of higher order cerebral functions such as memory, cognition, visual perception, language, and executive function. However, the onset of AD is insidious, and diagnosing the very early stages may be precluded as such tests may lack the necessary sensitivity and specificity. This, together with the potential for similar shortcomings in relation to assessing disease progression and response to treatment, has prompted the search for disease markers based on abnormalities in additional aspects of brain processing. One area receiving increasing investigation is the integrity of visual and visual-attention-related processing.

Another older work, published in 1995 in the *International Journal of Psychophysiology*, studied fixation and rapid (saccadic) eye movement in 62 patients, of whom 31 had some level of dementia [103]. "Changes in measures of fixation, but not saccade measurements, correlated with changes in Mini Mental State Exam scores over testing sessions. These data suggest that fixation is more sensitive than are saccades to the progression of AD."

Definition of Saccade Abrupt rapid small movements of both eyes, such as when the eyes scan a line of print. The saccades can be divided into two distinct groups: the major saccades that are easily observed with the naked eye and the minor saccades that are virtually unobservable without special instrumentation.

A diverse North American group of researchers recently reported, "AD patients were slower to reorient their attention to an uncued location. Also, AD patients showed a greater cost for switching attention to invalid left targets" [104]. Thus this study appears to corroborate the earlier findings that Alzheimer's patients do lack ability to fixate. Instruments to measure fixation are readily available, and the tests are inexpensive, reproducible, and noninvasive. Like many other tests discussed so far, these types of tests contribute to an overall disease diagnosis and understanding of the disease process. Tests like these do not measure "why," in other words, the root cause(s) of the disease, but give a definitive baseline that can be used to measure the course of disease in response to therapy and other changes. However, finding a clinic that measures fixation is challenging.

EYE STRUCTURES AND ALZHEIMER'S DISEASE

Microtubules

Together we have reviewed many ocular diagnostic techniques that have relevance to the early detection and screening for AD. Are you old enough to remember "Carnac the Magnificent" on the Johnny Carson Show? Presented here is the last eye diagnostic for AD (although there are many more, these are the most important to date). OCT is the most precise ocular tool for measuring Alzheimer's by studying the decay of the optic nerve and retinal nerve fiber layer. However, the "next big thing" in Alzheimer's research appears to be the protein "tau." In some respects, then, the best is saved for last.

The second hallmark of AD (first is beta-amyloid), described by Dr. Alzheimer, is neurofibrillary tangles. Note that Dr. Alzheimer did not create the current hierarchy. In fact, he felt neurofibrillary tangles were more important to the senile dementia, now called AD, compared to beta-amyloid. Tangles are abnormal collections of twisted protein threads found inside nerve cells. The chief component of tangles is a protein called tau. Healthy neurons are internally supported in part by structures called microtubules, which help transport nutrients and other cellular components, such as neurotransmitter-containing vesicles, from the cell body down the axon.

Tau, which usually has a certain number of phosphate molecules attached to it, binds to microtubules and appears to stabilize them. In AD, an abnormally large number of additional phosphate molecules attach to tau. As a result of this "hyperphosphorylation," tau disengages from the microtubules and begins to come together with other tau threads. These tau threads form structures called paired helical filaments, which can become enmeshed with one another, forming tangles within the cell. The microtubules can disintegrate in the process, collapsing the neuron's internal transport network. This collapse damages the ability of neurons to communicate with each other.

Microtubules (MTs) have significant roles in a broad range of biological functions, including shaping the neuronal structure and transporting intracellular cargoes. They are often referred to as the railway of the brain. It is reasonable to speculate that MT disruption profoundly affects neuronal architecture and function. Neurofibrillary tangles (NFTs) are a major neuropathological hallmark in brains affected by AD and related diseases. The central nervous system diseases in which NFT formation is predominant are categorized as tauopathies.

Since abnormally phosphorylated tau is a major component of neurofibrillary tangles, this is a possible link between so-called "phospho-tau accumulation" and neurodegeneration. NFT-bearing neurons often accompany loss of MTs. Interestingly, MT loss has also been found even in non–NFT-bearing neurons in the AD brain, suggesting that MT loss may precede tau phosphorylation and accumulation in brains affected by tauopathy. Thus, detection of microtubule loss is likely a very early sign of neurodegeneration, including AD.

A Japanese team wrote, "Microtubule destruction induces tau liberation and its subsequent phosphorylation," in 2010 [105]. According to the authors,

Newly expressed tau, which cannot bind to MTs, is then phosphorylated and accumulated in the neuronal cell body. If such an abnormality would be sustained over a long period, the accumulated tau may form NFT, which accompanies complete MT disruption as seen in brains affected by tauopathy. In fact, MT loss has also been found in non-NFT-bearing neurons in the brains of AD patients, and phosphorylated tau may gradually accumulate in the neuronal cell body for months. Thus changes to microtubules precede the development of the AD hallmark neurofibrillary tangles.

The authors go on to surmise the following: "Furthermore, amyloid deposition, which is the other pathological hallmark of AD, may also lead to MT destabilization, suggesting that MT loss may be an intermediate between amyloid-beta deposition and NFT formation in AD." Microtubule disruption may well indicate the formation of the two key hallmarks of AD. Measuring microtubule health is available today, as a noninvasive ocular examination.

Measuring Microtubules in the RNFL of the Eye

Microtubules are found in the retinal nerve fiber layer of the eye. The retinal nerve fiber layer (RNFL) in humans consists of bundles of ganglion cell axons running just under the surface of the retina. It is damaged in glaucoma and other diseases of the optic nerve and even other neurodegenerative diseases such as AD as we learned earlier. Assessment of RNFL structure is attractive for early diagnosis of Alzheimer's because decay of this tissue often precedes detectable vision loss or significant decay of tissue including neurons associated with AD.

RNFL structures reflect light. Reflectance arises from light scattering by microtubules, which are cylindrical structures oriented parallel to ganglion cell axons. Microtubules also cause the RNFL to exhibit birefringence. Birefringence is an optical property of materials that arises from the interaction of light with oriented molecular and structural components. For example, polarized sunglasses are "birefringent" and cause the orientation of light. The dominant source of RNFL birefringence is the array of approximately parallel microtubules in ganglion cell axons.

Optical measurements of the RNFL directly indicate the number of microtubules intersected by the measuring beam. The reasoning is as follows. In a birefringent material, light polarized in one direction travels more slowly than light polarized in the perpendicular direction. The delay experienced by the slower component is called "retardance" and, for a homogenous material, is proportional to thickness. The behavior of the light interacting with the microtubules is directly related to their health, which is a measure of their uniformity. Unhealthy microtubules lose their linear orientation and thus change the birefringence, a very measurable change.

Clinical studies have correlated the decrease of RNFL retardance to glaucoma damage, and damage includes a decrease in the number of microtubules. Microtubules are expected to disappear when axons die for glaucoma and other neurodegenerative diseases including AD. Although not definitively proven, death of axons and changes in microtubules' structure and integrity likely occur at the same time. Thus, measurement of microtubule health offers a window to the "tauopathy" process that is emerging as a major area of study into the causes of neurodegenerative disease.

The very recent research literature points to a shift away from amyloid and toward tau as a therapeutic target. Recent research articles sport titles like:

2012: Microtubule stabilizing agents as potential treatment for Alzheimer's disease and
    related neurodegenerative tauopathies [106]

2012: The microtubule-stabilizing agent, epothilone D, reduces axonal dysfunction, neurotoxicity, cognitive deficits, and Alzheimer-like pathology in an interventional study with aged tau transgenic mice [107]

2012: Hyperdynamic microtubules, cognitive deficits, and pathology are improved in tau transgenic mice with low doses of the microtubule-stabilizing agent BMS-241027 [108]

2013: Beyond taxol: microtubule-based treatment of disease and injury of the nervous system [109]

Let's hope that those studying tau catch on to this simple eye technique for measuring their biomarker so they can quickly assess the function of their new drugs.

SUMMARY AND CONCLUSIONS

Ocular biomarkers provide a means for early detection of AD. The Frost review article on this topic from 2010 supports the case made here [50]. The terrible impact of AD, both on those directly affected and on society in general, creates a pressing need for better treatments. By the time a person is diagnosed with "probable AD" using current techniques, significant irreversible neuronal degeneration has already occurred. Therefore, research into better treatments must be paralleled by research into technologies to screen populations for AD and to identify cases before cognitive symptoms arise.

Ocular changes reported in AD give hope for a noninvasive, cost-effective screening for AD. Evidence is accumulating in support of AD-related changes in the eye. Finding a sufficiently sensitive and specific ocular biomarker for AD is proving to be as far-fetched as is a single drug for the treatment of the disease. However, an ocular screening protocol for Alzheimer's, studying the constellation of impacted ocular tissue, and thus using multiple tests will benefit AD sufferers and researchers and possibly provide new insight into the molecular processes and genetic determinants of the disease today.

It is unlikely we will ever know when disease "officially" starts. It is known that young people who died of acute illness or unnatural causes show the earliest evidence for chronic diseases like Alzheimer's when autopsied. Thus our true goal is to determine when we can first observe disease. Hopefully detection can be pushed to as early a point in disease genesis as possible. The advantage of the eye is it can be investigated in great detail, inexpensively, and noninvasively. This is the utopia of chronic disease diagnostics and, particularly, AD.

References

[1] Gold DH, Weingeist TA, editors. The eye in systemic disease. Philadelphia: JB Lippincott; 1990. p. vii.
[2] Christensen C. The innovator's dilemma: when new technologies cause great firms to fail. Boston, MA: Harvard Business Press; 1997.
[3] Christensen CM, Bohmer R, Kenagy J. Will disruptive innovations cure health care? Harvard Business Rev 2000;78(5):102–12.
[4] Dowling JE. The retina: an approachable part of the brain. Cambridge, MA: Harvard University Press; 1987.
[5] Armstrong RA. Alzheimer's disease and the eye. J Optom 2009;2(3):103–11.
[6] Cogan DG. Alzheimer syndromes. Am J Ophthalmol 1987;104:183–4.

[7] Sadun AA, Borchert M, DeVita E, Hinton DR, Bassi CJ. Assessment of visual impairment in patients with Alzheimer's disease. Am J Ophthalmol 1987;104:113–20.

[8] Mendez MF, Tomsak RL, Remler B. Disorders of the visual system in Alzheimer's disease. Neurology 1990;40:439–43.

[9] Rizzo M, Nawrot M. Perception of movement and shape in Alzheimer's disease. Brain 1998;121:2259–70.

[10] Wood S, Mortel KF, Hiscock M, Bretmeyer BG, Caroselli JS. Adaptive and maladaptive utilization of color cues by patients with mild to moderate Alzheimer's disease. Arch Clin Neuropsychol 1997;12:483–9.

[11] Lakshminarayanan V, Lagrane J, Kean ML, Dick M, Shankle R. Vision in dementia: contrast effects. Neurol Res 1996;18:9–15.

[12] Trick GL, Trick LR, Morris P, Wolf M. Visual field loss in senile dementia of the Alzheimer's type. Neurology 1995;45(1):68–74.

[13] Kiyosawa M, Bosley TM, Chawluk J, et al. Alzheimer's disease with prominent visual symptoms; clinical and metabolic evaluation. Ophthalmology 1989;96:1077–85.

[14] Fletcher WA, Sharpe JA. Smooth pursuit dysfunction in Alzheimer's disease. Neurology 1988;38:272–7.

[15] Sadun AA, Bassi CJ. Optic nerve damage in Alzheimer's disease. Ophthalmology 1990;97:9–17.

[16] Zaccara G, Gangemi PF, Muscas GC, et al. Smooth-pursuit eye movements: alterations in Alzheimer's disease. J Neurol Sci 1992;112:81–9.

[17] Nissen MJ, Corkin S, Buoanno FJ, Growden JH, Wray SH, Baver J. Spatial vision in Alzheimer's disease; general findings and a case report. Arch Neurol 1985;42:667–71.

[18] Crow RW, Levin LB, LaBree L, Rubin R, Feldon SE. Sweep visual evoked potential evaluation of contrast sensitivity in Alzheimer's dementia. Invest Ophthalmol Vis Sci 2003;44:875–8.

[19] Schlotterer G, Mosovitch M, Crapper-McLachlan D. Visual processing deficits as assessed by spatial frequency contrast sensitivity and backward masking in normal ageing and Alzheimer's disease. Brain 1983;107:309–25.

[20] Ceccaldi M. Vision in Alzheimer's disease. Rev Neurol 1996;152:6–7.

[21] Katz B, Rimmer S. Ophthalmologic manifestations of Alzheimer's disease. Surv Ophthalmol 1989;34:31–43.

[22] Parisi V, Restuccia R, Fattapposta F, Mina C, Bucci M, Pierelli F. Morphological and functional retinal impairment in Alzheimer's disease patients. Clin Neurophysiol 2001;112(10):1860–7.

[23] Kergoat H, Kergoat MJ, Justino L, Chertkow H, Robillard A, Bergman H. Visual retinocortical function in dementia of the Alzheimer type. Gerontology 2002;48:197–203.

[24] Strenn K, Dalbianco P, Weghaupt H, Koch G, Vass C, Gottlob I. Pattern electroretinogram: luminance electroretinogram in Alzheimer's disease. J Neural Transm 1991;33:73–80.

[25] Prager TC, Schweitzer FC, Peacock LW, Garcia CA. The effect of optical defocus on the pattern electroretinogram in normal subjects and patients with Alzheimer's disease. Am J Ophthalmol 1993;116:363–9.

[26] Hinton DR, Sadun AA, Blancks JC, Miller CA. Optic nerve degeneration in Alzheimer's disease. N Engl J Med 1986;315:485–8.

[27] Philpot MP, Amin D, Levy R. Visual evoked potentials in Alzheimer's disease: correlations with age and severity. Electroencephalogr Clin Neurophysiol 1990;77:323–9.

[28] Harding GFA, Wright CE, Orwin A. Primary presenile dementia: the use of the visual evoked potential as a diagnostic indicator. Br J Psychol 1985;147:533–40.

[29] O'Neil D, Rowan M, Abrahams D, Feely JB, Walsh JB, Coakley D. The flash visual evoked potential in Alzheimer type dementia. Ir J Med Sci 1989;158:158.

[30] Bajalan AA, Wright CE, Van der Vliet VJ. Changes in the human visual evoked potential (VEP) caused by the anticholinergic agent hyoscine hydrobromide: comparison with results in Alzheimer's disease. J Neurol Neurosurg Psychiatr 1986;49:175–82.

[31] Coben LA, Danziger WL, Hughes CP. Visual evoked potentials in mild senile dementia of Alzheimer type. Electroencephalogr Clin Neurophysiol 1983;55:121–30.

[32] Fujimori M, Imamura T, Yamashita H, Hirono N, Mori E. The disturbances of object vision and spatial vision in Alzheimer's disease. Dement Geriatr Cogn Disord 1997;8:228–31.

[33] Geldmacher DS. Visuospatial dysfunction in the neurodegenerative diseases. Front Biosci 2003;8:E428–36.

[34] Glosser G, Baker KM, de Vries JJ, Alavi A, Grossman M, Clark CM. Disturbed visual processing contributes to impaired reading in Alzheimer's disease. Neuropsychologia 2002;40:902–9.

[35] Nakano N, Hatakeyama Y, Fukatsu R, et al. Eye–head coordination abnormalities and regional cerebral blood flow in Alzheimer's disease. Prog Neuropsychopharmacol Biol Psychiatry 1999;23:1053–62.

[36] Chapman FM, Dickinson J, McKeith I, Ballard C. Association among visual associations, visual acuity, and specific eye pathologies in Alzheimer's disease: treatment implications. Am J Psychiatry 1999;156:1983–5.

[37] Armstrong RA, Syed AB. Alzheimer's disease and the eye. Ophthal Physiol Opt 1996;16:S2–8.

[38] Suggestions for maintaining independent living. Real Health Systems; October 6, 2014. Available from: http://www.realhealth120.com/#!health-longevity-by-prevention/c1sig.

[39] Goldstein L, Muffat J, Cherny R, et al. Cytosolic β-amyloid deposition and supranuclear cataracts in lenses from people with Alzheimer's disease. Lancet 2003;361:1258–65.

[40] Hedges TR, Galves AP, Speigelman D, Barbas NR, Peli E, Yardley CJ. Retinal nerve fibre layer abnormalities in Alzheimer's disease. Acta Ophthalmol Scand 1996;74:271–5.

[41] Tsai CS, Ritch R, Schwartz B, et al. Optic nerve head and optic nerve fibre layer in Alzheimer's disease. Arch Ophthalmol 1991;109:199–204.

[42] Syed AB, Armstrong RA, Smith CUM. A quantitative analysis of optic nerve axons in elderly control subjects and patients with Alzheimer's disease. Folia Neuropathol 2005;43:1–6.

[43] Ronch JL. Alzheimer's disease: a practical guide to those who help others. New York: Continuum Books; 1989.

[44] Hof PR, Morrison JH. Quantitative analysis of a vulnerable subset of pyramidal neurons in Alzheimer's disease. II. Primary and secondary visual cortex. J Comp Neurol 1990;301:55–64.

[45] Armstrong RA, Nochlin D, Sumi SM, Alvord EC. Neuropathological changes in the visual cortex in Alzheimer's disease. Neurosci Res Commun 1990;6:163–71.

[46] Armstrong RA. Visual field defects in Alzheimer's disease patients may reflect differential pathology in the primary visual cortex. Optom Vis Sci 1996;73:677–82.

[47] Baker DR, Mendez MF, Townsend JC, Ilsen PF, Bright DC. Optometric management of patients with Alzheimer's disease. J Am Optom Assoc 1997;68:483–94.

[48] Berisha F, et al. Retinal abnormalities in early Alzheimer's disease. Invest Ophthalmol Vis Sci 2007;48(5):2285–9.

[49] Frost S, Martins RN, Yogesan K. Retinal screening for early detection of Alzheimer's disease. Digital teleretinal screening. Berlin, Heidelberg: Springer; 2012. p. 91–100.

[50] Frost S, Martins RN, Kanagasingam Y. Ocular biomarkers for early detection of Alzheimer's disease. J Alzheimers Dis 2010;22(1):1–16.

[51] Meier MH, et al. Microvascular abnormality in schizophrenia as shown by retinal imaging. Am J Psychiatry 2013;170(12):1451–9.

[52] Craddock TJA, et al. The zinc dyshomeostasis hypothesis of Alzheimer's disease. PLoS One 2012;7(3):e33552.

[53] Trempe CL. Zinc and macular degeneration. Arch Ophthalmol 1992;110(11):1517.

[54] Newsome DA, et al. Oral zinc in macular degeneration. Arch Ophthalmol 1988;106(2):192.

[55] Broun ER, et al. Excessive zinc ingestion. JAMA 1990;264(11):1441–3.

[56] Frederriske PH, et al. Oxidative stress increases production of beta-amyloid precursor protein and beta-amyloid (A-beta) in mammalian lens, and A-beta has toxic effects on lens epithelial cells. J Biol Chem 1996;271:10169–74.

[57] Clemons TE, Kurinij N, Sperduto RD. Associations of mortality with ocular disorders and an intervention of high-dose antioxidants and zinc in the Age-Related Eye Disease Study: AREDS report no. 13. Arch Ophthalmol 2004;122(5):716–26.

[58] Trempe C. Ranibizumab for age-related macular degeneration. N Engl J Med 2011;364:581–2.

[59] Folk JC, Stone EM. Ranibizumab therapy for neovascular age-related macular degeneration. N Engl J Med 2010;363:1648–55. [Erratum, N Engl J Med 2010;363:2474].

[60] Barbazetto IA, Saroj N, Shapiro H, Wong P, Ho AC, Freund KB. Incidence of new choroidal neovascularization in fellow eyes of patients treated in the MARINA and ANCHOR trials. Am J Ophthalmol 2010;149:939–46.

[61] Baird PN, Robman LD, Richardson AJ, et al. Gene–environment interaction in progression of AMD: the CFH gene, smoking and exposure to chronic infection. Hum Mol Genet 2008;17:1299–305.

[62] Vine AK, Stader J, Branham K, Musch DC, Swaroo PA. Biomarkers of cardiovascular disease as risk factors for age-related macular degeneration. Ophthalmology 2005;112:2076–80.

[63] Witmer MT. Treatment of exudative AMD: data from the CATT and IVAN trials. Ophthalmology 2012;9.

[64] Lee AJ, et al. Open-angle glaucoma and cardiovascular mortality: the Blue Mountains Eye Study. Ophthalmology 2006;113(7):1069–76.

[65] Cugati S, et al. Visual impairment, age-related macular degeneration, cataract, and long-term mortality: the Blue Mountains Eye Study. Arch Ophthalmol 2007;125(7):917.

[66] Wang JJ, Mitchell P, Smith W. Vision and low self-rated health: the Blue Mountains Eye Study. Invest Ophthalmol Vis Sci 2000;41(1):49–54.

[67] Wang JJ, et al. Visual impairment, age-related cataract, and mortality. Arch Ophthalmol 2001;119(8):1186.

[68] Younan C, et al. Cardiovascular disease, vascular risk factors and the incidence of cataract and cataract surgery: the Blue Mountains Eye Study. Neuroophthalmology 2003;10(4):227–40.

[69] Karpa MJ, et al. Direct and indirect effects of visual impairment on mortality risk in older persons: the Blue Mountains Eye Study. Arch Ophthalmol 2009;127(10):1347.

[70] Lee DJ, et al. Visual acuity impairment and mortality in US adults. Arch Ophthalmol 2002;120(11):1544.

[71] Klein R, Klein BEK, Moss SE. Age-related eye disease and survival: the Beaver Dam Eye Study. Arch Ophthalmol 1995;113(3):333.

[72] Klein R, et al. The association of cardiovascular disease with the long-term incidence of age-related maculopathy: the Beaver Dam Eye Study. Ophthalmology 2003;110(4):636–43.

[73] Knudtson MD, Klein BEK, Klein R. Age-related eye disease, visual impairment, and survival: the Beaver Dam Eye Study. Arch Ophthalmol 2006;124(2):243.

[74] Wong TY, et al. Retinal microvascular abnormalities and 10-year cardiovascular mortality: a population-based case–control study. Ophthalmology 2003;110(5):933–40.

[75] McCarty CA, Nanjan MB, Taylor HR. Vision impairment predicts 5 year mortality. Br J Ophthalmol 2001;85(3):322–6.

[76] Hennis A, et al. Lens opacities and mortality: the Barbados Eye Studies. Ophthalmology 2001;108(3):498–504.

[77] Xu L, et al. Mortality and ocular diseases: the Beijing Eye Study. Ophthalmology 2009;116(4):732–8.

[78] Borger PH, et al. Is there a direct association between age-related eye diseases and mortality? The Rotterdam Study. Ophthalmology 2003;110(7):1292–6.

[78a] Khanna RC, et al. Cataract, visual impairment and long-term mortality in a rural cohort in India: the Andhra Pradesh Eye Disease Study. PLoS One 2013;8(10):e78002.

[79] Klaver CCW, et al. Is age-related maculopathy associated with Alzheimer's disease? The Rotterdam Study. Am J Epidemiol 1999;150(9):963–8.

[80] Whitson HE, et al. Prevalence and patterns of comorbid cognitive impairment in low vision rehabilitation for macular disease. Arch Gerontol Geriatr 2010;50(2):209–12.

[81] Whitson HE, et al. The combined effect of visual impairment and cognitive impairment on disability in older people. J Am Geriatr Soc 2007;55(6):885–91.

[82] Kaarniranta K, et al. Age-related macular degeneration (AMD): Alzheimer's disease in the eye? J Alzheimers Dis 2011;24(4):615–31.

[83] Soscia SJ, et al. The Alzheimer's disease-associated amyloid β-protein is an antimicrobial peptide. PLoS One 2010;5(3):e9505.

[84] Pache M, Flammer J. A sick eye in a sick body? Systemic findings in patients with primary open-angle glaucoma. Surv Ophthalmol 2006;51(3):179–212.

[85] Fanelli JL. The dark side of the nerve—what's the link between glaucoma and Alzheimer's or Parkinson's? Can we use one to predict the other? Rev Optom 2011;148(7):53.

[86] Hinton DR, Sadun AA, Blanks JC, Miller CA. Optic-nerve degeneration in Alzheimer's disease. N Engl J Med 1986;315(8):485–7.

[87] Blanks JC, Hinton DR, Sadun AA, Miller CA. Retinal ganglion cell degeneration in Alzheimer's disease. Brain Res 1989;501(2):364–72.

[88] Parisi V. Correlation between morphological and functional retinal impairment in patients affected by ocular hypertension, glaucoma, demyelinating optic neuritis and Alzheimer's disease. Semin Ophthalmol 2003;18(2):50–7.

[89] Guo L, Salt TE, Luong V, Wood NE, Cheung W, Maass A, et al. Targeting amyloid-beta in glaucoma treatment. Proc Natl Acad Sci USA 2007;104:13444–9.

[90] Bizrah M, Guo L, Cordeiro MF. Glaucoma and Alzheimer's disease in the elderly. Aging Health 2011;7(5):719–33.

[91] Wostyn P, Audenaert K, Paul De Deyn P. Alzheimer's disease: cerebral glaucoma? Med Hypotheses 2010;74(6):973–7.

[92] Wostyn P. Intracranial pressure and Alzheimer's disease: a hypothesis. Med Hypotheses 1994;43:219–22.

[93] van Koolwijk LME, et al. Association of cognitive functioning with retinal nerve fiber layer thickness. Invest Ophthalmol Vis Sci 2009;50(10):4576–80.

[94] Jindahra P, Plant GT. Retinal nerve fibre layer thinning in Alzheimer disease. In: De La Monte S, editor. The clinical spectrum of Alzheimer's disease—the charge toward comprehensive diagnostic and therapeutic strategies. Rijeka, Croatia: InTech; 2011.

[95] Jefferis JM, Mosimann UP, Clarke MP. Cataract and cognitive impairment: a review of the literature. Br J Ophthalmol 2011;95(1):17–23.

[96] Harding JJ. Cataract, Alzheimer's disease, and other conformational diseases. Curr Opin Ophthalmol 1998;9(1):10–3.

[97] Krasodomska K, et al. Pattern electroretinogram (PERG) and pattern visual evoked potential (PVEP) in the early stages of Alzheimer's disease. Doc Ophthalmol 2010;121(2):111–21.

[98] Coburn KL, et al. Diagnostic utility of visual evoked potential changes in Alzheimer's disease. J Neuropsychiatry Clin Neurosci 2003;15(2):175–9.

[99] Tartaglione A, et al. Resting state in Alzheimer's disease: a concurrent analysis of flash-visual evoked potentials and quantitative EEG. BMC Neurol 2012;12(1):145.

[100] Brewer A, Barton B. Aging and dementia in human visual cortex: visual field map organization and population receptive fields. J Vis 2011;11(15):28.

[101] Kosnik W, Fikre J, Sekuler R. Visual fixation stability in older adults. Invest Ophthalmol Vis Sci 1986;27(12):1720–5.

[102] Tales A, Porter G. Visual attention-related processing in Alzheimer's. Rev Clin Gerontol 2008;18:229–43.

[103] Bylsma FW, et al. Changes in visual fixation and saccadic eye movements in Alzheimer's disease. Int J Psychophysiol 1995;19(1):33–40.

[104] Vasquez BP, et al. Visual attention deficits in Alzheimer's disease: relationship to HMPAO SPECT cortical hypoperfusion. Neuropsychologia 2011;49(7):1741–50.

[105] Miyasaka T, et al. Microtubule destruction induces tau liberation and its subsequent phosphorylation. FEBS Lett 2010;584(14):3227–32.

[106] Ballatore C, et al. Microtubule stabilizing agents as potential treatment for Alzheimer's disease and related neurodegenerative tauopathies. J Med Chem 2012;55(21):8979–96.

[107] Zhang B, et al. The microtubule-stabilizing agent, epothilone D, reduces axonal dysfunction, neurotoxicity, cognitive deficits, and Alzheimer-like pathology in an interventional study with aged tau transgenic mice. J Neurosci 2012;32(11):3601–11.

[108] Barten DM, et al. Hyperdynamic microtubules, cognitive deficits, and pathology are improved in tau transgenic mice with low doses of the microtubule-stabilizing agent BMS-241027. J Neurosci 2012;32(21):7137–45.

[109] Baas PW, Ahmad FJ. Beyond taxol: microtubule-based treatment of disease and injury of the nervous system. Brain 2013;136(10):2937–51.

Image Credit: *An elderly man with his hand to his forehead, a derivative of http://commons.wikimedia.org/wiki/File:India_-_An_old_man_-_0800.jpg. Available at Wikimedia Commons: https://commons.wikimedia.org/w/index.php?curid=25188098*

Inflammation Friend or Foe?

What role does inflammation play in Alzheimer's disease (AD)? The Alzheimer's Association is apparently on board with a belief that inflammation is connected to the disease based on extensive research funding provided in this area. They are not alone, as researchers all over the globe are pursuing the connection. A scholar.google search yields 300 research papers with Alzheimer's and inflammation in the title. Alzheimer's is not alone as a disease connected to inflammation. Search for any chronic disease and inflammation in scholar.google and your browser will fill with references. Arguably the strongest inflammation/disease connection is for cardiovascular diseases, despite the medical industry's attempt to focus just on cholesterol.

Where is the best place to screen for early signs of inflammation? Yes, the answer is the eye. The aging diseases common to the eye that have a systemic origin including macular degeneration, glaucoma, and cataracts, all have a strong inflammatory component. But there is a little-known twist to why the eye is such a good biomarker for inflammation. This research comes mainly from a small, boutique eye research center affiliated with the Harvard Medical School, namely the Schepens Eye Research Institute (Schepens). Here, a now deceased researcher, J. Wayne Streilein, delved into the concept of "ocular immune privilege," which could explain why aging diseases with inflammation as a component appear in the eye first. This topic will be explained more thoroughly later in this chapter, after we appreciate the breadth and depth of inflammation in chronic diseases that afflict our society.

INFLAMMATION AND DISEASE

Persistent, systemic inflammation is implicated as the root of practically all known chronic health conditions, including everything from rheumatoid arthritis and high cholesterol to dementia and cancer. But is inflammation a friend to our bodies or a foe? Clearly inflammation is a treasure of nature as it reflects the efforts of our immune system to protect us from disease. Now medical science is suggesting that inflammation can backfire and cause harm, but what if the real culprit is the cause of the inflammation. That is, what if the real problem is the underlying cause as to why our immune system is active in the

The End of Alzheimer's. http://dx.doi.org/10.1016/B978-0-12-812112-2.00007-0

first place? Then is inflammation ever a foe, or are we just not digging deep enough to find the real causes of chronic diseases? It may be that inflammation is both friend and foe. Consider two cases:

1. Due to a lack of "milieu interieur," or homeostasis, our bodies develop an "inflammatory" physiology. This weakened physiology allows the proliferation of opportunistic conditions (insults) that then leads to a vicious cycle of more inflammation and more insults.
2. An "insult(s)" attacks us, leading to an immune response, the result of which is inflammation. Depending upon our state of health, the inflammation may pass or appear as an inflammatory cascade described in case 1 above.

Definition of Insult In medical terms, an insult is the cause of some kind of physical or mental injury. For example, a burn on the skin (the injury) may be the result of a thermal, chemical, radioactive, or electrical event (the insult). Likewise, sepsis and trauma are examples of foreign insults, and Alzheimer's, multiple sclerosis (MS), and brain tumors are examples of insults to the brain.

In both cases, the insult is implied to be what causes damage to our bodies, but the inflammatory immune response may have a deleterious effect on us while affording protection. This is akin to a case of "friendly fire."

There are literally hundreds of illnesses associated with chronic inflammation that medicine has classified as unique and unrelated, when in fact they are all products of the same underlying imbalances inside our bodies. When the root causes of these imbalances are properly addressed, the chances of disease taking hold and producing inflammation is significantly reduced. Paul Clayton, author of *Out of the Fire*, nicely summarizes the evidence that this is true [1]. Clayton writes,

> Analysis of the mid-Victorian period in the United Kingdom reveals that life expectancy at age five was as good or better than exists today, and the incidence of degenerative disease was 10% of ours. Their levels of physical activity and hence calorific intakes were approximately twice ours. They had relatively little access to alcohol and tobacco; and due to their correspondingly high intake of fruits, whole grains, oily fish and vegetables, they consumed levels of micro- and phytonutrients at approximately 10 times the levels considered normal today. This paper relates the nutritional status of the mid-Victorians to their freedom from degenerative disease and extrapolates recommendations for the cost-effective improvement of public health today.

Most articles on the topic of diseases of inflammation go something like this:

> The modern epidemic of chronic, low-grade inflammation destroys the balance in your body. When your body's systems experience a constant inflammatory response, you become more susceptible to aging and disease.

This statement is likely to be the cart-before-the-horse. The real issue is that we, in modern society, have created the imbalance that makes us susceptible to disease. The imbalance creates inflammation first, followed by disease, then followed by more inflammation. When we understand cause/effect, then we know what to do. This false dogma about inflammation has been proven wrong so many times. When we squelch inflammation, disease gets worse more quickly. Sure, there can be short-term symptomatic relief, but that is not a proper approach to the management of the root causes of disease.

Here is another example of a misplaced interpretation:

And what are some of the primary causes of chronic inflammation? Excessive stress, poor diet that lacks vitamins and minerals, environmental toxicity, not drinking enough clean water, lack of sleep, and lack of exercise all contribute to low levels of chronic inflammation that often go undetected for many years until disease finally emerges.

Most of what is stated earlier is true. Indeed stress, poor diet, and other factors are deleterious, but how? These strains on our bodies weaken our immune systems and make us more susceptible to disease. Once we have disease, our immune system, as measured by inflammation, goes to work on our behalf. Don't blame inflammation. It is a treasure of nature. Respect inflammation as a warning sign for something deeper in the body. Perform a differential diagnosis, find the causes, and treat them, not the inflammation. Better yet, modulate your behavior to avoid inflammation in the first place. Paul Clayton's book can help you in this regard.

In AD, inflammation is known to be present. The two best-known signs of Alzheimer's, in the brains of its victims, are the plaques of amyloid-beta protein and the tangles of tau protein. Inflammation is also present as shown by the presence of microglia, neural cousins of pathogen-eating macrophages of the bloodstream. These brain immune system cells swarm around amyloid plaques and dying, tangle-ridden neurons. They seem helpful, eating up amyloid-beta as well as damaged cells. Many researchers question if their immunological enthusiasm also causes harm to healthy cells and accelerate the disease or even help to initiate it. In other words, does the immune system create collateral damage in the brain while trying to save neurons? Scientists have debated these questions for more than 2 decades without any firm resolution.

Now a burst of new research suggests that inflammation does, indeed, play a major role in Alzheimer's, and that targeting specific elements of that inflammation could be useful in treating or preventing the disease. We will explore the research and keep a keen eye on their conclusions as we seek the truth. Here, the truth means identification of the right culprit. Or, in the case of cholesterol and beta-amyloid, do we still need to dig deeper? As Dr. Craig Atwood said about beta-amyloid, "Is it the culprit or the only one that didn't get away?"

It may be true that, wherever it occurs in the body, chronic inflammation is a double-edged sword. The initial inflammatory response is meant to defend tissues against molecular foes, such as viruses, bacteria, cancerous cells, and harmful amyloid protein aggregates. But the longer it lasts, the more this inflammation stresses and kills healthy, "innocent bystander" cells. The immune system is at war with disease, and like in any conflict, collateral damage is possible. Over time, as in rheumatoid arthritis, for example, the inflammation can appear to become self-sustaining. This is a clear example of case 1 earlier, or is it?

Since the late 1980s, various studies have found hints that the chronic inflammation found in Alzheimer's hastens the disease process, and some believe it may even be a disease trigger. A review article from 1994 studied the results regarding inflammation and Alzheimer's over the previous decade, thus from 1984 to 1994 [2]. They determined the following:

The purpose of this article is to review evidence that inflammatory and immune mechanisms are important in the pathophysiology of Alzheimer's disease and to suggest new treatment strategies. METHOD: The authors review the English-language literature of the last 10 years pertaining to the pathophysiology of Alzheimer's disease. RESULTS: There is ample evidence supporting the hypothesis that inflammatory and immune mechanisms are involved in tissue destruction in Alzheimer's disease. Acute phase proteins are elevated in the serum and are deposited in amyloid plaques, activated microglial cells that stain for inflammatory cytokines accumulate around senile plaques, and complement components including the membrane attack complex are present around dystrophic neurites and neurofibrillary tangles. CONCLUSIONS: Clinical trials of anti-inflammatory/immunosuppressive drugs are necessary to determine whether alteration of these inflammatory mechanisms can slow the progression of Alzheimer's disease.

Fast-forwarding 20 years shows that antiinflammatory/immunosuppressive drugs do not slow Alzheimer's progression. In fact, just the opposite occurs—patients get worse more quickly. Is inflammation a friend or foe? This chapter endeavors to provide clarity, but one fact is clear, inflammation is a diagnostic for Alzheimer's risk.

A history of serious head injury, which typically causes brain inflammation, is known to be a risk factor for Alzheimer's. Systemic infection, another cause of inflammation, also appears to accelerate the disease. Several epidemiological studies have found, or at least alleged, that older people who use antiinflammatory drugs regularly appear to have significantly lower incidences of Alzheimer's. The value of those epidemiological studies came into question several years ago, when more rigorous placebo-controlled clinical trials on antiinflammatory drugs, ibuprofen, naproxen, and celecoxib, for example, failed to show signs of helping people who already have Alzheimer's dementia or early cognitive impairment. In some cases, these drugs apparently accelerated the course of the disease. This makes sense because the intended purpose of inflammation is to work in our favor. Inflammation is the result of our immune system activity.

"We really need to understand the [Alzheimer's] inflammation reaction better, what is good and what is bad," says Irene Knuesel, a researcher at the University of Zurich. Her laboratory and others have been discovering important clues about which is which. In all the diseases that show misfolded or unfolded proteins, which are proteins with unexpected conformational changes, researchers almost always associate an inflammatory response occurring at the same time. Thus it may be the inflammation or the causes (antecedents) of the inflammatory process that leads to the protein changes and the protein deposits. Inflammation is quickly becoming the "next big thing" in medicine.

We know that inflammation is a natural part of our body's protective response to "insult" (injury, trauma, infection, or irritation). Thus inflammation is a sign that our body is working to protect us. Is inflammation, like many medications and treatments, sometimes worse than the disease and causes harm to our bodies? Or are we misinterpreting the impact of inflammation and not appreciating that there are other forces that come before the inflammation that lead to disease. Let's explore the entire concept of inflammation with an open mind. What is known about inflammation, how it impacts our bodies, and the root causes of inflammation? Simply put, let's begin to understand the "why" of inflammation.

INFLAMMATION DEFINED

Inflammation (Latin, īnflammō, "I ignite, set alight") is part of the complex biological response of lymphoid and vascular tissues to harmful stimuli, such as pathogens, damaged cells, or irritants. Inflammation is a protective attempt by the body to remove the cause of the attack on the body and to initiate the healing process. Inflammation is not a synonym for infection, but it is very often associated with infection. Although a microorganism causes infection, inflammation is one of the responses we have, to the pathogen. Inflammation is considered a mechanism of innate immunity, as compared to adaptive immunity, which is specific for each pathogen. There are two classes of inflammation that can have similar or the same causes: acute inflammation and chronic inflammation.

Inflammation is intended to protect our body. It is our defense system, like the force fields on the USS Enterprise of Star Trek. No wonder strategies to halt inflammation are unable to arrest inflammatory diseases. Halting inflammation may have some palliative affects, but this is not a root cause strategy. Medicine must work to support our immune system as it wages war with the root cause(s) of illness.

Acute inflammation is the initial response of the body to harmful and sudden stimuli, such as trauma and is achieved by the increased movement of immune system molecules from the blood into the injured tissues. A cascade of biochemical events propagates the inflammatory response, involving the local vascular system, the immune system, and various cells within the injured tissue. Prolonged inflammation, known as chronic inflammation, leads to a progressive shift in the type of cells present at the site of inflammation and is characterized by simultaneous destruction (possibly from the insult or from the inflammation process) and healing of the tissue from the inflammatory process.

It is important to note that the inflammatory response emanates from the vascular system. Thus the vascular or circulatory system is intimately involved in inflammation and often holds clues as to the cause and certainly effects of inflammation. Any disease associated with an inflammatory response is, at least in part, a disease of the vascular system. And this clearly includes AD. We continue to see information that reinforces the concept that Alzheimer's is very much a vascular and not just a neurological disorder.

CHRONIC INFLAMMATION

Of the 10 leading causes of mortality in the United States, chronic, low-level inflammation contributes to the pathogenesis of at least seven. These include heart disease, cancer, chronic lower respiratory disease, stroke, AD, diabetes, and nephritis (inflammation of the kidneys). In general, the suffix "itis" infers some type of inflammatory disorder. This information was obtained from the Centers for Disease Control and Prevention 2011 (CDC).

Inflammation has classically been viewed as an acute (short term) response to tissue injury that produces characteristic symptoms and usually resolves spontaneously. More contemporary revelations show chronic inflammation to be a major factor in the development of degenerative disease and loss of youthful functions. Dr. Barry Sears, the renowned author of the *Zone* diet books, coined the term "silent inflammation" to better explain the insidious nature of chronic inflammation because it can smolder in the background and slowly erode our health, often without us being aware [3]. This is true of AD, as we know those people with symptoms already have substantial brain atrophy that is most likely caused by the long and stealth processes involved in chronic or silent inflammation.

The danger of chronic, low-level inflammation is that its silent nature belies its destructive power. It is now clear that the destructive capacity of chronic inflammation, including its causes, is unprecedented among physiologic processes. Not just in AD but also in other diseases, chronic inflammation once triggered can persist undetected for years or even decades, propagating cell death throughout the body. The process of inflammation, including all its causes, contributes so greatly to deterioration associated with the aging process, that this silent state of chronic inflammation has been coined "inflammaging."

Italian researchers first used the term "inflammaging" in 2000 [4]. A subsequent paper published in 2007 had the provocative title: *Inflammaging and anti-inflammaging: A systemic perspective on aging and longevity emerged from studies in humans* [5]. The abstract of that paper is included here:

> A large part of the aging phenotype, including immunosenescence (slow deterioration of the immune system), is explained by an imbalance between inflammatory and anti-inflammatory networks, which results in the low grade chronic pro-inflammatory status we proposed to call inflammaging. Within this perspective, healthy aging and longevity are likely the result not only of a lower propensity to mount inflammatory responses but also of efficient anti-inflammatory networks, which in normal aging fail to fully neutralize the inflammatory processes consequent to the lifelong antigenic burden and exposure to damaging agents. Such a global imbalance can be a major driving force for frailty and common age-related pathologies, and it should be addressed and studied within an evolutionary-based systems biology perspective. Evidence in favor of this conceptualization largely derives from studies in humans. We thus propose that inflammaging can be flanked by anti-inflammaging as major determinants not only of immunosenescence but eventually of global aging and longevity.

Chronic low-level inflammation may be threatening your health at this very moment, without you realizing it. The good news is that there are low-cost tests that can assess the inflammatory state within your body. The previous chapter on the eye provides the first line of diagnostic evaluation. Other tests that involve evaluation of your blood offer more detailed information. We are all pretty frightened by this amorphous concept of inflammaging. Statements like, "We thus propose that inflammaging can be flanked by anti-inflammaging," do not exactly provide a road map to a solution. However, Paul Clayton's works spell out an antiinflammaging recipe. Can you follow the recipe? The poorest of mid-Victorian England (1870) were able to, so you can, too, if you have the resolve.

Four signs characterize the classic manifestation of acute inflammation: Redness and heat result from the increased blood flow to the site of injury. Swelling results from the accumulation of fluid at the injury site, a consequence of the increased blood flow. Finally, swelling can compress nerve endings near the injury, causing the characteristic pain associated with inflammation. Pain is also important to make us aware of the tissue damage. Additionally, inflammation in a joint usually results in a fifth sign, impairment of function, which has the effect of limiting movement and forcing the rest of the injured joint to aid in healing. Keep in mind these signs and symptoms are seldom present in inflammaging.

One of the key contributors to inflammation is cellular stress created in part by the very process of life through the production of energy. In particular, aging is associated with declining mitochondrial efficiency and increased production of free radical molecules. Recent research identifies this age-associated aberration of mitochondrial function as a principle actuator of chronic inflammation. Mitochondrial dysfunction brings about inflammation that starts with the accumulation of free radicals that causes the mitochondrial membrane to become more permeable.

Free radicals are not the only cause of inflammatory cell death. Circulating sugars, primarily glucose, and fructose, are culprits as well. When these blood sugars come in contact with proteins and lipids (fats), a damaging reaction occurs that eventually leads to the activation of numerous inflammatory genes.

Additional biochemical inducers of a chronic inflammatory response include:

- Uric acid (urate) crystals, which can be deposited in joints during gouty arthritis (elevated levels are a risk factor for kidney disease, hypertension, and metabolic syndrome);
- Oxidized lipoproteins (such as LDL), a significant contributor to atherosclerotic plaques; and
- Homocysteine, a nonprotein-forming amino acid that is a marker and a risk factor for cardiovascular disease.

Together, these proinflammatory instigators promote a perpetual low-level chronic inflammatory state called parainflammation (parainflammation, silent inflammation, and inflammaging all refer to essentially the same processes). Medzhitov from Yale first presented this concept in the prestigious journal *Nature* in 2008 [6]. The abstract for that paper titled, *Origin and physiological roles of inflammation*, is included here.

> Inflammation underlies a wide variety of physiological and pathological processes. Although the pathological aspects of many types of inflammation are well appreciated, their physiological functions are mostly unknown. The classic instigators of inflammation—infection and tissue injury—are at one end of a large range of adverse conditions that induce inflammation, and they trigger the recruitment of leukocytes and plasma proteins to the affected tissue site. Tissue stress or malfunction similarly induces an adaptive response, which is referred to here as para-inflammation. This response relies mainly on tissue-resident macrophages and is intermediate between the basal homeostatic state and a classic inflammatory response. Para-inflammation is probably responsible for the chronic inflammatory conditions that are associated with modern human diseases.
>
> Although it progresses silently, para-inflammation, either as a sign or a cause, presents a major threat to the health and longevity of all aging humans. Chronic, low-level inflammation is associated with common diseases including cancer, type 2 diabetes, osteoporosis, cardiovascular diseases, and others. Thus, by targeting the myriad physiological variables that can create an inflammatory response, one can effectively temper chronic inflammation and reduce the risk for inflammatory diseases. However, first we must be aware of its presence.

IMMUNOSENESCENCE

Immunosenescence refers to the gradual deterioration of our immune system as we get older. It involves our capacity to respond to infections and maintain our long-term immune memory that was acquired (usually in our early life) either by infection or vaccination. A good example is chicken pox: we get infected during our childhood and develop a "lifelong" immunity until we develop shingles in old age as our long-term immune memory fails. Having shingles is a painful reminder that our immune system is failing. This age-associated immune deficiency is a major contributory factor to the increased frequency of morbidity (disease) and mortality (death) among the elderly.

As we age, our bodies produce more inflammation as a protective mechanism to counteract a failing immune system. This simple fact is not well understood by doctors of pharmaceutical companies that keep trying to use antiinflammatory medications in the treatment of AD and the other age related diseases. Antiinflammatory and antioxidant all make AD worst. Prednisone was tried to treat AD in one of the Harvard affiliated hospital years ago. The results were not good and the study was never reported as it is usual when treatment fail.

INFLAMMATION MARKERS

What are the methods for detecting inflammation, preferably at its earliest stages? The body, through the immune system, triggers a range of chemical and biochemical reactions to the causes or triggers of inflammation. It is the markers of inflammation that help us diagnose and understand diseases, such as Alzheimer's. It is very clear that, since Alzheimer's progresses for years to decades before showing clinical symptoms, and silent inflammation parallels disease, these markers appear in the blood and elsewhere in the body long before clinical disease. However, even though Alzheimer's is considered associated with inflammation, few if any neurologists order tests for these markers. Insurance companies are not covering tests for inflammation under standard-of-care diagnostic codes for Alzheimer's and other neurodegenerative diseases. Patients remain essentially undiagnosed or certainly under diagnosed.

What follows is a summary of some of the most prominent markers of inflammation used in research and diagnosis. Many can be detected by simple blood tests and are relatively inexpensive. In modern and traditional medicine, tests cannot be ordered without a diagnosis. Without the diagnosis, insurance will not cover tests. Prevention is still a concept in medicine, especially in America. And in some states, an individual cannot order their own blood tests to evaluate their own health under any circumstances. That presents a significant problem when it comes to measuring chronic low-grade (silent or para) inflammation, as most of us with this affliction have no symptoms upon which to create a diagnosis. However, there are ways to have these tests done in the United States. Be aware of new laws that are likely to be promoted in the future that further reduce the rights of individuals to access their own well-being with or without the help of a physician.

One organization that has taken a lead in providing diagnostic access to individuals is Life Extension Foundation. At their website (www.lef.org), an individual can go to "Products," then "Blood Testing," and choose from a wide variety of tests to evaluate his or her own health, for a fee. A test for evaluating silent inflammation is Item# LC100008. When you order this test, you are informed that it is not available to you in Massachusetts, New York, Rhode Island, and Pennsylvania. These are states with high regulatory burdens that result in individuals losing their freedoms and rights. If your doctor cannot order tests due to insurance payments, and you cannot order tests because big brother will not let you, then what?

Very few of the tests discussed later have associated diagnostic codes. A diagnosis of any of the major chronic diseases impacting our society also does not trigger these tests. Inflammation, per se, is not considered a disease or a treatable condition, which leaves the millions of patients with inflammation-based diseases in quite a quandary.

C-Reactive Protein

C-Reactive Protein (CRP) is a protein found in the blood, the levels of which rise in response to inflammation. It is one of several proteins rapidly produced by the liver during an inflammatory response. Its primary goal is thought to be to coat damaged cells to make them easier to recognize by other immune cells. Thus CRP helps with the destruction, assimilation, and secretion of bacteria, dead cells, and small mineral particles. Molecules like CRP are involved in inflammation and immune response and are often regarded in three different ways:

- having some type of protective action,
- being an innocent bystander, or
- being involved in the physiological activity in some type of deleterious way.

Thus, just like beta-amyloid in Alzheimer's and cholesterol in cardiovascular disease, there are arguments on either side of whether CRP is friend or foe.

The connection between CRP in the blood and AD is strong and growing. A search for "C-reactive protein" and Alzheimer's yields 30 references, almost half of which have been published since 2009. One such paper is titled, *A Clinical Significance of High-Sensitivity C-reactive Protein Level in Alzheimer's Disease and Vascular Dementia* [7]. Some goodies from this paper include:

> Background: There is increasing evidence about inflammatory processes in the development of dementia. Therefore, inflammation has been believed to play a pivotal role in cognitive decline, Alzheimer's disease (AD), and vascular dementia. High-sensitivity C-reactive protein (hs-CRP) is a sensitive systemic marker of inflammation, and increased levels of hs-CRP are associated with inflammatory reactions. It is important to identify modifiable risk factors, which could be used in preventing or delaying the onset of dementia. Therefore, we studied to clarify a clinical role of hs-CRP in AD and VaD (vascular dementia).
>
> Conclusions: The result of our study suggests the presence of inflammatory activity is related with dementia, not only AD, but also VaD associated with cerebrovascular disease.

CRP elevation above base levels is not definitively diagnostic on its own, as it can rise in several cancers, rheumatologic, gastrointestinal, and cardiovascular conditions, and infections, not to mention acute events like trauma. And, because it is so-called "nonspecific," testing for CRP has not become a standard or recognized test in baseline health assessments. However, measuring core body temperature with a thermometer is also nonspecific, yet it provides health care a great deal of information about cause, effect, and treatment. CRP can be viewed as a measure of a patient's temperature for chronic inflammation and disease.

Despite its presumed lack of specificity, elevation of CRP (as determined by a high-sensitivity CRP assay or hs-CRP) has a strong association with elevated risk of cardiovascular disease and stroke. Even as cardiovascular mortality trends ever so slightly downward, the current standard-of-care for medical diagnosis is woefully deficient in the early detection of cardiovascular disease. The Tim Russert story provides poignant evidence that modern medicine still does not have an adequate diagnostic and treatment approach to most cardiovascular afflictions.

A test for CRP is simple and is performed by drawing blood from a vein, usually from the inside of the elbow. After the puncture site is cleaned, a blood pressure cuff is placed around the upper arm, causing the veins below the band to swell with blood. A needle is then inserted into the vein, and the blood is collected. Antiserum is then used to detect CRP levels.

While this is a fairly accurate test, a low CRP level does not mean that there is no acute or chronic inflammation.

Explanations for a low CRP in the presence of chronic inflammatory disease may include that CRP is not involved in all the mechanisms associated with inflammation. Also, consider the case of Lyme disease. The Lyme bacteria (note the plural, as there are as many as 18 different types of Lyme bugs) are not steadily and consistently active. These microbes are often able to hide inside of cells (intracellular) and go into a quiet or spore phase. Under these circumstances, CRP may not be produced even though the underlying cause of the chronic inflammation, the Lyme bacteria, is still present. Levels of CRP reduce rapidly when the body does not detect the insult.

Details on C-Reactive Protein

The CRP acute phase response develops in a wide range of acute and chronic inflammatory conditions like bacterial, viral, or fungal infections; rheumatic, and other inflammatory diseases; malignancy; and tissue injury or necrosis. These conditions cause a release of interleukin-6 and other cytokines that trigger the synthesis of CRP and fibrinogen by the liver.

Definition of Cytokines Small proteins released by cells that have a specific effect on the interactions between cells, on communications between cells, or on the behavior of cells. The cytokines include the interleukins, lymphokines, and cell signal molecules such as tumor necrosis factor and the interferons, which trigger inflammation and respond to infections.

During the acute phase response, levels of CRP rapidly increase within 2 hours of acute insult including infection. It can rise up to 50,000-fold, reaching a peak at 48 hours. With resolution of the acute phase response, CRP declines with a relatively short half-life of 18 hours. Measuring CRP level is a screen for infectious and inflammatory diseases. Rapid, marked increases in CRP occur with inflammation, infection, trauma and tissue necrosis, malignancies, and autoimmune disorders. Because there are a large number of disparate conditions that can increase CRP production, an elevated CRP level does not diagnose a specific disease. An elevated CRP level can provide support for the presence of an inflammatory disease, such as rheumatoid arthritis, polymyalgia rheumatica or giant-cell arteritis, cardiovascular disease, macular degeneration, AD, and many other inflammation-based diseases including type 2 diabetes.

C-Reactive Protein in Disease Diagnosis

The CRP level is very good evidence of inflammation, as the only known factor to interfere with CRP production is liver failure. Measuring CRP levels therefore is useful in determining how a disease is progressing, and whether or not treatments given for the disease are working. For a comprehensive explanation of CRP and how its levels correspond to disease risk, consider reading *C-reactive protein and cardiovascular disease*, by Paul M. Ridker and Nader Rifai. Dr. Ridker is a Harvard Medical School Professor of Epidemiology, and Dr. Rifai has contributed to the science linking inflammation and macular degeneration [8].

A high sensitivity test called an hs-CRP test may be used to check CRP levels; this test will pick up even trace amounts of CRP that a regular blood test would not find. In healthy persons CRP levels are less than 10 mg/L and increases slightly as one ages; higher levels are found in women during late pregnancy, in women taking oral contraceptives, and in cases of mild inflammation and viral infections. A CRP level corresponds well with risk of heart disease and the myriad other diseases of aging and inflammation. Low hs-CRP indicates low current or future risk, while high hs-CRP infers high risk.

Since increased CRP levels may be influenced by acute (immediate) factors, a single measurement is not enough to predict a person's risk of chronic disease. As such, diagnosing chronic issues is done by doing at least two separate CRP tests 2 weeks apart and using the average number of both readings to assess a person's likelihood of having or getting disease. Since this test is akin to measuring your chronic disease temperature, the prudent should monitor CRP on an annual basis, following the same prescription of two tests over a 2-week period each year.

In cases of elevated CRP, enlightened medicine does not "treat" or lower the elevated CRP in itself. CRP is best used as a diagnostic tool to understand if inflammation is present. The key is to perform further tests to identify the underlying condition(s) causing the abnormal elevation of this marker. If CRP were an indicator for all chronic inflammatory diseases, then one simple annual test would be all that was necessary to measure the state of our overall chronic health. As we continue to spend one trillion dollars each year on medical and related research, a more exact cause/effect between CRP and inflammation will emerge. However, the CRP test is a very valuable tool but does not stand alone.

In AD, some research is finding low plasma (blood) levels of CRP associated with a significantly more rapid cognitive decline. Even under this circumstance, the authors believe, "Plasma markers of CRP may be associated with the rate of cognitive and functional decline in patients with Alzheimer's disease [9]." A group of Texas researchers followed up these results with another study that concluded,

These findings, together with previously published results, are consistent with the hypothesis that midlife elevations in CRP are associated with increased risk of AD development, though elevated CRP levels are not useful for prediction in the immediate prodromal (early symptom) years before AD becomes clinically manifest. However, for a subgroup of patients with AD, elevated CRP continues to predict increased dementia severity suggestive of a possible pro-inflammatory endophenotype in AD [10].

For those still clinging to the Amyloid Cascade Hypothesis, a group from China attempted to understand the connection between beta-amyloid formation and CRP in AD [11]. They concluded, "These results suggest that CRP cytotoxicity is associated with Aβ formation and Aβ-related markers expressions; CRP and Aβ were relevant in early-stage AD; CRP may be an important trigger in AD pathogenesis." The question left unanswered is: what is causing the elevated CRP (thus the beta-amyloid formation) in these Alzheimer's cases? Chapter 9 is revealing.

Interleukin-6

Interleukin-6 (IL-6) is a cytokine (a chemical which enables communication between cells) that is secreted by T cells and macrophages as part of the immune inflammation response to trauma. The body makes CRP from IL-6, thus the levels of these two substances should be closely related. Elevated levels of IL-6, then, are a reliable indicator of the amount of inflammation occurring in the body. A blood test is used to ascertain the IL-6 level in the body.

Elevated Interleukin-6, as with CRP, is associated with AD, and the blood is a good place to perform diagnostic evaluation. A team from Scotland and Hungary wrote, *Serum interleukin-6 levels correlate with the severity of dementia in Down syndrome and in Alzheimer's disease*, in 1997 [12]. Their research helps us to understand the cause/effect relationship.

> Inflammatory processes are suspected in the pathomechanism of Alzheimer's dementia (AD) but the serum and cerebrospinal fluid (CSF) levels of inflammatory cytokines are not yet determined in the different forms of the disorder. Results: Normal serum IL-6 levels were found in the mild-moderate stage, but significantly increased levels were found in the severe stage of both dementia groups. The CSF concentrations remained within the normal range in all groups. Positive correlations between the serum IL-6 levels and age and the severity of the disease were present.
> Conclusion: These findings suggest a disease stage dependent general activation of the immune system both in sporadic AD and in DS with AD.

Adipose tissues make large amounts of IL-6. Increased blood sugar levels also lead to more manufacturing of IL-6. Many in the medical community believe that the inflammation resulting from enhanced IL-6 production is a major contributor to cardiovascular problems and inflammatory diseases of aging (inflammaging). High IL-6 levels have also been found in people with AD, as this disease is increasingly being viewed as an inflammatory brain disorder. It is important to note that when IL-6 is ascribed to Alzheimer's, it is through a systemic blood measurement and is not a brain-specific measurement. Results like these show how Alzheimer's is not a brain-only disease.

Different from CRP, IL-6 acts as both a proinflammatory and antiinflammatory cytokine. It is secreted by T cells and macrophages to stimulate immune response, for example, during infection and after trauma, especially burns or other tissue damage leading to inflammation. IL-6 has been shown to be required in mice for resistance against the bacterium Streptococcus pneumoniae. IL-6 is also a "myokine," a cytokine produced from muscle, and it is elevated in response to muscle contraction. It is significantly elevated with exercise, and precedes the appearance of other cytokines in the circulation. During exercise, it is thought to act in a hormone-like manner to mobilize extracellular substrates and/or augment substrate delivery. Smooth muscle cells of many blood vessels also produce IL-6 as a proinflammatory cytokine. It may be that IL-6, due to its both pro- and antiinflammatory activity, is less specific as a marker for inflammation compared to CRP.

IL-6 is one of the most important mediators of fever and of the acute phase response. It is capable of crossing the blood brain barrier and initiating processes that change the body's temperature set point. In muscle and fatty tissue, IL-6 stimulates energy mobilization, which leads to increased body temperature. IL-6 can be secreted by macrophages in response to specific microbial molecules. This leads to a process that induces intracellular signaling cascades that give rise to inflammatory cytokine production.

IL-6 is also produced by fat cells and is thought to be a reason obese individuals have higher levels of CRP. However, a new belief is that IL-6 is a response, not a cause. Therefore, obesity is probably a manifestation of an inflammatory physiology created by an imbalance caused by a gross excess of omega-6 fatty acids based on our current corn-based diet. The inflammation created by the omega-6 excess makes us vulnerable to pathogens that hijack our metabolism and create insulin resistance. IL-6 is responsible for stimulating acute phase protein synthesis, as well as the production of neutrophils (immune cells that can destroy pathogens) in the bone marrow. It supports the growth of B cells and is antagonistic to regulatory T cells.

IL-6 is relevant to many diseases, such as diabetes, atherosclerosis, depression, AD, systemic lupus erythematosus, prostate cancer, and rheumatoid arthritis. Thus, IL-6 is an important marker for inflammation and should be monitored in the same way, and in parallel with CRP.

Tumor Necrosis Factor Alpha

Tumor Necrosis Factor Alpha (TNF-α) is an intercellular signaling protein, a cytokine, which can be released by multiple types of immune cells in response to cellular damage, stress, or infection. Originally identified as an antitumor compound produced by macrophages (immune cells), TNF-α is required for proper immune surveillance and function. Acting alone or with other inflammatory mediators, TNF slows the growth of many pathogens. It activates the bactericidal effects of neutrophils (white blood cells), and is required for the replication of several other immune cell types. Excessive TNF, however, is thought to lead to a chronic inflammatory state, but a more likely scenario is that it is a marker for an underlying process, including "subclinical" infection. That is, infection that is not easily detected with available tests. It is believed that TNF can increase thrombosis (blood clotting), decrease cardiac contractility, and may be implicated in tumor initiation and promotion. Thus, like other cytokines, this substance may be friend, foe, or both. But, based on our belief about inflammation, it is mostly a friend, and the foe lie under the radar.

Dysregulation of TNF production has been implicated in a variety of human diseases, including AD, cancer, major depression, and inflammatory bowel disease (IBD). While still controversial, studies of depression and IBD are currently being linked by TNF levels. TNF may be produced primarily by macrophages, but it is produced also by a broad variety of cell types including lymphoid cells, mast cells, endothelial cells, cardiac myocytes, adipose tissue, fibroblasts, and neuronal tissue. Large amounts of TNF are released in response to lipopolysaccharide (LPS = dead bacteria cell membranes), other bacterial products, and Interleukin-1 (IL-1). In the skin, mast cells appear to be the predominant source of preformed TNF, which can be released upon inflammatory stimulus (e.g., LPS).

It has a number of actions on various organ systems, generally together with IL-1 and IL-6. Much of this is beyond the scope of this book. However, some of the actions include: suppressing appetite, fever, and increasing insulin resistance. Again, is the TNF a cause or a signal to a deeper cause? A local increase in concentration of TNF will cause the signs of inflammation to occur: heat, swelling, redness, pain, and loss of function. High concentrations of TNF induce shock-like symptoms, whereas, the prolonged exposure to low concentrations of TNF can result in cachexia (physical wasting away with loss of weight and muscle mass). This can be found, for example, in cancer patients and certainly in Alzheimer's patients who frequently are reported to lose muscle but the presumption continues to be a loss of appetite.

There are many examples connecting TNF to Alzheimer's. Consider the following article titled, *Detection of interleukin-1, interleukin-6, and tumor necrosis factor in serum in patients with Alzheimer's disease* [13]. In this Chinese study, 26 patients with confirmed dementia of the

Alzheimer's type and healthy controls were analyzed. The study authors showed that all three of these inflammatory markers were elevated in the Alzheimer's patients compared to the healthy people. They concluded, "Inflammatory cytokines IL-1, IL-6, and TNF play important roles in the neurodegeneration process associated with AD, and the inflammation may contribute to the pathogenesis of AD." Clearly TNF is a systemic blood marker for Alzheimer's but it is probably a friend and not a foe as the authors suggest.

Fibrinogen

Fibrinogen is a protein that makes the blood sticky. During the blood clotting process, the blood vessel walls or the clotting factors in the blood release a chemical into the bloodstream. This causes fibrinogen, an inert protein found in blood plasma, to be converted into fibrin. The fibrin molecule is unique in its ability to link together, forming long threads that wrap around the platelet plug. The threads act much like a spider-web, catching more platelets, red blood cells, and other substances to form a clot. Fibrinogen levels also become elevated with tissue inflammation or tissue destruction. Thus, the fibrinogen test is a reliable measure of the amount of inflammation occurring in the body. High fibrinogen levels may also be an indicator of an increased risk of heart or circulatory disease. The changing levels of fibrinogen can also be used to monitor the course of an ongoing inflammation; decreased levels indicate an improvement, while increases are indicative of a worsening condition. Recent research has shown that fibrin plays a key role in the inflammatory response and development of rheumatoid arthritis.

According to the medical literature, fibrinogen is associated with an increased risk of AD and vascular dementia. In fact, a paper with that exact title was published by researchers at Erasmus University Medical Center, Rotterdam, the Netherlands [14]. The summary of that article is provided here:

> Vascular and inflammatory factors may play an important role in the pathogenesis of dementia. Studies reported an association between plasma levels of inflammation markers and the risk of dementia. Both fibrinogen and C-reactive protein are considered inflammatory markers. Fibrinogen also has important hemostatic properties. We investigated the association of fibrinogen and C-reactive protein with dementia.
>
> Conclusions: High fibrinogen levels were associated with an increased risk of both Alzheimer's disease and vascular dementia.

Unlike CRP, fibrinogen is believed to be a direct contributor to atherosclerosis. The combination of elevated fibrinogen with other vascular disease risk factors can substantially increase disease potential. However, there are currently no medical treatments available to lower fibrinogen levels, according to the standard thinking in medicine. However, when fibrinogen is viewed as a messenger of disease rather than a cause requiring lowering, then the causes of inflammation that contribute to its elevation can be treated and likely lower its levels.

Fibrinogen levels can be measured in venous blood. Normal levels are about 1.5–2.77 g/L, depending on the method used. In typical circumstances, fibrinogen is measured in citrated plasma samples in the laboratory. Higher levels are, among others, associated with cardiovascular disease (>3.43 g/L). It may be elevated in any form of inflammation, as it is an acute-phase protein; for example, it is especially apparent in human gingival tissue during the initial phase of periodontal disease.

Lipoprotein (a)

Lipoprotein(a) [Lp(a)] an inherited abnormal protein attached to LDL, the so-called bad cholesterol. Lp(a) increases coagulation and triples cardiovascular disease risk. Lp(a) may be involved in the recruitment of inflammatory cells. Individuals without Lp(a) or with very low Lp(a) levels seem to be healthy. Thus plasma Lp(a) is certainly not vital, at least under normal environmental conditions. Its function might not be vital but just evolutionarily advantageous under certain environmental conditions, for example, in case of exposure to certain infectious diseases.

Lp(a) is suggested to have a role in AD, in that increased serum concentrations increase the risk for cerebrovascular disease, which may be one avenue to Alzheimer's [15]. Japanese researchers state that Lp(a) is seen to have and expanding role in the development of both vascular and neurological diseases [16]. They site articles in which Lp(a) serum concentrations have been shown to be related to vascular dementia but also site that the relationship to AD remains controversial.

High levels of Lp(a), sometimes referred to as the "bad" form of cholesterol, have been found to be an independent risk factor for heart disease and stroke. This form of "bad" cholesterol can be inherited and can be difficult to lower with diet alone. What's more important is that other cholesterol numbers can be just fine, while the Lp(a) level is still very high. The only way to know is to have blood level of Lp(a) checked.

An ideal Lp(a) level is considered less than 30. Intermediate levels are between 30 and 60, and high levels are considered over 60. It's not known yet what causes a high Lp(a) levels, but genetic factors may be most important. Lp(a) levels may rise after menopause in women. Levels are also increased by the "male" hormone testosterone (although women make testosterone, too—just less of it). Some reports suggest that soy supplementation may also boost Lp(a) levels. Is an appropriate therapy for the lowering of Lp(a) or finding the cause of the elevated Lp(a) and treating that root cause?

Homocysteine

Progressively elevated blood levels of homocysteine are a documented risk marker for cardiovascular events and other chronic diseases of aging. If you want to know more about homocysteine, read some of the writings of Kilmer McCully. Also, consider reading his memoirs [17]. Dr. McCully put forth the homocysteine theory of cardiovascular disease in the 1970s while at Harvard Medical School. Unfortunately for Dr. McCully, timing is everything, and Harvard was already starting to ride the cholesterol train. The result is that Dr. McCully's work was dismissed, as was he. In fact, he was blacklisted and struggled to find his next job. Now with cholesterol diminishing in importance, Dr. McCully is back, and homocysteine is considered the single best biomarker for cardiovascular disease and accelerated aging. However, it is difficult for a doctor to order a homocysteine test for anything but cardiovascular disease within the standard-of-care.

Homocysteine helps illuminate the strong connection between cardiovascular disease and Alzheimer's. A simple google.scholar search of the key words "Alzheimer's" and "homocysteine" yields 21,300 research articles while a search including "cardiovascular" yields 88,300 results. Homocysteine is detected in the systemic blood, illustrating that AD is not just isolated in the brain. The first "hit" in the Google search yields, *Plasma homocysteine as a risk factor for dementia and Alzheimer's disease,* by scientists at BU and Tufts University [18]. The information in this paper comes from the world-famous Framingham Heart Study. The conclusion of this article is that "increased plasma homocysteine level is a strong, independent risk factor for the development of dementia and AD."

If you have a concern about your future risk of AD, please get a homocysteine test. But how? Your doctor will not order it because it is not covered for a presumption of Alzheimer's, and you cannot order it yourself in several states in the United States. You may have to travel abroad just to determine your risk for AD.

More studies demonstrate the connection between elevated levels of homocysteine (hyperhomocysteinemia) and occurrence of AD besides other cognitive impairments. Researchers suggest that B-group vitamin supplementation (including folate) may possibly decrease chances to develop AD. A recent Oxford University Study found that high doses of folic acid, as well as vitamins B6 and B12 helped lower the blood levels of homocysteine and further reduced the associated brain shrinkage by up to 90% in the areas of the brain most affected in Alzheimer's patients [19].

The Oxford study seems a bit too good to be true. Thus, according to the Mayo Clinic, "Vitamin B-12 helps maintain healthy nerve cells and red blood cells. A vitamin B-12 deficiency—most common in older adults and vegetarians—can cause various signs and symptoms, including memory loss. In such cases, vitamin B-12 supplements can help improve memory. In the absence of a vitamin B-12 deficiency, there's no evidence that vitamin B-12 supplements enhance memory for people who have AD. Still, vitamin B-12 remains an important part of a healthy diet." Is the Oxford study wrong? This is a case where more research will help clarify this quandary.

A high level of blood serum homocysteine is a powerful risk factor for cardiovascular disease. However, studies that preceded the Oxford work, which attempted to decrease the risk of cardiovascular disease (not Alzheimer's) by lowering homocysteine, was not fruitful [20–22]. A study was conducted on nearly 5,000 Norwegian heart attack survivors who already had severe, late-stage heart disease. No study has been conducted yet in a preventive capacity on subjects who are in a relatively good state of health. However, Dr. McCully has shown in several research studies that the development of arteriosclerosis requires elevated levels of homocysteine in the blood.

Giving vitamins, in particular, folic acid, B6, and B12, to reduce homocysteine levels may not quickly offer benefit. However, a significant 25% reduction in stroke was found in the "HOPE-2" study, even in patients mostly with existing serious arterial decline, although the overall death rate was not significantly changed by the intervention in the trial [23]. Reducing homocysteine does not quickly repair existing structural damage of the artery architecture. However, science strongly supports the supposition that homocysteine degrades and inhibits formation of three main structural components of the artery, collagen, elastin, and the proteoglycans. Alternatively, could homocysteine be yet another marker rather than a participant in the disease process? Few, if any have tried to truly understand the "first cause" of homocysteine elevation and attempt to do clinical trials on the root cause. And, considering that homocysteine is intimately connected to inflammation, it certainly could be friend and foe.

A possible explanation for this homocysteine conundrum may be provided by research from Duke, titled, *Novel risk factors for stroke: Homocysteine, inflammation, and infection* [24]. The author indicates that the evidence between elevated levels of homocysteine and stroke is fairly strong. They also indicate that patients with chronic inflammation, as well as those with chronic or acute infection, are at elevated stroke risk. If we take these two statements and combine them, then the thesis looks something like this: "Homocysteine is a marker for inflammation that is caused by chronic or acute infection."

The statement earlier is not accepted by mainstream medicine. Instead, they offer their own set of hypotheses to address the failure of homocysteine-lowering therapies to reduce cardiovascular event frequency. One suggestion is that folic acid may directly cause an increased build-up of arterial plaque, independent of its homocysteine-lowering effects. Alternatively, folic acid and vitamin B12 may cause an overall change in gene methylation levels in vascular cells, which may also promote plaque growth. Finally, altering methylation activity in cells might increase methylation of l-arginine to asymmetric dimethylarginine, which can increase the risk of vascular disease. Thus alternative homocysteine-lowering therapies may yet be developed which show greater effects on development and progression of cardiovascular disease. Somehow the concept of chronic low-grade infection at the root of the inflammation process is just not in the current medical vernacular.

The infection hypothesis, seeming overlooked in the medical literature, infers that homocysteine has concomitant beneficial as well as detrimental effects. And homocysteine is as much a marker for other processes as well as being physiologically active. Thus, lowering homocysteine without determining the root cause(s) of homocysteine elevation is likely to have no impact on disease, or worse, have a negative effect. For example, if homocysteine is actually protective in some capacity, then the insult for which homocysteine is providing protection must be treated first before homocysteine-lowering therapy stands a chance of showing health benefits.

Like other tests for inflammation, the homocysteine tests involve obtaining a blood sample. Levels of homocysteine above 11 infer a higher risk of both cardiovascular mortality and AD. A level of 14 is associated with a doubling of the risk for AD. Indeed, homocysteine is considered to be the single best marker for future risk of chronic diseases of aging. Consider reading further about homocysteine, including the book titled, *The H Factor*.

More detailed information on homocysteine and Alzheimer's is found in Chapter 8.

Ceramides

These fatty molecules, particularly when phosphorylated, are plentiful when inflammation is present in the body. They may not be directly tied into the physiology of inflammation, but they are at least like inflammation paparazzi. Diseases with excess ceramides include: Alzheimer's, type 2 diabetes, and insulin resistance. All these diseases have strong inflammatory components.

Ceramides are a family of waxy lipid molecules. A ceramide is composed of sphingosine and a fatty acid. They are found in high concentrations within the cell membrane of cells.

They are one of the component lipids that make up sphingomyelin, one of the major lipids in the lipid bilayer that forms a continuous barrier around cells. Ceramides participate in a variety of cellular signaling; examples include regulating differentiation, proliferation, and programmed cell death (apoptosis) of cells. They also play an important role in inflammatory responses [25]. The so-called C1P (phosphorylated ceramide) is considered a potent proinflammatory agent [26].

Serum ceramides increase the risk of Alzheimer's disease," according to a Mayo Clinic group [27]. Here is an excerpt from their results and conclusions:

> Compared to the lowest tertile, the middle and highest tertiles of ceramide were associated with a 10-fold and 7.6-fold increased risk of AD respectively. Total and high-density lipoprotein cholesterol and triglycerides were not associated with dementia or AD.
>
> Results from this preliminary study suggest that particular species of serum ceramides are associated with incident AD.

Ceramides appear to be emerging as an early diagnostic marker for AD. According to the Mayo team, pathways and products of metabolism of ceramides are altered early in the course of AD.

Some interesting titles linking ceramides and Alzheimer's include:

2010: Alterations of the sphingolipid pathway in Alzheimer's disease: new biomarkers and treatment targets? [28]
2010: Serum sphingomyelins and ceramides are early predictors of memory impairment. [29]
2010: Plasma ceramides are altered in mild cognitive impairment and predict cognitive decline and hippocampal volume loss. [30]
2012: Blood serum miRNA: non-invasive biomarkers for Alzheimer's disease. [31]
2012: The role of ceramides in selected brain pathologies: ischemia/hypoxia, Alzheimer's disease. [32]

Ceramides are fatty molecules. Another word for fat is lipid. Lipids are a group of naturally occurring molecules that include fats, waxes, sterols, fat-soluble vitamins (such as vitamins A, D, E, and K), monoglycerides, diglycerides, triglycerides, phospholipids, and others. The function of lipids includes storing energy, signaling, and acting as structural components of cell membranes. This classic view is now expanding, and the role of lipids in metabolism, not just in energy storage, is emerging. Here are some quick facts on the brain:

- The weight of your brain is about 3 pounds.
- Your skin weighs twice as much as your brain.
- Your brain is made up of about 75% water.
- Your brain consists of about 100 billion neurons.
- There are anywhere from 1,000 to 10,000 synapses for each neuron.
- There are no pain receptors in your brain, so your brain can feel no pain.
- There are 100,000 miles of blood vessels in your brain.
- Your brain is the fattest organ in your body and may consist of at least 60% fat.
- Your brain uses 20% of the total oxygen in your body. For a 150-pound person, that implies that it uses 10 times the energy compared to the average bodily tissue.

Fats play a role in both creating and quelling inflammation, depending upon the type. Balance and moderation is the key to maintain appropriate physiology. The question becomes, if fats are restricted, what takes their place? It is usually carbohydrates of some type, and too much may be contributing to the chronic diseases with which our society is plagued. David Perlmutter, MD is a progressive neurologist who offers a solution through his book, *The Better Brain Book: The Best Tools for Improving Memory and Sharpness and for Preventing Aging of the Brain* [33].

Nuclear Factor Kappa-Beta

Nuclear factor kappa-beta (NF-kB) is important in the initiation of the inflammatory response. When cells are exposed to damage signals (such as TNF-α or oxidative stress), they activate NF-kB, which turns on the expression of over 400 genes involved in the inflammatory response. This staggering number of genes helps us appreciate the complexity of a seemingly simple concept—inflammation. The triggered inflammatory response includes other inflammatory cytokines, as well as proinflammatory enzymes including cyclooxygenase-2 (COX-2) and lipoxygenase. COX-2 is the enzyme responsible for synthesizing proinflammatory prostaglandins, and it is the target of nonsteroidal antiinflammatory drugs (ibuprofen, aspirin) and COX-2 inhibitors (Celebrex and others).

Interleukins and Cytokines

Interleukins and cytokines have many functions in the promotion and resolution of inflammation. Proinflammatory interleukins that have been the subject of most research include IL-1β, IL-6, and IL-8. IL-1β helps immune cells to move out of blood vessels and into damaged or dysfunctional tissues. We already reviewed IL-6 in detail. Recall that it has both proinflammatory and antiinflammatory roles, and it coordinates the production of compounds required during the progression and resolution of acute inflammation. IL-8 is expressed by both immune and nonimmune cells and helps to attract neutrophils (immune cells that can destroy pathogens) to sites of injury.

The cytokine factors mentioned earlier (interleukins, TNF-α) are "long-distance messages." They are produced by cells at the site of inflammation and released into the blood, carrying information about the inflammatory response throughout the body. In contrast, eicosanoids are "local" messages. They are produced by cells that are at the site of inflammation and travel short distances (locally within the same organ, to neighboring cells, or sometimes only to different parts of the same cell, where they elicit immune defenses).

Eicosanoids

There are several families of eicosanoids (including prostaglandins, prostacyclins, leukotrienes, and thromboxanes) that are created by most cell types in all major organ systems. Aside from their roles in inflammation (and antiinflammation), prostaglandins have a variety of functions in cell growth, kidney function, digestion, and the constriction and dilation of blood vessels. Thromboxanes are important mediators of the blood clotting process. Proinflammatory leukotrienes are important for recruiting and activating white blood cells during inflammation, and they are best studied for their role in airway constriction and anaphylaxis.

Cells produce eicosanoids using unsaturated fatty acids that are part of their cell membranes. The fatty acid starting materials for eicosanoid synthesis are the essential fatty acids linoleic acid (omega-6) and its derivative arachidonic acid (AA), as well as alpha-linolenic acid (an omega-3) and its derivatives eicosapentaenoic acid (EPA) and docosahexaenoic acid (DHA). While generalizations about roles of these fatty acids in eicosanoid synthesis should be approached cautiously, the most potent inflammatory eicosanoids are produced from omega-6 fatty acids (linoleic and arachidonic acids). Diets high in omega-3 fatty acids are associated with lower biomarkers of inflammation and cardiovascular disease risk; proposed mechanisms include the production of less inflammatory or antiinflammatory eicosanoids and through the cyclooxygenase and lipoxygenase enzymes (see later).

Cyclooxygenases and Lipoxygenases

The eicosanoids (earlier) require several enzymatic steps to be synthesized from unsaturated fatty acids; the cyclooxygenase (COX) and lipoxygenase (LOX) enzymes catalyze the first steps in these reactions. Cyclooxygenases initiate the conversion of omega-3 and omega-6 derivatives into one of the many prostaglandins or thromboxanes. The interest in COX enzyme metabolism comes from the fact that its inhibition leads to decreased prostaglandin synthesis and therefore a reduction in inflammation, fevers, and pain. The analgesic and antiinflammatory activity of aspirin and the nonsteroidal antiinflammatory drugs (NSAIDS, like ibuprofen and naproxen) is due to their inhibition of COX enzymes.

There are two COX enzymes with well-defined roles in humans (COX-1 and COX-2). COX-2 has the most relevance to the inflammatory process: it is normally inactive, but is turned on during inflammation and stimulates this process of inflammation by creating proinflammatory prostaglandins and thromboxanes.

Lipoxygenases convert fatty acids into proinflammatory leukotrienes, which are important local mediators of inflammation. Several potent inflammatory leukotrienes are produced by 5-LOX in mammals. Lipoxygenase enzymes and the proinflammatory factors they produce have a fundamental role in the inflammatory process by aiding in the recruitment of white blood cells to the site of inflammation. They also stimulate local cells to produce cytokines, which amplifies the inflammatory response. Thus, LOX enzymes may be involved in a wide variety of inflammatory conditions, and represent an additional target for antiinflammatory therapy and diagnosis.

While COX and LOX enzymes are most often associated with proinflammatory processes, it is important to remember that both enzymes also produce factors that inhibit or resolve inflammation and promote tissue repair (including the prostacyclins and lipoxins). The proper transition from the pro- to antiinflammatory activities of the COX and LOX enzymes is important for the progression of a healthy inflammatory response.

Cortisol (Hydrocortisone)

Cortisol is a steroid hormone, more specifically a glucocorticoid, produced in the adrenal gland. It is released in response to stress and a low level of blood glucocorticoids. Its primary functions are to increase blood sugar through gluconeogenesis; suppress the immune system; and aid in fat, protein, and carbohydrate metabolism. It also decreases bone formation. During pregnancy, increased production of cortisol between weeks 30 and 32 initiates production of fetal lung surfactant to promote maturation of the lungs. Various synthetic forms of cortisol are used to treat a variety of diseases.

Cortisol prevents the release of substances in the body that cause inflammation. Thus its presence is indicative of a latent state of inflammation. It stimulates gluconeogenesis (the breakdown of protein and fat to provide metabolites that can be converted to glucose in the liver), and it activates antistress and antiinflammatory pathways.

Alpha 1-Antichymotrypsin

This substance, alpha 1-antichymotrypsin (ACT), is considered as an inflammatory protein. Its association with AD dates back to the 1980s. It is a member of the serine proteinase inhibitor (serpin) family. Alpha 1-antichymotrypsin has been identified as a major constituent of the neurofibrillary plaques associated with AD, and in vitro studies have shown that it enhances the rate of amyloid-fibril formation. These observations and recent genetic evidence suggest that alpha 1-antichymotrypsin is important in the pathogenesis of AD, either as a target for therapy or as a diagnostic marker.

In 1990, Japanese researchers showed how ACT has a strong association with Alzheimer's [34]. The conclusion from the 1990 paper reads, "We concluded that the measurement of serum levels of α1-antichymotrypsin could be useful as a screening marker for Alzheimer's-type dementia. In addition, CSF levels also could be a useful marker for Alzheimer's-type dementia, because they might reflect the state of dementia."

N-terminal Pro-Brain Natriuretic Peptide

N-terminal pro-brain natriuretic peptide (NT-proBNP) is an important cardiovascular risk marker with powerful independent prognostic value for detection of clinical and subclinical cardiac dysfunction. Elevated levels indicate the presence of ongoing myocardial (heart) stress and potentially an underlying cardiac disorder. Since evidence is accumulating that connects inflammation to cardiovascular disease, NT-proBNP could well emerge as a key measure of inflammation and diseases of aging associated with cardiovascular diseases like AD.

NT-proBNP is primarily used as an aid in the diagnosis of congestive heart failure. The test is further indicated for evaluating the level of risk of individuals with acute coronary syndrome and congestive heart failure. The test may also serve as an aid in the assessment of increased risk of cardiovascular events and mortality in individuals at risk for heart failure who have stable coronary artery disease.

Thyroid-Stimulating Hormone

Thyroid-stimulating hormone (TSH or thyrotropin) is a peptide hormone synthesized and secreted by thyrotrope cells in the pituitary gland, which regulates the endocrine function of the thyroid gland. Inflammation of the thyroid gland may occur after a viral illness (subacute thyroiditis). This condition is associated with a fever and a sore throat that often makes it painful to swallow. The thyroid gland is also tender to touch. There may be generalized neck aches and pains. Inflammation of the gland with an accumulation of white blood cells known as lymphocytes (lymphocytic thyroiditis) may also occur. In both of these conditions, the inflammation leaves the thyroid gland "leaky," so that the amount of thyroid hormone entering the blood is increased.

Myeloperoxidase

Myeloperoxidase (MPO) is a peroxidase enzyme. MPO is most abundantly expressed in neutrophil granulocytes (a subtype of white blood cells). It is a lysosomal protein stored in the neutrophil. MPO has a heme (blood like) pigment, which causes its green color in secretions rich in neutrophils (immune cells that can destroy pathogens), such as pus and some forms of mucus.

Recent studies have reported an association between myeloperoxidase levels and the severity of coronary artery disease. It has been suggested that myeloperoxidase plays a significant role in the development of the atherosclerotic lesion and rendering plaques unstable. Again, inflammation is implicated in atherosclerosis and plaque buildup, thus a connection between inflammation and myeloperoxidase clearly exists.

A Cleveland Clinic Foundation 2003 study suggested that MPO could serve as a sensitive predictor for myocardial infarction (heart attack) in patients having chest pain [35]. Since then, there have been over 100 published studies documenting the utility of MPO testing. Most recently, a Canadian group reported that elevated MPO levels more than doubled the risk for cardiovascular mortality over a 13-year period, and measuring both MPO and CRP provided added benefit for risk prediction than just measuring CRP alone [36].

Alzheimer's, Parkinson's, and MS all have a MPO connection. In an article by Texas scientists titled, *Microglia and myeloperoxidase: A deadly partnership in neurodegenerative disease*, the authors reiterate the connection between these diseases and inflammation [37]. "The presence of [MPO] in these diseased brains has been reported by a number of investigators," they stated. They also believe that MPO is involved in the perpetuation of inflammation and is a likely target for treatment. This has yet to be proven, and MPO should be viewed as a marker for inflammation, rather than a target, pending more definitive research. However, with an association between MPO, cardiovascular disease and Alzheimer's illuminated, it makes sense to perform both the MPO and CRP tests together, to get a better picture of current or future disease risk. If an MPO and CRP combination test is more instructive compared to a single test, imagine the diagnostic power of combining more of these tests into your health portfolio.

F2-Isoprostanes

F2-Isoprostanes (F2-IsoPs) are the "gold-standard" for quantifying oxidative stress. Increased free radical-mediated injury to the brain is proposed to be an integral component of several neurodegenerative diseases, including AD. Lipid peroxidation is a major outcome of free radical-mediated injury to brain, where it directly damages membranes and generates a number of oxidized products. F2-IsoPs, one group of lipid peroxidation products derived from arachidonic acid, are especially useful as biomarkers of lipid peroxidation. F2-IsoP concentration is selectively increased in diseased regions of brain from patients who died from advanced AD, where pathologic changes include beta-amyloid, amyloidogenesis, neurofibrillary tangle formation, and extensive neuron death, says a team from U. Penn [38]. There is broad agreement that increased cerebrospinal fluid (CSF) levels of F2-IsoPs also are present in patients with early AD. These results indicate that in CSF, F2-IsoPs may aid in the assessment of laboratory diagnosis of AD.

There are several favorable attributes that make measurement of F2-IsoPs attractive as a reliable indicator of oxidative stress: (1) F2-IsoPs are specific products of lipid peroxidation; (2) they are stable compounds; (3) levels are present in detectable quantities in all normal biological fluids and tissues, allowing the definition of a normal range; (4) their formation increases dramatically in a number of animal models of oxidant injury; (5) their formation is changed by antioxidant status; and (vi) their levels are not affected by lipid content of the diet. Measurement of F2-IsoPs in plasma can be utilized to assess total production of F2-IsoPs.

The U. Penn authors show a profound understanding for AD. They state:

> Alzheimer's disease (AD) includes a group of dementing neurodegenerative disorders that have diverse etiologies but the same hallmark brain lesions. Since oxidative stress may play a role in the pathogenesis of AD and isoprostanes are chemically stable peroxidation products of arachidonic acid, we measured both iPF2α-III and iPF2α -VI using gas chromatography-mass spectrometry in AD and control brains.
>
> The levels of both isoprostanes ... were markedly elevated in both frontal and temporal poles of AD brains compared to the corresponding cerebella. ... These data suggest that specific isoprostane analysis may reflect increased oxidative stress in AD.
>
> Oxidative stress is implicated as one of the major underlying mechanisms behind many acute and chronic diseases and is involved in normal aging. However, the measurement of free radicals or their end products is complicated. Thus, proof of association of free radicals in disease conditions has been absent. Isoprostanes are now considered to be reliable biomarkers of oxidative stress, as evidenced by a study organized by the National Institutes of Health. A number of these compounds have potent biologic activities including certain inflammatory properties. Isoprostanes are involved in many human diseases.

Measurement of bioactive F2-isoprostanes in body fluids offers a unique analytic test to study the role of free radicals in physiology, oxidative stress–related diseases, acute or chronic inflammatory conditions, and also in the assessment of various antioxidants, radical scavengers, and drugs.

Be careful how you interpret the conclusions about oxidative stress. Currently, antioxidant therapy is very much in vogue. However, our body is an engine, not unlike an automobile. Life is an oxidative process. The first line of defense our immune system presents is white blood cells that produce peroxide to kill invaders. What is peroxide? It is oxygen with a free radical. If you overdue antioxidant supplement, you may inhibit your immune system, with potentially dire consequences. Please consider an antiinflammation, as opposed to an antioxidant program. This strategy will reduce your oxidative stress burden without compromising your immune system. Antiinflammation does not mean NSAIDS and other antiinflammatory drugs either. Read Chapter 11, the work of Paul Clayton, and Dr. Sears's book, *The Inflammation Zone*, to better understand an antiinflammation approach.

Adiponectin

The traditional role attributed to fat tissue is energy storage. However, a change in perspective has risen since the discovery of the adipokines (cytokines secreted by adipose tissue). Advances in the biology of the adipose tissue has revealed that it is not simply an energy storing organ but also secretes a variety of molecules, termed adipokines, that affect processes beyond metabolic regulation.

Adiponectin is a protein hormone that is involved in a number of metabolic processes, including glucose regulation and fatty acid metabolism. Adiponectin is exclusively secreted from adipose tissue into the bloodstream and is very abundant in plasma relative to many hormones. Levels of the hormone are inversely correlated with body fat percentage in adults. The hormone plays a role in the suppression of the metabolic processes that may result in type 2 diabetes, obesity, atherosclerosis, and an independent risk factor for metabolic syndrome, all of which have inflammatory components. Adiponectin in combination with leptin has been shown to completely reverse insulin resistance in mice. An emerging view is that insulin resistance is an inflammatory disorder, not a sugar regulation disorder as it is currently managed under the standard-of-care.

Circulating levels of adiponectin decrease with increasing obesity and are lower in patients with type 2 diabetes, metabolic syndrome, and cardiovascular disease compared with controls matched by body mass index. Several reports demonstrated antiinflammatory effects of adiponectin. Because increased adipose tissue is associated with low-grade chronic inflammation, and proinflammatory factors inhibit adiponectin production, the current hypothesis states that chronic inflammation associated with visceral (fat around internal organs) obesity inhibits production of adiponectin, perpetuating inflammation. The negative correlation between adiponectin and markers of inflammation supports this idea.

In contrast, adiponectin levels are elevated rather than decreased in inflammatory conditions not associated with excess weight. This occurs in classic chronic inflammatory/autoimmune diseases that are unrelated to increased adipose tissue, such as rheumatoid arthritis, inflammatory bowel disease, type 1 diabetes, and cystic fibrosis. In these patients, adiponectin levels positively, rather than negatively, correlate with inflammatory markers. Furthermore, proinflammatory effects of adiponectin have been reported in tissues, such as joint synovium and colonic epithelium. Thus, adiponectin is regulated in the opposite direction and may exert different functions in classic versus obesity-associated inflammatory conditions.

Adiponectin exerts some of its effects via the brain. Consider a paper by Japanese scientists published in 2011 titled, *Adiponectin in plasma and cerebrospinal fluid in MCI and Alzheimer's disease* [39]. The authors sought to study the connection between molecules released from adipose (fat) tissue and AD. Specifically, they investigated the following:

> Life style-related disorders such as hypertension, diabetes, dyslipidemia, and obesity are reported to be a great risk of dementia. Adipocytokines released from adipose tissue are thought to modulate some brain functions including memory and cognition. We here analyzed adiponectin, one of the most important adipocytokines, in plasma and cerebrospinal fluid (CSF) from cognitive normal controls (NC), mild cognitive impairment (MCI) subjects, and patients with Alzheimer's disease (AD) and discussed if/how adiponectin could relate to the pathogenesis of AD.

Results: The levels of adiponectin in plasma and in CSF showed a positive correlation. Plasma adiponectin was significantly higher in MCI and AD compared to NC (normal controls), whereas CSF adiponectin was significantly higher in MCI compared to NC.

Conclusion: It is possible that the level of adiponectin in plasma reflects its level in CSF. The tendency to have higher adiponectin in plasma and CSF from MCI and AD suggests that this molecule plays a critical role in the onset of AD.

The lesson learned is that fat tissue is not just an energy storage reserve. In this tissue are some potentially destructive molecules that are associated with diseases of aging, such as AD. What is yet to be established is the possibility that these molecules may have protective as well as deleterious properties, like many of the inflammatory molecules that have been discussed so far. A test for adiponectin involves taking a blood sample drawn from a vein in the patient's arm. A single test, thus a single value of adiponectin level, like any of the markers of inflammation presented here, is of little absolute diagnostic value. This parameter should be measured annually, or more frequently in anyone suffering from chronic disease.

β2-Microglobulin

β2-Microglobulin (β2-M) is a measure of the activity of the acquired immune system and can provide information about infection and inflammation. It is a protein and was originally purified from the urine of a patient with renal disease. β2-M is present in particularly high concentrations on the surface of lymphocytes (small white blood cells that play a large role in defending the body against disease). Serum levels of β2-M are a function of its rate of synthesis or release into the serum pool and of its rate of clearance. In inflammatory disorders (e.g., rheumatoid arthritis, systemic lupus erythematosus, Sjögren's Syndrome, and Crohn's disease) and in lymphoproliferative diseases (e.g., multiple myeloma, β cell lymphoma, and chronic lymphocytic leukemia), serum β2-M levels are frequently elevated, reflecting an increased rate of synthesis. The link between β2-M and Alzheimer's is a bit of a stretch, at the moment. But we cannot exclude any parameter from consideration owing to how little we know about AD. Since it is tied to inflammation and immune response, it is worth considering.

There is precedence in the literature for an association with AD. A British and Finnish team wrote, *Evaluation of CSF cystatin C, beta-2-microglobulin, and VGF as diagnostic biomarkers of Alzheimer's disease using SRM*, in the *Journal of the Alzheimer's Association* [40]. The conclusion of that research is: "Application of the SRM assay supports the initial discovery of B2M, VGF, and cystatin-C as potential CSF biomarkers of AD." That is, β2-M is a marker for Alzheimer's, at least in the cerebral spinal fluid.

White Blood Cells

White blood cells are part of your immune system that fights infections and diseases. Abnormal white blood cell levels may be a sign of infection, blood cancer, or an immune system disorder. Today in medicine, ranges of acceptable white blood cell counts on the upper end are 10,000. Many studies indicate that levels of 7,800 or even lower in women, for example, represent a high risk for cardiovascular mortality [41,42]. This fact is largely ignored in the standard-of-care and helps explain why cardiovascular disease is our #1 killer.

A 2013 paper explains the connection between AD, inflammation, and the various components of our blood that comprise the innate immune system [43]. The abstract brilliantly captures the truth about how a simple blood test can tell us much more than fancy instrumentation.

> Inflammation is part of the complex biological response of vascular tissues to harmful stimuli, such as pathogens, damaged cells, or irritants. This is a mechanism of innate immunity, which may cause an increase in the number of monocytes (a type of white blood cell) and neutrophils (the most abundant type of white blood cells of the innate immune system) circulating in the blood. Literature indicated that chronic inflammation might be a factor in developing neurological problems, including Alzheimer's, Parkinson's, and other similar illnesses.
>
> Our main objective is to identify peripheral markers of Alzheimer's disease and for that purpose, we are looking at the profile of white blood cells focusing on monocytes, neutrophils, lymphocytes (white blood cells—adaptive immune system) and basophils (more rare white blood cell type).
>
> 27 patients of Alzheimer's disease (AD) diagnosed by magnetic resonance imaging and neuropsychological tests were observed for their blood profile. Key observations during this study were that the levels of monocytes in the blood of the diagnosed AD patients were high regardless of their age and sex. For those patients, whose monocytes were in normal range, their neutrophil levels were significantly high, whereas blood levels of lymphocytes and basophils were found to be constantly low. Escalated levels of monocytes and neutrophils are hallmarks of chronic inflammation and may be a precursor to AD.
>
> A low lymphocyte count specifies that the body's resistance to fight infection is substantially reduced, whereas low basophil levels indicate their over utilization due to chronic allergic inflammatory condition. Likewise, blood glucose and creatinine levels were high whereas calcium ions were low. Our studies indicated that white blood cells along with other inflammatory byproducts may act as peripheral markers for early diagnosis of Alzheimer's disease.

SELAH! (Note: Wikipedia indicates this biblical word is hard to translate. In this context, using modern vernacular it simply means, "boom!")

Eosinophil granulocytes, usually called eosinophils or eosinophiles (or, less commonly, acidophils), are white blood cells that are one of the immune system components responsible for combating multicellular parasites and certain infections in vertebrates. Along with mast cells, they also control mechanisms associated with allergy and asthma. They are granulocytes that develop during hematopoiesis (formation of blood cells) in the bone marrow before migrating into blood.

Natural Killer (NK) Cells: NK cells are a part of innate immune system and play a major role in defending the host from both tumors and virally infected cells. NK cells distinguish infected cells and tumors from normal and uninfected cells by recognizing changes of a surface molecule called MHC (major histocompatibility complex) class I. NK cells are activated in response to a family of cytokines called interferons. Activated NK cells release cytotoxic (cell-killing) granules, which then destroy the altered cells. They were named "natural killer cells" because of the initial notion that they do not require prior activation in order to kill cells which are missing MHC class I.

T cells (Thymus cells) and B cells (Bone cells) are the major cellular components of the adaptive immune response. T cells are involved in cell-mediated immunity whereas B cells are primarily responsible for humoral immunity (relating to antibodies). The function of T cells and B cells is to recognize specific "nonself" antigens, during a process known as antigen presentation. Once they have identified an invader, the cells generate specific responses that are tailored to maximally eliminate specific pathogens or pathogen infected cells. B cells respond to pathogens by producing large quantities of antibodies, which then neutralize foreign objects like bacteria and viruses.

In response to pathogens some T cells, called T helper cells, produce cytokines that direct the immune response while other T cells, called cytotoxic T cells, produce toxic granules that contain powerful enzymes which induce the death of pathogen infected cells. Following activation, B cells and T cells leave a lasting legacy of the antigens they have encountered, in the form of memory cells. Throughout the lifetime of an animal, these memory cells will "remember" each specific pathogen encountered, and they are able to mount a strong and rapid response if the pathogen is detected again.

Blood Glucose

Glucose is a type of sugar that the body uses for energy. Abnormal glucose levels in blood may be a sign of diabetes. Diabetes is an inflammatory disorder that can be crudely described as an energy crisis in the body (and the brain). Diabetes has been known for a long time but it is now at pandemic levels. For example, recent estimates put the level of diabetes sufferers in China at 115,000,000. Many of the researchers whose ideas are expressed in this chapter indicate that all these chronic inflammatory diseases are interconnected. Thus, Alzheimer's has roots in diabetes, insulin resistance, and metabolic syndrome.

Many researchers are labeling Alzheimer's disease type 3 diabetes. A scientific review by Brown University titled, *Alzheimer's Disease Is Type 3 Diabetes–Evidence Reviewed*, shed light on the connection [44]. The authors state, "Currently, there is a rapid growth in the literature pointing toward insulin deficiency and insulin resistance as mediators of AD-type neurodegeneration..." Based on the body of evidence, the authors ultimately concluded, "the term 'type 3 diabetes' accurately reflects the fact that AD represents a form of diabetes that selectively involves the brain and has molecular and biochemical features that overlap with both type 1 diabetes mellitus and T2DM (type 2 diabetes mellitus)." This concept is covered in more detail later.

Insulin and Insulin Resistance

Insulin is a hormone produced by the pancreas, which is central to regulate carbohydrate and fat metabolism in the body. Insulin causes cells in the liver, muscle, and fat tissue to take up glucose from the blood, storing it as glycogen in the liver and muscle.

Insulin stops the use of fat as an energy source by inhibiting the release of glucagon. With the exception of the metabolic disorder diabetes mellitus and metabolic syndrome, insulin is provided within the body in a constant proportion to remove excess glucose from the blood, which otherwise would be toxic. When blood glucose levels fall below a certain level, the body begins to use stored sugar as an energy source through glycogenolysis, which breaks down the glycogen stored in the liver and muscles into glucose, which can then be utilized as an energy source. The level of insulin is central to metabolic control mechanisms. It is also used as a control signal to other body systems (such as amino acid uptake by body cells). In addition, it has several other anabolic effects throughout the body.

When control of insulin levels fails, diabetes mellitus will result. As a consequence, insulin is used medically to treat some forms of diabetes mellitus. Patients with type 1 diabetes depend on external insulin (most commonly injected subcutaneously) for their survival because the hormone is no longer produced internally. Patients with type 2 diabetes are often insulin resistant and, because of such resistance, may suffer from a "relative" insulin deficiency. Some patients with type 2 diabetes may eventually require insulin if other medications fail to control blood glucose levels adequately. Over 40% of those with type 2 diabetes require insulin as part of their diabetes management plan. Once insulin enters the human brain, it enhances learning and memory and benefits verbal memory in particular. Enhancing brain insulin signaling by means of intranasal insulin administration enhances temperature regulation and glucose levels in response to food intake. This suggests that central nervous insulin contributes to the control of whole-body energy regulation. In layman's terms, this means that the control center for insulin is the brain.

> Is it any wonder, considering the brain is but 2% of body mass while consuming at least 20% of all body energy? When our bodies are insulin resistant, the brain is severely impacted.

Insulin resistance normally refers to reduced glucose-lowering effects of insulin. However, other functions of insulin can also be affected. For example, insulin resistance in fat cells reduces the normal effects of insulin on lipids and results in reduced uptake of circulating lipids and increased hydrolysis of stored triglycerides. Increased mobilization of stored lipids in these cells elevates free fatty acids in the blood plasma. Elevated blood fatty-acid concentrations (associated with insulin resistance and diabetes mellitus type 2), reduced muscle glucose uptake, and increased liver glucose production all contribute to elevated blood glucose levels. High plasma levels of insulin and glucose due to insulin resistance are a major component of the metabolic syndrome. If insulin resistance exists, more insulin needs to be secreted by the pancreas. If this compensatory increase does not occur, blood glucose concentrations increase and type 2 diabetes occurs.

Inflammation is now implicated in causing insulin resistance. A 2008 UC San Diego review titled, *Inflammation and Insulin Resistance* is just one of many papers with the exact same title [45]. The abstract of that paper is as follows:

> Obesity-induced chronic inflammation is a key component in the pathogenesis of insulin resistance and the metabolic syndrome. In this review, we focus on the interconnection between obesity, inflammation, and insulin resistance... While the initiating factors of this inflammatory response remain to be fully determined, chronic inflammation in these tissues could cause localized insulin.

Even though this paper focuses on obesity induced insulin resistance, the culprits come from the usual list of proinflammatory cytokines. This infers that any process of inflammation may contribute to insulin resistivity assuming enough triggering factors are also involved. Proof of the broad connection between inflammation and insulin resistance is illustrated through a scholar.google search. The key words "insulin resistance" and "inflammation" yields 196,000 hits ($98,000,000,000 spent on this research). With a "title only" search using the same keywords, 539 articles are found.

It's no surprise that there is research titled, *Insulin resistance and Alzheimer's disease* [46]. Insulin resistance is clearly a key link between Alzheimer's and inflammation.

RISK FACTORS FOR CHRONIC INFLAMMATION

There are several risk factors that increase the likelihood of creating and perpetuating a low-level inflammatory response.

Age: In contrast to younger individuals, older adults can have consistently elevated levels of several inflammatory molecules, especially IL-6, CRP, and TNF-α. Our youth is rapidly catching up to their parents and grandparents as inflammatory diseases like type 2 diabetes expands. These elevations are observed even in healthy older individuals. While the reasoning for this age-associated increase in inflammatory markers is not thoroughly understood, it may reflect cumulative mitochondrial dysfunction and oxidative damage or may be the result of other risk factors associated with age, such as increases in visceral (abdominal) body fat, increases in infectious burdens from periodontal, and other smoldering disease, or reductions in sex hormones.

Obesity: Fat tissue is an endocrine organ, storing and secreting multiple hormones and cytokines into circulation and affecting metabolism throughout the body. For example, fat cells produce and secrete both TNF-α and IL-6, and visceral (abdominal) fat can produce these inflammatory molecules at levels sufficient to induce a strong inflammatory response (see Insulin Resistance earlier). Visceral fat cells can produce 3 times the amount of IL-6 as fats cells elsewhere, and in overweight individuals, may be producing up to 35% of the total IL-6 in the body. Fat tissue can also be infiltrated by macrophages, which secrete proinflammatory cytokines. This accumulation of macrophages appears to be proportional to body mass index (BMI) and appear to be a major cause of low-grade, systemic inflammation, and insulin resistance in obese individuals.

Diet: If this section was written 10 years ago, it might have stated, "A diet high in saturated fat is associated with higher pro-inflammatory markers, particularly in diabetic or overweight individuals." This is simply false. A real culprit driving inflammation is an overabundance of omega-6 fatty acids in our diets. We evolved based on a marine diet, and to be optimally healthy the "right" omega-6/omega-3 ratio is about 3 to 1. Although few of us know our ratio, the average American is probably at 10 to 1 or higher. The average unhealthy American is likely over a 20 to 1 ratio. The omega-6 acids are extremely important to life. However, at >5 to 1 relative to omega-3, omega-6 acids promote chronic inflammation. Diets high in synthetic trans-fats (such as those produced by hydrogenation) have been associated with increases in inflammatory markers (IL-6, TNF-α, IL-8, CRP). The increases in markers of inflammation due to synthetic trans-fats may be more pronounced in individuals that are also overweight.

General dietary over-consumption is a major contributor to inflammation and other detrimental age-related processes in the modern world. Therefore, eating a calorie-restricted diet is an effective means of relieving physiologic stressors. Indeed, several studies show that calorie restriction provides powerful protection against inflammation. However, the choice of calories is of utmost important. Our society currently puts an overemphasis on trans-fat and "low fat" for fear of gaining "fat." However, both proteins and carbohydrates are more proinflammatory compared to healthy fats, thus fats must not be avoided.

Foods rich in micronutrients are truly missing in our diets. Appreciating the diet of the mid-Victorian British, as explained by Paul Clayton, provides the information we all need to live Alzheimer's-free [1].

In their book *Ultra-Prevention*, the Drs. Liponis and Hyman recommend a 6-week program to optimal health that includes a major focus on diet [47]. They state that the starting point of a diet-focused approach to health is to "clear the sludge." They recommend:

> Remove anti-nutrients to overcome malnutrition and give your body the support it needs. Anti-nutrients are compounds in food that can cause more harm than good when we eat them. Examples of anti-nutrients include hydrogenated and saturated fats, sugar, refined grains and starches, caffeine, alcohol, and carcinogens. The above is just the introduction to the section and, of course, contains valuable details worth reading (minus the statement about saturated fats).

Low sex hormones. Among their many roles in biology, sex hormones also modulate the immune/inflammatory response. The cells that mediate inflammation (such as neutrophils and macrophages) have receptors for estrogens and androgens that enable them to selectively respond to sex hormone levels in many tissues.

Exhaustive research has shown that testosterone and estrogen can repress the production and secretion of several proinflammatory markers, including IL-1β, IL-6, TNF-α, and the activity of NF-kb. These observations have been corroborated by observational studies that have linked lower testosterone levels in elderly men to increases in inflammatory markers. Several studies have shown an increase in inflammatory IL-1β, IL-6, and TNF-α following surgical or natural menopause.

Conversely, the preservation of sex hormone levels is associated with reductions in the risk of several inflammatory diseases, including atherosclerosis, asthma in women, and rheumatoid arthritis in men. Hormone replacement therapy (HRT) may partially exert its protective effects through an attenuation of the inflammatory response. Reductions in the risks of coronary heart disease and inflammatory bowel disease in some individuals, as well as levels of some circulating inflammatory cytokines (including IL-1B, IL-8, and TNF-α) is observed in some studies of women on HRT.

Atwood and coworkers put forth a comprehensive theory on aging and age-related diseases, such as AD where hormone imbalance and deficiency is at the root [48]. Consider reading his article with the following provocative title, *Living and Dying for Sex: A Theory of Aging Based on the Modulation of Cell Cycle Signaling by Reproductive Hormones* [49].

Smoking: Cigarette smoke contains several inducers of inflammation, particularly reactive oxygen species. Chronic smoking increases production of several proinflammatory cytokines (TNF-α, IL-1β, IL-6, IL-8), while simultaneously reducing production of antiinflammatory molecules. Smoking also increases the risk of periodontal disease, an independent risk factor for increasing systemic inflammation.

Sleep Disorders: Production of inflammatory cytokines (TNF-α and IL-1β) appears to follow a circadian rhythm and may be involved in the regulation of sleep. Disruption of normal sleep can lead to daytime elevations of these proinflammatory molecules. Plasma levels of TNF-α and/or IL-6 were elevated in patients with excessive daytime sleepiness, including those with sleep apnea and narcolepsy. These elevations in cytokines were independent of body mass index or age, although persons with higher visceral body fat were more likely to have sleep disorders.

Infection: Periodontal disease is a major preventable cause of inflammatory diseases. Pathogens can produce a systemic inflammatory response that may affect several other systems, such as the heart and kidneys. It is by this mechanism that periodontal disease is thought to be a risk factor for cardiovascular diseases. Suggested reading is an excellent treatise called *Oral and Whole Body Health,* produced and published by *Scientific American Magazine* as a separate publication [50]. This document can be found by Googling the title. The oral cavity, it turns out, is an incubator for disease like no other place in the body. The kitchen is arguably the dirtiest place in your home, not the bathroom. So too, the mouth the dirtiest place in your body because of the presence of decaying food, oxygen, and water. It is the perfect environment to foster bacterial growth. To compound the problem, 93% of Americans do not get a regular dental cleaning.

Infections associated with periodontal disease are just the tip of the iceberg, and this topic is discussed in detail in a later chapter. Toxoplasmosis is another infectious species and may be the #1 most under diagnosed health problem in the United States based on information provided by the Centers for Disease Control. If it is in the United States, it is likely prevalent around the globe. Indeed, a Swedish group wrote a paper titled, *Toxoplasmosis. The most common parasitic infection in Europe, but not fully understood and probably underdiagnosed* [51]. Should your loved one with diseases of inflammation be tested for Toxoplasmosis?

Stress (both physical and emotional) can lead to inflammatory cytokine release (IL-6); stress is also associated with decreased sleep and increased body mass (stimulated by release of the stress hormone cortisol), both of which are independent causes of inflammation.

PUFA 3, PUFA 6, and PUFA6/3 Ratio: Omega-3 fats, also known as PUFA 3 (polyunsaturated fatty acids), are a small group of molecules technically known as fatty acids. The key clinical omega-3 fats are EPA and DHA, which are found largely in cold-water fish. It is well established in current literature that a higher blood level of these important fats may help to reduce the risk of heart disease and stroke.

As discussed earlier, the PUFA 6/3 ratio is important as PUFA 6 tends to promote proinflammatory molecules while PUFA 3 does just the opposite. High ratios (more PUFA 6) (>5) are associated with chronic silent inflammation. The following is a very concise abstract that discusses the history of our diets with respect to PUFAs in general and provides information on diseases impacted by an "imbalance" of PUFAs. The paper was published in 2008 by The Center for Genetics, Nutrition and Health, Washington, DC and was titled, *The Importance of the Omega-6/Omega-3 Fatty Acid Ratio in Cardiovascular Disease and Other Chronic Diseases* [52].

Several sources of information suggest that human beings evolved on a diet with a ratio of omega-6 to omega-3 essential fatty acids (EFA) of ~1 whereas in Western diets the ratio is 15/1–16.7/1. Western diets are deficient in omega-3 fatty acids and have excessive amounts of omega-6 fatty acids compared with the diet on which human beings evolved and their genetic patterns were established. Excessive amounts of omega-6 polyunsaturated fatty acids (PUFA) and a very high omega-6/omega-3 ratio, as is found in today's Western diets, promote the pathogenesis of many diseases, including cardiovascular disease, cancer, and inflammatory and autoimmune diseases, whereas increased levels of omega-3 PUFA (a lower omega-6/omega-3 ratio) exert suppressive effects.

In the secondary prevention of cardiovascular disease, a ratio of 4/1 was associated with a 70% decrease in total mortality. A ratio of 2.5/1 reduced rectal cell proliferation in patients with colorectal cancer, whereas a ratio of 4/1 with the same amount of omega-3 PUFA had no effect. The lower omega-6/omega-3 ratio in women with breast cancer was associated with decreased risk. A ratio of 2–3/1 suppressed inflammation in patients with rheumatoid arthritis, and a ratio of 5/1 had a beneficial effect on patients with asthma, whereas a ratio of 10/1 had adverse consequences.

These studies indicate that the optimal ratio may vary with the disease under consideration. This is consistent with the fact that chronic diseases are multigenic and multifactorial. Therefore, it is quite possible that the therapeutic dose of omega-3 fatty acids will depend on the degree of severity of disease resulting from the genetic predisposition. A lower ratio of omega-6/omega-3 fatty acids is more desirable in reducing the risk of many of the chronic diseases of high prevalence in Western societies, as well as in the developing countries.

A relatively simple blood test for lipids provides the information to determine the PUFA 6/3 ratio. For example, Life Extension Foundation provides an "omega score" test that provides the PUFA ratio (catalog number LCOMEGA).

INFLAMMATION AND CARDIOVASCULAR DISEASE

Cardiovascular disease information deserves a prominent position on any book about Alzheimer's because:

1. The two diseases have many of the same root causes (and many of the references included in this book corroborate this); and
2. AD is following the same (futile) track that cardiovascular disease followed (and continues to follow) in the standard-of-care. Therefore, there are many lessons to be learned by understanding the history of cardiovascular disease. Hopefully medicine will correct the errant path it is currently pursuing for AD.

What follows is an admission by a cardiac surgeon. Do you see a neurologist making the same kind of admission 10–20 years from now about AD? You can almost take out any reference to cardiovascular disease in the following "admission" and replace it with AD, and the message would not be altered significantly.

Dr. Dwight Lundell is the past Chief of Staff and Chief of Surgery at Banner Heart Hospital, Mesa, AZ. He is a controversial figure for a variety of reasons, one of them being that he speaks out against the standard-of-care. His summary is reproduced here because much of what he says is corroborated by the research presented in this chapter. Please note that the authors do not completely agree with this description of the action of chronic inflammation in heart disease provided by Dr. Lundell. However, this explanation appears substantially correct and is helpful in understanding cardiovascular disease. For a comprehensive physiological explanation of cardiovascular disease published in a peer-reviewed journal, please read the work of Dr. Kilmer McCully [53,53a], and the article from 2015 posted in FineForme:

> We physicians with all our training, knowledge, and authority often acquire a rather large ego that tends to make it difficult to admit we are wrong. So, here it is. I freely admit to being wrong. As a heart surgeon with 25 years of experience, having performed over 5,000 open-heart surgeries, today is my day to right the wrong with medical and scientific fact.

I trained for many years with other prominent physicians labeled "opinion makers." Bombarded with scientific literature, continually attending education seminars, we opinion makers insisted heart disease resulted from the simple fact of elevated blood cholesterol.

The only accepted therapy was prescribing medications to lower cholesterol and a diet that severely restricted fat intake. The latter of course we insisted would lower cholesterol and heart disease. Deviations from these recommendations were considered heresy and could quite possibly result in malpractice.

It is not working!

These recommendations are no longer scientifically or morally defensible. The discovery a few years ago that inflammation in the artery wall is the real cause of heart disease is slowly leading to a paradigm shift in how heart disease and other chronic ailments will be treated.

The long-established dietary recommendations have created epidemics of obesity and diabetes, the consequences of which dwarf any historical plague in terms of mortality, human suffering, and dire economic consequences.

Despite the fact that 25% of the population takes expensive statin medications and despite the fact we have reduced the fat content of our diets, more Americans will die this year of heart disease than ever before.

Statistics from the American Heart Association show that 75 million Americans currently suffer from heart disease, 20 million have diabetes and 57 million have pre-diabetes. These disorders are affecting younger and younger people in greater numbers every year.

Simply stated, without inflammation being present in the body, there is no way that cholesterol would accumulate in the wall of the blood vessel and cause heart disease and strokes. Without inflammation, cholesterol would move freely throughout the body as nature intended. It is inflammation that causes cholesterol to become trapped.

Inflammation is not complicated—it is quite simply your body's natural defense to a foreign invader such as a bacteria, toxin, or virus. The cycle of inflammation is perfect in how it protects your body from these bacterial and viral invaders. However, if we chronically expose the body to injury by toxins or foods the human body was never designed to process, a condition occurs called chronic inflammation. Chronic inflammation is just as harmful as acute inflammation is beneficial.

Dr. Dwight Lundell continues.

What thoughtful person would willfully expose himself repeatedly to foods or other substances that are known to cause injury to the body? Well, smokers perhaps, but at least they made that choice willfully.

The rest of us have simply followed the recommended mainstream diet that is low in fat and high in polyunsaturated fats and carbohydrates, not knowing we were causing repeated injury to our blood vessels. This repeated injury creates chronic inflammation leading to heart disease, stroke, diabetes, and obesity.

Let me repeat that: The injury and inflammation in our blood vessels are caused by the low-fat diet recommended for years by mainstream medicine.

What are the biggest culprits of chronic inflammation? Quite simply, they are the overload of simple, highly processed carbohydrates (sugar, flour, and all the products made from them) and the excess consumption of omega-6 vegetable oils like soybean, corn, and sunflower that are found in many processed foods.

Take a moment to visualize rubbing a stiff brush repeatedly over soft skin until it becomes quite red and nearly bleeding. If you kept this up several times a day, every day for five years. If you could tolerate this painful brushing, you would have a bleeding, swollen infected area that became worse with each repeated injury. This is a good way to visualize the inflammatory process that could be going on in your body right now.

Regardless of where the inflammatory process occurs, externally or internally, it is the same. I have peered inside thousands upon thousands of arteries. A diseased artery looks as if someone took a brush and scrubbed repeatedly against its wall. Several times a day, every day, the foods we eat create small injuries compounding into more injuries, causing the body to respond continuously and appropriately with inflammation.

While we savor the tantalizing taste of a sweet roll, our bodies respond alarmingly as if a foreign invader arrived declaring war. Foods loaded with sugars and simple carbohydrates, or processed with omega-6 oils for long shelf life have been the mainstay of the American diet for six decades. These foods have been slowly poisoning everyone.

How does eating a simple sweet roll create a cascade of inflammation to make you sick?

Imagine spilling syrup on your keyboard and you have a visual of what occurs inside the cell. When we consume simple carbohydrates such as sugar, blood sugar rises rapidly. In response, your pancreas secretes insulin whose primary purpose is to drive sugar into each cell where it is stored for energy. If the cell is full and does not need glucose, it is rejected to avoid extra sugar gumming up the works.

When your full cells reject the extra glucose, blood sugar rises producing more insulin and the glucose converts to stored fat.

What does all this have to do with inflammation? Blood sugar is controlled in a very narrow range. Extra sugar molecules attach to a variety of proteins that in turn injure the blood vessel wall. This repeated injury to the blood vessel wall sets off inflammation. When you spike your blood sugar level several times a day, every day, it is exactly like taking sandpaper to the inside of your delicate blood vessels.

Dr. Dwight Lundell continues.

While you may not be able to see it, rest assured it is there. I saw it in over 5,000 surgical patients spanning 25 years who all shared one common denominator—inflammation in their arteries.

Let's get back to the sweet roll. That innocent looking goody not only contains sugars, it is baked in one of many omega-6 oils such as soybean. Chips and fries are soaked in soybean oil; processed foods are manufactured with omega-6 oils for longer shelf life. While omega-6s are essential—they are part of every cell membrane controlling what goes in and out of the cell — they must be in the correct balance with omega-3's.

If the balance shifts by consuming excessive omega-6, the cell membrane produces chemicals called cyto-kines that directly cause inflammation.

Today's mainstream American diet has produced an extreme imbalance of these two fats. The ratio of imbalance ranges from 15:1 to as high as 30:1 in favor of omega-6. That's a tremendous amount of cytokines causing inflammation. In today's food environment, a 3:1 ratio would be optimal and healthy.

To make matters worse, the excess weight you are carrying from eating these foods creates overloaded fat cells that pour out large quantities of pro-inflammatory chemicals that add to the injury caused by having high blood sugar. The process that began with a sweet roll turns into a vicious cycle over time that creates heart disease, high blood pressure, diabetes and finally, Alzheimer's disease, as the inflammatory process continues unabated.

There is no escaping the fact that the more we consume prepared and processed foods, the more we trip the inflammation switch little by little each day. The human body cannot process, nor was it designed to consume, foods packed with sugars and soaked in omega-6 oils.

There is but one answer to quieting inflammation, and that is returning to foods closer to their natural state. To build muscle, eat more protein. Choose carbohydrates that are very complex, such as colorful fruits and vegetables. Cut down on or eliminate inflammation- causing omega-6 fats like corn and soybean oil and the processed foods that are made from them.

One tablespoon of corn oil contains 7,280 mg of omega-6; soybean contains 6,940 mg. Instead, use olive oil or butter from grass-fed beef.

Animal fats contain less than 20% omega-6 and are much less likely to cause inflammation than the supposedly healthy oils labeled polyunsaturated. Forget the "science" that has been drummed into your head for decades. The science that saturated fat alone causes heart disease is non-existent. The science that saturated fat raises blood cholesterol is also very weak. Since we now know that cholesterol is not the cause of heart disease, the concern about saturated fat is even more absurd today.

The cholesterol theory led to the no-fat, low-fat recommendations that in turn created the very foods now causing an epidemic of inflammation. Mainstream medicine made a terrible mistake when it advised people to avoid saturated fat in favor of foods high in omega-6 fats. We now have an epidemic of arterial inflammation leading to heart disease and other silent killers.

What you can do is choose whole foods your grandmother served and not those your mom turned to as grocery store aisles filled with manufactured foods. By eliminating inflammatory foods and adding essential nutrients from fresh unprocessed food, you will reverse years of damage in your arteries and throughout your body from consuming the typical American diet. —*Dr. Dwight Lundell*

INFLAMMATION AND DISEASES OF AGING

There is no single biological process, as part of the immune response, which triggers inflammation that is definitive for complex diseases of aging like Alzheimer's. However, taken together with all the other tests for the disease, these tests help us paint a more comprehensive diagnostic picture and provide clues for treatments, both palliative and possibly those that lead to a cure or at least a delay of its development.

Consider the research from the University of British Columbia that is published in the very prestigious *Annals of the New York Academy of Sciences* [54]. This is one of 198,000 records on the topic of inflammation and aging and is titled, *Inflammation and the degenerative diseases of aging*.

> Chronic inflammation is associated with a broad spectrum of neurodegenerative diseases of aging. Included are such disorders as Alzheimer's disease (AD), Parkinson's disease (PD), amyotrophic lateral sclerosis, the Parkinson-dementia complex of Guam, all of the tauopathies, and age-related macular degeneration. Also included are such peripheral conditions as osteoarthritis, rheumatoid arthritis, atherosclerosis, and myocardial infarction.
>
> Inflammation is a two-edged sword. In acute situations, or at low levels, it deals with the abnormality and promotes healing. When chronically sustained at high levels, it can seriously damage viable host tissue. We describe this latter phenomenon as autotoxicity to distinguish it from autoimmunity. The latter involves a lymphocyte-directed attack against self-proteins. Autotoxicity, on the other hand, is determined by the concentration and degree of activation of tissue-based monocytic phagocytes (immune system).
>
> Microglial cells are the brain representatives of the monocyte phagocytic system. Biochemically, the intensity of their activation is related to a spectrum of inflammatory mediators generated by a variety of local cells. The known spectrum includes, but is not limited to, prostaglandins, pentraxins, complement components, anaphylotoxins, cytokines, chemokines, proteases, protease inhibitors, adhesion molecules, and free radicals. This spectrum offers a huge variety of targets for new anti-inflammatory agents.

The McGeers, from the University of British Columbia, are giants in the inflammation area. In their 2004 work, cited earlier, they indicate that antiinflammatory agents appear to work for treating such inflammatory diseases. What these authors were not privy to is a multitude of trials and studies subsequent to 2004 that showed that, in most cases, dementia and Alzheimer's patients suffer from pure antiinflammatory treatments. Inflammation is, for the most part, a treasure of our immune system as it fights off insults of all types. Sure, there may be a "double-edged sword," but suppressing the immune response before it has done its job has negative consequences. As with many things in life, timing is everything.

Importantly, this work shows the fundamental connection between inflammation and AD and a wide variety of other diseases. The usual suspects, the markers of inflammation for the other diseases of inflammation, show up in Alzheimer's patients. They also did show that inflammation is beneficial.

An international team put together another significant study in 2009 [55].

> Recent scientific studies have advanced the notion of chronic inflammation as a major risk factor underlying aging and age-related diseases. In this review, low-grade, unresolved, molecular inflammation is described as an underlying mechanism of aging and age-related diseases, which may serve as a bridge between normal aging and age-related pathological processes. This new view on the role of molecular inflammation as a mechanism of aging and age-related pathogenesis can provide insights into potential interventions that may affect the aging process and reduce age-related diseases, thereby promoting healthy longevity.

Alzheimer's is one of the diseases of aging. Aging, or certainly the acceleration of the aging process, is well recognized as the primary risk factor. A search of inflammation and AD yields 123,000 references (approximately $62,000,000,000 in research money). What is inflammation called when it is associated with the brain, the nervous system, and neurons?

NEUROINFLAMMATION

Neuroinflammation is a hot topic in modern neuroscience. A relatively new open-access journal, the *Journal of Neuroinflammation*, focuses on this field. Neuroinflammation is inflammation of parts of the nervous system. This subject allows neurology to cast an impression that inflammation of the nervous system is a special case and isolated from general inflammation. This is decidedly not true because of the connection between the circulatory system and inflammation. The inflammatory markers discussed to this point are all found in the circulatory system when associated with Alzheimer's. However, neuroinflammation is potentially a "gateway" subject that will eventually unite neurology with the rest of medicine when it comes to investigating AD.

In an article from 2000 titled, *Alzheimer's disease and neuroinflammation*, the McGeer team continues to contribute to the inflammation dialog [56]. They state that it is now generally accepted that the lesions of AD are associated with a host of inflammatory molecules. Specifically, "The upregulations for CRP in the AD hippocampus is comparable to those in osteoarthritic joints. This lends further support to the hypothesis that chronic inflammation may be causing neuronal death in AD." More research points to the overlap of inflammation and Alzheimer's and diseases outside of the brain.

Indiana University contributed to the inflammation/AD discussion dating way back to 1997 [57]. "CRP is a plasma acute-phase protein, normally not found in the brain. Previous studies have demonstrated the presence of CRP in the senile plaques of AD." The authors go on to show that CRP was found associated with neurofibrillary tangles, and their data support an involvement of inflammatory processes in the etiology (cause) of AD.

HISTORY OF INFLAMMATION AND ALZHEIMER'S DISEASE

In the clinical setting, the connection between inflammation and AD has been largely ignored or overlooked. The connection between inflammation and AD has a robust history in research that extends back to the 1980s and really came of age in the 1990s through the work of McGeer and Rogers [56]. Interestingly, the connection between what may be the cause of inflammation and its association with AD extends back even further, but that is the subject of Chapter 9.

Following the work of McGeer and Rogers, it is slowly becoming accepted that cellular and molecular components of immune system reactions (inflammation) are associated with AD. The list presented here gives an idea of the depth and breadth of their studies.

1990: Anti-inflammatory drugs and Alzheimer's disease. [57a]
1992: Complement activation by beta-amyloid in Alzheimer's disease. [57b]

1992: Anti-inflammatory agents as a therapeutic approach to Alzheimer's disease. [58]
1992: Immune-related mechanisms of Alzheimer's disease pathogenesis. [58a]
1993: Does anti-inflammatory treatment protect against Alzheimer's disease? [58b]
1993: Clinical trial of indomethacin in Alzheimer's disease. [58c]
1994: Neuroimmune mechanisms in Alzheimer's disease pathogenesis. [58d]
1996: Inflammation and Alzheimer's disease pathogenesis. [58e]
1997: Molecular and cellular characterization of the membrane attack complex, C5b-9, in Alzheimer's disease. [58f]
2000: Key issues in Alzheimer's disease inflammation. [58g]
2006: Inflammation, anti-inflammatory agents, and Alzheimer's disease: the last 12 years. [58h]
2007: Therapeutic approaches to inflammation in neurodegenerative disease. [58i]
2008: Type 2 diabetes: local inflammation and direct effect of bacterial toxic components. [58j]
2010: Inflammation in transgenic mouse models of neurodegenerative disorders. [58k]
2016: Common mechanisms involved in Alzheimer's disease and type 2 diabetes: a key role of chronic bacterial infection and inflammation. [58l]
2016: Inflammation, antiinflammatory agents, and Alzheimer's disease: the last 22 years. [58m]

In the 2006 article, the McGeer team noted [59]:

1. "Activated microglia" are associated with the Alzheimer's lesions. Microglia are a type of glial (brain) cell that are the resident macrophages of the brain and spinal cord and thus act as the first and main form of active immune defense in the central nervous system (CNS). Microglia constitute 20% of the total glial cell population within the brain. Macrophages function in both nonspecific defense (innate immunity) as well as help initiate specific defense mechanisms (adaptive immunity).
2. Rheumatoid arthritics were relatively spared from AD. Even though numerous studies have been conducted on the antiinflammatory drugs used for rheumatoid arthritis patients, the results with AD have not been very good. They state, "In vitro, activated microglia release factors which are toxic to neurons, and these can be partially blocked by NSAIDs." "In vitro" means literally in glass, as in a test tube. A test that is performed in vitro is one that is done in glass or plastic vessels in the laboratory. In vitro is the opposite of in vivo (in a living organism). This brings up the entire topic of disease models. That is, can research simulate AD in a test tube or a rat? The answer is substantially no.

In the in vitro study, the actual disease is not present. It is an oversimplification of the complexity of the disease and is subject to failure for many different reasons. The literature is full of examples of test tube experiments that don't translate to us humans. In fact, at latest count, 236 out of 236 AD clinical trials have failed. However, they were all successful in the "animal models" or else they would not have advanced to the clinical trial phase.

The Inflammation/Alzheimer's Diseases Association gained a strong scientific basis in the early 1990s when two key discoveries were made. The first discovery was that immune competent cells (activated microglia and astrocytes) and inflammatory proteins (cytokines and complement proteins) were found in the vicinity of the amyloid plaques and neurofibrillary tangles. Many of the earliest results were at first dismissed as inaccurate given the perception of the brain as "immune privileged," meaning that the brain does not elicit inflammation in response to antigens or damage. However, there is now an abundant literature on the presence of acute phase proteins in amyloid plaques, which are activated microglial cells where inflammatory cytokines and other inflammatory markers were identified in Alzheimer's patients.

Since the initial discovery of a potential inflammatory component to Alzheimer's, studies have broadened to look at a variety of inflammation-associated factors for dementias and other titles ascribed to dementia, like cognitive decline. Researchers have looked in a variety of places, not just the brain, for these markers, including: the circulatory system, cerebral spinal fluid, in genes that involve or regulate the immune system, and other parts of the body known to contribute to inflammatory load like the mouth looking at gingivitis.

A central question is how do these "nonbrain" indicators of inflammation relate to inflammation in the brain and the process of neurodegeneration? Even though it is clear that the circulatory system courses through the brain, the mechanism(s) by which systemic inflammation affects the development of AD are not specifically known. Some hypothesize that so-called peripheral inflammation may be contributing to neuron degeneration by either directly causing the disease or by lowering the threshold for the disease by weakening the general immune system.

The concept of peripheral inflammation being a contributor to Alzheimer's became more credible when the concept of brain "immune privilege" was dispelled by research that showed blood borne cytokines (inflammation markers) do cross the blood-brain barrier, especially at specific sites where the barrier is damaged. In these instances, there was a clear cause/effect relationship between those cytokines and the development of Alzheimer's plaques. The Japanese and McGeer published the first paper on this topic in 1991 and 1992, respectively [58,60]. A definitive work was authored in the very prestigious journal, *Nature*, in 2001 by the Karolinska Institute, Sweden [61]. This important work opened the door to investigate Alzheimer's beyond the brain. However, over a decade later, medicine still largely rebuffs this idea. The authors provided the following insight:

> Inflammatory reactions against invaders in the body call upon cytokine molecules that elicit systemic responses, such as fever, fatigue, increased pain sensitivity and appetite loss, mediated by the central nervous system. But how cytokines can induce these effects has been a mystery as they are unlikely to cross the blood–brain barrier.
>
> Here we show that cerebral vascular cells express components enabling a blood-borne cytokine to stimulate the production of prostaglandin E2, an inflammatory mediator whose small size and lipophilic properties allow it to diffuse into the brain parenchyma. As receptors for this prostaglandin are found on responsive deep neural structures, we propose that the activated immune system controls central reactions to peripheral inflammation through a prostaglandin-dependent, cytokine-mediated pathway.

It makes sense to conclude that low-grade systemic inflammation constitutes a common denominator in neurodegenerative and vascular diseases, possibly via detrimental effects on the vasculature (blood vessels). And, the actual mechanism of the disease may not be strictly tied to the crossing of the blood-brain barrier by inflammatory molecules. If the inflammatory molecules are able to pass, it is also quite possible that the insult(s) (pathogens, for example) that lead to the development of the blood-borne inflammation can also pass through the barrier. This being the case, the inflammation can initiate in the brain, rather than being simply transported to the brain. In either case, the likely result, which is neuroinflammation and excessive cell death, develops.

It's unlikely that any single mechanism ever occurs in isolation in AD since patients present so much variability, including age of onset, severity, and many other parameters. There are consideration for the role of inflammation and development of disease. One possibility is that inflammation-associated genes are also related to dementia. An individual with a proinflammatory genotype would manifest proinflammation in the blood and other peripheral areas and the brain. Elevated peripheral inflammation could also affect brain inflammation by making neurons more prone to inflammation and tissue damage since general inflammation in the body weakens the person's immune system and renders them more susceptible to disease. In this way, chronic systemic inflammation affectively lowers the threshold for development and expansion of AD. It is this possibility that encourages us to believe that, if we can manage inflammation and its causes, we don't need to "cure" AD because its onset can be delayed by many years. This will significantly reduce the number of people who ever come down with the disease.

THE FRAMINGHAM HEART STUDY AND ALZHEIMER'S DISEASE

In 1948, the Framingham Heart Study, under the direction of the National Heart Institute (now known as the National Heart, Lung, and Blood Institute or NHLBI), embarked on an ambitious project in health research. At the time, little was known about the general causes of heart disease and stroke, but the death rates for cardiovascular diseases (CVD) had been increasing steadily since the beginning of the century and had become an American epidemic (it continues to be the #1 killer of Americans). The Framingham Heart Study became a joint project of the National Heart, Lung, and Blood Institute and Boston University.

The objective of the Framingham Heart Study was to identify the common factors or characteristics that contribute to CVD by following its development over a long period of time in a large group of participants who had not yet developed overt symptoms of CVD or suffered a heart attack or stroke. Many research papers came and continue to propagate from the Framingham Study. One such paper is titled, *Inflammatory markers and the risk of Alzheimer's disease, The Framingham Study* [62]. The objective of this particular study was to examine whether serum cytokines and spontaneous production of peripheral blood mononuclear cell (PBMC) cytokines (markers for inflammation) are associated with the risk of incident AD. The researcher concluded that higher spontaneous production of interleukin 1 (IL-1) or tumor necrosis factor α (TNF-α) by peripheral blood mononuclear cells may be a marker of future risk of AD in older individuals. These data strengthen the evidence for a role of inflammation in the development of clinical AD.

MULTIPLE MARKERS OF INFLAMMATION PREVALENT IN ALZHEIMER'S

A recent paper titled, *A meta-analysis of cytokines in Alzheimer's disease*, confirms the strong connection between inflammation and the disease [63]. This article by a Canadian team is extremely important because it looks objectively at a wealth of previously published works and performs a "metaanalysis" on the combination of all the data. A result of a metaanalysis study is a probability or "P-value."

Definition of Meta-Analysis Quantitative statistical analysis that is applied to separate but similar experiments of different and usually independent researchers and that involves pooling the data and using the pooled data to test the effectiveness of the results.

Definition of Cytokine The term "cytokine" is derived from a combination of two Greek words: "cyto" meaning cell and "kinos" meaning movement. Cytokines are cell-signaling molecules that aid cell-to-cell communication in immune responses and stimulate the movement of cells toward sites of inflammation, infection, and trauma.

Definition of P-value The statistical probability of the occurrence of a given finding by chance alone in comparison with the known distribution of possible findings, considering the kinds of data, the technique of analysis, and the number of observations. The P value may be noted as a decimal: P <.01 means that the likelihood that the phenomena tested occurred by chance alone is less than 1%. The lower the P value, the less likely the finding would occur by chance alone.

The **Canadian team (1)** stated, "The complex neurodegenerative cascade leading to AD neuropathology and symptoms is characterized by altered production, aggregation, and clearance of the amyloid-β peptide deposited in plaques, and hyperphosphorylation of the tau protein forming neurofibrillary tangles [64–67]. Interventions targeted at this cascade to date have been unsuccessful [68,69], while current pharmacotherapy, which is only modestly effective [70–73], does not act on these mechanisms. Therefore, other etiologic hypotheses are needed to help guide the development of alternative or adjunctive treatment strategies with disease-modifying potential."

Interpretation: The authors agree that beta-amyloid and modified tau are part of the Alzheimer's process. They assert that treating beta-amyloid and/or tau is an ineffective means to change the course of the disease. They indicate that alternative root cause (etiologic) ideas must be pursued if Alzheimer's is to be effectively treated.

Canadian team (2): "There is considerable evidence to suggest that an inflammatory response may be involved in the AD neurodegenerative cascade [74,75]. For example, pathology studies have shown that the pro-inflammatory cytokine interleukin (IL)-1β is overexpressed six-fold in the brains of AD patients compared with control subjects [76], especially in the vicinity of amyloid plaques [77]."

Interpretation: Inflammation is very prevalent in the blood of AD patients, and inflammatory factors are found in the vicinity of the AD plaques. In particular, interleukin (IL)-1β is present at 6 times the level expected. Our world is not linear, thus a sixfold increase in concentration may mean a 36 times higher risk of disease. Chapter 11 discusses the significance of inflammation and recovery of the brain (neurogenesis). Interleukin (IL)-1β plays a key role in the ability of the brain to recover from disease.

Canadian team (3): Clinically, associations between AD and many inflammatory biomarkers, including the cytokines IL-1β, IL-2, IL-4, IL-6, IL-8, IL-10, IL-12, IL-18, interferon (IFN)-γ, tumor necrosis factor (TNF)-α, transforming growth factor (TGF)- β, and the acute phase reactant protein CRP have been documented. However, these associations are often inconsistent between studies [78]. Moreover, cytokines have been sampled from peripheral blood and from cerebrospinal fluid (CSF), sometimes with discordant results [79]. For example, some studies have found higher peripheral blood concentrations of IL-6 in AD patients compared with control subjects, while CSF IL-6 concentrations did not differ between the same AD and control subjects [80]. Thus, the clinical literature remains to identify a pattern of immune activation associated with AD.

Interpretation: Like most studies in medicine, due to the complexities of the human body and disease and conditions of studies, results often vary. However, the body of evidence points strongly to a connection between inflammation and Alzheimer's, particularly in the peripheral blood, but not always in the cerebral spinal fluid (CSF).

Canadian team (4): Results from individual studies can be combined quantitatively using meta-analytical techniques to improve the strength of evidence. Therefore, this study reports the results of a meta-analysis conducted to determine whether the concentrations of specific cytokines differ quantitatively between patients diagnosed with AD and control subjects as measured from peripheral blood and CSF.

Interpretation: The best way to draw a conclusion is to combine all the studies and look at the results from a statistical viewpoint.

Canadian team (5): This metaanalysis reports significantly higher concentrations of the pro-inflammatory cytokines IL-6, TNF-α, IL-1β, IL-12, and IL-18 in the peripheral blood of AD subjects compared with control subjects. Evidence was particularly strong for IL-6, IL-12, and IL-18, which were significantly associated with AD in subgroups of studies that matched control subjects for age, in which subjects were free of inflammatory comorbidity and that used comparable assay techniques. While both positive and negative results have been reported in individual studies, these metaanalytic results strengthen the clinical evidence that AD is accompanied by a peripheral inflammatory response.

Interpretation: General body-wide inflammation is associated with AD.

Canadian team (6): In peripheral blood, there were significantly higher concentrations of:

- Interleukin (IL)-6; *P* value: <0.00001 (0.001% that this is by chance)
- Tumor necrosis factor (TNF)-α; *P* value = 0.01 (1% that this is by chance)
- IL-1β; *P* value < 0.00001 (0.001% that this is by chance)
- Transforming growth factor (TGF)-β; *P*-value = 0.0006 (0.06% that this is by chance)
- IL-12; *P*-value <0.00001 (0.001% that this is by chance)
- IL-18; *P*-value = 0.03 (3% that this is by chance)

There were significantly higher concentrations of TGF-β in the cerebral spinal fluid of AD subjects compared with control subjects. *P*-value = 0.006 (0.6% that this is by chance)

Interpretation: The presence of inflammation in Alzheimer's patients is not by chance.

Canadian team (7): Conclusions: These results strengthen the clinical evidence that AD is accompanied by an inflammatory response, particularly higher peripheral concentrations of IL-6, TNF-α, IL-1β, TGF-β, IL-12, and IL-18, as well as higher CSF concentrations of TGF-β.

MICROGLIA

Microglia is a type of glial cell that is the resident macrophage of the brain and spinal cord, and thus acts as the first and main form of active immune defense in the central nervous system (CNS). Microglia constitutes 20% of the total glial cell population within the brain. Microglia (and astrocytes) is distributed in large nonoverlapping regions throughout the brain and spinal cord. Microglia is constantly scavenging the CNS for damaged neurons, plaques, and infectious agents.

The brain and spinal cord were considered "immune privileged" organs in that they are separated from the rest of the body by a series of endothelial cells known as the blood-brain barrier, which prevents most infections from reaching the vulnerable nervous tissue. This notion is now dispelled. In the case where infectious agents are directly introduced to the brain or cross the blood–brain barrier, microglial cells must react quickly to decrease inflammation and destroy the infectious agents before they damage the sensitive neural tissue.

Due to the lack of availability of antibodies from the rest of the body (few antibodies are small enough to cross the blood brain barrier), microglia must be able to recognize foreign bodies, swallow them, and act as antigen-presenting cells activating T-cells. Since this process must be done quickly to prevent potentially fatal damage, microglia is extremely sensitive to even small pathological changes in the central nervous system, especially the brain. This sensitivity is achieved in part by having unique potassium channels that respond to even small changes in extracellular potassium.

Macrophages are cells produced by the differentiation of monocytes in tissues. Macrophages function in both nonspecific defense (innate immunity) as well as help initiate specific defense mechanisms (adaptive immunity) of vertebrate animals. Their role is to phagocytize (engulf and then digest) cellular debris and pathogens, either as stationary or as mobile cells. They also stimulate lymphocytes and other immune cells to respond to pathogens. They are specialized phagocytic cells that attack foreign substances, infectious microbes, and cancer cells through destruction and ingestion.

Monocytes are a type of white blood cell and are part of the innate immune system of vertebrates including all mammals (including humans), birds, reptiles, and fish. Monocytes play multiple roles in immune function. Such roles include replenishing resident macrophages and dendritic cells under normal states, and in response to inflammation signals, monocytes can move quickly (approximately 8–12 hours) to sites of infection in the tissues and divide/differentiate into macrophages and dendritic cells to elicit an immune response. Half of them are stored in the spleen.

So let's review what microglias (macrophages and monocytes) do to determine if we can better understand inflammation, its causes, diagnosis, and treatment in AD. Microglias go after damaged neurons, plaques, and infectious agents. Macrophages go after foreign substances, cellular debris, infectious microbes, and cancer cells. Question: which of these are root causes?

- Damaged neurons: This is a "what" not a "why," as damage occurs after disease.
- Plaques: Again, this is a "what" not a "why." Plaques may be or are toxic but they appear to be part of the disease process, not the root of the disease.
- Cellular debris, foreign substances: Again, this is after the fact.
- Cancer cells: There may be a correlation between brain cancer or other cancers and AD, but it is not well documented. Growth of cancer cells is a symptom, not the disease.
- Infectious microbes: Could these be (one of) the roots of inflammation and AD?

Is the relationship between inflammation and AD becoming clear? Is it apparent, based on Chapter 2, that inflammation is much more important compared to beta-amyloid? Is it also fairly certain that inflammation is on our side and we have to dig deeper, say, into a differential diagnosis process, to ferret out the real root causes of AD?

ALZHEIMER'S DISEASE AND INFLAMMATION: THE ROTTERDAM STUDY

To finish the concept of inflammation and AD, let's look to the Rotterdam Study. This extensive study of humans and their diseases and causes of death helps tie together this entire concept of inflammation, AD, and the eye, namely:

- Not only are markers of inflammation associated with AD, but also they often precede the disease clinically. Thus, markers of inflammation are an early diagnostic for Alzheimer's.
- Eye diseases, through inflammation, are connected to Alzheimer's. Thus, eye diseases are a message to dig deeper into your physiology through blood tests for inflammation to determine your risk for the future development of AD.

> The eye and the blood provide a perfect 1-2 punch
> for the early diagnosis of Alzheimer's disease.

The Rotterdam Study was introduced in Chapter 6 and provided proof that eye diseases are systemic diseases that lead to high, early mortality. It is a forward looking (prospective), population-based study. The aim of the Rotterdam Study was to investigate factors that determine the occurrence of cardiovascular, neurological, ophthalmological, endocrinological, and psychiatric diseases in elderly people. The study was established in 1990 by Professor Albert Hofman of the Department Epidemiology & Biostatistics at the Erasmus Medical Center in Rotterdam, the Netherlands. Inhabitants of Ommoord, a suburb of Rotterdam, were invited to participate on a regular basis.

The Rotterdam Study comprised three different groups over a period of 15 years. The initial group started in 1990 with 7,983 men and women aged 55 years and over. Follow-up visits were held in 1994–95, in 1997–99, and in 2002–04. As of 2008, 14,926 subjects aged 45 years or over comprised the Rotterdam Study. The findings of the Rotterdam Study have been presented in over a 1,000 research articles and reports.

In 2000–01 second group was established. Another 3,011 inhabitants of Ommoord aged 55 years and over agreed to participate. The partakers of this second group underwent follow-up examinations in 2004–05. A third group of the Rotterdam Study started in 2006, this time with inhabitants aged 45 years and over. Almost 4,000 participants have been included in this third study.

A typical examination at the Rotterdam Study included an extensive home interview and two visits for clinical examinations. Possible risk factors were measured at the research center. Clinical outcomes were continuously monitored throughout the study period for all participants of the three groups. Data on illness (morbidity) and mortality were collected at general practitioners' practices and hospitals.

Many studies from Rotterdam speak to the topic of AD. The most poignant of these is titled, *Inflammatory Proteins in Plasma and the Risk of Dementia, The Rotterdam Study* [81]. The goal of the study was to determine whether high levels of inflammatory proteins in plasma samples are associated with an increased risk of dementia. Their result tells us what we need to know. They conclude: "Plasma levels of inflammatory proteins are increased before clinical onset of dementia, AD, and vascular dementia." We now know we can look to the blood to determine those who are at risk of Alzheimer's and dementias. The specific biomarkers they mentioned included: CRP, IL-6, and ACT [34].

As a reminder from the previous chapter, essentially the same result found for AD was found for macular degeneration in the Rotterdam Study. In the article, *C-reactive protein level and risk of aging macula disorder: The Rotterdam Study*, the researchers show that elevated baseline levels of CRP were associated with the development of early and late age-related macular degeneration (AMD) in this large population-based cohort [34]. There you go. Exchange the term AMD with AD and the conclusions are essentially the same. There are many more papers that independently arrive at the same conclusion that it is hard to ignore the truth. AMD and AD are connected through inflammation and, most likely, through the causes of the inflammation.

INFLAMMATION, BLOOD VESSEL DAMAGE, AND ALZHEIMER'S DISEASE

In Alzheimer's, we now know that inflammation is present. But what does the chronic inflammation (or its causes) do from a physiological point that leads to AD? We know that inflammation signals that the immune system is activated, or possibly over-activated. But where does the damage start, and can we prevent or reverse this process? A Yale study titled, *Alzheimer's dementia begins as a disease of small blood vessels, damaged by oxidative-induced inflammation and dysregulated amyloid metabolism: implications for early detection and therapy*, helps us understand that it is the vessels that are damaged first [82]. The vessels, of course, supply oxygen, nutrients, and remove waste. As a general rule of thumb, a healthy capillary must be within three cell diameters of any living cell for the cell to survive. Thus the health of our small vessels, in the brain and elsewhere, is paramount to our health.

Marchesi, the Yale author, describes what is known and unknown about the earliest stages of AD, focusing on the possibility that the initial pathological changes involve oxidative-induced inflammatory damage to small blood vessels. The resulting ischemia (loss of blood flow, in this context at a cellular level) activates amyloid-processing enzymes and other pro-inflammatory factors that eventually compromise neuronal functions, leading, over time, to the complex lesions that characterize advanced disease. He states, "The idea that blood vessel damage is primary (to the cause of Alzheimer's disease) has a long history and many prior advocates."

What (really) damages small vessels? Keep reading and do not skip Chapters 8 or 9.

INFLAMMATION AND THE STANDARD-OF-CARE

We are now aware of many diseases, mainly chronic diseases of accelerated aging, in which chronic low-grade inflammation plays a role. We also learned that inflammation markers are in the blood long before diseases, such as Alzheimer's become clinically relevant (more important to us humans). Does modern medicine view inflammation as a disease? Sadly, it is not classified as a disease. There are no codes or treatments for chronic inflammation. Where does that leave you, the patient? You get to wait until you become sufficiently ill for (insurance coded) symptoms to develop that medicine is then very willing to treat. Even in this case, does inflammation get treated in the standard-of-care? Here, the answer is again no, at least not at its root. You will be treated for the symptoms of the disease and not the causes. An internist at a prestigious Boston hospital stated to a colleague who asked for a CRP test, "Why do you want that test? Even if is elevated, we do not know how to treat it."

Did he really mean that "we don't know how to treat" or, in the upside-down world of medicine where we confuse health insurance with health care, did he really mean, "we don't know how to get paid for treating it?"

INFLAMMATION: WHY THE EYE?

The eye is a transparent piece of tissue. Using a microscope and other means, doctors can look into the eye and observe structures to assess their relative level of health or disease. Assuming that AD and macular degeneration are very similar diseases, which one starts first? The eye, due to its easy access, allows observation of the disease, thus even if the diseases start at the same time, the eye provides the earliest indications of disease genesis. In addition to its easy access, it appears that the eye may be more susceptible to inflammatory diseases compared to other tissue. This being the case, the transparency and susceptibility of the eye make it panacea for detecting processes that damage tissue.

J. Wayne Streilein, formerly of the Schepens Eye Research Institute and now part of the Harvard Medical School facilities, was the world expert on inflammation and the eye. In a lengthy article in the Karger Gazette, Dr. Steilein stated [83]:

> Not surprisingly, inflammation, if it occurs within the eye, is a profound threat to vision. In an inflamed eye, light transmission through the visual axis can be impeded and diffracted by leukocytes and plasma proteins, and the visual axis itself can be distorted, causing the focused light image to fall away from the photoreceptor outer segments. Thus, the dilemma! Inflammation is one of the most important pathways by which immune mechanisms protect a tissue against pathogens. It is this dilemma—the need for immune protection, and the vulnerability to the consequences of inflammation—that lies at the heart of immune privilege in the eye. Through adaptation, evolution has devised a special form of immune protection (we call it immune privilege) that enables the eye to resist the vast majority of pathogens by using processes largely devoid of inflammation, thereby avoiding loss of vision. We should remember that adaptations of this type represent biologic compromises, and in the case of ocular immune privilege, the compromise renders the eye vulnerable to those organisms whose pathogenicity and virulence can only be eliminated with the aid of overt inflammation.

Thus the eye is both assessable for observation, and for (some? all? many?) chronic diseases of inflammation, it has apparent vulnerability such that these diseases may start in the eye before other tissue.

If you are concerned about inflammaging, find an eye doctor who specializes in eye pathology and ask for a thorough assessment. That doctor may not interpret your pathology results as presented in this book, but now you can.

References

[1] Clayton P. Out of the fire. Why chronic inflammation is the root of all disease and HOW to put out the flames. Hong Kong: PharmacoNutrition Press; 2013.

[2] Aisen PS, Davis KL. Inflammatory mechanisms in Alzheimer's disease. Am J Psychiatry 1994;151:1105.

[3] Sears B. Silent Inflammation. Nutraceuticals World 2005.

[4] Franceschi C, et al. Inflammaging: an evolutionary perspective on immunosenescence. Ann N Y Acad Sci 2000;908(1):244–54.

[5] Franceschi C, et al. Inflammaging and anti-inflammaging: a systemic perspective on aging and longevity emerged from studies in humans. Mech Ageing Dev 2007;128(1):92–105.

[6] Medzhitov R. Origin and physiological roles of inflammation. Nature 2008;454(7203):428–35.

[7] Wang MJ, et al. A clinical significance of high-sensitivity C-reactive protein level in Alzheimer's disease and vascular dementia. Dementia Neurocog Disorders 2012;11(4):131–5.

[8] Seddon JM, et al. Association between C-reactive protein and lutein/zeaxanthin, fish intake, body mass index and other age-related macular degeneration risk factors. Invest Ophtalmol Vis Sci 2005;46(5):2380.

[9] Locascio JJ, et al. Plasma amyloid {beta}-protein and C-reactive protein in relation to the rate of progression of Alzheimer disease. Arch Neurol 2008;65(6):776.

[10] O'Bryant SE, et al. Decreased C-reactive protein levels in Alzheimer disease. J Geriatr Psychiatry Neurol 2010;23(1):49–53.

[11] Bi BT, et al. Promotion of β-amyloid production by C-reactive protein and its implications in the early pathogenesis of Alzheimer's disease. Neurochem Int 2012;60(3):257–66.

[12] Kalman J, et al. Serum interleukin-6 levels correlate with the severity of dementia in Down syndrome and in Alzheimer's disease. Acta Neurol Scand 1997;96(4):236–40.

[13] Yang R, et al. Detection of interleukin-1, interleukin-6 and tumor necrosis factor in serum in patients with Alzheimer disease. Mod Practical Med 2010;6:005.

[14] van Oijen M, et al. Fibrinogen is associated with an increased risk of Alzheimer disease and vascular dementia. Stroke 2005;36(12):2637–41.

[15] Solfrizzi V, et al. Lipoprotein (a), apolipoprotein E genotype, and risk of Alzheimer's disease. J Neurol Neurosurg Psychiatry 2002;72(6):732–6.

[16] Iwamoto T, et al. Dual inverse effects of lipoprotein(a) on the dementia process in Japanese late-onset Alzheimer's disease. Psychogeriatrics 2004;4:64–71.

[17] McCully, K. Pioneer of the homocysteine theory. Nova Biomedical; 2013.

[18] Seshadri S, et al. Plasma homocysteine as a risk factor for dementia and Alzheimer's disease. N Engl J Med 2002;346(7):476–83.

[19] Douaud G, et al. Preventing Alzheimer's disease-related gray matter atrophy by B-vitamin treatment. Proc Natl Acad Sci 2013;110(23):9523–8.

[20] de Craen AJM, Stott DJ, Westendorp RG. Homocysteine, B vitamins, and cardiovascular disease. N Engl J Med 2006;355(2):205–11.

[21] Lonn E, et al. Homocysteine lowering with folic acid and B vitamins in vascular disease. N Engl J Med 2006;354(15):1567–77.

[22] Bonaa KH, Njolstad I, Ueland PM, et al. Homocysteine lowering and cardiovascular events after acute myocardial infarction. N Engl J Med 2006;354:1578–88.

[23] Saposnik G, et al. Homocysteine-lowering therapy and stroke risk, severity, and disability additional findings from the HOPE 2 trial. Stroke 2009;40(4):1365–72.

[24] Goldstein LB. Novel risk factors for stroke: homocysteine, inflammation, and infection. Curr Atheroscler Rep 2000;2(2):110–4.

[25] Arana L, et al. Review ceramide and ceramide 1-phosphate in health and disease. Lipids Health Dis 2010;9:15.

[26] Lamour NF, Chalfant CE. Ceramide-1-phosphate: the "missing" link in eicosanoid biosynthesis and inflammation. Mol Interven 2005;5(6):358.

[27] Mielke MM, et al. Serum ceramides increase the risk of Alzheimer disease: the Women's Health and Aging Study II. Neurology 2012;79(7):633–41.

[28] Mielke MM, Lyketsos CG. Alterations of the sphingolipid pathway in Alzheimer's disease: new biomarkers and treatment targets? Neuromol Med 2010;12(4):331–40.

[29] Mielke MM, et al. Serum sphingomyelins and ceramides are early predictors of memory impairment. Neurobiol Aging 2010;31(1):17–24.

[30] Mielke MM, et al. Plasma ceramides are altered in mild cognitive impairment and predict cognitive decline and hippocampal volume loss. Alzheimers Dement 2010;6(5):378–85.

[31] Geekiyanage H, et al. Blood serum miRNA: non-invasive biomarkers for Alzheimer's disease. Exp Neurol 2012;235(2):491–6.

[32] Car H, et al. The role of ceramides in selected brain pathologies: ischemia/hypoxia, Alzheimer disease. Postepy Hig Med Dosw 2012;66:295.

[33] Perlmutter D, Colman C. The better brain book: the best tools for improving memory and sharpness and for preventing aging of the brain. New York: Penguin; 2005.

[34] Matsubara E, et al. α1-Antichymotrypsin as a possible biochemical marker for Alzheimer-type dementia. Ann Neurol 1990;28(4):561–7.

[35] Brennan ML, Penn MS, Van Lente F, Nambi V, Shishehbor MH, Aviles RJ, et al. Prognostic value of myeloperoxidase in patients with chest pain. N Engl J Med 2003;349(17):1595–604.

[36] Heslop CL, Frohlich JJ, Hill JS. Myeloperoxidase and C-reactive protein have combined utility for long-term prediction of cardiovascular mortality after coronary angiography. J Am Coll Cardiol 2010;55(11):1102–9.

[37] Lefkowitz DL, Lefkowitz SS. Microglia and myeloperoxidase: a deadly partnership in neurodegenerative disease. Free Radic Biol Med 2008;45(5):726–31.

[38] Praticò D, et al. Increased F2-isoprostanes in Alzheimer's disease: evidence for enhanced lipid peroxidation in vivo. FASEB J 1998;12(15):1777–83.

[39] Une K, et al. Adiponectin in plasma and cerebrospinal fluid in MCI and Alzheimer's disease. Eur J Neurol 2011;18(7):1006–9.

[40] Ward MA, et al. Evaluation of CSF cystatin C, beta-2-microglobulin, and VGF as diagnostic biomarkers of Alzheimer's disease using SRM. Alzheimer's Dementia 2011;7(4):S150–1.

[41] Kojima S, et al. The white blood cell count is an independent predictor of no-reflow and mortality following acute myocardial infarction in the coronary interventional era. Ann Med 2004;36(2):153–60.

[42] Do Lee C, et al. White blood cell count and incidence of coronary heart disease and ischemic stroke and mortality from cardiovascular disease in African-American and White men and women: atherosclerosis risk in communities study. Am J Epidemiol 2001;154(8):758–64.

[43] Shad KF, et al. Peripheral markers of Alzheimer's disease: surveillance of white blood cells. Synapse 2013;67(8):541–3.

[44] Suzanne M, Wands JR. Alzheimer's disease is type 3 diabetes–evidence reviewed. J Diabetes Sci Technol 2008;2(6):1101.

[45] de Luca C, Olefsky JM. Inflammation and insulin resistance. FEBS Lett 2008;582(1):97–105.

[46] de la Monte, Suzanne M. Insulin resistance and Alzheimer's disease. BMB Rep 2009;42(8):475–81.

[47] Hyman M, Mark L. Ultraprevention: The 6-week plan that will make you healthy for life. New York: Simon and Schuster; 2003.

[48] Atwood CS, et al. Dysregulation of the hypothalamic-pituitary-gonadal axis with menopause and andropause promotes neurodegenerative senescence. J Neuropathol Exp Neurol 2005;64(2):93–103.

[49] Bowen RL, Atwood CS. Living and dying for sex. Gerontology 2004;50(5):265–90.

[50] Abbate J, Guynup S, Genco R. Scientific American presents: oral and whole body health. 2006, 6–49. http://media.dentalcare.com/media/en-US/products/owbh.pdf.

[51] Evengård B, Forsgren M, Uggla A. Toxoplasmosis. The most common parasitic infection in Europe, but not fully understood and probably underdiagnosed. Läkartidningen 1997;94(38):3249.

[52] Simopoulos, Artemis P. The importance of the omega-6/omega-3 fatty acid ratio in cardiovascular disease and other chronic diseases. Exp Biol Med 2008;233(6):674–88.

[53] Ravnskov, Uffe, Kilmer S, McCully. Vulnerable plaque formation from obstruction of vasa vasorum by homocysteinylated and oxidized lipoprotein aggregates complexed with microbial remnants and LDL autoantibodies. Ann Clin Lab Sci 2009;39(1):3–16.

[53a] Lundell, D. World Renowned Heart Surgeon Speaks Out On What Really Causes Heart Disease. FineForme, July 2015. Available from: http://www.fineforme.net/world-renowned-heart-surgeon-speaks-out-on-what-really-causes-heart-disease/#more-189.

[54] McGeer PL, McGeer EG. Inflammation and the degenerative diseases of aging. Ann NY Acad Sci 2004;1035(1):104–16.

[55] Chung HY, et al. Molecular inflammation: underpinnings of aging and age-related diseases. Ageing Res Rev 2009;8(1):18–30.

[56] McGeer PL, McGeer EG, Yasojima K. Alzheimer disease and neuroinflammation. Advances in dementia research. Vienna: Springer; 2000. p. 53–57.

[57] Duong T, Nikolaeva M, Acton PJ. C-reactive protein-like immunoreactivity in the neurofibrillary tangles of Alzheimer's disease. Brain Res 1997;749(1):152–6.

[57a] Martyn C. Anti-inflammatory drugs and Alzheimer's disease. BMJ 2003;327(7411):353–4.

[57b] Rogers J, et al. Complement activation by beta-amyloid in Alzheimer disease. Proc Natl Acad Sci USA 1992;89(21):10016–20.

[58] McGeer PL, Joseph R. Anti-inflammatory agents as a therapeutic approach to Alzheimer's disease. Neurology 1992;42(2):447–9.

[58a] Rogers J, et al. Immune-related mechanisms of Alzheimer's disease pathogenesis. In: Khachaturian ZS, Blass JP, editors. Alzheimer's Disease: New Treatment Strategies. New York: Marcel Dekkar Inc; 1992. p. 147–63.

[58b] McGeer PL, et al. Does anti-inflammatory treatment protect against Alzheimer's disease? In: Khachaturian ZS, Blass J, editors. Alzheimer's Disease—New Treatment Strategies. New York: Marcel Dekker Inc; 1993. p. 165–71.

[58c] Rogers J, et al. Clinical trial of indomethacin in Alzheimer's disease. Neurology 1993;43(8):1609–11.

[58d] McGeer PL, et al. Neuroimmune mechanisms in Alzheimer disease pathogenesis. Alzheimer Dis Assoc Disord. 1994;8(3):149–58.

[58e] Rogers J, et al. Inflammation and Alzheimer's disease pathogenesis. Neurobiol Aging. 1996;17(5):681–6.

[58f] Webster S, et al. Molecular and cellular characterization of the membrane attack complex, C5b-9, in Alzheimer's disease. Neurobiol Aging. 1997;18(4):415–21.

[58g] Cooper NR, et al. Key issues in Alzheimer's disease inflammation. Neurobiol Aging. 2000;21(3):451–3.

[58h] McGeer PL, et al. Inflammation, anti-inflammatory agents and Alzheimer disease: the last 12 years. J Alzheimers Dis. 2006;9(3 Suppl.):271–6.

[58i] Klegeris A, McGeer EG, McGeer P. Therapeutic approaches to inflammation in neurodegenerative disease. Curr Opin Neurol 2007;20(3):351–7.

[58j] Miklossy J, Martins R, Darbinian N, et al. Type 2 diabetes: local inflammation and direct effect of bacterial toxic components. Open Pathol J 2008;2:86.

[58k] Schwab C, Klegeris A, McGeer PL. Inflammation in transgenic mouse models of neurodegenerative disorders. Biochim Biophys Acta. 2010;1802(10):889–902.

[58l] Miklossy J, McGeer PL. Common mechanisms involved in Alzheimer's disease and type 2 diabetes: a key role of chronic bacterial infection and inflammation. Aging 2016;8(4):575–88.

[58m] McGeer PL, Rogers J, McGeer EG. Inflammation, antiinflammatory agents, and Alzheimer's disease: the last 22 years. J Alzheimers Dis 2016;54(3):853–7.

[59] McGeer PL, Rogers J, McGeer EG. Inflammation, anti-inflammatory agents and Alzheimer disease: the last 12 years. J Alzheimer's Dis 2006;9:271–6.

[60] Itagaki S, et al. Characteristics of reactive microglia in Alzheimer's and Parkinson's disease brain tissue. Basic, clinical, and therapeutic aspects of Alzheimer's and Parkinson's diseases. New York: Springer; 1991. 381–384.

[61] Ek M, et al. Inflammatory response: pathway across the blood–brain barrier. Nature 2001;410(6827):430–1.

[62] Tan ZS, et al. Inflammatory markers and the risk of Alzheimer disease: the Framingham Study. Neurology 2007;68(22):1902–8.

[63] Swardfager W, et al. A meta-analysis of cytokines in Alzheimer's disease. Biol Psychiatry 2010;68(10):930–41.

[64] Bharadwaj PR, Dubey AK, Masters CL, Martins RN, Macreadie IG. Abeta aggregation and possible implications in Alzheimer's disease pathogenesis. J Cell Mol Med 2009;13:412–21.

[65] Karlnoski RA, Rosenthal A, Kobayashi D, Pons J, Alamed J, Mercer M, et al. Suppression of amyloid deposition leads to long-term reductions in Alzheimer's pathologies in Tg2576 mice. J Neurosci 2009;29:4964–71.

[66] Cataldo AM, Paskevich PA, Kominami E, Nixon RA. Lysosomal hydrolases of different classes are abnormally distributed in brains of patients with Alzheimer disease. Proc Natl Acad Sci USA 1991;88:10998–1002.

[67] Bancher C, Brunner C, Lassmann H, Budka H, Jellinger K, Wiche G, et al. Accumulation of abnormally phosphorylated tau precedes the formation of neurofibrillary tangles in Alzheimer's disease. Brain Res 1989;477:90–9.

[68] Salloway S, Sperling R, Gilman S, Fox NC, Blennow K, Raskind M, et al. A phase 2 multiple ascending dose trial of bapineuzumab in mild to moderate Alzheimer disease. Neurology 2009;73:2061–70.

[69] Aisen PS, Saumier D, Briand R, Laurin J, Gervais F, Tremblay P, et al. A phase II study targeting amyloid-beta with 3APS in mild-tomoderate Alzheimer disease. Neurology 2006;67:1757–63.

[70] Kaduszkiewicz H, Zimmermann T, Beck-Bornholdt HP, van den Bussche H. Cholinesterase inhibitors for patients with Alzheimer's disease: systematic review of randomised clinical trials. BMJ 2005;331:321–7.

[71] McShane R, Areosa SA, Minakaran N. Memantine for dementia. Cochrane Database Syst Rev 2006;2. CD003154.

[72] Courtney C, Farrell D, Gray R, Hills R, Lynch L, Sellwood E, et al. Long-term donepezil treatment in 565 patients with Alzheimer's disease (AD 2000): randomised double-blind trial. Lancet 2004;363:2105–15.

[73] Gill SS, Anderson GM, Fischer HD, Bell CM, Li P, Normand SL, Rochon PA. Syncope and its consequences in patients with dementia receiving cholinesterase inhibitors: a population-based cohort study. Arch Intern Med 2009;169:867–73.

[74] Wyss-Coray T. Inflammation in Alzheimer disease: driving force, bystander or beneficial response? Nat Med 2006;12:1005–15.

[75] Griffin WS, Mrak RE. Interleukin-1 in the genesis and progression of and risk for development of neuronal degeneration in Alzheimer's disease. J Leukoc Biol 2002;72:233–8.

[76] Griffin WS, Stanley LC, Ling C, White L, MacLeod V, Perrot LJ, et al. Brain interleukin 1 and S-100 immunoreactivity are elevated in Down syndrome and Alzheimer disease. Proc Natl Acad Sci U S A. 1989;86(19):7611–5.

[77] Griffin WS, Sheng JG, Roberts GW, Mrak RE. Interleukin-1 expression in different plaque types in Alzheimer's disease: significance in plaque evolution. J Neuropathol Exp Neurol 1995;54:276–81.

[78] Perry VH, Cunningham C, Holmes C. Systemic infections and inflammation affect chronic neurodegeneration. Nat Rev Immunol 2007;7:161–7.

[79] Rota E, Bellone G, Rocca P, Bergamasco B, Emanuelli G, Ferrero P. Increased intrathecal TGF-beta1, but not IL-12, IFN-gamma and IL-10 levels in Alzheimer's disease patients. Neurol Sci 2006;27:33–9.

[80] Tarkowski E, Blennow K, Wallin A, Tarkowski A. Intracerebral production of tumor necrosis factor-alpha, a local neuroprotective agent, in Alzheimer disease and vascular dementia. J Clin Immunol 1999;19:223–30.

[81] Engelhart MJ, et al. Inflammatory proteins in plasma and the risk of dementia: the Rotterdam study. Arch Neurol 2004;61(5):668.

[82] Marchesi VT. Alzheimer's dementia begins as a disease of small blood vessels, damaged by oxidative-induced inflammation and dysregulated amyloid metabolism: implications for early detection and therapy. FASEB J 2011;25(1):5–13.

[83] Streilein JW. Ocular immune privilege–protection that preserves sight! Karger Gaz 2001;64:5–8.

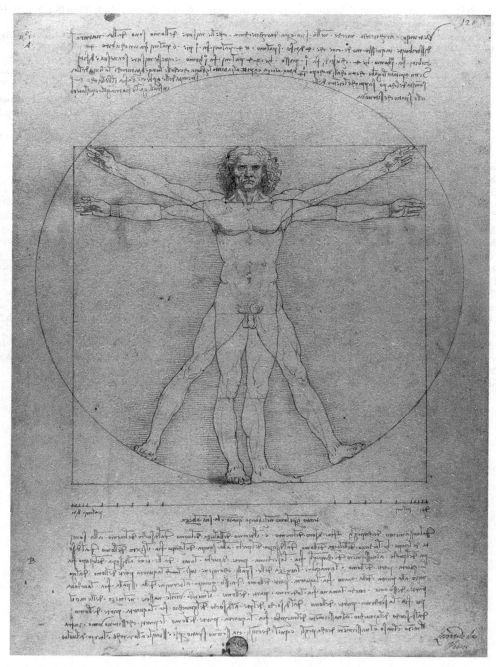

Image Credit: Leonardo Da Vinci's Vitruvian Man - Photo by Luc Viatour (www.lucnix.be). Available at Wikimedia Commons: https://commons.wikimedia.org/wiki/File:Da_Vinci_Vitruve_Luc_Viatour.jpg

Alzheimer's: Beyond the Brain

Human life, particularly in health and disease, is the result of countless independent forces impinging simultaneously on the total organism and setting in motion a multitude of interrelated responses.
—*Rene Dubos, French-American bacteriologist and author of the slogan "Think globally, act locally"*

There is profound evidence that the Alzheimer's disease originates systemically, as systemic treatments for non-brain related illnesses clearly change the course of the disease.

—*P.J. Seberger, MD, PhD*

"The key to combating the Alzheimer's epidemic is to focus on diagnosing the disease long before it ravages the brain. And the best place to start is the heart," says Jack C. de la Torre, Adjunct Professor of Psychology at The University of Texas at Austin and prolific author on Alzheimer's. Now we know that that best place for early diagnosis of chronic inflammatory diseases, including those of the heart, is the eye.

According to de la Torre's research, the opportune window for diagnosis is middle age, when vascular risk factors, such as type 2 diabetes, hypertension, and heart disease, are strongly linked with Alzheimer's. "Vascular risk factors to Alzheimer's disease (AD) offer the possibility of markedly reducing dementia by early identification and treatment," says de la Torre. "Improved understanding coupled with preventive strategies could be a monumental step forward in reducing worldwide prevalence of AD, which is doubling every 20 years."

"Reducing AD prevalence by focusing right now on vascular risk factors, even with our limited technology, is not a simple or easy task," De la Torre says. "But the task must not be delayed because time is running out for millions of people whose destiny with dementia may start sooner rather than later."

Solutions suggested by Dr. de la Torre make sense, particular in light of the shortcomings of explanations and solutions that are "brain only." Thankfully he is not the only researcher to explore beyond the brain into systemic (in the entire body) and overlapping causes of Alzheimer's. The evidence is mounting that Alzheimer's originates, and very often has pathologies, that link the disease to systemic (body-wide) adverse physiological changes. Most in neurology will argue that these are so-called comorbidities, and thus are more coincidental than a cause. However, since medicine has dedicated over 100 years to the "brain-only" hypothesis, it's now time to put considerable effort toward exploring elsewhere.

Inflammation (Chapter 7) is now being considered a major component of AD. Can inflammation occur solely in the brain? Hardly. Inflammation is usually tied to the vascular (circulatory) system. This system is so highly connected that it's difficult to imagine that a disease

The End of Alzheimer's. http://dx.doi.org/10.1016/B978-0-12-812112-2.00008-2

presenting with markers in the circulatory system is not body wide. Consider the circulatory system of a famous athlete, Lance Armstrong. While racing up the French Alps, the same red blood cell passes through his heart every 10 seconds or so. And surely that blood cell takes a path through the brain now and then. Now that's a highly connected system! Sure neurons of the brain do not circulate, but it is the circulatory system that energizes and nourishes those neurons.

ALZHEIMER'S AND CARDIOVASCULAR DISEASES SHARE RISK FACTORS

We know the brain is not an organ in isolation. There are 100,000 ft. of blood vessels in the brain (almost a marathon in distance). This link between cardiovascular diseases and AD is not new, having been discovered over 20 years ago during a series of autopsies on AD patients. While working in the Kentucky medical examiner's office, Larry Sparks was checking brain tissues looking for early signs of AD. He noticed that those who had the telltale plaques of Alzheimer's had one thing in common. He took the slides and put them into two piles, those with heart disease and those without heart disease. All the brain plaques and tangles showed up in the pile with heart disease. Dr. Sparks went on to be a scientist with Sun Health Research Institute in Arizona before his passing. That finding corroborated other results suggesting heart disease could be a forerunner of brain disease.

"The link between heart disease and AD is growing in strength every few months," said Bill Thies, a scientific director of the Alzheimer's Association. "And we predict it will continue to grow. I'm not surprised that there's a relationship," he said. "The heart is the organ that supplies essential elements to many parts of the body, and the brain is just one of the first" [1].

In one proposed mechanism, atherosclerosis due to long-standing hypertension leads to poor blood flow to the brain, which may contribute to the development of dementia by causing neurons supplied by those blood vessels to die due to hypoxia. Low diastolic blood pressure in late-life may further worsen blood flow to the brain. Peripheral arterial disease (narrowed arteries leading to poor circulation in the arms or legs) has also been associated with a twofold increase in the risk for AD, further supporting the theory that the factors that are associated with or cause inadequate blood circulation are the same as those that contribute to AD.

These cardiovascular risk factors appear to have an additive effect. Therefore, if you have two of these risk factors, your risk of AD increases by about four times, and if you have three risk factors, your risk for AD increases by about six times (maybe nine times), and so on. Some of these risk factors also affect the rate of progression of AD. Specifically, atrial fibrillation, hypertension, and angina have been associated with a greater rate of decline among people with AD.

It is important to recognize the link between cardiovascular disease and AD, because cardiovascular disease is often preventable or treatable. By managing cardiovascular disease, it may be possible to decrease the risk of developing AD and even impact the course of the disease even if it has taken hold. Research into the effect of treating cardiovascular risk on the risk of AD is ongoing. But based on the available evidence, managing cardiovascular risk factors should be an integral part of managing AD.

The key question is, however, the following: how should cardiovascular risk be managed? If you look at the field of cardiology, unfortunately, like neurology, the ability to treat the underlying causes of vascular disease is limited. For example, they use blood pressure–lowering medication for high blood pressure. They use cholesterol-lowering medication for high cholesterol. They use surgical procedures to open up clogged vessels. Do any of these methods attack a root cause? Does lowering blood pressure help a brain starving for oxygen? Statins for lowering cholesterol were to be a miracle cure for cardiovascular disease, but as shown in Chapter 5, that approach is an abject failure even though over 21 million Americans still are led to believe that therapy is worth the substantial risk associated with those drugs.

The true answer for managing cardiovascular disease risk is properly provided by scientific giants like Claude Bernard, Louis Pasteur, and Kilmer McCully. We can also learn a great deal from the writings of Paul Clayton, cited in Chapter 7. Before advancing into a complete rewrite of the field of cardiology, let's look at the scientific evidence connecting Alzheimer's and cardiovascular diseases.

A New York University School of Medicine study in 2004 implies that higher levels of "good" cholesterol in middle age may help preserve brain function in later years [2]. This finding was reported at the 9th International Conference on Alzheimer's disease and Related Disorders (ICAD), presented by the Alzheimer's Association. Researchers at Harvard University and Brigham and Women's Hospital in Boston reported on the relationship between cardiovascular risk factors and cognitive function in 4081 women aged 65 years and older [3]. The researchers correlated performance on a battery of neuropsychological tests with levels of cholesterol and triglycerides determined several years earlier. The only factor that influenced cognitive performance was HDL (or "good") cholesterol. That is, high LDL cholesterol, the alleged culprit in cardiovascular disease, did not impact Alzheimer's. The researchers found a consistent increase in cognitive health paralleling higher levels of HDL. A less robust correlation was seen for lower levels of LDL (or "bad") cholesterol and better cognitive performance. Triglyceride levels did not appear to influence cognition. These results suggest the possibility that simple, well-established lifestyle modifications to increase HDL levels—increased physical activity, moderate alcohol intake, and high intake of mono- and polyunsaturated fatty acids (omega 3 fatty acids)—could have a substantial public health impact beyond heart disease.

In a dementia study published in 2004, a large group of people who had mild cognitive impairment were examined and followed for several years [4]. Subjects who had high blood pressure or other vascular problems at the beginning of the study were found to be twice as likely to develop dementia, compared with those without these risks, the researchers found. Half of those with vascular risks progressed to Alzheimer's, compared with only 36% of those without. That represents a 50% increase in risk for those suffering from certain types of cardiovascular problems.

Various factors that contribute to diabetes mellitus and cardiovascular disease also appear to promote a decline in mental function in the elderly. Jacob Mintzer, MD, of the Medical University of South Carolina in Charleston presented data from the Charleston Heart Study [5]. In a sample of more than 700 men and women, Mintzer measured diabetes and cardiovascular disease risk factors and correlated them with cognitive health several years after the measurements. Mintzer and colleagues found that the presence of high cardiovascular risk factors increased the risk of later cognitive decline, with a particularly strong effect in African

Americans. Specifically, the presence of coronary heart disease, excessively high cholesterol, or hypertension also led to more rapid cognitive decline during aging. Interestingly, they found that although the diagnosis of diabetes increased the risk of cognitive decline by as much as twofold, the presence of high levels of fasting glucose substantially decreased the risk of cognitive decline in diabetic patients. Why? The most reasonable answer, based on simple understanding of brain physiology, may be related to the high-energy demand of the brain. If glucose is too low, especially in those with some level of insulin resistance, then the brain suffers an energy crisis.

Today, we view diabetes as a disease of excess sugar because that is what is detected in the blood of patients suffering from the disease. The ACCORD Study (Action to Control Cardiovascular Risk in Diabetes) puts that simplistic notion to bed [6]. Actually, the elevation of blood glucose and serum insulin is a normal response to the lack of glucose in tissue and is one of the causes for a very significant increase in mortality when doctors try to strictly control serum glucose in type 2 diabetics. Vessel and general inflammation and insulin resistance is the problem, not the blood glucose. Older people suffering from mild memory and cognition problems may be less likely to progress to full-blown AD if they receive the proper treatment for medical conditions like diabetes. Directly lowering blood glucose levels is not the right approach, however.

A group of researchers from China began following 837 residents aged 55 and older who had mild cognitive impairment but not dementia [7]. Of these, 414 had at least 1 detectable medical condition that can impair blood flow to the brain. After 5 years, 298 of the participants had developed Alzheimer's. Subjects who had high blood pressure or other vascular problems at the beginning of the study were twice as likely to develop the dementia, compared with those without these risks, the researchers found. Among those with vascular problems, those who received treatment were almost 40% less likely to develop Alzheimer's than those who did not.

Many studies like this one add to mounting evidence that some of the health and risk factors that contribute to cardiovascular disease, and resultant strokes and heart attacks, also increase the risk of Alzheimer's. Some studies showed that cardiovascular risk factors are associated directly with cognitive decline because the risk factors overlap. These data suggest that tests and diagnostics for cardiovascular disease have excellent predictive power toward Alzheimer's. Thus we should be rethinking and enhancing diagnostics for latent cardiovascular diseases beyond the "standard-of-care," and this should be used as an initial screening for AD risk. The connection goes beyond Alzheimer's and heart disease. More than 50 chronic degenerative diseases have a common disease pathway, and inflammation plays a key role as we now know. The good news is that this is the first set of risk factors identified that people can actually do something about. Medications and lifestyle changes that have been used to promote heart health may actually protect brain health as well.

Research data leave little doubt that we should focus on helping people to manage their numbers, but not the ones you might think. Blood pressure is important but simply lowering it puts the brain at risk. Cholesterol is of lower importance with emphasis on HDL and not LDL. As with blood pressure, blood sugar levels are telling, but simply lowering them directly, with medication, again puts the brain at risk. However, with a surging epidemic of disease associated with these three "numbers," it is evident that just controlling the number is not the solution. Thus, it is not the numbers themselves, but the methods used in their control that make the difference. Chapter 7 tells us that we need to focus on a new set of numbers that are associated with inflammation, and when it is controlled, our "numbers" will probably be optimal. All these studies help to confirm the idea that cardiovascular health and brain health are closely related. Controlling your cholesterol number, for example, with cholesterol-modifying therapies, does not get at the core of disease. It is probably harmful. We must dig deeper.

There are two scientists who have taken a lead in deepening the connection between Alzheimer's and cardiovascular disorders: Jack C. de la Torre and Kilmer McCully. Dr. de la Torre is brilliant in his persistence at showing the Alzheimer's/heart connection. Dr. McCully is equally brilliant, as a pathologist, at getting to the root of the causes that appear common to both diseases.

DR. JACK C. DE LA TORRE, MD, PhD

To say he is prolific in his study of Alzheimer's and cardiovascular diseases is an understatement. What follows is a shockingly long list of publications by Dr. de la Torre. Why shocking? Because despite the well-formulated connection between "treatable" cardiovascular diseases and "untreatable" AD, medicine continues to avoid this approach. Thus all 50 references are included here to show the profundity of the connection:

1993: Brain blood flow restoration 'rescues' chronically damaged rat CA1 neurons. [8]
1993: Can disturbed brain microcirculation cause Alzheimer's disease? [9]
1996: Chronic reduction of cerebral blood flow in the adult rat: late-emerging CA1 cell loss and memory dysfunction. [10]
1997: Cerebromicrovascular pathology in Alzheimer's disease compared to normal aging [11]
1997: Cerebrovascular changes in the aging brain. [12]
1997: Reduced cytochrome oxidase and memory dysfunction after chronic brain ischemia in aged rats. [13]
1998: Reversal of ischemic-induced chronic memory dysfunction in aging rats with a free radical scavenger–glycolytic intermediate combination. [14]
1999: Cerebral hypoperfusion yields capillary damage in the hippocampal CA1 area that correlates with spatial memory impairment. [15]
1999: Cerebrovascular pathology in Alzheimer's disease. [16]
1999: Critical threshold cerebral hypoperfusion causes Alzheimer's disease? [17]
2000: Chronic cerebrovascular ischemia in aged rats: effects on brain metabolic capacity and behavior. [18]
2000: Critically attained threshold of cerebral hypoperfusion: can it cause Alzheimer's disease? [19]
2000: Critically attained threshold of cerebral hypoperfusion: the CATCH hypothesis of Alzheimer's pathogenesis. [20]
2000: Evidence that Alzheimer's disease is a microvascular disorder: the role of constitutive nitric oxide. [21]
2000: Impaired brain microcirculation may trigger Alzheimer's disease. [22]
2000: Vascular pathophysiology in Alzheimer's disease. [23]

2001: Effects of aging on the human nervous system. [24]
2002: Alzheimer's disease as a vascular disorder nosological evidence. [25]
2002: Alzheimer's disease: vascular etiology and pathology. [26]
2003: Hippocampal nitric oxide upregulation precedes memory loss and Aβ 1-40 accumulation after chronic brain hypoperfusion in rats. [27]
2004: Alzheimer's disease is a vasocognopathy: a new term to describe its nature. [28]
2004: Drug therapy in Alzheimer's disease. [29]
2004: Is nitric oxide a key target in the pathogenesis of brain lesions during the development of Alzheimer's disease? [30]
2004: The role of nitric oxide in the pathogenesis of brain lesions during the development of Alzheimer's disease. [31]
2004: Vascular dynamics in Alzheimer and vascular dementia. [32]
2005: Cerebrovascular gene linked to Alzheimer's disease pathology. [33]
2005: Inhibition of vascular nitric oxide after rat chronic brain hypoperfusion: spatial memory and immunocytochemical changes. [34]
2005: Is Alzheimer's disease preceded by neurodegeneration or cerebral hypoperfusion? [35]
2005: Mitochondria DNA deletions in atherosclerotic hypoperfused brain microvessels as a primary target for the development of Alzheimer's disease. [36]
2006: How do heart disease and stroke become risk factors for Alzheimer's disease? [37]
2006: The deadly triad: heart disease, stroke and Alzheimer's disease. [38]
2008: Alzheimer's disease prevalence can be lowered with non-invasive testing. [39]
2008: Pathophysiology of neuronal energy crisis in Alzheimer's disease. [40]
2009: Carotid artery ultrasound and echocardiography (CAUSE) testing to lower the prevalence of Alzheimer's disease. [41]
2009: Cerebral and cardiac vascular pathology in Alzheimer's disease. [42]
2009: Cerebrovascular and cardiovascular pathology in Alzheimer's disease. [43]
2010: Alzheimer's disease is incurable but preventable. [44]
2010: Basics of Alzheimer's disease prevention. [45]
2010: The vascular hypothesis of Alzheimer's disease: bench to bedside and beyond. [46]
2010: Vascular risk factor detection and control may prevent Alzheimer's disease. [47]
2011: Three postulates to help identify the cause of Alzheimer's disease. [48]
2012: A tipping point for Alzheimer's disease research. [49]
2012: A turning point for Alzheimer's disease? [50]
2012: Alzheimer's disease: how does it start? [51]
2012: Cardiovascular risk factors promote brain hypoperfusion leading to cognitive decline and dementia. [52]
2012: Cerebral hemodynamics and vascular risk factors: setting the stage for Alzheimer's disease. [53]
2012: Physiopathology of vascular risk factors in Alzheimer's disease. [54]
2013: Cardiovascular disease promotes cognitive dysfunction that can signal Alzheimer's and vascular dementia. [55]
2013: Alzheimer's disease is associated with increased risk of haemorrhagic stroke. [56]
2014: Detection, prevention, and pre-clinical treatment of Alzheimer's disease. [57]
2016: Cerebral perfusion enhancing interventions: a new strategy for the prevention of Alzheimer dementia. [58]

The best way to honor Dr. de la Torre for his considerable work on our behalf is to read his summation on Alzheimer's and then assist him and other like-minded researchers and clinicians move forward toward a preventative solution. Here are the words of Dr. de la Torre from his paper titled "A tipping point for Alzheimer's disease research" [49]:

A drunkard loses his keys on a dark street and is looking for them under a lamp-post. A policeman comes over and asks what he's doing. 'I'm looking for my keys' he says. 'I lost them over there.' The policeman looks puzzled. 'Then why are you looking for them over here?' 'Because,' replies the drunk, 'the light is so much better.' There is undeniable logic to look for things lost in the dark where the light is better. But when the search under the light consistently yields nothing, it is wiser to move on and look elsewhere, even where the light is dim. This has been the Alzheimer's Disease (AD) paradox for the last two decades, as the search for the cause of this disorder has been carried out under a very limited spotlight that has produced no clear benefits for patients at-risk or those already diagnosed with AD.

This trackless clinical progress is all the more baffling in view of over 75,000 scientific papers published on the subject of AD during the last two decades. A considerable amount of the Alzheimer's research done during this period has focused its efforts in attempting to validate that AD is triggered by the excessive accumulation in the brain, by a sticky substance called amyloid-β-peptide (Abeta) [59]. Although several versions of the Abeta hypothesis have been put forward, its main thesis is that Abeta is responsible for the neurodegeneration and cognitive loss seen in AD. This canonical view has engendered more than 30,000 papers dealing mainly with the biochemistry, genetics, pathology, molecular activity, and clinical effects of Abeta. The Abeta hypothesis has influenced and guided the research in the field of Alzheimer's dementia that many workers regard it as the gold standard of scientific investigation. Despite the monumental bibliography that Abeta research has generated, patients today are no better off, medically speaking, than was described by Alois Alzheimer's in 1907 [60].

Dr. de la Torre continues:

Recently, the hypothesis that Abeta deposition in the brain is the cause of sporadic AD was put to the test in a series of pharmaceutical sponsored clinical trials using immunization against Abeta42, the reputed culprit, in order to reduce Abeta aggregation, or clear it completely from the brain. Patients with mild to moderate AD were treated with anti-Abeta compounds such as AlzheMed [tramiprosate], flurizan [tarenflurbill], semagacestat, AN-1792, solanezumab, Elan 301 and 302 [bapineuzumab].

Despite virtual clearance of Abeta from the brain or limiting its aggregation, none of the anti-Abeta vaccines used were able to improve cognitive function or slow down the neurodegenerative progress characteristic of AD [61,62]. Some of the treated patients developed adverse effects to the amyloid immunotherapy, the most serious being the development of meningoencephalitis [63].

These compounds did what they were supposed to do, to clear Abeta from the brain, but not what they were expected to do, make patients even slightly better.

Abeta's contribution to neuritic plaque development likely plays a secondary role in AD neurodegeneration but studies consistently show it does not correlate with the brain cell, synaptic and metabolic loss, or to the initiation of cognitive impairment leading to Alzheimer's dementia [64–67]. The relentless clinical trial failures pointedly indicate that pharmacotherapeutic elimination of Abeta from the brain does not improve any of the pathologic features that characterize AD.

FUTURE OF THE ALZHEIMER'S/SYSTEMIC DISEASE CONNECTION

More from Dr. de la Torre:

Getting back to the drunk under the lamp-post, common sense dictates that confining the search for the key to any disease process to a single spot weakens our confidence that good science is being practiced. Consequently, another approach to combat AD is slowly gaining ground. Prevention, rather than a nostrum for sporadic AD, appears as a more realistic strategy to impact the catastrophic consequences of this dementia. Prevention should aim at preserving normal cognitive function as long as possible. The best time to apply preventive measures is during middle age in asymptomatic individuals when early mechanism-based interventions can be most effective. Prevention of AD is possible for three reasons:

1. Mounting evidence indicates that AD is a heterogeneous disorder evolving mainly in aging people with preventable vascular risk factors (VRF) that are reported to progressively lower cerebral blood flow to a level where mild cognitive impairment and AD appear [54].
2. VRF such as cardiovascular disorders, hypertension, dyslipidemia, atherosclerosis, diabetes type 2, cerebrovascular disease, and others [68] are readily diagnosed, can be treated early, and could be clinically monitored for long-term changes.
3. Diagnostic tools that can identify VRF prior to the onset of AD and interventions presently available to limit VRF activity are needed to be assessed in randomized clinical trials to determine their value in preventing cognitive decline.

Assuming VRF prevention can lower AD prevalence, medical centers staffed with multi-disciplinary specialists composed of neurologists, psychologists, cardiologists, gerontologists, and technical personnel, trained in neurodegenerative, vascular disorders and dementia, should be established to identify and treat when appropriate and follow-up patients who are at-risk of developing dementia. This approach would result in a cost-effective measure that should taper the devastating socio-economic calamity that is anticipated will increase from $200 billion in direct costs in 2012 to $1.1 trillion in 2050. More importantly, prevention can dramatically reduce new cases of AD from a present 5.4 million to 16 million people in the US alone during the next 40 years [69]. The time has come to start a plan that will vigorously reduce AD prevalence world-wide or we will all surely pay the colossal medical and economic price for the failure to act.

DR. KILMER McCULLY

What Dr. McCully lacks in volume of research, he more than makes up for in content. This is not to disparage Dr. de la Torre, who solidifies the AD/CVD connection. Where Dr. de la Torre is broad, Dr. McCully is deep. They are the perfect 1–2 punch against Alzheimer's. Dr. McCully gets to the root of the problem. And the root of the disease is deep down in the vessels. If you are allowed to read but one research paper on chronic diseases, then Dr. McCully's paper with the following very long title is it!: "Vulnerable plaque formation from obstruction of vasa vasorum by homocysteinylated and oxidized lipoprotein aggregates complexed with microbial remnants and LDL autoantibodies" [70].

There is not much that Dr. McCully and Uffe Ravnskov leave to chance as they brilliantly develop their thesis about how LDL cholesterol is not "bad" but an important part of our body's defense mechanism. They guide us to an understanding about how cholesterol, in all its forms, is very important to human function, especially immune response. Be aware that 25% of all cholesterol is found in the brain, yet the brain is only 2% of the mass of the body. That cholesterol is up there for a reason. If it were so evil, we would be DOA. They also point us to a vast body of research that explains the underpinnings of cardiovascular disease and, by extension, AD. This is a new view compared to conventional wisdom, and their teaching is seldom if ever to be found in the standard-of-care. I know as my 91-year-old mother has aortic stenosis, and her cardiologist tried to prescribe statin.

Here are some key points from the Dr. McCully article:

There is general agreement that atherosclerosis begins as an inflammatory process in the arterial wall, and also that rupture of a vulnerable plaque is the starting point for the creation of the occluding thrombus in myocardial infarction and ischemic stroke (various cardiovascular disease events). Therefore, any hypothesis about the cause of atherosclerosis and its consequences must necessarily be able to point to the origin of the inflammation and to explain how a vulnerable plaque is created.

Dr. McCully continued by explaining the current accepted view of cardiovascular disease and then stated, "However, it conflicts with many clinical, epidemiological, pathological, and experimental observations. There are in particular six disturbing facts:

1. The concept that high LDL cholesterol causes endothelial dysfunction (breakdown of the blood vessel walls) is unlikely because there is no association between the concentration of LDL cholesterol in the blood and the degree of endothelial dysfunction.
2. The concept that endothelial damage leads to influx of LDL cholesterol is unlikely as well, because the atherosclerotic plaques seen in extreme hyper-homocysteinemia caused by inborn errors of methionine metabolism do not contain any lipids in spite of pronounced endothelial damage.
3. No study of unselected individuals has found an association between the concentration of LDL or total cholesterol in the blood and the degree of atherosclerosis at autopsy.
4. In studies of women and the elderly, hypercholesterolemia (high cholesterol) is a weak risk factor for cardiovascular disease, or, in most cases, not a risk factor at all, although the large majority of cardiovascular deaths occur in people above 65 years of age. [And people over the age of 70 have a lower total mortality rate if their cholesterol is elevated (above 250).]
5. Among individuals with familial hypercholesterolemia (FH, genetically high cholesterol) there is no association between LDL-cholesterol and the prevalence or the progress of cardiovascular disease. The higher coronary mortality in young people with FH may instead be due to inherited abnormalities of the coagulation system, often seen in FH and a strong risk factor for coronary heart disease in this population.
6. With one exception, an occluding coronary thrombus has never been produced experimentally in rodents by hypercholesterolemia (high cholesterol) alone, indicating that the pathological process in these models may differ from that in human beings."

If you are compelled by Dr. McCully's arguments against LDL (bad) cholesterol, then what is the cause of cardiovascular diseases? He provides a detailed explanation that is well worth reading. Here is just the introduction to his elegant explanation:

In the following discussion we present a new interpretation of the origin of vulnerable plaques (the cause of cardiovascular events) that we think is in better agreement with presently available evidence. This interpretation is based on the fact that the lipoproteins (cholesterol) function as a nonspecific immune system that binds and inactivates microorganisms and their toxins by complex formation (and by the bactericidal effects of cholesterol). In the case of a massive microbial invasion, these complexes may aggregate, in particular in the presence of hyperhomocysteinemia, because homocysteine thiolactone causes aggregation and precipitation of thiolated LDL. Complex formation and aggregation may also be enhanced by autoantibodies against thiolated LDL and oxidized LDL. Because of high extra-capillary tissue pressure, the aggregates may be trapped in arterial vasa vasorum, resulting in local vascular ischemia, intramural cell death, and the creation of vulnerable plaques.

Such plaques have many characteristics of a micro-abscess, which, by rupturing, initiates the occluding thrombosis (closing of the vessel with a blood clot) and releases its content of infectious material into the circulation and the myocardium. This suggested chain of events explains why many of the clinical symptoms and laboratory findings in acute myocardial infarction are similar to those seen in infectious diseases. It also explains the frequent presence of microbial remnants in atherosclerotic plaques, the many associations between infections and cardiovascular disease, the similarities between myocarditis and myocardial infarction, and why cholesterol accumulates in the arterial wall.

The simple summary is the following: The vulnerable plaques that are fundamental to acute cardiovascular disease events are caused by the interaction of the immune system, by way of LDL cholesterol, and pathogens. Yes, cholesterol is there to protect, not to hurt. No wonder people over the age of 70 with high cholesterol (>250) live much longer compared to their age mates with cholesterol levels below 160 [71].

It is interesting that our ancestors often knew more about medicine compared to our collective intelligence today. For example, talk to any cardiologist and they will tell you something like: "bad cholesterol (from eating too much fat) collects on the inside of the walls of your arteries, causing a blockage, that then leads to heart attack or stroke." However, not one part of that explanation is accurate. In fact, the true mechanism for cardiovascular disease was elucidated in "63." The actual date was 1863 by a German doctor named Koester [72]. The reference for that work is included here:

Koester W. Endarteritis and arteritis. Berl Klin Wochenschr 1876;13:454–5.

Koester explained that the disease of the vessels starts from the outside, at the vasa vasorum, which is the outermost layer of the blood vessel architecture, and works its way inside. The vasa vasorum is the layer of small vessels that nurture the wall of larger arteries. Plaques are the result of abnormal neovascularization (new vessel formation) that derive from this vasa vasorum vascular network. Serum proteins and lipids leak from these abnormal vessels and accumulate in the extravascular space. When this accumulation is suddenly released, inside your blood vessel, you are highly likely to have an immediate cardiovascular event like a stroke or heart attack. Before it bursts you may experience symptoms of angina or ischemia as this sack swells in the vessel wall into your vessel—from the outside, thus narrowing your lumen.

This type of abnormal neovascularization is seen in many diseases, for example, in the eye in wet age-related macular degeneration (AMD), in the joint of rheumatoid arthritis, in the lesions of psoriasis, in wound healing, and in amyloid angiopathy in the brain, to name a few.

Eat your cholesterol and other healthy and essential fats. Your brain needs the nourishment. But listen to Dr. McCully and protect yourself against pathogens by building your immune system.

Here is a selected reading list of Dr. McCully's publications:

1993: Chemical pathology of homocysteine I. Atherogenesis. [73]
1994: Chemical pathology of homocysteine II. Carcinogenesis and homocysteine thiolactone metabolism. [74]
1994: Chemical pathology of homocysteine. III. Cellular function and aging. [75]
2004: Homocysteine and heart disease. [76]
2005: Hyperhomocysteinemia and arteriosclerosis: historical perspectives. [77]
2007: Homocysteine, vitamins, and vascular disease prevention. [78]
2009: The heart revolution. [79]
2009: Chemical pathology of homocysteine. IV. Excitotoxicity, oxidative stress, endothelial dysfunction, and inflammation. [80]
2011: Chemical pathology of homocysteine. V. Thioretinamide, thioretinaco, and cystathionine synthase function in degenerative diseases. [81]
2012: How macrophages are converted to foam cells. [82]

2012: Vegetarianism produces subclinical malnutrition, hyperhomocysteinemia and atherogenesis. [83]

2014: Biofilms, lipoprotein aggregates, homocysteine, and arterial plaque rupture. [84]

2015: Homocysteine metabolism, atherosclerosis, and diseases of aging. [85]

2015: Homocysteine and the pathogenesis of atherosclerosis. [86]

2015: The active site of oxidative phosphorylation and the origin of hyperhomocysteinemia in aging and dementia. [87]

2016: Lack of an association or an inverse association between low-density-lipoprotein cholesterol and mortality in the elderly: a systematic review. [88]

2016: Homocysteine, infections, polyamines, oxidative metabolism, and the pathogenesis of dementia and atherosclerosis. [89]

HOMOCYSTEINE: BRIEF HISTORY AND CARDIOVASCULAR DISEASE

In 1969, Dr. McCully published an unorthodox conclusion in the *American Journal of Pathology* regarding a new possible cause of heart disease. This move soon cost him his job at Harvard University, he says. Dr. McCully proposed that a substance called homocysteine could, when allowed to accumulate to toxic levels, degenerate arteries and produce heart disease. Homocysteine, an amino acid, is a normal by-product of protein metabolism (specifically, of the amino acid methionine), which does not create a problem when present in small amounts.

However, Dr. McCully observed that children with elevated levels of homocysteine showed signs of blood vessel degeneration similar to these observed in middle-aged adults with heart disease. He next demonstrated that when rabbits were injected with homocysteine, they developed arterial plaques within 3–8 weeks. Homocysteine apparently curtails the ability of blood vessels to expand, keeping them restricted and narrow. It accomplishes this by increasing connective tissue growth and by degenerating the elastic tissue in the arterial walls. Dr. McCully argued that high-protein diets, more than fats and cholesterol, seem to be a prime cause of heart disease. After publication of his novel theory, subsequently backed by considerable clinical support, Harvard denied him tenure, and effectively fired him.

The evidence continues to mount in support of Dr. McCully's homocysteine theory. In 1992, researchers at Harvard University School of Public Health showed that men with homocysteine levels only 12% higher than average had 3.4 times greater risk of heart attack than those with normal levels. Also that year, the *European Journal of Clinical Investigation* showed that 40% of stroke victims have elevated homocysteine levels compared to only 6% of controls [90]. Compare this to a study reported in the *American Heart Journal*, which found that out of 137,000 people admitted to the hospital in the United States with heart attacks, nearly 75% had "normal" LDL cholesterol levels. You can see quite clearly here that high homocysteine levels are a bigger risk factor for heart disease compared to high cholesterol.

A University of Washington study published in 1995 in the *Journal of the American Medical Association* reviewed numerous studies linking homocysteine with heart disease and concluded that homocysteine represents a strong independent risk factor for stroke even after adjustment of other risk factors [91]. According to Dr. McCully, heart disease is, in part, attributed to "abnormal processing of protein in the body because of deficiencies of B vitamins in the diet." It is this B-vitamin deficiency that allows homocysteine, normally converted to a harmless substance, to accumulate to dangerous levels. "Protein intoxication," characterized by excess homocysteine, contributes to damaging the cells and tissues of arteries, "participating in the many processes that lead to loss of elasticity, hardening and calcification, narrowing of the lumen (arterial passageway), and formation of blood clots within arteries." This older view on homocysteine metabolism and disease still holds some validity, but lowering homocysteine without understanding and treating the cause of the disease and the reason for the elevation in homocysteine is now known to be a contraindicated therapeutic strategy.

Elevated homocysteine has the potential of displacing (the misguided) high cholesterol levels as the major dietary factor in heart disease. A German study (1991) looked at the coronary arteries of 163 males with chest pain and concluded that the arterial narrowing was due more to blood levels of homocysteine than to cholesterol. In 1997, Dr. McCully declared: "Elevated blood homocysteine is estimated to account for at least 10% of the risk of coronary heart disease in the US population." Now that is a reasonable and not overly dramatized conclusion.

The current tests and diagnoses for asymptomatic cardiovascular disease are woefully inadequate. The classic risk factors such as the "Framingham score" and stress testing are not particularly predictive (recall the Tim Russett story). Some of the recently considered inflammatory markers are much more telling but tend to lose predictive power with advanced age. However, if you could have only one test, opt for homocysteine. Quoting a recent study, "a model based on homocysteine concentration alone was a better predictor of cardiovascular mortality in very old people with no history of cardiovascular disease than were models based on classic risk factors. These preliminary findings call for validation in a separate cohort and, if confirmed, could eventually lead to a revision of current guidelines and corresponding indicators of quality of care" [92].

Do you understand the need for a new guideline? Why don't doctors just do the homocysteine test? Remember, only a few ICD-9 or ICD-10 diagnostic codes justify the reimbursement for measuring homocysteine in people, and it will take at least 10 years for the test to be included in the guidelines. How many unnecessary lives will be lost to cardiovascular and AD over this period of time due to inadequate diagnosis?

HOMOCYSTEINE AND ALZHEIMER'S

We all need to appreciate that, in chronic disease, there is never one cause, diagnosis, or cure. Dr. McCully indicated that homocysteine accounts for 10% of cardiovascular disease and, by extension, that includes Alzheimer's, but what about the other 90%? The most astute researchers in any given field of chronic disease discovery often use the term "cascade." If you are inclined to reading long, complicated, and pedantic definitions, go to Wikipedia and absorb the definition of "root cause." Multiple coincidental are often required to trigger complex reactions. This is not just true in medicine but extends into every other field of science and technology.

Definition of Cascade A succession of actions, processes, or operations, as of a physiological process. Alternatively, a series that once initiated continues to the end, each step being triggered by the preceding one, sometimes with cumulative effect.

Many more studies than cited here show a correlation between the amount of homocyste-ine in blood and the extent of cardiovascular disease risk. There is no such thing as a one-for-one correlation in chronic disease; therefore statistics are used to describe the level of risk. It takes massive studies involving large populations to confidently pin down statistical risk for complex diseases with a myriad of different and potentially interrelated risk factors. Homo-cysteine has emerged as one of the most consistent predictors of disease and early mortality. The predictive powers of homocysteine extend into AD and dementias.

Given the connection between cardiovascular diseases and AD, there is a strong, but not surprising, connection to AD. In fact, a search of scholar.google for homocysteine and AD yields 17,500 records ($8,750,000,000 in research). A short list of article titles is presented at the end of this section.

Speaking of statistics, the BU and Tufts University authors of the following 2002 study made some rather bold statistical statements about levels of homocysteine and risk of AD [93]. The paper titled "Plasma homocysteine as a risk factor for dementia and Alzheimer's disease" concludes:

> The results of our prospective, observational study indicate that there is a strong, graded association be-tween plasma total homocysteine levels and the risk of dementia and Alzheimer's disease. An increment in the plasma homocysteine level of 5 micromol per liter increased the risk of Alzheimer's disease by 40%.
>
> A plasma homocysteine level in the highest age-specific quartile doubled the risk of dementia or Alzheim-er's disease. A similar result was found when the single criterion of hyperhomocysteinemia (base-line plasma homocysteine, >14 micromol per liter) was used. The magnitude of this effect is similar to the magnitude of the increases in the risks of death from cardiovascular causes and stroke associated with a similar increment in the plasma homocysteine level, which have been previously described in the Framingham cohort.

This research is full of rich information on the predictive power of homocysteine. This research concludes the following:

- Increase homocysteine value by 5, and increase the Alzheimer's risk by 40%.
- Homocysteine value >14, risk of Alzheimer's increases by 40% compared to healthy levels (some consider 12 and others 9 as a healthy level).
- Homocysteine-related risks for Alzheimer's and cardiovascular diseases are the same.

> The prospective nature of this study and the strong association between newly diagnosed dementia and Alzheimer's disease and plasma homocysteine levels measured eight years before base line suggest that the elevation in the homocysteine level preceded the onset of dementia. Finally, subjects with a sustained eleva-tion of plasma homocysteine had the greatest risk of dementia.

Elevated homocysteine is detected well before clinical symptoms of Alzheimer's. Ho-mocysteine is not a definitive diagnosis for Alzheimer's but these data suggest that people should get their homocysteine levels checked routinely in a prevention mode and not just as a determinant once disease is known. Sustained elevated homocysteine presents the greatest statistical risk of developing symptoms of chronic disease. The solution is regular (annual?) monitoring of homocysteine levels.

The Kentucky nuns study conducted a good many years ago showed that nuns who have the most mini-strokes showed the symptoms of AD, while many with lots of beta-amyloid did not have signs of that disease [94]. The study also showed that nuns who were most likely to suffer AD had low blood levels of the vitamin folic acid and high levels of homocysteine.

Tufts researchers, as early as 1996, showed that those with high homocysteine levels performed poorly on cognitive tests, compared with those who had low homocysteine levels [95]. In addition, low levels of vitamin B12 and folic acid were also associated with low cognitive test scores. A 6-year study, concluded in 1997, found that people who took vitamin B6 and B12 supplements performed better on cognitive tests, including recall ability [96]. Interestingly, in 2013, vitamin B12 supplementation and improvement in cognition once again made headlines through major new sources. And these outlets presented these findings as "new." It's sad how we are letting trillions of dollars' worth of fine medical literature lay fallow.

Caution—simply lowering homocysteine may not impact AD, or cardiovascular disease for that matter. This is similar to lowering cholesterol. It has an impact on a very small slice of the severely ill population only despite what advertisements lead us to believe. Homocysteine is an important part of a diagnostic profile. However, some studies suggest that just lowering homocysteine may not improve Alzheimer's and may actually complicate the disease. A deep and broad root cause analysis followed by treatments based on comprehensive results, rather than a "one-off" treatment, will likely lead to better outcomes. When the basics of disease are understood and treated, the homocysteine-lowering interventions may be more successful.

Australians published a review paper in 2012 that provided reasonable proof that simply lowering homocysteine with vitamin therapy does not change the course of cognitive impairment [97]. They concluded, "B-vitamin supplementation did not show an improvement in cognitive function for individuals with or without significant cognitive impairment. Supplementation of vitamins B12, B6, and folic acid alone or in combination does not appear to improve cognitive function in individuals with or without existing cognitive impairment. It remains to be established if prolonged treatment with B-vitamins can reduce the risk of dementia in later life." Thus we must regard homocysteine and associated vitamin deficiency as a very important diagnostic parameter first, and as a treatable target as secondary upon establishing a more complete cause/effect profile.

These authors make some extremely important points. They also raise some very important questions. Why, in some cases, does B vitamin supplementation appear to work, while in others it does not? Is this a case of research fraud? In this case, fraud or misinterpreted data are probably not at issue; rather it is likely due to patient variability. Here, the issue is the complexity of disease. Homocysteine may be elevated as either or both a "friend and/or foe," depending upon the underlying etiology and physiology. This is akin to inflammation that may have multiple causes. Homocysteine may rise, for example, in the face of infection. Treatment with B vitamins will lower homocysteine but will not impact infection. A search of the medical literature shows a strong connection between elevated homocysteine and intracellular infection like *H. pylori* [98]. Just taking B vitamins is not an adequate treatment for *H. pylori* infections. This provides one plausible explanation for when homocysteine lowering with vitamins does not impact disease.

Several of the first papers that included that term Alzheimer's and homocysteine in the title of the article were published in 1998 [99–102]. Their thesis was that homocysteine and AD are connected, and this argument has withstood the test of time. The first one was published in the *International Journal of Geriatric Psychiatry* with the title "Total serum homocysteine in senile dementia of Alzheimer's type." The conclusion of that paper was that "SDAT (senile dementia, Alzheimer's Type) patients have significantly elevated tHcy (homocysteine)." In another paper, published in *Archives of Neurology*, the authors recognized that "recent studies suggest that vascular disease may contribute to the cause of Alzheimer's disease (AD). Since elevated plasma total homocysteine (tHcy) level is a risk factor for vascular disease, it may also be relevant to AD." They concluded, "Low blood levels of folate and vitamin B12, and elevated tHcy levels were associated with AD."

According to Dr. Martha Savaria Morris from Tufts University [103],

> A high circulating concentration of the amino acid homocysteine is an independent risk factor for stroke. Alzheimer's disease (AD) commonly co-occurs with stroke. Epidemiological studies found associations between hyperhomocysteinaemia and both histologically confirmed AD and disease progression and revealed that dementia in AD was associated with evidence of brain infarcts (strokes) on autopsy. Thus, hyperhomocysteinaemia (high homocysteine levels) and AD could be linked by stroke or microvascular disease. However, given known relations between B-group-vitamin deficiency and both hyperhomocysteinaemia and neurological dysfunction, direct causal mechanisms are also plausible.
>
> 25% of dementia cases are attributed to stroke. The possibility that some of the other 75% might be prevented by the lowering of homocysteine concentrations greatly increases the hope of maintaining self-sufficiency into old age.

Dr. Morris goes on to say that there is an expectation that AD levels could diminish in the United States and Canada due to folate supplementation. However, this does not reflect the probable cause/effect relationships between homocysteine and disease. True, homocysteine is toxic, but it is also a marker of disease. Interestingly, 25% of dementia is attributed to stroke. Microvessel disease is not classified as stroke and may be a substantial part of the "75%" outside the purview of stroke. Thus, maybe almost all of dementias are modifiable by lowering homocysteine at a root cause level! Finding the cause of the elevated homocysteine and controlling those that result in elevation of this biomarker is important if we want to decrease age-related chronic diseases.

Here is a collection of some titles relating homocysteine and AD:

1998: Folate, vitamin B12, and serum total homocysteine levels in confirmed Alzheimer's disease. [101]

1998: Total serum homocysteine in senile dementia of Alzheimer's type. [99]

1999: Homocysteine and Alzheimer's disease. [104]

2000: Homocysteine, Alzheimer's disease, and cognitive function. [105]

2001: Homocysteine potentiates copper- and amyloid-beta peptide-mediated toxicity in primary neuronal cultures: possible risk factors in the Alzheimer's-type neurodegenerative pathways. [106]

2002: Plasma homocysteine as a risk factor for dementia and Alzheimer's disease. [93]

2003: Homocysteine and Alzheimer's disease. [103]

2004: Homocysteine, Alzheimer's genes and proteins, and measures of cognition and depression in older men. [107]

2005: Homocysteine and folate as risk factors for dementia and Alzheimer's disease. [108]

2006: Elevated plasma homocysteine levels: risk factor or risk marker for the development of dementia and Alzheimer's disease? [109]

2007: Relations between homocysteine, folate and vitamin B12 in vascular dementia and in Alzheimer's disease. [110]

2008: Relationship between genetic polymorphism, serum folate and homocysteine in Alzheimer's disease. [111]

2009: Homocysteine a prognostic marker for cognitive decline in Alzheimer's patients. [112]

2010: Homocysteine as a predictor of cognitive decline in Alzheimer's disease. [113]

2011: Plasma homocysteine level and apathy in Alzheimer's disease. [114]

2011: Homocysteine: a biomarker in neurodegenerative diseases—review. [115]

2012: Effect of homocysteine lowering treatment on cognitive function: a systematic review and meta-analysis of randomized controlled trials. [97]

2013: Plasma homocysteine, Alzheimer's and cerebrovascular pathology: a population-based autopsy study. [116]

2013: Betaine attenuates Alzheimer's-like pathological changes and memory deficits induced by homocysteine. [117]

2013: Hydroxysafflor yellow A ameliorates homo-cysteine induced Alzheimer's-like pathologic dysfunction and memory/synaptic disorder. [118]

2014: Homocysteine exacerbates β-amyloid pathology, tau pathology, and cognitive deficit in a mouse model of Alzheimer's disease with plaques and tangles. [119]

2014: Correlation between behavioural and psychological symptoms of Alzheimer's type dementia and plasma homocysteine concentration. [120]

A review by German researchers sums up the value of homocysteine as a diagnostic marker for AD. Their paper was published in 2011 with the title "Homocysteine: a biomarker in neurodegenerative diseases" [115].

> Diseases of the central nervous system are found in patients with severe hyperhomocysteinemia (HHcy) (elevated homocysteine). Epidemiological studies show a positive, dose-dependent relationship between mild-to-moderate increases in plasma total homocysteine concentrations (Hcy) and the risk of neurodegenerative diseases, such as Alzheimer's disease, vascular dementia, cognitive impairment, or stroke … Therefore, recent evidence supports the role of Hcy (homocysteine) as a potential biomarker in age-related neurodegenerative diseases.

The complete homocysteine story continues to unfold. Stating a now familiar theme, is homocysteine friend or foe or both? There is no question about the detrimental effects of homocysteine. But are these a necessary evil of some underlying protective process as is the case (some now think) for beta-amyloid? The answer to one question is unequivocal. Should you have your homocysteine level routinely checked? Yes!

The connection between homocysteine and infection is documented, but it is not strong yet. It is likely to continue to develop. As it does, we will gain a better understanding why simple homocysteine lowering does not always lead to improvement in cognition, for example. Here is an example of research that documents the connection between infection, homocysteine, and brain vascular disease, "Association of Helicobacter pylori infection and serum homocysteine level in patients with cerebral infarction" [121]. The authors conclude, "*Helicobacter pylori* infection may increase the risk of cerebral infarction, which might be associated with the increased serum homocysteine level."

Most studies in medicine focus on just one parameter. For example, a study will measure the impact of just homocysteine and attempt to factor out all other variables. What often is lacking is the evaluation of overlapping parameters. Do two parameters increase an effect by a factor of two or by more or less? Right now we do not know due to the lack of studies, but the world generally follows statistics, that is, the bell curve. Most things do not progress linearly, they progress geometrically. Consider the following number sequences:

Linear: 1, 2, 3, 4, 5, 6, 7, 8, 9, 10
Geometric: 1, 2, 4, 8, 16, 32, 64, 128, 256, 512

Thus, if two health factors advance or recede geometrically as opposed to linearly, the impact on health or disease is tremendous. In considering drug–drug interactions, the same math probably applies.

Here is a study from the Netherlands that looks at two health factors, homocysteine and inflammation (IL-6 and CRP). The title of the study is "Homocysteine and inflammation: predictors of cognitive decline in older persons?" [122]. The authors state, "Higher homocysteine at baseline was negatively associated with prolonged lower cognitive functioning and a faster rate of decline in information processing speed and fluid intelligence. The negative association between higher homocysteine and immediate recall was strongest in persons with a high level of IL-6."

The authors use the term "strongest," but we don't know if the impact is additive (linear) or synergistic (geometric). However, this type of study shows the power of a broad and deep differential diagnosis that leads to a multifactorial treatment. Clearly, a multifactorial treatment will have at least a linear, and hopefully geometric, impact on recovery.

OTHER DISEASES OF AMYLOID BUILDUP

Beta-amyloid is found in the senile plaques of dementia brains of the Alzheimer's type. Are you aware that the same "misfolded protein" process occurs beyond the brain, in other parts of the body, even when dementia or Alzheimer's is not present?

Dr. Christopher M. Dobson of Cambridge University, England, is one of the foremost experts on "amyloidosis" and he, in a review article from 2006, provided the following insight [123]:

> Two greatly debilitating illnesses—Alzheimer's disease, which is largely associated with aging, and variant Creutzfeldt–Jakob disease (vCJD), which is linked with the recent bovine spongiform encephalopathy (BSE) epidemic in Britain—have had a huge impact on public consciousness. Despite their very different origins, these diseases are closely related to each other and also to the 20 or so other 'amyloidoses'—so called because they involve the aberrant deposition of proteins in the form of amyloid fibrils or plaques. Other diseases in this group, which are less publicized but no less devastating to those affected, are Parkinson's disease, type II (late-onset) diabetes and rare conditions such as familial insomnia.
>
> Amyloidoses thus include some of the most feared and costly diseases in the Western world. Alzheimer's may very soon be the most prevalent and socially disruptive illness in the ageing populations of all developed countries. It is remarkable, therefore, that most of these diseases were virtually unknown until relatively recently. Indeed, the first detailed description of an amyloid pathology was made less than 100 years ago, when Alzheimer's identified the form of dementia that bears his name. As with many amyloidoses, Alzheimer's disease is usually sporadic, although there are less common hereditary forms that can afflict relatively young people—Alzheimer's most famous patient was in her early 50s.

> As mentioned above, all of these diseases, whether sporadic, familial or transmissible, are associated with deposits in tissue of proteins that are normally soluble. .... A major puzzle was that although the proteins involved have different structures in their soluble forms, the fibrils look remarkably similar, suggesting that the molecular structures in the aggregated forms are essentially the same.

Proteins are continually damaged and repaired or replaced throughout our life (wear and tear). With aging, the damaged proteins are not repaired or degraded quickly and start to accumulate. These proteins are the biomarker of diseases related to aging. The strength of the immune system is central to proper protein processing. Important factors like nutrition and exercise prop up the immune system. Infection plays a major role in aging morbidity (disease) and mortality (death). Controlling infection through immune health and treatment reduces protein buildup and contributes to healthy longevity.

From a diagnostic perspective, any appearance anywhere in the body of these misfolded amyloid proteins may be an indication of an underlying, or root cause, process that may result in AD. Certainly any amyloid-type disease is a chronic debilitating condition. It gives further possible credence to the concept that disease is an upside-down pyramid, that many diseases may develop from a single or just a few causes, all of which are related to accelerated aging.

Dr. Dobson provides a comprehensive list of diseases of amyloidosis. However, there are two very important forms of amyloidosis that connect back to AD. The first is the connection between amyloid protein and the heart, and its (now obvious?) connection to AD. This will be covered at the end of this chapter. The second is "Alzheimer's disease of the muscle," also known as Inclusion Body Myositis (IBM). Interestingly, news out of the International Alzheimer's Association meeting referred to muscle loss as being part of the symptoms of Alzheimer's, but no one in popular media indicated a connection between Alzheimer's and IBM.

ALZHEIMER'S DISEASE OF THE MUSCLE

Weight loss is well known to be associated with AD. Between 20% and 40% of patients suffering from mild to moderate forms of AD are affected by weight loss, irrespective of where they live. This weight loss increases as the disease develops and represents a predictive factor of mortality. Certain studies also show that loss of weight can precede the diagnosis of the disease. In this case, it could be one of the early signs of the pathological process and should be included in the entire differential diagnosis. Clinical practice shows that weight loss is accompanied by a series of complications including alterations of the immune system, muscular atrophy, falls, fractures, dependence, and other afflictions leading to a worsening of the patient's health and to an increased risk of institutionalization and mortality.

Inclusion Body Myositis (IBM) is a disease where the amyloid is found in the muscle. Although not well correlated at present, it makes sense that AD sufferers, with a burden of amyloid in the brain and in the eye, also have amyloid in the muscle and thus have muscular disease comorbidity. Although it is assumed that AD sufferers simply lose their appetite due to the state of their deteriorating brain, that may only be part of the explanation for weight loss. Amyloid in the muscle can lead to atrophy of the muscle tissue just as amyloid in the brain leads to atrophy there.

Definition of Inclusion Body Myositis (IBM) IBM is an inflammatory muscle disease, characterized by slowly progressive weakness and wasting of both distal and proximal muscles, most apparent in the muscles of the arms and legs. There are two types: sporadic inclusion body myositis (sIBM) and hereditary inclusion body myopathy (hIBM).

Let's do a quick exercise. In the above definition, take out the word "muscle" and replace it with "brain." Sounds like IBM and AD are the same disease? Note, too, that there are two types of AD, both sporadic and familial (hereditary). Again, the connection to the two diseases is more than just interesting. The next paragraph presents the biochemistry of IBM and its connection to AD excerpted from an article titled "Inclusion-body myositis A myodegenerative conformational disorder associated with Aβ, protein misfolding, and proteasome inhibition" [124].

An intriguing feature is the accumulation within sIBM (sporadic Inclusion Body Myositis) muscle fibers of amyloid-beta (Aβ), phosphorylated tau protein, and at least 20 other proteins that are also accumulated in the brain of Alzheimer's disease patients.

The connection, at least in the halls of research, between IBM and AD is not new. A 1994 paper set the stage for classifying these two diseases as the same, or, at least, very similar. The paper holds the title "Twisted tubulofilaments of inclusion body myositis muscle resemble paired helical filaments of Alzheimer's brain and contain hyperphosphorylated tau" [125].

We know the definition of IBM, and it is as close as Wikipedia. We also know the connection between IBM and AD as it is as close as scholar.google.com. Yet many in the Alzheimer's field say: "Weight loss is often (always?) a sign of an insufficient calorie intake, which must therefore be adjusted on an individual basis." However, the reality is that weight loss may well be associated with IBM that, once diagnosed, may indicate that a disease process similar to (or identical to) Alzheimer's is in progress in muscle. For patients with IBM who are asymptomatic for dementia, this might be their early warning diagnosis for the disease. Do not adjust the diet of these patients; instead provide them with a differential diagnosis!

IBM results in general, progressive muscle weakness. The muscles in the thighs called the quadriceps and the muscles in the arms that control the making of a fist are usually affected early on. Common early symptoms include frequent tripping and falling (you will see the significance of this highlight shortly), weakness going up stairs, and trouble manipulating the fingers, for example, turning doorknobs and gripping keys. Foot drop in one or both feet is a symptom of IBM.

Here are a dozen or so news reports from the Paris International Alzheimer's Association conference of 2011 discussing trips/falls:

- Falls an early clue to Alzheimer's. www.webmd.com
- Falls may indicate earliest stages of Alzheimer's. http://edition.cnn.com/
- Falls linked to early Alzheimer's disease. www.usatoday.com
- Falls, eye tests may detect early Alzheimer's. http://news.abs-cbn.com/
- Are falls a harbinger of Alzheimer's disease? www.medscape.com
- Falls may be early sign of Alzheimer's. www.sciencedaily.com
- Falls and eye test may give clues to Alzheimer's. www.hurriyetdailynews.com
- Falls could signal early Alzheimer's disease. www.everydayhealth.com.
- Falls more common in early Alzheimer's. www.medicalnewstoday.com

- Study: Falls may be a sign of Alzheimer's disease. www.washingtontimes.com
- Frequent falls may be early Alzheimer's sign. www.consumeraffairs.com
- Falls could show early signs of Alzheimer's disease. www.thesuitmagazine.com
- Falls may indicate earliest stages of Alzheimer's and need for further evaluation. Alzheimer's Association. www.prnewswire.com
- Falls an early sign for Alzheimer's. www.thirdage.com
- The researchers noted that Alzheimer's has been linked to balance. www.tcmwell.com
- Falls may be early sign of Alzheimer's. www.alzfamserv.org
- Falls may be early sign of Alzheimer's. http://esciencenews.com/
- Falls, retina changes could point to Alzheimer's. www.redorbit.com

In all these news flashes, not one referred to Inclusion Body Myositis. The question is the following: who didn't make the connection, the media or the researchers at the conference? Isn't it more likely that people develop coordination issues and fall more frequently due to a combination of factors including IBM and some sort of neurodegenerative process? Would it make sense that those with Alzheimer's who fall more frequently also have the comorbidity of IBM? IBM is the most common degenerating muscle disease in the elderly.

Both AD and IBM are amyloid diseases; thus it is likely that the amyloid is not just in the brain but also in the muscle and impacts not just cognitive functioning but also muscle functioning. Excessive falling is therefore a sign of a generalized amyloid disease that can impact many systems. Falls and loss of muscle mass are then an earlier indicator for Alzheimer's and also a more generalized systemic amyloid disease.

Let's take a look at one of these reports in detail to see what was noted. This is one as reported by WebMD: "Falls may be an early sign of Alzheimer's disease, researchers report. [126]"

In a study of 125 older adults who appeared physically and cognitively healthy, two-thirds of those with large deposits of Alzheimer's-associated plaque in their brains suffered falls. In contrast, only one-third of those with little or no plaque experienced falls. "This is a really important finding. We didn't expect to see such a significant increase—a doubling—of falls [in people with a lot of plaque]," says Susan Stark, PhD, of Washington University in St. Louis.

Is this a really important finding or a lack of a thorough diagnosis for IBM that has as symptoms, reduced muscle mass, trips and falls? But the authors and pundits persist ….

"To our knowledge, this is the first study to identify a risk of increased falls related to a diagnosis of pre-clinical Alzheimer's disease," she tells WebMD. "Preclinical Alzheimer's disease is used to describe people with large deposits of Alzheimer's-associated plaque in their brains, despite appearing cognitively normal." The findings were presented at the 2011 Alzheimer's Association International Conference.

"Falls are a leading cause of disability, premature nursing home placement, and death among older adults. Older people with Alzheimer's suffer more than double the rate of falls as people without the disorder, because of problems with balance, gait disorders, and visual and spatial perception," Stark says.

Stark notes that previous research has shown that weight loss may be another sign of preclinical Alzheimer's disease. "We're trained to look for cognitive symptoms, but we're finding there may be other symptoms, in this case, motor symptoms, as well," she says. "If the findings hold up, older patients who fall a lot may want to be examined for memory and other cognitive problems."

William E. Klunk, MD, PhD, Professor of Psychiatry and Neurology at the University of Pittsburgh School of Medicine and a member of the Alzheimer's Association's Medical & Scientific Advisory Council, says, "The study is important because it may help pave the way for earlier detection of Alzheimer's disease, before problems with memory emerge."

So this is what Alzheimer's evaluation has come to. The Paris 2011 conference represented 5000 of the brightest researchers presenting their latest findings, representing billions of your research dollars at work. People who fall more often may be on the path to AD. And those with amyloid deposits in their eye may also be heading toward AD (Chapter 6). Are you going to be able to determine if your loved one needs further evaluation based on a fall? How many falls are enough to warrant further investigation? Could the falls be related to blood pressure–lowering medication or some other confounding factor instead? The choice is yours to wait for an incident, much like a heart attack with cardiovascular disease. Or, you might decide to be proactive and have a broad array of more definitive tests, for example, for inflammation and eye biomarkers, performed. If you have a family member or close friend who falls more, what do you do next? What does the standard-of-care really provide? Insinuating that falls can be used as an early diagnostic is truly a raising of the white flag on understanding Alzheimer's. However, if you have an elderly parent who has fallen once or more, take action and seek a deeper understanding.

Clinically it may be easy to identify those predisposed to falls through simple observations. Those individuals with a slow gait, having difficulty standing without using their arms for support, or having a weak handshake are those likely with muscle disease. They are candidates for more broad and deep diagnostic testing.

ALZHEIMER'S DISEASE OF THE HEART

A group of North Carolina researchers may have closed the loop circling Alzheimer's and cardiovascular diseases. They published a paper in 2013 titled "Proteotoxicity and cardiac dysfunction—Alzheimer's disease of the heart?" [127]. Here is some key language from their abstract that explains the connection between these two diseases:

> Protein homeostasis plays a role in the development of numerous disorders. Misfolded proteins (beta-amyloid, for example) are central in the pathophysiology of neurodegenerative diseases such as Huntington's disease, Parkinson's disease, and Alzheimer's disease. In the past several years, (the same type of) misfolded proteins have been found to play a role in the pathophysiology of common human cardiac diseases such as a pathological cardiac hypertrophy (increase in volume) and dilated and ischemic (loss of blood flow) cardiomyopathies (heart muscle disease), leading to the suggestion that protein misfolding is a key contributor to the progression of heart failure.
>
> In this review, we explore the contribution of protein misfolding to the pathophysiology of cardiac disease, describing why these proteins become misfolded and how the innate systems that usually dispose of them break down. We then discuss how the knowledge obtained from studying protein misfolding in other diseases, such as Alzheimer's disease, may aid us in understanding the pathophysiological mechanisms of cardiac diseases and developing new treatments that focus on preventing or reversing protein misfolding in the heart.

What that last statement infers is rather amazing. They appear to be saying that we will learn how to treat heart disease by treating AD!

To summarize the work of Dr. de la Torre and this (new) group from North Carolina (and their contributors):

1. The heart is a major risk factor for AD but not the only one. AD can develop when vascular risk factors unfold during aging, a complex process that promotes brain hypoperfusion (decreased blood flow), progressive cognitive decline, and neurodegenerative changes.
2. Misfolded proteins occur in both diseases and offer an explanation as to the root cause(s) that are common and overlap in Alzheimer's and cardiovascular diseases.

Recall from Chapter 2 that many scientists, including the Kennedy Chaired Professor of Neurology at Harvard Medical School, indicate that the misfolded proteins are actually protective and part of our immune response. Thus beta-amyloid or misfolded proteins are likely a symptom, not a cause. It is Dr. McCully, even though he does not specifically refer to misfolded proteins, who describes the biology and physiology of the disease process that underlies these diseases. Dr. Christopher M. Dobson reminds us that accumulation of misfolded proteins is not exclusive to AD. However, the accumulation of beta-amyloid 1-42 protein in particular, in various tissues of our body, is a hallmark of the abnormal breakdown of a genetically normal protein that is associated with many of the chronic inflammatory diseases related to aging. In the context of AD, key inflammatory diseases are as follows:

- Alzheimer's disease: Brain
- Cortical Cataract: Lens of Eye
- Drusen: Age-related Macular Degeneration (AMD)
- Optic Nerve: Glaucoma

AMYLOIDOSIS

Diseases of misfolded proteins, also referred to as amyloidosis, are a group of diseases that have in common the abnormal accumulation of misfolded proteins that are often genetically abnormal, as opposed to beta-amyloid 1-42, and are prone to accumulate. Aging may be a combination of increased formation of these proteins along with a decreased ability to recognize and remove these proteins.

Here is a partial list of diseases that have amyloidosis as part of their pathology, not including those listed earlier:

Aortic medial amyloid aorta: The largest artery in the body, the aorta, arises from the left ventricle of the heart, goes up (ascends) a little ways, bends over (arches), and then goes down (descends) through the chest and through the abdomen to where it ends by dividing into two arteries called the common iliac arteries that go to the legs.

Atherosclerosis (also known as arteriosclerotic vascular disease or ASVD) is a condition in which an artery wall thickens. It is a syndrome affecting arterial blood vessels, a chronic inflammatory response in the walls of arteries that is characterized by inflammation and proliferation of the vasa vasorum (angiogenesis) and accumulation of inflammatory cells and lipid deposits.

Cardiac arrhythmias, isolated atrial amyloidosis (also known as arrhythmia and irregular heartbeat), are any of a large and heterogeneous group of conditions in which there is abnormal electrical activity in the heart. The heartbeat may be too fast or too slow, and may be regular or irregular.

Cerebral amyloid angiopathy (CAA), also known as congophilic angiopathy, is a form of angiopathy in which amyloid deposits form in the walls of the blood vessels of the central nervous system. The term congophilic is used because the presence of the abnormal aggregations of amyloid can be demonstrated by microscopic examination of brain tissue after application of a special stain called Congo red.

Diabetes mellitus type 2, formerly non–insulin-dependent diabetes mellitus (NIDDM) or adult-onset diabetes, is a metabolic disorder that is characterized by high blood glucose in the context of insulin resistance and relative insulin deficiency.

Dialysis-related amyloidosis is a form of amyloidosis associated with chronic renal failure.

Parkinson's disease (also known as Parkinson disease, Parkinson's, idiopathic parkinsonism, primary parkinsonism, PD, or paralysis agitans) is a degenerative disorder of the central nervous system. The motor symptoms of PD result from the death of dopamine-generating cells in the substantia nigra, a region of the midbrain; the cause of this cell death is unknown.

Rheumatoid arthritis (RA) is a chronic, systemic inflammatory disorder that may affect many tissues and organs, but principally attacks flexible (synovial) joints. The process produces an inflammatory response of the capsule around the joints (synovium) secondary to swelling (hyperplasia) of synovial cells, excess synovial fluid, and the development of fibrous tissue (pannus) in the synovium. The pathology of the disease process often leads to the destruction of articular cartilage and ankylosis of the joints.

Inclusion Body Myositis (IBM) is an inflammatory muscle disease, characterized by slowly progressive weakness and wasting of both distal and proximal muscles, most apparent in the muscles of the arms and legs. There are two types: sporadic inclusion body myositis (sIBM) and hereditary inclusion body myopathy (hIBM). In sporadic inclusion body myositis muscle, two processes, one autoimmune and the other degenerative, appear to occur in the muscle cells in parallel.

Systemic AL amyloidosis or just primary amyloidosis is the most common form of systemic amyloidosis in the United States. The disease is caused when a person's antibody-producing cells do not function properly and produce abnormal protein fibers made of components of antibodies called light chains. These light chains come together to form amyloid deposits in different organs, which can cause serious damage to these organs.

Transmissible spongiform encephalopathies (TSEs), also known as prion diseases, are a group of progressive conditions that affect the brain and nervous system of many animals, including humans. According to the most widespread hypothesis, they are transmitted by prions, although some other data suggest an involvement of a *Spiroplasma* infection. Mental and physical abilities deteriorate, and myriad tiny holes appear in the cortex causing it to appear like a sponge (hence "spongiform") when brain tissue obtained at autopsy is examined under a microscope. The disorders cause impairment of brain function, including memory changes, personality changes, and problems with movement that worsen over time. Prion diseases of humans include classic Creutzfeldt–Jakob disease, new variant Creutzfeldt–Jakob disease (nvCJD, a human disorder related to mad cow disease), Gerstmann–Sträussler–Scheinker syndrome, fatal familial insomnia, and kuru. These conditions form a spectrum of diseases with overlapping signs and symptoms.

Could it be that all these diseases stem from the same origin? If this is true, then one treatment or prevention method has the potential to alleviate all these dreaded diseases.

Dr. Dobson concludes his review on the process of amyloid and general misfolding of proteins on an optimistic note [123]:

Despite the complexity of the protein aggregation process, the findings described above show that dramatic progress in its elucidation has been made in recent years. This progress relates particularly to our understanding of the nature and significance of amyloid formation and to how this process relates to the normal and aberrant behavior of living organisms. Increasingly sophisticated techniques are now being applied to elucidate the "amyloid phenomenon" in ever-greater detail. Of special significance is the manner in which a wide variety of ideas from across the breadth of the biological, physical, and medical sciences is being brought together to probe important unifying principles.

… even our present understanding of the mechanism of amyloid formation is leading to more reliable methods of early diagnosis and to more rational therapeutic strategies that are either in clinical trials or approaching such trials. Thus, despite the rapidity with which diseases of the type discussed here are increasingly afflicting the human populations of the modern world, there are grounds for optimism that present progress in understanding their nature and origins will lead, in the not too distant future, to the beginnings of widely applicable and effective means to combat their spread and their debilitating consequences.

Let's all share in Dr. Dobson's excitement and become aware of the importance of this amyloid process as contributing to the overall process of aging, the acceleration of aging, and AD. Let's also learn from the mistakes of the past where too much emphasis and hope was placed on one factor of a disease that is of great concern. Alzheimer's is a multifactorial disease, and any solution to diagnosis and treatment will require looking at amyloid, the causes of the formation of amyloid, inflammation, cardiovascular diseases, and numerous other root causes.

DO RESEARCHERS KNOW THE CAUSE OF AMYLOID-BASED DISEASES?

Is there any research that exists today that might provide insight into the questions raised by Dr. Dobson? Is there a unifying theory of amyloid-based diseases?

Dr. Judith Miklossy of the International Alzheimer's Research Association in Switzerland provides an answer, at least in part, as to why misfolded proteins occur in Alzheimer's, and likely all other amyloid diseases. One of her many research articles in this area is titled "Chronic inflammation and amyloidogenesis in Alzheimer's disease—role of spirochetes" [128]. Dr. Miklossy suggests that microorganisms (bacteria) may be involved in the genesis of misfolded proteins and Alzheimer's. The abstract to her paper includes these thoughts:

Alzheimer's disease (AD) is associated with dementia, brain atrophy and the aggregation and accumulation of a cortical amyloid-β peptide (beta-amyloid). Chronic bacterial infections are frequently associated with amyloid deposition. It had been known for a century that the spirochete *Treponema pallidum* (bacteria with oral cavity origin) can cause a dementia that looks like Alzheimer's disease. Bacteria or their poorly degradable debris (dead bacterial cells) have a number of deleterious effects on the body including triggering inflammation. All these processes induced by the dead bacteria, when exposed to the immune system, are involved in the pathogenesis (development of disease) of AD. Bacteria, or their debris, highly prevalent in the population at large may initiate a cascade of events leading to chronic inflammation and amyloid deposition in AD.

Curiously, Dr. Alzheimer's initial hunch over 100 years ago has merit based on this emerging bacterial hypothesis. He suggested that microorganisms might cause the "senile plaques" now known as AD.

Taken together, the works of Dr. Miklossy, Dr. McCully, Dr. de la Torre, and Claude Bernard likely provide the necessary information to push Alzheimer's back at least 5 years, if not completely off the map, like it was at the turn of the 20th century. The work of Dr. Miklossy is highlighted in Chapter 9.

DIABETES AND ALZHEIMER'S

Our bodies and the mechanisms of disease are all connected. Another piece of circumstantial evidence is that, to this point in this book, we have mentioned the term "diabetes" and "diabetic" 200 times, yet have not written a section on diabetes. The connection between diabetes and eye diseases and vascular diseases is without question and we have tied these two diseases back to Alzheimer's. Dr. Perlmutter, through his famous book, "*Grain brain*," discusses the Alzheimer's/diabetes connection well and we recommend you read his work [129].

DIABETES, ALZHEIMER'S, AND WOMEN

Women suffer from Alzheimer's at a much higher frequency compared to men. In fact, approximately 50% more women than men have Alzheimer's. One reason often suggested is the extra life span of women, and this no doubt plays a role in Alzheimer's, which is a disease of aging. However, age alone does not fully explain the gross excess of women with the disease.

A very recent review tells us that diabetes is associated with a 60% increased risk of any dementia in both sexes [130]. This conclusion was based on 14 studies, 2,310,330 individuals, and 102,174 cases of dementia. Yes, these data are statistically significant. Digging into the details of this review, we find that the diabetes-associated "risk ratio" (research-speak for simply "risk") for vascular dementia is 2.34 in women and a much lower 1.73 in men. For nonvascular dementia, the risk is 1.53 in women and 1.49 in men. Overall, women with diabetes had a 19% greater risk for the development of dementia compared to men.

The increase in dementia for women compared to that for men, in diabetics, is for vascular dementia and not Alzheimer's specifically. However, we believe that the difference between vascular dementia and Alzheimer's is subtle at best and subject to both interpretation and misinterpretation.

AD has not been considered a vascular disease but new studies are slowly changing the consensus toward the truth—that Alzheimer's is a vascular disease. Here is a point to ponder. A large vessel (one involved in a stroke, e.g.) is >1 mm in diameter. The size of blood vessels varies enormously, from a diameter of about 25 mm (1 in.) of the aorta to only 8 μm in the capillaries. This is a 3000-fold range. But that is just in diameter. The capillary has a cross-sectional area (think of a water pipe) 100,000 times smaller compared to the aorta (area = 3.14159 times the square of the radius). This is the important measurement because pipes are round. Also, note that a capillary is too small to be detected and studied by MRI—the major technique used to look for brain changes associated with Alzheimer's. Thus medicine cannot see the cause—assuming it is a vascular blockage or other change at the capillary level.

Does it make sense that if large vessels can clog (and cause a stroke, e.g.), a capillary with a size up to 100,000 times smaller can also clog—and maybe even clogs first? We all probably suffer from some clogged capillaries. As more and more capillaries in our brain lose the ability to transport sufficient blood, neurons begin to die along with our memory. It is likely that vascular dementia and Alzheimer's are essentially the same diseases but, in Alzheimer's, the majority of the vascular issues stem from capillaries.

CONCLUSIONS

Understanding Alzheimer's requires an understanding of our whole body. No body system or disease mechanism can be overlooked. In this chapter we point mainly at vascular disease and its association with Alzheimer's. However, a thorough understanding of the true causes of vascular disease—including inflammation discussed in Chapter 7 and Infection discussed in Chapter 9—reveals the fundamentals of Alzheimer's.

The solution is to optimize your whole-body health. Be careful about the likely first line of defense of modern medicine for vascular diseases, for example, the lowering of blood pressure. Elevated blood pressure is not the disease; it is simply a symptomatic treatment to avoid problems associated with weak vessels. The combination of weak vessels and artificially lowered blood pressure will accelerate Alzheimer's/dementia. The real problem is the weak vessels and the underlying causes of this condition. Modern medicine does not have a solution for vascular disease causes and, by extension, Alzheimer's. Statins, beta-blockers, ACE inhibitors, blood pressure meds, and the like are exactly what you DON'T want to lower your AD risk. They treat symptoms, often at the expense of true brain health.

The infection theory of Alzheimer's is emerging. Many infectious species, notably *Chlamydia pneumoniae*, live in, weaken vessels, and disrupt and diminish blood flow. This pathogen has been found in many many Alzheimer's brains—by no coincidence. Our clinics look for and treat vascular diseases caused by infection as part of a whole-body approach to Alzheimer's. Our hope is that the standard-of-care (modern medicine) will catch up and start properly treating Alzheimer's sufferers.

References

[1] Sadly this guy is just another shill for the drug companies.
[2] Reiss AB, et al. Cholesterol in neurologic disorders of the elderly: stroke and Alzheimer's disease. Neurobiol Aging 2004;25(8):977–89.
[3] Stampfer MJ. Cardiovascular disease and Alzheimer's disease: common links. J Intern Med 2006;260(3):211–23.
[4] Ganguli M, et al. Mild cognitive impairment, amnestic type an epidemiologic study. Neurology 2004;63(1): 115–21.
[5] Sambamurti K, et al. Cholesterol and Alzheimers disease: clinical and experimental models suggest interactions of different genetic, dietary and environmental risk factors. Curr Drug Targets 2004;5(6):517–28.
[6] Cushman WC, et al. Effects of intensive blood-pressure control in type 2 diabetes mellitus. N Engl J Med 2010;362(17):1575.
[7] Li J, et al. Vascular risk factors promote conversion from mild cognitive impairment to Alzheimer disease. Neurology 2011;76(17):1485–91.
[8] de la Torre JC, et al. Brain blood flow restoration 'rescues' chronically damaged rat CA1 neurons. Brain Res 1993;623(1):6–15.
[9] de la Torre JC, Mussivand T. Can disturbed brain microcirculation cause Alzheimer's disease? Neurol Res 1993;15(3):146.
[10] Pappas BA, De La Torre JC, Davidson CM, Keyes MT, Fortin T. Chronic reduction of cerebral blood flow in the adult rat: late-emerging CA1 cell loss and memory dysfunction. Brain Res 1996;708(1):50–8.

[11] de la Torre JC. Cerebromicrovascular pathology in Alzheimer's disease compared to normal aging. Gerontology 1997;43(1–2):26–43.

[12] de la Torre JC. Cerebrovascular changes in the aging brain. Adv Cell Aging Gerontol 1997;2:77–107.

[13] de la Torre JC, et al. Reduced cytochrome oxidase and memory dysfunction after chronic brain ischemia in aged rats. Neurosci Lett 1997;223(3):165–8.

[14] de la Torre JC, et al. Reversal of ischemic-induced chronic memory dysfunction in aging rats with a free radical scavenger–glycolytic intermediate combination. Brain Res 1998;779(1):285–8.

[15] De Jong GI, Farkas E, Stienstra CM, Plass JRM, Keijser JN, de la Torre JC, et al. Cerebral hypoperfusion yields capillary damage in the hippocampal CA1 area that correlates with spatial memory impairment. Neuroscience 1999;91(1):203–10.

[16] de la Torre JC, Hachinski V. Cerebrovascular pathology in Alzheimer's disease. Ann NY Acad Sci 1997;826:1–523.

[17] de la Torre JC. Critical threshold cerebral hypoperfusion causes Alzheimer's disease? Acta Neuropathol 1999;98(1):1–8.

[18] Cada A, de la Torre JC, Gonzalez–Lima F. Chronic cerebrovascular ischemia in aged rats: effects on brain metabolic capacity and behavior. Neurobiol Aging 2000;21(2):225–33.

[19] de la Torre JC. Critically attained threshold of cerebral hypoperfusion: can it cause Alzheimer's disease? Ann NY Acad Sci 2000;903(1):424–36.

[20] de la Torre JC. Critically attained threshold of cerebral hypoperfusion: the CATCH hypothesis of Alzheimer's pathogenesis. Neurobiol Aging 2000;21(2):331–42.

[21] de la Torre JC, Stefano GB. Evidence that Alzheimer's disease is a microvascular disorder: the role of constitutive nitric oxide. Brain Res Rev 2000;34(3):119–36.

[22] de la Torre JC. Impaired brain microcirculation may trigger Alzheimer's disease. Neurosci Biobehav Rev 1994;18(3):397–401.

[23] Laske C, Stellos K. Vascular Pathophysiology of Alzheimer's disease. Curr Alzheimer Res 2014;11(1):1–3.

[24] de la Torre JC, Fay LA. Effects of aging on the human nervous system. Principles and practice of geriatric surgery. New York: Springer; 2001. p. 926–48.

[25] de la Torre JC. Alzheimer disease as a vascular disorder nosological evidence. Stroke 2002;33(4):1152–62.

[26] de la Torre JC, et al. Alzheimer's disease: vascular etiology and pathology. Ann NY Acad Sci 2002;977:1–526.

[27] de la Torre JC, et al. Hippocampal nitric oxide upregulation precedes memory loss and Aβ 1-40 accumulation after chronic brain hypoperfusion in rats. Neurol Res 2003;25(6):635–41.

[28] de la Torre JC. Alzheimer's disease is a vasocognopathy: a new term to describe its nature. Neurol Res 2004;26(5):517–24.

[29] de la Torre J, et al. Drug therapy in Alzheimer's disease. N Engl J Med 2004;351:1911–3.

[30] Aliyev A, Seyidova D, Rzayev N, Obrenovich ME, Lamb BT, Chen SG, et al. Is nitric oxide a key target in the pathogenesis of brain lesions during the development of Alzheimer's disease? Neurol Res 2004;26(5):547–53.

[31] Seyidova D, Aliyev A, Rzayev N, Obrenovich M, Lamb BT, Smith MA, et al. The role of nitric oxide in the pathogenesis of brain lesions during the development of Alzheimer's disease. In Vivo 2004;18(3):325–34.

[32] Goldsmith HS, de La Torre JC. Vascular dynamics in Alzheimer and vascular dementia. Neurol Res 2004;26(5):453.

[33] de la Torre JC. Cerebrovascular gene linked to Alzheimer's disease pathology. Trends Mol Med 2005;11(12):534–6.

[34] de la Torre JC, Aliev G. Inhibition of vascular nitric oxide after rat chronic brain hypoperfusion: spatial memory and immunocytochemical changes. J Cereb Blood Flow Metab 2005;25(6):663–72.

[35] de la Torre JC. Is Alzheimer's disease preceded by neurodegeneration or cerebral hypoperfusion? Ann Neurol 2005;57(6):783–4.

[36] Aliyev A, Chen SG, Seyidova D, Smith MA, Perry G, de la Torre J, et al. Mitochondria DNA deletions in atherosclerotic hypoperfused brain microvessels as a primary target for the development of Alzheimer's disease. J Neurol Sci 2005;229:285–92.

[37] de la Torre JC. How do heart disease and stroke become risk factors for Alzheimer's disease? Neurol Res 2006;28(6):637–44.

[38] de la Torre JC. The deadly triad: heart disease, stroke and Alzheimer's disease. Neurol Res 2006;28(6):577–8.

[39] de la Torre JC. Alzheimer's disease prevalence can be lowered with non-invasive testing. J Alzheimers Dis 2008;14(3):353–9.

[40] de la Torre JC. Pathophysiology of neuronal energy crisis in Alzheimer's disease. Neurodegener Dis 2008;5(3–4):126–32.

[41] de la Torre JC. Carotid artery ultrasound and echocardiography (CAUSE) testing to lower the prevalence of Alzheimer's disease. J Stroke Cerebrovasc Dis 2009;18:319–28.

[42] de la Torre JC. Cerebral and cardiac vascular pathology in Alzheimer's disease. In: Maccione R, Perry G, editors. Current hypotheses and research milestones in Alzheimer's disease. New York, Heidelberg: Springer; 2009. p. 159–69.

[43] de la Torre JC. Cerebrovascular and cardiovascular pathology in Alzheimer's disease. Int Rev Neurobiol 2009;84:35–48.

[44] de la Torre JC. Alzheimer's disease is incurable but preventable. J Alzheimers Dis 2010;20(3):861–70.

[45] de la Torre JC. Basics of Alzheimer's disease prevention. J Alzheimers Dis 2010;20:687–8.

[46] de la Torre JC. The vascular hypothesis of Alzheimer's disease: bench to bedside and beyond. Neurodegener Dis 2010;7(1–3):116–21.

[47] de la Torre JC. Vascular risk factor detection and control may prevent Alzheimer's disease. Ageing Res Rev 2010;9(3):218–25.

[48] de la Torre JC. Three postulates to help identify the cause of Alzheimer's disease. J Alzheimers Dis 2011;24(4):657–68.

[49] de la Torre JC. A tipping point for Alzheimer's disease research. J Alzheimers Dis Parkinsonism 2012;2:e120.

[50] de la Torre JC. A turning point for Alzheimer's disease? Biofactors 2012;38(2):78–83.

[51] de la Torre JC. Alzheimer's disease: how does it start? J Alzheimers Dis 2002;4(6):497–512.

[52] de la Torre JC. Cardiovascular risk factors promote brain hypoperfusion leading to cognitive decline and dementia. Cardiovasc Psychiatry Neurol 2012;2012:367516.

[53] de la Torre JC. Cerebral hemodynamics and vascular risk factors: setting the stage for Alzheimer's disease. J Alzheimers Dis 2012;32(3):553–67.

[54] de la Torre JC. Physiopathology of vascular risk factors in Alzheimer's disease. J Alzheimers Dis 2012;32(3):517–8.

[55] de la Torre JC. Cardiovascular disease promotes cognitive dysfunction that can signal Alzheimer's and vascular dementia. In: Waldstein S, Elias M, editors. Neuropsychology of cardiovascular disease. 2nd ed. New York: Taylor & Francis; 2015.

[56] Stone J, De La Torre JC. Alzheimer's disease is associated with increased risk of haemorrhagic stroke. Evid Based Ment Health 2013;16:88.

[57] De la Torre JC. Detection, prevention, and pre-clinical treatment of Alzheimer's disease. J Alzheimers Dis 2014;42(Suppl. 4):S327–8.

[58] de la Torre JC. Cerebral perfusion enhancing interventions: a new strategy for the prevention of Alzheimer dementia. Brain Pathol 2016;5(2016):618–31.

[59] Hardy J, Selkoe DJ. The amyloid hypothesis of Alzheimer's disease: progress and problems on the road to therapeutics. Science 2002;297:353–6.

[60] Alzheimer A, Stelzmann RA, Schnitzlein HN, Murtagh FR. An English translation of Alzheimer's 1907 paper, "Uber eine eigenartige Erkankung der Hirnrinde". Clin Anat 1995;8:429–31.

[61] Kambhampaty A, Smith-Parker J. Eli Lilly's solanezumab faces grim prospects of attaining conditional FDA approval in mild Alzheimer's. Financial Times; 2012.

[62] Holmes C, Boche D, Wilkinson D, Yadegarfar G, Hopkins V, et al. Long-term effects of Abeta42 immunisation in Alzheimer's disease: follow-up of a randomised, placebo-controlled phase I trial. Lancet 2008;372:216–23.

[63] Aisen PS. Alzheimer's disease therapeutic research: the path forward. Alzheimers Res Ther 2009;1:2.

[64] Arriagada PV, Growdon JH, Hedley-Whyte ET, Hyman BT. Neurofibrillary tangles but not senile plaques parallel duration and severity of Alzheimer's disease. Neurology 1992;42:631–9.

[65] Bouras C, Kövari E, Herrmann FR, Rivara CB, Bailey TL, et al. Stereologic analysis of microvascular morphology in the elderly: Alzheimer disease pathology and cognitive status. J Neuropathol Exp Neurol 2006;65:235–44.

[66] Pimplikar SW. Reassessing the amyloid cascade hypothesis of Alzheimer's disease. Int J Biochem Cell Biol 2009;41:1261–8.

[67] Fjell AM, Walhovd KB. Neuroimaging results impose new views on Alzheimer's disease—the role of amyloid revised. Mol Neurobiol 2012;45:153–72.

[68] Deschaintre Y, Richard F, Leys D, Pasquier F. Treatment of vascular risk factors is associated with slower decline in Alzheimer disease. Neurology 2009;73:674–80.

[69] Alzheimer's Association. 2012 Alzheimer's disease facts and figures. Alzheimers Dement 2012;8:131–68.

[70] Ravnskov U, McCully KS. Vulnerable plaque formation from obstruction of vasa vasorum by homocysteinylated and oxidized lipoprotein aggregates complexed with microbial remnants and LDL autoantibodies. Ann Clin Lab Sci 2009;39(1):3–16.

[71] Thelle DS, Tverdal A, Selmer R. Is the use of cholesterol in mortality risk algorithms in clinical guidelines valid? Ten years prospective data from the Norwegian HUNT 2 study. J Eval Clin Pract 2012;18(6):1219.

[72] Koester W. Endarteritis and arteritis. Berl Klin Wochenschr 1876;13:454–5.

[73] McCully KS. Chemical pathology of homocysteine. I. Atherogenesis. Ann Clin Lab Sci 1993;23(6):477–93.

[74] McCully KS. Chemical pathology of homocysteine. II. Carcinogenesis and homocysteine thiolactone metabolism. Ann Clin Lab Sci 1994;24(1):27–59.

[75] McCully KS. Chemical pathology of homocysteine. III. Cellular function and aging. Ann Clin Lab Sci 1994;24(2):134–52.

[76] McCully KS. Homocysteine and heart disease. 2004;81:81–83.

[77] McCully KS. Hyperhomocysteinemia and arteriosclerosis: historical perspectives. Clin Chem Lab Med 2005;43(10):980–6.

[78] McCully KS. Homocysteine, vitamins, and vascular disease prevention. Am J Clin Nutr 2007;86(5):1563S–8S.

[79] McCully K, McCully M. The heart revolution. New York: HarperCollins; 2009.

[80] McCully KS. Chemical pathology of homocysteine. IV. Excitotoxicity, oxidative stress, endothelial dysfunction, and inflammation. Ann Clin Lab Sci 2009;39(3):219–32.

[81] McCully KS. Chemical pathology of homocysteine. V. Thioretinamide, thioretinaco, and cystathionine synthase function in degenerative diseases. Ann Clin Lab Sci 2011 Fall;41(4):301–14.

[82] Ravnskov U, McCully KS. How macrophages are converted to foam cells. J Atheroscler Thromb 2012;19:949–50.

[83] Ingenbleek Y, McCully KS. Vegetarianism produces subclinical malnutrition, hyperhomocysteinemia and atherogenesis. Nutrition 2012;28(2):148–53.

[84] Ravnskov U, McCully KS. Biofilms, lipoprotein aggregates, homocysteine, and arterial plaque rupture. MBio. 2014;5(5):e01717–e1814.

[85] McCully KS. Homocysteine metabolism, atherosclerosis, and diseases of aging. Compr Physiol 2015;6(1):471–505.

[86] McCully KS. Homocysteine and the pathogenesis of atherosclerosis. Expert review of clinical pharmacology 2015;8(2):211–9.

[87] McCully KS. The active site of oxidative phosphorylation and the origin of hyperhomocysteinemia in aging and dementia. Ann Clin Lab Sci 2015;45(2):222–5.

[88] Ravnskov U, et al. Lack of an association or an inverse association between low-density-lipoprotein cholesterol and mortality in the elderly: a systematic review. BMJ Open 2016;6(6):e010401.

[89] McCully KS. Homocysteine, infections, polyamines, oxidative metabolism, and the pathogenesis of dementia and atherosclerosis. J Alzheimers Dis 2016;54(4):1283–90.

[90] Brattström L, et al. Hyperhomocysteinaemia in stroke: prevalence, cause, and relationships to type of stroke and stroke risk factors. Eur J Clin Invest 1992;22(3):214–21.

[91] Boushey CJ, et al. A quantitative assessment of plasma homocysteine as a risk factor for vascular disease. JAMA 1995;274(13):1049–57.

[92] De Ruijter W, et al. Use of Framingham risk score and new biomarkers to predict cardiovascular mortality in older people: population-based observational cohort study. BMJ 2009;338:a3083.

[93] Seshadri S, et al. Plasma homocysteine as a risk factor for dementia and Alzheimer's disease. N Engl J Med 2002;346(7):476–83.

[94] The nun study. Washington, DC: National Institutes of Health; 1997.

[95] Riggs KM, et al. Relations of vitamin B-12, vitamin B-6, folate, and homocysteine to cognitive performance in the Normative Aging Study. Am J Clin Nutr 1996;63(3):306–14.

[96] La Rue A, et al. Nutritional status and cognitive functioning in a normally aging sample: a 6-y reassessment. Am J Clin Nutr 1997;65(1):20–9.

[97] Ford AH, Almeida OP. Effect of homocysteine lowering treatment on cognitive function: a systematic review and meta-analysis of randomized controlled trials. J Alzheimers Dis 2012;29(1):133–49.

[98] Dierkes J, et al. *Helicobacter pylori* infection, vitamin B12 and homocysteine. Dig Dis 2003;21(3):237–44.

[99] McCaddon A, et al. Total serum homocysteine in senile dementia of Alzheimer type. Int J Geriatr Psychiatry 1998;13(4):235–9.
[100] Clarke R, et al. Plasma homocysteine, folate and vitamin B-12 as risk factors for confirmed Alzheimer's disease. Arch Neurol 1998;55(1449):55.
[101] Clark R, et al. Folate, vitamin B12, and serum total homocysteine levels in confirmed Alzheimer's disease. Arch Neurol 1998;55:1449–55.
[102] Doyle RM, et al. Exploring the homocysteine pathway in Alzheimer's disease. Arch Neurol 1998;55:1449–55.
[103] Morris MS. Homocysteine and Alzheimer's disease. Lancet Neurol 2003;2(7):425–8.
[104] Miller JW. Homocysteine and Alzheimer's disease. Nutr Res 1999;57(4):126–9.
[105] Miller JW. Homocysteine, Alzheimer's disease, and cognitive function. Nutrition. 2000;16:675–7.
[106] White AR, et al. Homocysteine potentiates copper-and amyloid-beta peptide-mediated toxicity in primary neuronal cultures: possible risk factors in the Alzheimer's-type neurodegenerative pathways. J Neurochem 2001;76(5):1509–20.
[107] Flicker L, Martins RN, Thomas J, et al. Homocysteine, Alzheimer genes and proteins, and measures of cognition and depression in older men. J Alzheimers Dis. 2004;6(3):329–36.
[108] Ravaglia G, Forti P, Maioli F, et al. Homocysteine and folate as risk factors for dementia and Alzheimer disease. Am J Clin Nutr 2005;82(3):636–43.
[109] Seshadri S. Elevated plasma homocysteine levels: risk factor or risk marker for the development of dementia and Alzheimer's disease? J Alzheimers Dis. 2006;9(4):393–8.
[110] Köseoglu E, et al. Relations between homocysteine, folate and vitamin B12 in vascular dementia and in Alzheimer disease. Clin Biochem 2007;40(12):859–63.
[111] Kageyama M, et al. Relationship between genetic polymorphism, serum folate and homocysteine in Alzheimer's disease. Asia Pac J Public Health 2008;(20 Suppl.):111–7.
[112] Homocysteine a prognostic marker for cognitive decline in Alzheimer's patients. The Brown University Geriatric Psychopharmacology Update, 2009;13(8):1-8.
[113] Oulhaj A, et al. Homocysteine as a predictor of cognitive decline in Alzheimer's disease. J Geriatr Psychiatry 2010;25(1):82–90.
[114] Chen C-S, et al. Plasma homocysteine level and apathy in Alzheimer's disease. J Am Geriatr Soc 2011;59(9):1752–4.
[115] Herrmann W, Obeid R. Homocysteine: a biomarker in neurodegenerative diseases. Clin Chem Lab Med 2011;49(3):435–41.
[116] Hooshmand B, et al. Plasma homocysteine, Alzheimer and cerebrovascular pathology: a population-based autopsy study. Brain 2013;136(Pt 9):2707–16.
[117] Chai GS, et al. Betaine attenuates Alzheimer-like pathological changes and memory deficits induced by homocysteine. J Neurochem 2013;124(3):388–96.
[118] Ya-Qin L, et al. Hydroxysafflor yellow a ameliorates homocysteine-induced Alzheimer-like pathologic dysfunction and memory/synaptic disorder. Rejuvenation Res. 2013;16(6):446–52.
[119] Li JG, et al. Homocysteine exacerbates β-amyloid pathology, tau pathology, and cognitive deficit in a mouse model of Alzheimer disease with plaques and tangles. Ann Neurol. 2014;75(6):851–63.
[120] Zheng Z, et al. Correlation between behavioural and psychological symptoms of Alzheimer type dementia and plasma homocysteine concentration. Biomed Res Int 2014;2014:383494.
[121] Wu HQ, Zhang X, Tang Y. Association of *Helicobacter pylori* infection and serum homocysteine level in patients with cerebral infarction. Zhejiang Da Xue Xue Bao Yi Xue Ban 2012;41(1):89–92.
[122] Van den Kommer TN, et al. Homocysteine and inflammation: predictors of cognitive decline in older persons? Neurobiol Aging 2010;31(10):1700–9.
[123] Chiti F, Dobson CM. Protein misfolding, functional amyloid, and human disease. Annu Rev Biochem 2006;75:333–66.
[124] Askanas V, Engel WK. Inclusion-body myositis A myodegenerative conformational disorder associated with Aβ, protein misfolding, and proteasome inhibition. Neurology 2006;66(1 Suppl. 1):S39–48.
[125] Askanas V, et al. Twisted tubulofilaments of inclusion body myositis muscle resemble paired helical filaments of Alzheimer brain and contain hyperphosphorylated tau. Am J Pathol 1994;144(1):177.
[126] Laino C. Falls an early clue to Alzheimer's. WebMD, 2011. Available from: http://www.webmd.com/alzheimers/news/20110718/falls-an-early-clue-to-alzheimers#1.

[127] Willis MS, Patterson C. Proteotoxicity and cardiac dysfunction—Alzheimer's disease of the heart? N Engl J Med 2013;368(5):455–64.

[128] Miklossy J. Chronic inflammation and amyloidogenesis in Alzheimer's disease—role of spirochetes. J Alzheimers Dis 2008;13(4):381–91.

[129] Perlmutter D. "Grain brain." The surprising truth about wheat, carbs, and sugar—your brain's silent killers. New York: Little, Brown and Company; 2013.

[130] Chatterjee S, et al. Type 2 diabetes as a risk factor for dementia in women compared with men: a pooled analysis of 2.3 million people comprising more than 100,000 cases of dementia. Diabetes Care 2016;39(2):300–7.

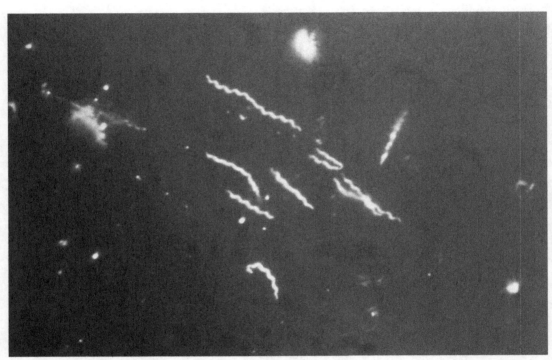

Image Credit: Borrelia burgdorferi. Available at Wikimedia Commons: https://upload.wikimedia.org/wikipedia/commons/f/f3/Borrelia_burgdorferi_%28CDC-PHIL_-6631%29_lores.jpg

CHAPTER

9

Does Infection Cause Alzheimer's?

The textbooks say in 1900 most people died of infectious diseases, and today most people don't die of infectious disease; they die of cancer and heart disease and Alzheimer's and all these things. Well, in ten years I think the textbooks will need to be rewritten to say, 'Throughout history most people have died of infectious disease, and most people continue to die of infectious disease.' It appears that many diseases we didn't think were infectious may be caused by infectious agents after all.

—*Paul Ewald, Professor of Biology, University of Louisville*

Based on medical research dating back over 100 years, there is little doubt about the connection between infection, Alzheimer's disease (AD), and other neurodegenerative diseases. The purpose of this chapter is to provide sufficient evidence to convince you about this connection. Now, assuming you appreciate this infection/Alzheimer's connection: what promotes infection leading to AD in some, while others are apparently protected? This question is as fundamental as is the question about the causes of aging. Ultimately, we find that it relates to our overall health and the changing health of our immune system with age (immunosenescence). However, as you will discover in this chapter, the infection/Alzheimer's connection provides a target for both diagnosis and treatment for those seeking prevention and those already afflicted with the disease.

Does infection cause AD? The answer is a decided "no," but infection is very much involved in the disease process. An apt way to describe the involvement of infection is it "strikes the final blow." Infection amplifies and exacerbates the effects of the other neurodegenerative processes. Examining and understanding other chronic infectious diseases sheds light on the cause/effect of infection in AD.

Toxoplasmosis: The Center for Disease Control (CDC) is concerned about Toxoplasmosis and its health consequences. Toxoplasmosis is an infection due to the parasite *Toxoplasma gondii*. According to the CDC, Toxoplasmosis is considered to be a leading cause of death attributed to foodborne illness in the United States. More than 60 million men, women, and children in the United States carry the Toxoplasma parasite (and it likes brain tissue), but very few have symptoms because the immune system usually keeps the parasite from causing illness.

However, women newly infected with Toxoplasma during pregnancy and anyone with a compromised immune system should be aware that toxoplasmosis could have severe consequences, including birth defects and death. Toxoplasmosis is considered one

The End of Alzheimer's. http://dx.doi.org/10.1016/B978-0-12-812112-2.00009-4

of the Neglected Parasitic Infections, a group of five parasitic diseases that have been targeted by CDC for public health action.

So, what does this information mean in the context of Alzheimer's? Potentially deleterious infectious species are present in many (if not all) of us. Toxoplasmosis is present in one-sixth of American, and there are many more "latent" bacteria like Toxoplasmosis. The bacteria are not active, or at least are not present at clinically significant levels in most people because they have healthy immune systems battling to suppress it. Chapter 7 discussed Alzheimer's and inflammation. People with inflammation have immune systems that are active, and possibly compromised. These individuals are the ones susceptible to the adverse consequences of bacterial infection. It is in these inflamed "hosts" that the bacteria are opportunistic and proliferate, causing disease.

Tuberculosis: One-third of the world's population is thought to have been infected with *Mycobacterium tuberculosis* [1], with new infections occurring in about 1% of the population each year. In 2007, an estimated 13.7 million chronic cases were active globally [2]. The rate of tuberculosis in different areas varies across the globe; about 80% of the population in many Asian and African countries test positive in tuberculin tests, while only 5%–10% of the United States population test positive [3]. More people in the developing world contract tuberculosis because of a poor nutrition and impaired immune systems that have been accentuated by high rates of HIV infection and the corresponding development of AIDS. Again we note that many people have the tuberculosis bacteria, but relatively few show clinical symptoms of the diseases of tuberculosis. In fact, only 0.21% of those infected with tuberculosis actually show disease symptoms.

Rickettsia bacteria: Here is another example of a bug that contributes to Alzheimer's or Alzheimer's-like symptoms. It is present in many but expressed in few. For those with clinically relevant levels of rickettsias, the consequences are severe. Consider this definition of the disease caused by rickettsias. "Typhus (Richettsia Typhii) is any of several similar diseases caused by Rickettsia bacteria. The name comes from the Greek typhos meaning smoky or hazy, describing the state of mind of those affected with typhus. The causative organism Rickettsia is an obligate intracellular parasite bacterium [4]." Smoky or hazy mind certainly sounds like a dementia and possibly AD.

By extrapolation, this helps us appreciate the cause/effect between Alzheimer's and infection. As you will see in this chapter, infection is present in Alzheimer's, and it likely participates in the disease process. However, many people without Alzheimer's likely harbor the infectious species, too. What prevents them from having AD is a younger and healthier immune system. Those with the disease have an older immune system, that has suffered immune system decay (immunosenescence) and/or suppression, and have some or all the risk factors leading to that situation. These people have more activation of their innate immune system to control infection, as seen through an increase in inflammation. These are the true root causes of Alzheimer's.

INFECTION AND ALZHEIMER'S DISEASE

Alzheimer's disease (AD), the most frequent cause of dementia, is a form of amyloidosis. It has been known for a century that dementia, brain atrophy and amyloidosis can be caused by chronic bacterial infections, namely by Treponema pallidum (bacterium with spiral shape). Bacteria and viruses are powerful stimulators of inflammation. It was suggested by Oscar Fischer in 1907 and discussed by Aloïs Alzheimer and colleagues that microorganisms might be contributors in the generation of senile plaques in AD [5,6].

—ScienceDaily, May 25, 2008

Microbes, including bacteria, virus, and fungi, even when dead (their cellular remnants) are powerful stimulators of the innate and acquired immune system. The connection between microorganisms and the immune system is apparent in essentially any definition of the immune system and its action.

> The immune system is a system of biological structures and processes within an organism that protects against disease. In order to function properly, an immune system must detect a wide variety of agents, from viruses to parasitic worms, and distinguish them from the organism's own healthy tissue. If you survive childhood, your immune system does this well and only starts to decline at age 55 or so.

Amyloid deposits in the brain, the so-called senile plaques, are the hallmark of AD. Much of the research on Alzheimer's is conducted on animals. Do you know how the amyloid deposits are often produced in models of AD in animals used to study the disease? These amyloid deposits are induced by injecting living or dead bacteria (LPS) into the experimental animals. The result is a strong immune response and inflammation [7].

The other hallmark of AD, the "tangles," are also formed in animals in the presence of bacteria and the cell membranes of dead bacteria (LPS). Judith Miklossy from Switzerland along with researchers from the United States, Australia, and Canada showed these tangles, known medically as hyperphosphorylated tau, are tied to bacteria. Tau is a very important component of the AD process. The specific type of bacteria used in the study is called a spirochete. The bacteria associated with Lyme disease and several predominant periodontal bacteria are part of a family of bacteria known as spirochetes.

Definition of Lipopolysaccharides (LPS) also known as lipoglycans, are large molecules consisting of a lipid and a polysaccharide joined by a bond; they are found in the outer membrane of Gram-negative bacteria, act as endotoxins, and cause strong immune responses in animals.

Definition of Endotoxin toxin that is a structural molecule of the bacteria that is recognized by the immune system.

Definition of Spirochetes belong to a phylum of distinctive diderm (double-membrane) bacteria, most of which have long, helically coiled (spiral-shaped) cells.

The LPS of bacteria is a powerful inflammatory factor and also drives the production of amyloid deposits as seen in both in vitro (outside the body) and in vivo (in the body) animal models. Either dormant bacterium poised to reactivate or bacterial remnants of dead bacteria may persist indefinitely in the affected tissues. They may thus act as a chronic "insult" inducing chronic inflammation.

Definition of Insult any stressful stimulus which, under normal circumstances, does not affect the host organism, but which may result in morbidity (disease), when it occurs in a background of preexisting compromising conditions or in an aging immune system.

Diseases of aging, including AD, do not set in quickly. They develop slowly over time (silent inflammation, parainflammation, and inflammaging). The question becomes: is it possible that bacterial infection is present for the duration as the disease incubates or is it there just at the end of the disease and strikes the final blow? What we do know is that certain types of bacterial cell walls are highly resistant to degradation by human enzymes and thus may be highly persistent, creating a continuous or intermittent, but unrelenting attack that causes our immune system to trigger processes that show up as chronic inflammation. The robust nature of infectious agents was shown in a University of North Carolina paper published in 1967 titled, *Persistence of group A streptococcal cell walls related to chronic inflammation of rabbit dermal connective tissue* [8].

> It has been proposed that the chronic inflammatory process is the direct result of the persistence in tissue of complexes from streptococcal cell walls... Thus, the cell wall fragments can persist in tissue in a relatively innocuous state, and as the polysaccharide is gradually removed by tissue enzymes, the mucopeptide is exposed to produce chronic irritation.

Researchers south of the border in South Carolina showed that bacteria might infect a remote site in the body, which on interaction with the immune system, leads to chronic general inflammation. The now classic paper by Fox, titled, *Role of bacterial debris in inflammatory diseases of the joint and eye*, sheds light on the elusiveness of bacteria [9]. The abstract is provided here:

> Several distinct rheumatic conditions (including Lyme arthritis, Reiter's syndrome, and rheumatic fever) as well as certain forms of the blinding disease, uveitis, may share a common etiology. In each instance specific bacterial pathogens may infect a distant site, which on interaction with the immune system, leads to a sterile inflammation in the joint or eye. These 'reactive' conditions may result, in some cases, from prior localization of non-viable bacterial remnants (including the cell wall or peptidoglycan (LPS)) or alternatively 'dormant' fastidious bacteria in the affected joint or eye where they act as persisting antigens.
>
> Classical diagnostic (cell culture) techniques would not detect the presence of these alleged microbial antigens. Alternative approaches for detection of ubiquitous components of bacteria in the host (using appropriate chemical, molecular, and immunological techniques) are discussed.

A lesson learned? Even if it is believed that infectious species are involved in a chronic inflammatory disease, it may be difficult to prove. Thus a very robust process of diagnosis is required. A controversial approach is to assume their presence, based on inferential evidence, such as inflammatory markers, and consider treatment options, as long as such treatments "do no harm." The stealth nature of certain bacteria creates a conundrum. Within the standard-of-care, if its not detected, it is not there, and thus not treated. As you will see, these bacteria are there and need to be treated. Here is a quote from Claude Bernard, one of the top medical scientists of the 19th century:

> The experimenter who does not know what he is looking for will not understand what he finds.

Lyme disease is arguably the most well-known of the chronic diseases of bacteria. The cause of Lyme disease, discovered in 1982, is a class of bacteria known as spirochetes [10]. In particular, Lyme is caused by the spirochete know as Borrelia. *Borrelia burgdorferi* is the pathogen associated with Lyme disease in North America. There are now recognized to be 18 genomic species of Borrelia that may contribute to disease, one of which we refer to as Lyme disease. Despite these numbers, just *B. burgdorferi* is subject of testing and diagnosis in the United States. In Europe, testing for "Lyme" includes at least three forms of Borrelia.

To exacerbate the diagnostic gap for Borrelia, research papers over a large span of time show that FDA approved tests for "Lyme" often give variable, thus unreliable results. A study from 2011 titled, *Large differences between test strategies for the detection of anti-Borrelia antibodies are revealed by comparing eight ELISAs and five immunoblots* [11].

> In conclusion, ELISAs (enzyme-linked immunosorbent assay) and immunoblots for detecting anti-Borrelia antibodies have widely divergent sensitivity and specificity, and immunoblots for detecting anti-Borrelia antibodies have only limited agreement. Therefore, the choice of ELISA–immunoblot combination severely influences the number of positive results, making the exchange of test results between laboratories with different methodologies hazardous. The widespread availability of more specific and sensitive assays for the detection of anti-Borrelia antibodies will open the way for a reappraisal of the two-tier testing system.

Another study on the precision and accuracy of Lyme bacteria tests dates back to 1993 [12], and estates that "the results demonstrate the variable performance of commercial serologic kits for detection of antibodies to Borrelia burgdorferi."

These simple studies explain that infectious species that cause chronic illnesses can be very low grade and hide, and that even the more well-known of these bacteria are difficult to identify with a broad range of approved tests. These simple truths underscore that it is highly presumptuous to assume they are not present and not a potential root cause or exacerbator of AD.

Prior to the discovery of Alzheimer's, a common disease of senility was referred to as "general paresis." General paresis of the insane was first described as a distinct disease in 1822. Originally, the cause was believed to be an inherent weakness of character or constitution. While Esmarch and Jessen had asserted as early as 1857 that syphilis caused general paresis, progress toward the general acceptance by the medical community of this idea was only accomplished later [13].

Definition of Syphilis a sexually transmitted infection caused by the spirochete bacterium *Treponema pallidum* subspecies pallidum.

Finally, in 1913, all doubt about the syphilitic nature of general paresis was finally eliminated when researchers definitively found syphilitic spirochetes in the brains of the afflicted. Is history repeating itself in AD? Consider the following quote from an article titled, *On Some Clinical Manifestations of Congenital Syphilis* by Hubert Armstrong, MD in 1914 [14].

> In 1914 a bewildered doctor noted, "The discovery of the spirochete and of the Wassermann [test] ... has thrown the whole subject of syphilis into the melting-pot, from which new conceptions of the disease are emerging ... These are still malleable, and fresh investigations are daily entailing their remodeling..."

Dr. Armstrong astutely indicated that the diagnosis (Wassermann test) was changing the face of medicine.

Today, when Dr. Judith Miklossy of Switzerland discusses the connection between spirochetes and AD she is often met with bewilderment! Her first paper on this subject was in 1993 titled, *Alzheimer's disease—A spirochetosis*? [15] Notice the 80-year time gap between the assertion and acceptance of a connection between general paresis and bacteria. Can your loved one with Alzheimer's wait another 80 years for the standard-of-care to catch up to research and provide bona fide diagnoses and treatments?

Fischer put forth the first suggestion that bacteria, other than syphilis, may be involved in dementia in 1907. He suggested that senile plaques might correspond to colonies of microorganisms. Shortly thereafter, in 1911 Dr. Alois Alzheimer made the same suggestion. A few years later, in 1913, researchers showed that a bacterium associated with syphilis was directly linked to dementia. Based on this 1913 study, it is becoming accepted that the bacterium is a cause for brain (cortical) atrophy, activation of microglial cells (the brain immune defense system), and amyloid deposits. That's over 100 years ago!

At the time (1913), the syphilis bacterium was identified to cause general paresis. We now know that general paresis and AD clinically and biochemically look like the same disease. Alzheimer's himself noted the clinical similarity of his now-famous patient to the paresis. In many early studies through the 1920s, authors described the syphilis bacterium colonies as confined to the brain in patients with general paresis, and the distribution of these colonies were identical to those of the senile plaques that we now refer to as the Alzheimer's hallmark beta-amyloid plaques. The other AD hallmark, the neurofibrillary tangles, have also been described in general paresis as early as 1908 [16] and later described in 1910 [17] and 1958 and are thus also present with bacteria [18].

SMOLDERING CHRONIC INFLAMMATION, INFECTION, AND ALZHEIMER'S DISEASE

Consider the possibility that slow acting unconventional infectious agents, ones that may even look like a virus or hide inside cells, requiring decades to become active, may be involved in AD. Maybe these pathogens are present at birth and slowly become more active as we age and our immune system weakens? What if these disease-causing bugs are opportunistic and proliferate when our immune system becomes involved in fighting off other disease from as broad as tooth decay to concussion? This is not a new concept, having been proposed, at least in part, by several authors going back to the 1970s. H. M. Wisniewski from the New York State Institute for Basic Research in Developmental Disabilities wrote in 1978, *Possible viral etiology of neurofibrillary changes and neuritic plaques* [19].

Several authors crossed the meridian into the bacteria/chronic disease and AD camp from 1987–96. A scholar.google search for "infection" and "Alzheimer's" anywhere in a medical journal yields over 90,000 records. Refining the search such that the keywords are only in the title yielded 17 hits (be careful because scholar.google does not accommodate wild cards so the spelling must be an exact match to the title, for example, there are 7 hits with "bacteria" but 10 hits with "bacterial." If the search is broadened to include dementia, 167 records are noted. Some of the earliest titles include, *Cytomegalovirus infection and Alzheimer's disease* [20], and *Recent topics on Alzheimer's disease. The possibility of Alzheimer's disease as a delayed-type infection--study on prions of proteinaceous infective particles* [21].

The underlying bacteria responsible for Lyme disease is the difficult-to-detect bug, *B. burgdorferi*. Lyme is not just a disease of the joints. Dementia associated with brain (cortical) atrophy and activation of microglia (brain immune response) occurs in late stages of chronic Lyme infection. The Lyme bacteria are from the class called spirochetes. There are striking similarities between neurosyphilis and neuroborreliosis (Lyme disease) with respect to the clinical and pathological aspects. Fallon from Columbia University established the parallels in 1994 [22]. Excerpts from the abstract titled, *Lyme Disease: A Neuropsychiatric Illness*, follow:

> Lyme disease is a multisystemic illness that can affect the central nervous system (CNS), causing neurologic and psychiatric symptoms. The goal of this article is to familiarize psychiatrists with this spirochetal illness.
>
> Up to 40% of patients with Lyme disease develop neurological involvement of either the peripheral or central nervous system. Dissemination to the CNS can occur within the first few weeks after skin infection. Like syphilis, Lyme disease may have a latency period of months to years before symptoms of late infection emerge...A broad range of psychiatric reactions have been associated with Lyme disease including... dementia.
>
> The microbiology of Borrelia burgdorferi sheds light on why Lyme disease can be relapsing and remitting and why it can be refractory (not yielding, at least not readily) to normal immune surveillance and standard antibiotic regiments. Once in the central nervous system, B. burgdorferi may remain latent, only to cause illness months to years later. Several features are known to contribute to an organism's resistance to standard lengths of antibiotic treatment. These features include an intracellular location, long replication time, genetic variability, and the ability to become sequestered in difficult-to-penetrate sites. B. burgdorferi appears to possess all of these characteristics.

Fallon (and coauthor Nields) concludes with, "Psychiatrists who work in endemic areas need to include Lyme disease in the differential diagnosis of any atypical psychiatric disorder." This idea by Fallon is a good start. However, all areas and specialties in medicine need to consider the wide range of spirochetes, not just Lyme, for just about any chronic disease for which they have no explanation or solution.

Fallon makes a very important point that was first expressed by Dr. Miklossy—that the Lyme bug, *B. Burgdorferi*, is not the only spirochete that contributes to dementia or AD.

Periodontal spirochetes are highly predominant in the population at large. *B. burgdorferi* infection alone cannot explain AD. The title of her initial article clearly shows this as the title is, *Alzheimer's disease—A spirochetosis?* [15], and not "Lyme neuroborreliosis."

Lyme disease is difficult but not impossible to treat. Fallon reported that Lyme, like other bacteria, have more than one form. A dormant form can persist and avoid therapeutic agents. Yet, within the standard-of-care, patients are administered one antibiotic for 10 days. As you will see, this is a woefully inadequate treatment protocol and completely ignores the "relapsing and remitting" aspect of the pathogen. This is regardless of the symptoms of Lyme disease. That is, a patient with joint issues from Lyme or a patient with "Alzheimer's" and Lyme require the same robust treatment regimen that is unknown within the standard-of-care. Ask your doctor to treat you for 1 year and watch the response.

To further understand the connection between AD and Lyme disease, we contacted an assistant dean of a prestigious teaching and research institution who is an expert on Lyme disease. We discussed various cause/effect connections associated with Lyme disease. When we asked him about the Lyme/Alzheimer's connection, he responded by saying, "You have lost all credibility with me as there is no connection."

Later that day, we stopped by our clinic that specializes in neurodegenerative disease diagnosis and treatment. I (TJL) posed what the assistant dean had said to the person responsible for populating the electronic medical records system. She was working on four random patient records at the time. She said, "Oh yeah, have the dean come over and review two of these cases and he might change his mind." This occurred in 2011, 17 years after the paper by Fallon describing the connection between Lyme disease and dementia.

The most recently and widely studied bacterium associated with AD is *Chlamydia pneumoniae (Chlamydophila pneumoniae. C. pneumoniae,* CP). This microbe is not a sexually transmitted bacterium, but others in the "chlamydia" class are. A search of "chlamydia" and "Alzheimer's" yields more than 5,000 records. *C. pneumoniae* and the Lyme bacteria are probably just the "tip of the iceberg" of bacteria to be implicated in AD. Stealth intracellular bacteria are likely the missing link between AD and cardiovascular disease. In the previous chapter, much emphasis was placed on the work of Dr. Kilmer McCully and his description of the mechanism by which infectious species are involved in cardiovascular disease. In his landmark paper, he referenced work by a German group that evaluated both diseased and healthy tissue, and they found pathogens in the diseased tissue [23].

Ott et al. [24] identified fragments from >50 different microbial species within atherosclerotic plaques (with an average of 12), but not a single one in normal arterial tissue.

No one is really interested in having enough blood drawn to search for over 50 microbes propagating in their vessels. And 50 might well be a considerable underestimate of the breadth of microbial possibilities. However, doesn't your family member with Alzheimer's deserve to be evaluated for at least one? A test for inflammation might justify searching for these bacteria, but remember, they can evade testing and the immune system while causing disease.

It is not the intent of this book to cover every single microbe that may be involved in AD. The take-home lesson is that these bacteria can initiate persistent "silent" inflammation that may be a significant contributor to the disease. Identifying the causing bacteria is important. However, it is possible to determine their presence by searching for and diagnosing the chronic inflammatory process by way of markers of inflammation already discussed in Chapter 7. This is accomplished by measuring the patient's blood and associated techniques that measure health and disease in your entire system, not just in the brain.

CHLAMYDIA PNEUMONIAE

C. (more recently being classified as Chlamydophila) *pneumoniae* is an intracellular pathogen responsible for a number of different acute and chronic infections. It is estimated that CP may infect more than 50% of the world population, most of whom have no symptoms and may never develop symptoms assuming their immune system stays strong and is able to keep the bug at bay. The recent deepening knowledge on the biology and the use of increasingly more sensitive and specific detection measures has allowed demonstration of CP in a large number of persons suffering from different diseases including cardiovascular (atherosclerosis and stroke), central nervous system (CNS) disorders, and dementias.

CP is most well-known as a common cause of human respiratory disease. It was first isolated from the conjunctiva of a child in Taiwan in 1965, but it was not until the early 1980s that it was scientifically identified as a distinct chlamydia species. [Be aware there are three common types of chlamydia, *Chlamydia trachomatis* (the STD), *Chlamydia psittaci*, and *C. pneumoniae*.] It was established as a major respiratory pathogen in 1983 when it was isolated from the throat of a college student at the University of Washington. Most likely, the respiratory tract primarily transmits CP from human to human, and infection spreads slowly. The incubation period is several weeks, which is longer than that for many other respiratory pathogens.

CP accounts for 6%–20% of community acquired pneumoniae (CAP) in adults, but participates in coinfection involving other bacterial agents in approximately 30% of adult cases of CAP. Here we see the opportunistic ability of infectious species. When the host (us) is weakened by something else, another infection, for example, a "latent" bug like CP multiplies geometrically. Some studies have suggested a possible association between CP infection and acute exacerbations of asthma and chronic obstructive pulmonary diseases (COPD). In recent years, however, in addition to respiratory diseases, an increasing number of publications report detection of CP in chronic diseases beyond the respiratory tract. In fact, specific molecular diagnostic techniques have allowed research to demonstrate the presence of CP DNA in a large number of persons suffering from different diseases other than cardiovascular (atherosclerosis and stroke), such as osteoarthritis and CNS disorders.

Clearly there is much to learn, and many important issues remain unanswered with regard to the role that CP may play in initiating disease. A growing body of evidence concerns the involvement of this pathogen in cardiovascular disease and chronic neurological disorders and particularly in AD and multiple sclerosis (MS). The association between CP and MS is even stronger than the one between CP and AD. Both of these associations will be discussed here, shortly.

C. pneumoniae is not a sexually transmitted disease. The so-called STD is C. trachomatis, a common sexually transmitted disease. When you breathe in Chlamydia, it can infect the ciliated cells, the cells lining the airways. Ciliated cells are like an escalator that moves mucus along. Chlamydia can paralyze the ciliated cells because it steals their energy. When you have an active infection of any kind, the body has an immune response to it. Part of the innate immune response is that monocytes and macrophages try to engulf the pathogen and kill it. But they can't always kill Chlamydia because it is an intracellular (inside the cell) infection. Like most other intracellular infections, it is much more difficult to be eliminated by the acquired immune system. This explains why vaccination does not work against intracellular infections.

CP can silently infect other cells in the body. With these tricky infections, a patient doesn't have symptoms. All the while, Chlamydia is metabolizing, and growing. Bugs like Chlamydia are called "obligate intracellular." Obligate because they can't make their own energy; they have to steal energy from the human cell. This means the chlamydia-infected cell doesn't work the way it was designed. They are diseased. In some ways infected cells and the infected tissue appear insulin resistant due to the metabolic action of the microbes. No wonder some researchers are calling Alzheimer's type 3 diabetes. Maybe *Chlamydia pneumoniae* is the cause?

Chlamydia also likes to infect endothelial cells, the cells that line blood vessels. When there's inflammation in the body, there often is angiogenesis, meaning new blood vessel cells are formed. Chlamydia is drawn to those cells. So if Chlamydia is in the blood and there's an inflammation in your body, you are set up for the inflammation to increase. This is referred to as a secondary inflammatory response that leads to a chronic problem. The concept of secondary inflammation caused by Chlamydia is conjecture. It may be that the bug is primary in causing the inflammation and subsequent disease. When the infection is in the joints, the inflammatory process is referred to as Rheumatoid Arthritis. When the infection is in the brain, it causes MS, dementia, or AD, depending on where and when the infection and resulting inflammation settles, as well as immunosenescence (the changes in the immune system associated with age).

Most people generally believe that a bacterial infection is something that is over with quickly and is usually easily treated with any number of antibiotics. Historic exceptions are Leprosy, toxoplasmosis, active tuberculosis, and the herpes zoster virus that brings lingering misery that can last for years. These are intracellular infections that impact most of the world population. Diseases from these bugs hit us when we are old due to immunosenescence or other immune compromising events earlier in life. In general, bacterial diseases appear suddenly and can be dealt with by a highly tuned immune system, perhaps backed up with a dose of antibiotics. The bugs that apparently cause Alzheimer's are more like those that cause Leprosy and tuberculosis, in terms of resistance to treatment due to their intracellular nature.

We are now learning that asthma, arthritis, vascular diseases, sinus infection, and a number of other diseases that sometimes nag us for as long as a person lives may also be the result of bacterial infections. Exploring the role that bacteria play in these ailments is a new topic in medicine. That is why, although the diseases themselves are hardly novel, a recent International Conference on Emerging Infectious Diseases, held in Atlanta by America's Centers for Disease Control (CDC), devoted a session to them.

GERM THEORY OF DISEASE: BACK TO THE FUTURE MEDICINE

We could learn so much from our forebears if we would only listen. Did you listen to your parent's advice? Do your children listen to your advice? The answer is likely "no," until you or they achieved a higher level of "enlightenment." Modern medicine is no different. In some respects, like a teenager, medicine seems to chide, "How can scientists from the 19th century know more than us? After all, we continue to progress in the breadth and depth of our knowledge. And we have sophisticated tools that our ancestors could not even imagine." Yet, as discussed previously, Claude Bernard of 19th century from France provides us with lessons on homeostasis for which we clearly lack appreciation or understanding. The proof is rampant chronic diseases that did not exist during his time.

Another giant, who remains highly respected, but largely ignored, is Louis Pasteur. Pasteur, like Bernard, hailed from France and was at the pinnacle of his fame during the mid-to-late 1800s. His laboratory, at one time, commandeered 10% of the research budget of France. Pasteur is credited with developing the "Germ Theory" of disease. His name is certainly attached to this theory, but less significant names in history really developed the postulate. Regardless, there is much we can learn from the germ theory, in the modern context of chronic disease.

Germ theory states that many diseases are caused by the presence and actions of specific microorganisms within the body. The theory was developed and gained gradual acceptance in Europe and the United States from the mid-1800s. It eventually superseded existing miasma (bad air from rotting organisms) and contagion (one disease could change into another or might manifest itself differently in different people) theories of disease and in so doing radically changed the practice of medicine. It remains a guiding theory that underlies aspects of contemporary medicine.

Awareness of the physical existence of germs preceded the theory by more than 2 centuries. Discoveries made by several historical figures pointed the way to germ theory. On constructing his first simple microscope in 1677, Antoni van Leeuwenhoek was surprised to see tiny organisms, which he called "animalcules," in the droplets of water he was examining. He made no connection with disease, and although later scientists observed germs in the blood of people suffering from disease, they suggested that the germs were an effect of the disease, rather than the cause. This fit with the then popular theory of spontaneous generation. In the case of Alzheimer's, this may be a prevailing belief of which all of us need to be acutely (or chronically) aware.

The observations and actions of Ignaz Semmelweis, Joseph Lister, and John Snow would retrospectively be acknowledged as contributing to the acceptance of germ theory. But it was the laboratory research of Louis Pasteur in the 1860s and then Robert Koch in the following decades that provided the scientific proof for germ theory. Their work opened the door to research into the identification of disease-causing germs and potential life-saving treatments. Today we suffer from a myriad of diseases that may (surely) have a "germ" cause, but medicine often chooses to address symptoms rather than explore germs as a possible cause. A classic modern example is a story about a bug that causes stomach ulcers. It is an initially sad tale that has a happy ending, a Nobel Prize in Medicine.

HELICOBACTER PYLORI

One of the first chronic complaints found to have an unexpected bacterial cause was stomach ulceration. The evidence that a bug called *Helicobacter pylori* triggers ulcers had been accumulating since the 1970s. America's medical establishment (the National Institutes of Health) officially accepted the idea in 1994. This encouraged others who were looking for hitherto unsuspected connections between infections and disease.

Dr. Warren, a pathologist from Australia, made a revolutionary discovery during a routine diagnosis of diseased tissue [25–29]. He made extra effort in his diagnosis due to his nascent curiosity. Through his advanced diagnosis, he eventually was able to prove a cause/effect relationship that has profoundly impacted treatments of stomach ulcers. Warren, along with his colleague, Dr. Marshall, hypothesized that a specific bacteria known today as H-Pylori caused stomach ulcers. At that time (and inexplicably still by some medical professionals today), the belief was that these ulcers were caused mainly by stress and other causes of excess stomach acid. Marshall and Warren's suggestion that ulcers may be caused by bacteria was initially viewed by some researchers as absurd and outrageous. Martin Blaser of the Division of Infectious Diseases at the Vanderbilt University School of Medicine thought a 1983 talk by Marshall was "the most preposterous thing I'd ever heard; I thought, 'This guy is a madman.'" In fact, these two Australians, Warren and Marshall, were not even invited to present their data at gastroenterology society meetings for many years. Blaser has since become one of the leading researchers on *H. pylori*, having changed his belief 180 degrees in the face of overwhelming evidence.

Dr. David Forman of the Imperial Cancer Research Fund thought that Marshall's claim that bacteria are responsible for various stomach diseases including cancer was a "totally crazy hypothesis." But he thought it worth demolishing, and since has concluded that *H. pylori* infection is a major factor in gastric cancer as well as ulcers. Through the course of their trials, Warren and Marshall even infected themselves with H-pylori and then successfully treated themselves with appropriate and known medications. In 2005, the two were awarded the Nobel Prize in Medicine for their discovery and the work they did to prove their thesis.

Paul Thagard is Professor of Philosophy and Director of the Cognitive Science Program at the University of Waterloo. He has written extensively on the philosophy of acceptance of new discoveries. His paper, aptly titled, *Ulcers and Bacteria I: Discovery and Acceptance* is certainly worth reading to fully understand the complexities and pitfalls of new scientific discovery [30]. One of the pitfalls to any new idea is "Not Invented Here." According to Princeton University, "'Not Invented Here' is a term used to describe persistent social, corporate, or institutional culture that avoids using or buying already existing products, research, or knowledge because of their external origins. It is normally used in a pejorative sense, and may be considered an antipattern." Antipatterns are considered ineffective and/ or counterproductive in practice. Are there scientists, researchers, and physicians out there who, like Warren and Marshall regarding ulcers, have similar but ignored Alzheimer's discoveries?

The take-home message is: don't accept the standard-of-care when it fails to provide an acceptable solution. There is research occurring at Harvard and in your backyard by people committed to finding workable solutions for Alzheimer's and other "incurable" diseases. Search for knowledge and truths. Many people in medicine and other disciplines have sought and found solutions beyond the conventional wisdom. Also, consider the many examples of medical pioneers (Appendix 6). Mavericks often are the brunt of much criticism and disparagement prior to their theses being proven correct. Galileo was imprisoned for explaining that the earth is not the center of the universe. You owe it to yourself, your family, and your loved ones to pursue a quest for answers. Dig deep into discovery and seek to determine if alternative views are supported by valid science. Also consider the agenda of the establishment and those who seek to discredit alternatives.

The impact of H-pylori on the body was first thought confined to the stomach, but it is now emerging that H-Pylori may migrate into the brain and be involved in dementias and AD. A number of other bacterial suspects, including CP, appear to have broader effects. Interestingly, similar symptoms can be produced by a variety of organisms, as they all appear to stimulate the same immune response. The diseases appear to be a manifestation of that response connected to the tissue in which the bacterial infection proliferates.

CHLAMYDIA PNEUMONIAE AND CARDIOVASCULAR DISEASES

CP has a strong connection to atherosclerosis (hardening of the arteries). Over the past few years, research done in Finland, Italy, Britain, Argentina, and America confirmed that this bacterium likes to inhabit the fatty "plaques" which accumulate in blood-vessel walls when arteries harden. An alternate view is that these organisms cause the fatty plaques in the first place. The first report of an association between CP infection and atherosclerosis was made in 1992, "Chronic Chlamydia pneumoniae infection as a risk factor for coronary heart disease in the Helsinki Heart Study" [31], and its conclusion reads as follows:

> Conclusion: The results suggest that chronic C. pneumoniae infection may be a significant risk factor for the development of coronary heart disease.

Subsequently, many other reports concerning the association between CP and atherosclerosis at several arterial sites have confirmed a link between CP and atherosclerosis. How does this infectious species attack our vessels?

- First, CP gains access to the vascular system during local inflammation in lower respiratory tract infection.
- Second, the infected air passage immune system unwittingly transmits it through the mucosal barrier and gives the pathogen access to the lymphatic system, systemic circulation, and vessel walls.
- Third, CP can infect a variety of cells commonly found in cell walls of vessels, including coronary artery endothelial cells, macrophages, and aortic smooth muscle cells.
- Fourth, CP may influence vessel cell wall biology by changing the macrophage-lipoprotein interactions.

In this setting, the outer wall of chlamydia, LPS, may lead to inflammation and other deleterious processes that cause tissue damage, possibly by starving the tissue of oxygen (hypoxia). Of course, oxygen starved tissue quickly dies. Try holding your breath for 2 minutes. That is what is happening on a cellular level when CP infection invades tissue. These bugs are parasites, and they feed off our blood supply for energy.

This viewpoint is somewhat controversial and contrary to the traditional view about cholesterol and how a plaque develops and ruptures, thus causing a stroke or heart attack (myocardial infarction). The new hypothesis, presented in the previous chapters, ties together all known research and science. It explains, at great length, how the mechanism of cardiovascular disease has been misinterpreted and incorrectly made cholesterol, particularly LDL ("bad") cholesterol the villain. McCully et al. explained the role of CP and other bacteria in the cause of cardiovascular diseases [23]. Some excerpts from Dr. McCully's paper, not covered in Chapter 8, are reproduced here:

> The predilection for plaques within systemic arteries (the existing hypothesis within the standard-of-care) also contradicts the idea that microbes attack the endothelium directly, because if this were so, atherosclerosis would be just as common in veins. Also the focal occurrence of atherosclerotic lesions is in better accordance with a microbial genesis, because if elevated LDL cholesterol were the most important cause, atherosclerosis should be a more generalized disease.
> The increased incidence of cardiovascular events found after treatment with rofecoxib and other nonsteroidal anti-inflammatory drugs contradicts the idea that atherosclerosis is caused by the inflammation itself, but it is in accord with an infectious origin of atherosclerosis, where inflammation is a necessary step for healing.
> Chlamydia pneumoniae is not the only microbe that is found in atherosclerotic plaques. Ott et al identified fragments from >50 different microbial species within atherosclerotic plaques, but not a single one in normal arterial tissue [24]. On average, each patient had microbial remnants from 12 different species; some patients had more, some had fewer, and other investigators have found various virus species as well.
> It is highly unlikely that a single antibiotic could eliminate >50 different microbial species. It is not even likely that antibiotics could eliminate Chlamydia pneumoniae, because this species is able to survive inside

living cells, where they are resistant to the effects of antibiotics. Furthermore, antibiotics are generally ineffective against viral infections. Whether the total burden of multiple microbial invasions or the effect of a single pathogen is the key to progression remains to be determined.

McCully also points out that antibiotic treatment of cardiovascular disease has not been generally successful. He points out the following: "An apparent contradiction to our interpretation is that prevention of cardiovascular disease by antibiotics has been largely unsuccessful. However, in these trials patients have usually received a single antibiotic, chosen because it was effective against *C. pneumoniae*, the organism that has been studied most intensively, and the trials have been of relatively short duration."

J. Thomas Grayston, Professor Emeritus of Epidemiology at University of Washington School of Public Health, wrote an editorial in the journal *Circulation* in 2003, which also provides relevance to the argument that bacterial infection is a critical component of the cardiovascular disease process [32]. His comments are in regard to treatment of cardiovascular disease with antibiotics, presumed to be caused by microbes. He found that the clinical trials involving antibiotic treatment were poorly designed, thus doomed to failure. He said:

> I have previously described the errors in study design and the inadequate treatment course of this trial. The short antibiotic courses used in the London and ROXIS studies influenced the treatment course in a number of subsequent studies. This was despite efforts in cardiology journals in 1998 and 1999 to educate cardiologists about the microbiology of Chlamydia and treatment requirements for chronic Chlamydia infection [33,34].
>
> There is a large body of experience with antibiotic treatment of chronic Chlamydia trachomatis and Chlamydia psittaci infections, the other Chlamydia species that infect humans. Successful treatment has been uncommon and has required vigorous, long-term, carefully controlled antibiotic administration. The life cycle of Chlamydia explains why treatment is difficult. The infectious, extra cellular, nonreplicating form of the organism (elementary body) is not susceptible to antibiotics. It may remain viable in the body for weeks to months before reinfecting a susceptible cell. This is why eradication of the organism after acute infection is difficult. Furthermore, the intracellular replicating form (reticulate body) that is susceptible to antibiotics is capable of entering a "persistent" phase for an indeterminate time that is not susceptible to antibiotics.
>
> Based on experience treating chronic Chlamydia infections and knowledge of the life cycle of the organism, I recommended that the treatment course in clinical trials with CHD be for one year.

In summary, Drs. Grayston, McCully, Stephan (Ott), and many others recognize the association between CP and other bacteria and cardiovascular disease. Together they realize that treatment is difficult if only one bacterium is involved. However, it is likely that more than one may be involved regardless of how many or few are diagnosed. Therefore, treatment should continue for a minimum of 1 year and probably should involve a mixture of antibiotics (and antiinflammatory agents of a very specific nature). Successful treatment is far more complex than doling out antibiotic pills to a cardiovascular or AD sufferer for a year. A doctor must delve deep into human physiology and understand the individual's phenotype (everything that makes you, you), to develop a bona fide and comprehensive treatment approach for these serious diseases of aging.

There are at least two reasons to believe that bacteria are at the root cause of AD, stemming from our understanding about the relationship between CVD and AD. First, using antibiotics to treat an atherosclerotic patient who has had one heart attack reduces the risk of his suffering a second. Keep in mind that this result is based on a short-duration treatment. Longer duration studies will likely yield better results. Second, research has come up with a plausible mechanism that is described earlier. A recent experiment at the University of Louisville in

Kentucky illustrates that infecting endothelial cells taken from the walls of coronary arteries with Chlamydia stimulates the production of molecules called chemokines [35]. That is not surprising, since the role of chemokines is to attract disease-fighting white blood cells called neutrophils (immune cells that can destroy pathogens) and monocytes to the blood-vessel walls. Once there, however, these blood cells invade the endothelium, causing it to become inflamed.

Such inflammation is, in fact, a normal response to bacterial infection. The curious question is why the inflammation would become chronic in atherosclerosis, when in the case of most infections it is transient. The most likely explanation, as we have already noted through the work of Grayston is the variable life forms of the bugs resulting in their ability to go stealth from treatment, only to reappear later. But whatever the cause, since another effect of inflammation is to attract platelets to the area, the creation of a clot, with the attendant risk of a heart attack or stroke, is a common consequence of this continual inflammation caused by CP and similar bugs.

The main obstacles to support a definitive role of CP in chronic diseases are that few dependable methods exist to safely and confidently diagnose chronic infection. Recall the earlier discussion about the 13+ FDA methods to detect the Lyme bacteria and how they give inconsistent results. Because these bugs can go stealth, inflammation markers are not always present or telling. Also, in chronic chlamydial infections, the so-called "chlamydial persistent state" is resistant to conventional antibiotic agents, including those specifically designed to target the bug, much like in the case of Lyme disease.

CP has two forms that are quite distinct in both form and function. One form, referred to as the elementary body (EB) is both infectious and metabolically inactive. This nonreplicating infectious particle is released when infected cells rupture. The elementary body is responsible for the bacteria's ability to spread from person to person. This form is analogous to a spore.

Definition of Spore a refractile, oval body formed within bacteria, especially Bacillus and Clostridium, which is regarded as a resting stage during the life history of the cell, and is characterized by its resistance to environmental changes.

The reticulate body (RB) is a noninfectious and metabolically active intracellular (inside cells) form that replicates by binary fission, reorganizes into EBs, and then is released by cell death (lysis). In general, it is likely that this aberrant developmental step leads to the persistence of viable Chlamydia within infected cells over long periods. CP is able to create an intracellular niche from where it promotes host cell (our cells) survival or death, modulates regulatory host cell signaling pathways, and bypasses the host cell's defense mechanisms (hides from our normal defenses). Thus, CP can induce a persistent infection due to the inability of the host to completely eliminate the pathogen.

That we cannot eradicate CP from out bodies, at least not easily, especially because we seldom know it is there, and few attempts to diagnose for this bacterium. This leads to a state of chronic infection and inflammation that makes the individual susceptible to proliferation of CP and/or many other infectious species that may be smoldering as the body weakens and cause disease. Further, CP can enter into a state of quiescence with intermittent periods of replication characterized by the body's production of specific proteins and proinflammatory cytokines capable of triggering tissue damage. Thus, a one-time measurement of inflammation, the byproduct of infection, may not reveal anything unusual. Patients with developing chronic diseases and a one-time "good" blood test need to be reevaluated periodically.

Under pressure from a strong immune system in a healthy person, the metabolic processes of the CP are diminished, and the body controls the spread and development of disease. But in modern society, we lack homeostasis and a strong immune system. This is evidenced by an epidemic of chronic diseases attributable to poor diet and, most likely, proliferation of CP and its counterparts. These chronic infections, which can last for several decades, can also initiate autoimmunity and many of the so-called autoimmune diseases. To a large extent, the form of the disease may depend on the phenotype of the person, including, to some degree, their genetic makeup. It is thought that CP can be controlled through a combination of environmental (behavioral) factors that include augmenting immune health and reducing traditional risk factors. But, depending upon the extent to which CP has developed, these actions may not be sufficient to put the pathogen "at bay." Alzheimer's is a disease associated with advanced age. Our elderly have weakening immune systems. It all adds up.

Definition of Phenotype The observable physical or biochemical characteristics of an organism, as determined by both genetic makeup and environmental influence.

The prevalence of CP infection increases with age, just as does the diseases of aging like AD. Antibodies against CP begin to appear at school age for our young lacking homeostasis, but are rare in children under the age of five, except in developing and tropical countries. Prevalence of CP increases at ages 5–14, and it continues to increase slowly as we age, tracking the health of our immune system. This would seem to infer an epidemic or even a pandemic with regard to this type of chronic infection, and that most people are infected and reinfected for life. However, the adaptive immune system is able to keep the levels of CP below that which leads to clinical disease, at least in those who are proactive about their health.

Solving for the root causes of chronic inflammation that we more thoroughly understand, and a cure for atherosclerosis and AD will come closer. Chronic inflammation and infection is not restricted to the arterial walls. It appears to be the linking factor of many of these diseases. Arthritis is inflammation and likely infection of the joints. Crohn's disease (also suspected of being caused by bacteria) is an inflammation of the bowel. Ulcers are inflammations of the stomach caused by the bug H-pylori. However, the first step toward a cure is an understanding of the root causes of disease, the factors that allow its spread and proliferation. A differential diagnosis is necessary to identify specific disease targets and treatments.

As a wide range of diseases now put down to the general process of aging do turn out to be infections, a new field of treatment must open. Over the past few years, drug companies have been reluctant to invest in new antibiotics because of concern about their overuse and resistance development. Soon they may change their minds. Vaccines against germs, which have been regarded as unworthy of attention, might also be developed if such germs are acknowledged to cause serious diseases.

There is a hope that in the future it may be possible to pop a pill or have a shot to keep you both free of heart disease and sharpen the mind. But the drug taken will never dictate the measure of success in treatment. Personal responsibility, proper diet, exercise, and homeostasis are all crucial factors that improve the performance of drugs. In the case of antibiotics, the ultimate curative is your immune system. The drugs simply help lower the infectious burden to manageable levels, whereupon your immune system takes over. If your immune system is in poor health, expect a rapid relapse or minimal effect regardless of how a drug may be promoted.

CHLAMYDIA PNEUMONIAE AND THE BRAIN

CP is able to invade and infect a wide variety of human cells, such as epithelial, endothelial, myocardial, and smooth muscle cells as well as macrophages, monocytes, and lymphocytes. This bug is able to disperse through the entire body following exposure to a respiratory infection. Certainly, the presence of CP DNA in peripheral blood mononuclear cells strongly suggests that such dissemination can occur in a number of different tissues. More disturbing is that it appears that CP infection can pass through the blood-brain barrier by way of white blood cells, suggesting a mechanism by which the organism may enter the central nervous system (CNS). This may account for the delivery of the organism to the CNS and result in chronic injury and diseases like dementia and AD.

Infection is a trigger to inflammation, and many of the diseases of aging involve inflammation. But a key question is how can CP cross the blood-brain barrier and get into the brain? The brain was assumed to have a significant barrier to pathogens. Certain types of blood cells may traffic CP across the blood-brain barrier, shed the organism in to the vessels of the brain, and induce neuroinflammation. The demonstration of CP by histopathological (the study of diseased tissues at a microscopic level) in the late-onset Alzheimer's dementia suggests a relationship between central nervous system infection with CP and the signs and symptoms of AD.

Dr. Brian Balin of Department of Biomedical Sciences, Philadelphia College of Osteopathic Medicine, Philadelphia, Pennsylvania leads the way at identifying pathogens, particularly CP, in the brain of Alzheimer's sufferers. In his 2003 publication titled, *Chlamydia pneumoniae infection promotes the transmigration of monocytes through human brain endothelial cells*, Dr. Balin explains the migration of CP into the AD brain [36].

> We have investigated the effects of Chlamydia pneumoniae on human brain endothelial cells (the thin layer of cells that lines the interior surface of blood vessels) and human monocytes (white blood cells of the immune system) as a mechanism for breaching the blood-brain barrier (BBB) in Alzheimer's disease (AD). Human brain endothelial cells and peripheral blood monocytes may be key components in controlling the entry of C. pneumoniae into the human brain. Our results indicate that C. pneumoniae infects blood vessels and monocytes in AD brain tissues compared with normal brain tissue.... Thus, infection at the level of the vasculature (circulatory system) may be a key initiating factor in the pathogenesis of neurodegenerative diseases such as sporadic AD.

The next important question to be answered, after an understanding of how CP gets into the brain, is how does the disease spread across the brain? Several researchers have taken the baton and led this investigation. One article on this subject is titled, *Prion-like transmission of protein aggregates in neurodegenerative diseases* [37]. The authors from France, Sweden, and the United States state:

> Neurodegenerative diseases are commonly associated with the accumulation of intracellular or extracellular protein aggregates. Recent studies suggest that these aggregates are capable of crossing cellular membranes and can directly contribute to the propagation of neurodegenerative disease pathogenesis. We propose that, once initiated, neuropathological changes might spread in a 'prion-like' manner and that disease progression is associated with the intercellular transfer of pathogenic proteins. The transfer of naked infectious particles between cells could therefore be a target for new disease-modifying therapies.

The implication is that the disease spreads through the transfer of proteins, possibly by way of "naked" infectious particles. The presumption is these "prion" type particles are not living entities. However, could these presumed "naked" particles actually be difficult to detect intracellular infectious species like spirochetes or viruses? In which case, treatment would not be as implied in the above quote.

Definition of Prion an infectious agent composed of protein in a misfolded form. This is the central idea of the prion hypothesis, which remains debated. This would be in contrast to all other known infectious agents (virus/bacteria/fungus/parasite), which must contain nucleic acids (either DNA, RNA, or both).

Another important paper, from University College London Institute of Neurology in 2012, is titled, *The Spread of Neurodegenerative Disease* [38]. The authors confirm that, until recently, only in prion disease had the spread of disease been shown. They go on to say, "The first evidence that proteins other than prions could be pathologic was described in 1994, when the introduction of brain tissue from a patient with AD into the brains of aged marmosets was followed by the seeding of beta-amyloid plaques." The study continues to support their hypothesis with a variety of other important facts that appear to parallel the spread of "prion infectious agents."

Could it be that the disease spreads through proteins, or might there be an alternative explanation? Could it be that stealth pathogens are present in the protein material and that is the cause of the propagation of disease? Or could it be that these infectious species initiate the formation of the protein materials that then spread? Science probably does not offer a definitive answer yet, but let's explore these possibilities by returning our attention back to CP.

In recent years, a growing body of evidence concerning the involvement of CP in neurological diseases has been gradually increasing. This was supported in particular by the detection of genomic material of the microorganism into the cerebrospinal fluid (CSF) of patients with MS, AD, meningoencephalitis, and neurobehavioral disorders. The first paper that reported an association between CP infection and AD was published in 1998 by Balin and coauthors titled, *Identification and localization of Chlamydia pneumoniae in the Alzheimer's brain* [39]. The authors found that 90% of AD brains were positive for CP, and the organism was detected in various sections of brain (hippocampus, cerebellum, temporal cortex, and prefrontal cortex). CP spread throughout the brain! All these regions of the brain exhibited AD signs including tangles and plaques. Electron microscopic results revealed CP in the brain tissue, and wet chemistry techniques indicated strong "labeling" for CP in the sections of the brain most affected by AD, while no "labeling" was noted in the controls.

Importantly, CP was highly associated with the AD plaques and tangles, and these "colonies" of CP were shown to be metabolically active organisms. In the Balin report, a strong association of APOE-ε4 genotype and CP infection was found, suggesting, as shown in reactive arthritis, that the APOE-ε4 gene may promote some aspects of CP propagation in AD [40]. More importantly, in the context of the spreading of the disease, does it make sense that if proteins contained active CP bacteria, then it is the CP bacteria that is responsible for disease propagation, at least in part?

The Balin study received a great deal of public and scientific attention, and other laboratories replicated the findings. In 2000, two independent investigators found CP in brains of neurodegenerative disease patients validating Balin's results [41,42]. Another 2000 study titled, *Failure to detect chlamydia pneumoniae in brain sections of Alzheimer's disease patients* provides an opposing view [43]. However, what about the other 50+ infectious species that could lead to neurodegeneration? In 2006, Gerard demonstrated CP in 80% of AD samples and 11% of the control [44]. This later study again showed that the organism was active. In addition, they demonstrated that astrocytes, microglia, and neurons all served as host cells for CP in the AD brain, and that infected cells were found in close proximity to both plaques and tangles in the AD brain.

In the years following the initial studies by Balin, research has provided more insights into mechanisms of infection as a potential cause of AD. Here are some of the recognized insights:

- The amount of CP found in the AD brain appears to vary with the APOE genotype.
- Infection by CP of the vascular system facilitates access of CP to the nervous system and the brain.
- CP may be responsible for the deleterious loosening of "tight junctions."
- Infection with CP through the olfactory pathways promotes the production of extracellular amyloid-like plaques suggesting that this could be a primary trigger for AD. The nasal passages are a likely target for infection. Following entry, potential damage and/or cell death may occur in the main olfactory bulb and olfactory cortex, thereby setting the stage for further neuronal damage.

The scientific knowledge surrounding AD and infection by CP is still growing. Standardization of diagnostic techniques will certainly allow for better comparability of studies. However, other systemic infections as potential contributors to the pathogenesis of AD should be considered. That is, the lack of detection of *C. pneumoniae* does not imply that infection is not involved in the AD process. Lyme disease and several other stealth bacteria may be present with, or in the absence of, CP. In cardiovascular disease, Ott indicated that diseased tissue, on average, contains 8–12 separate microbial entities [24].

Balin et al. continued to take the lead connecting CP with AD. Their paper in the *Journal of Alzheimer's Disease*, titled, *Chlamydophila pneumoniae and the etiology of late-onset Alzheimer's disease* provides a plausible explanation of how CP, and possibly similar infectious species in general, contributes to the degradative aging process that is AD [45]. The abstract is reproduced here:

Sporadic, late-onset Alzheimer's disease (LOAD) is a non-familial, progressive neurodegenerative disease that is now the most common and severe form of dementia in the elderly. That dementia is a direct result of neuronal damage and loss associated with accumulations of abnormal protein deposits in the brain. Great strides have been made in the past 20 years with regard to understanding the pathological entities that arise in the AD brain, both for familial AD (~5% of all cases) and LOAD (~95% of all cases).

The neuropathology observed includes: neuritic senile plaques (NSPs), neurofibrillary tangles (NFTs), neuropil threads (NPs), and often deposits of cerebrovascular amyloid. Genetic, biochemical, and immunological analyses have provided a relatively detailed knowledge of these entities, but our understanding of the 'trigger' events leading to the many cascades resulting in this pathology and neurodegeneration is still quite limited. For this reason, the etiology of AD, in particular LOAD, has remained elusive. However, a number of recent and ongoing studies have implicated infection in the etiology and pathogenesis of LOAD. This review focuses specifically on infection with Chlamydophila (Chlamydia) pneumoniae in LOAD and how this infection may function as a "trigger or initiator" in the pathogenesis of this disease.

Balin in his abstract referred to a number of recent and ongoing studies. Here is a listing, by year and title, of reports and research papers that discuss the connection between AD and CP.

1998: Chlamydia pneumoniae: an emerging pathogen in Alzheimer's disease. [46]
1998: Identification and localization of Chlamydia pneumoniae in the Alzheimer's brain. [39]
1999: Ultrastructural Analysis of Chlamydia Pneumoniae in the Alzheimer's Brain. [47]
2000: Chlamydia pneumoniae in the Alzheimer's brain—Is DNA detection hampered by low copy number. [48]

2000: Identification of Chlamydia pneumoniae in the Alzheimer's brain. [49]

2001: What is the evidence for a relationship between Chlamydia pneumoniae and late-onset Alzheimer's disease? [50]

2002: Chlamydia Pneumoniae Infection in Human Monocytes and Brain Endothelial Cell: Initiating Factors in the Development of Alzheimer's Disease. [51]

2002: Pathology in the brains of young, non-transgenic, balb/c mice following infection with chlamydia pneumoniae: a model for late onset/sporadic Alzheimer's disease. [52]

2003: Chlamydia Pneumoniae in Cerebrospinal Fluid of Patients with Alzheimer's Disease and in Atherosclerotic Plaques. [53]

2004: Chlamydia pneumoniae in the Pathogenesis of Alzheimer's. Chlamydia pneumoniae. [54]

2004: Is Chlamydia associated with Alzheimer's? [55]

2004: The effects of the chlamydia pneumoniae infectious process on neuronal cells: implications for Alzheimer's disease. [56]

2004: The induction of Alzheimer's disease-like pathology and recovery of viable organism from the brains of BALB/c mice following infection with chlamydia. [57]

2005: Chlamydia pneumoniae as a potential etiologic agent in sporadic Alzheimer's disease. [58]

2005: The load of Chlamydia pneumoniae in the Alzheimer's brain varies with APOE genotype. [59]

2006: Caspase activity is inhibited in neuronal cells infected with Chlamydia pneumoniae: Implications for apoptosis in Alzheimer's disease. [60]

2006: Chlamydia pneumoniae isolates from the Alzheimer's brain are poorly related to other strains of this organism. [61]

2006: Chlamydophila (Chlamydia) pneumoniae in the Alzheimer's brain. [62]

2006: Chlamydophila (Chlamydia) pneumoniae infection of human astrocytes and microglia in culture displays an active, rather than a persistent, phenotype. [63]

2006: Isolates of Chlamydia pneumoniae from the Alzheimer's brain infecting cultured human astrocytes and microglia display an active, not a persistent, growth. [64]

2007: Evaluation of CSF-Chlamydia pneumoniae, CSF-tau, and CSF-Abeta42 in Alzheimer's disease and vascular dementia. [65]

2007: Method of diagnosing Alzheimer's disease by detection of chlamydia pneumoniae. [66]

2008: Induction of amyloid processing in astrocytes and neuronal cells coinfected with herpes simplex virus 1 and Chlamydophila (Chlamydia) pneumoniae: Linkage of an infectious process to Alzheimer's disease. [67]

2009: The use of magnetic resonance imaging to evaluate Chlamydia as an aetiological agent in Alzheimer's disease. [68]

2010: Analysis of Chlamydia pneumoniae-infected monocytes following incubation with a novel peptide, acALY18: A potential treatment for infection in Alzheimer's disease. [69]

2010: Chlamydia pneumoniae infection and Alzheimer's disease: a connection to remember? [70]

2010: Immunohistological detection of Chlamydia pneumoniae in the Alzheimer's disease brain. [71]

2011: Human monocytes, olfactory neuroepithelia, and neuronal cells with Chlamydia pneumoniae alters expression of genes associated with Alzheimer's. [72]

2012: Analysis of Inflammasome Gene Regulation Following Chlamydia pneumoniae Infection of THP1 Monocytes: Implications for Alzheimer's Disease. [73]

2012: Infection with Chlamydia Pneumoniae Alters Calcium-associated Gene Regulation and Processes in Neuronal Cells and Monocytes: Implications for Alzheimer's Disease. [74]

2013: Chlamydia pneumoniae infection of neuronal cells induces changes in calcium-associated gene expression consistent with Alzheimer's Disease. [75]

2013: Analysis of autophagy and inflammasome regulation in neuronal cells and monocytes infected with Chlamydia Pneumoniae: Implications for Alzheimer's disease. [76]

2014: Electron microscopy studies elucidate morphological forms of Chlamydia pneumoniae in blood samples from patients diagnosed with mild cognitive impairment (MCI) and Alzheimer's disease (AD)(LB81). [77]

HERPES SIMPLEX VIRUS LONG ASSOCIATED WITH ALZHEIMER'S DISEASE

Virus, in particular, herpes simplex virus (HSV), is considered by many researchers to be at the root of AD. To illustrate the history of the connection between HSV and Alzheimer's, we present two research papers, one published in 1983 and the other in 2009.

1983: *Neurological disease and herpes simplex virus*, from the University of Manchester, United Kingdom [78].

The brains of 43 patients, some with various neurological disorders, other controls, were examined for herpes simplex virus (HSV) antigen using immunoperoxidase technique.

The three patients with herpes simplex encephalitis shared a pattern of staining, consistent with that reported previously. However, of the other 40 patients, only two (one a patient with Alzheimer's disease, the other a control patient) showed areas of brain positive for HSV antigen (VA). In the patient with Alzheimer's disease VA (inferring herpes simplex virus) was present within nerve and glial cells of the amygdala, within oligodendrocytes of the optic and olfactory tracts and in macrophages within the temporal cortex hippocampus and cerebellum. In the control patient VA was seen only in oligodendrocytes of optic chiasma and olfactory tract.

The scarcity of these findings suggests coincidental disease processes within these two patients and means that any hypothesis implicating HSV as an etiological agent in degenerative disease must still remain extremely speculative.

2009: Alzheimer's Disease-Specific Tau Phosphorylation is Induced by Herpes Simplex Virus Type 1 [79].

Neurofibrillary tangles are one of the main neuropathological features of Alzheimer's disease (AD) and are composed of abnormally phosphorylated forms of a microtubule-associated protein called tau. What causes this abnormal phosphorylation is unknown. Our previous studies have implicated herpes simplex virus type 1 (HSV1) as an etiological agent in AD, and so we investigated whether infection with this virus induces AD-like tau phosphorylation. Here we demonstrate that HSV1 causes tau phosphorylation at several sites, including serine 202, threonine 212, serine 214, serine 396 and serine 404. In addition, we have elucidated the mechanism involved by showing that the virus induces glycogen synthase kinase 3β and protein kinase A, the enzymes that cause phosphorylation at these sites. Our data clearly reveal the importance of HSV1 in AD-type tau phosphorylation, and support the case that the virus is a cause of the disease. Together with our previous data, our results point to the use of antiviral agents to slow the progression of the disease.

The researchers make the very bold statement that HSV1 "is a cause of the disease." Their findings make sense physiologically. They very astutely indicated "a" rather than "the" cause of the disease. How do you hold a virus at bay? The best way is to stay healthy, reduce inflammatory factors, and bolster your immune system. Add HSV1 to your list of considerations when obtaining a differential diagnosis for AD. But you surely see that AD, if you believe in this infection theory, is not caused by one pathogen. Do not seek treatment for HSV1 if you are infected with H-pylori, CP, or the myriad of other pathogen that can lead to AD. If you are not testing, you are guessing.

OTHER VIRUSES

In recent years, we have come to learn that viruses are involved in many chronic diseases of aging. Extensive study of AIDS truly expanded our understanding of what can attack the body and how it responds to such attacks. Another example is HPV (human papilloma virus) and its connection to cervical cancer. The list goes on but few have studied the possible connection between AD and virus. An interesting paper on the connection between virus and Alzheimer's was published way back in 1985. It did not give any specific proof that unspecified viruses were involved in the genesis of Alzheimer's, but it opened the dialog. The paper was titled, *Viral Aetiology of Diseases of Obscure Origin* [80].

Here is the pithy abstract by the clever Brit, from the Department of Microbiology Guy's Hospital Medical School London, who forwarded the idea (translated into American English):

> Viruses have often been suggested as factors in the etiology of diseases of obscure origin. Much of the work however, has been unacceptable or irreproducible, and considerable skepticism is advocated. A short survey is given of the types of evidence for viruses in the etiology of these diseases. Modern techniques for detecting virus-specific antigens or nucleic acid sequences, and modern ideas about disease processes, make it likely that more definite information will be available before long. Presence of viruses, however, can be causal or merely casual, and the difficult question of proof is discussed, with reference to updated Koch's postulates.

The diseases surveyed in greater detail include cancer, neurological disease (MS, Guillain Barre syndrome, Parkinson's disease, AD), connective tissue disease (systemic lupus erythematosus, rheumatoid arthritis), Crohn's disease and ulcerative colitis, juvenile diabetes, autoimmune thyroiditis, Paget's disease of bone, and atherosclerosis.

Mims, the author, suggested possible mechanisms by which viruses cause such diseases, with special emphasis on the viral triggering of damaging autoimmune responses. This work did not make a clear connection between virus and Alzheimer's but certainly paved the way for more research. Mims brought up Koch's postulates that should be used in the evaluation of any infectious agent and AD. Koch's postulates are four criteria designed to establish a causal relationship between a causative microbe and a disease. The postulates were formulated by Robert Koch and Friedrich Loeffler in 1884 and refined and published by Koch in 1890 [81]. Koch applied the postulates to establish the etiology of anthrax and tuberculosis, but the postulates have been generalized to other diseases.

Koch's postulates are:

1. The microorganism must be found in abundance in all organisms suffering from the disease, but should not be found in healthy organisms.

2. The microorganism must be isolated from a diseased organism and grown in pure culture.
3. The cultured microorganism should cause disease when introduced into a healthy organism.
4. The microorganism must be reisolated from the inoculated, diseased experimental host and identified as being identical to the original specific causative agent.

Koch subsequently made postulate "1" less rigid because of the emerging understanding of stealth and latent microorganisms.

MULTIPLE SCLEROSIS AND ALZHEIMER'S DISEASE: A COMMON ROOT CAUSE?

There appears to be a connection between AD and MS, as inflammation appears to be a common factor in both diseases. A diagnosis of MS does not portend AD. However, since these diseases overlap, many of the diagnostic tests suggested here for AD may also help clarify MS. Also, more research is needed to show the connection between the antecedents of these two diseases. MS is a disease of the young to middle aged, while Alzheimer's is a disease of the elderly.

The first paper that contained both AD and MS in the title was published in 1990s. The first one connecting MS and AD and inflammation was published in 1995 by Paul Patterson from the California Institute of Technology, with the following title: *Cytokines in Alzheimer's disease and multiple sclerosis* [82]. The abstract read:

> Cytokines are well-known as mediators of inflammation, and recent work has highlighted the role of these agents and inflammatory events in Alzheimer's disease and multiple sclerosis. The discovery of subclasses of T-helper cells has provided a critical framework to aid in understanding how the cytokine network regulates these diseases.

Fast forward to 2009 and a paper titled, *Dementia in multiple sclerosis: is it Alzheimer's or a new entity?* [83] The authors state that MS is a chronic autoimmune multifocal inflammatory demyelinating disease with secondary neurodegeneration affecting predominantly females between 20–40 years of age. The clinical course over years is very variable. Milder cognitive dysfunctions in various conditions are very frequent in acute stage and in the further courses of the disease.

> Is the 'secondary neurodegeneration' of the Alzheimer's type in MS? We do know that dementia is relatively rare in MS, due in part by the young age of those afflicted. This study followed 400 MS patients for 10 years, tracking dementia. They found three types of dementia associated with MS, an early onset, an onset that occurs at the same time as loss of motor abilities, and a late onset dementia.

Even with this well-done study, there remains a lack of connection between MS and AD due to the "heterogeneity" of dementia with MS. That is, there was no consistent pattern of dementia onset with MS noted. The author was emphatic that dementia associated with MS is not as rare as described broadly in the medical literature.

Is there any evidence for a connection at a root cause level between AD and MS? The French published a study in 2003 titled, *Atherosclerosis, multiple sclerosis, and Alzheimer's disease: what role for Herpesviridae?* [84]

Definition of *Herpesviridae* a large family of DNA viruses that cause diseases in animals, including humans. The
members of this family are also known as herpes viruses. The family name is derived from the Greek word her-
pein ("to creep"), referring to the latent, recurring infections typical of this group of viruses.

The tenant of their study was that many authors suggest that these "very different" diseases may all, at times, see involvement from *Herpesviridae*, which are ubiquitous, and are
commonly involved in well-identified diseases. Their final conclusion was that "any formal
conclusion is impossible, and more extensive studies are warranted." However, they did
acknowledge that the scientific evidence connecting these diseases is growing.

Here is some of the research from around the globe connecting Alzheimer's and MS:

- Israeli's made a connection between the two seemingly unrelated diseases, in that they
  have been found to share a troublesome gene [85]. APOE4, a gene that predisposes
  people to Alzheimer's, is now shown to predict how quickly and severely MS progresses.
  The connection could yield insights into how both diseases damage the brain, and
  possibly inspire new treatments.
- In *Multiple Sclerosis and Alzheimer's Disease,* published by an Austrian team in 2008, the
  authors acknowledged that both diseases involve chronic inflammation with microglia
  activation (brain immune system activation) [86]. They concluded that microglia
  activation in the MS cortex alone has little or no influence on the development of cortical
  AD pathology.
- In 2011 in the *Journal of Neural Transmission,* another Austrian reviewed *Mechanisms of
  neurodegeneration shared between multiple sclerosis and Alzheimer's disease* [87]. In that paper
  the author asserted that MS and AD are fundamentally different diseases. However,
  recent data suggest that certain mechanisms of neurodegeneration may be shared
  between the two diseases. Inflammation drives the disease in MS. It is also present in
  AD lesions. In both diseases, degeneration of neurons, axons, and synapses occur on
  the background of profound mitochondrial injury. Thus even though the diseases are
  different, many of the physiological symptoms and causes overlap.

The diseases "are fundamentally different?" Does this statement refer to symptoms or
causes? As you will see, it is an accurate statement for symptoms but not so for causes.

A google.scholar search on terms "multiple sclerosis" and "infection" yields 300,000 hits.
The first hit presented in the search is titled, *Epidemiological evidence for multiple sclerosis as an
infection.* Since there is strong evidence for an association between Alzheimer's and infection,
maybe this is the root cause link between the two diseases. Any differential diagnosis of MS
should certainly consider evaluation of inflammation markers and possibly infectious species.

MULTIPLE SCLEROSIS AND
CHLAMYDIA PNEUMONIAE

Dr. David Wheldon of England is well-known for his study of MS. He has research articles
to his credit but he fancies himself a humble clinician. We need many, many more humble
clinicians like Dr. David Wheldon. He has teamed up with a research group out of Vanderbilt
University in the United States. One of their coauthored papers is titled, *Multiple sclerosis:*

an infectious syndrome involving Chlamydophila pneumoniae [88]. In addition to this article, he web-published a compelling summary on the connection between MS and infection titled, *Empirical antibacterial treatment of infection with Chlamydophila pneumoniae in Multiple Sclerosis.* Excerpts from his website are included here.

> After much controversy there is now powerful evidence for the respiratory pathogen Chlamydophila (Chlamydia) pneumoniae being a causal factor in some variants of the neurological illness multiple sclerosis. A series of remarkable studies finds:
>
> - The presence of C. pneumoniae gene sequences in the cerebrospinal fluid of patients who have the disease, and culture of the organism when sensitive cultural methods are used.
> - An association of new C. pneumoniae respiratory infections with episodes of clinical relapse.
> - A statistically significant elevation of C. pneumoniae-specific serum antibody levels when the disease shifts into the progressive form
> - Antibodies to C. pneumoniae in the cerebrospinal fluid of patients with the disease.
> - Intrathecal production of Chlamydia pneumoniae-specific high-affinity antibodies is significantly associated with a subset of multiple sclerosis patients with progressive forms.
> - Evidence of active C. pneumoniae protein synthesis in the central nervous system, with production of a bacterial protein evoking an antibody shown to cause death of oligodendrocyte precursor cells.
> - A peptide specific to C. pneumoniae causes inflammatory CNS disease (with some parallels to MS) in rats.
> - C. pneumoniae gene transcription in the CSF of patients with MS.
> - MRI improvement in antibiotic-treated patients with early disease in a small but fastidious double-blind trial of non-immunomodulatory antibiotics.
> - MRI improvement, with reduction of the number of Gd-enhancing lesions, in a second treatment study with minocycline.
> - An association of C. pneumoniae in the CNS with MS is demonstrated by immunohistochemical, molecular and ultrastructural methods.

He points out that the association between *C. pneumoniae* and MS has been shown by a "surprisingly diverse array of methods: cultural, molecular (both DNA and RNA based), immunohistological, serological (blood and CSF based), animal model, ultrastructural and therapeutic trial." He points out that it is the diversity of methods that truly support the conclusion that this infectious species are significantly involved in the development and progression of MS. Dr. Wheldon further notes that antichlamydial treatments have been very promising. Dr. Wheldon explains:

> The mechanism of Multiple Sclerosis as caused by C. pneumoniae and its parallels to Alzheimer's disease is striking. C pneumoniae is known to patchily parasitize the cells, which line small blood vessels, causing episodes of vasculitis (vessel inflammation). This is a local inflammatory process characterized by tiny punctures in the vessel walls and leakage of blood-components into the surrounding tissue space. It can be visualized directly in the retinal veins, where the vessels appear to be coated with a thin grayish sheath. This sheath is comprised of T lymphocytes. A very similar pathology takes place in the brain in early MS. The association between sheathing of retinal veins and MS was first made in 1944. The anatomical distribution of lesions within the brain in MS is often centered on small veins; elongated plaques may follow the sinuous curves of the vessels they surround.

Wheldon educates us on the history of MS and inflammation (caused by infection), and that a clinician, Rindfleisch, had recognized, over 130 years ago, that inflammation of small vessels (vasculitis) precedes neural damage.

Wheldon connects the eye and MS. "Examination of the eye reveals retinal vasculitis in about a third of persons with early MS, but it is probably present in far more. It is especially common following optic neuritis (a common precursor of MS), and is characterized by leakage of dye in a fluorescein dye test, blood cells, and cuffing of the vessel walls by inflammatory cells. Where it is seen, there is a raised likelihood that MS will follow." The eye thus shows up as a powerful diagnostician in two neurodegenerative diseases, Alzheimer's and MS.

MS is currently considered an autoimmune demyelinating disease. Myelin is an insulating lipoprotein; its sudden local loss causes the acute MS relapse. But Wheldon points out that this myelin loss may be secondary to infection. "The very fact that retinal vasculitis is commonly associated with MS casts considerable doubt on myelinopathy being the root cause of MS; myelin, and the oligodendrocyte cells which produce it, are not found in the retina, and the earliest pathological manifestations of MS are in blood-vessels, not nerves and glial cells."

Dr. Wheldon makes some important points about treatment for the infectious cause of MS. His suggestions apply for AD when the same infections species, CP, is found. Note that an emphasis is placed primarily on diagnosis because once true root causes are determined, a good doctor is able to carry out proper treatment. However, current medicine does not appear to completely appreciate the difficultly in treating intracellular infectious species like *C. pneumoniae*. Thus Dr. Wheldon's prescription is quite far removed for traditional clinical treatment regiments.

Visit Dr. Wheldon's website for a more comprehensive understanding about the connection between MS and infection, found by Googling "MS" and "David Wheldon." On his site, he gives much more proof connecting *C. pneumoniae* infection and MS. He also gives a detailed treatment prescription, assuming that CP is identified as part of the pathology of the disease, based on a differential diagnosis.

PERIODONTAL BACTERIA LINKED TO ALZHEIMER'S

What is the dirtiest room in your house? Most would suggest that the bathroom is the dirtiest. However, bacteria do not proliferate in the presence of other bacteria, generally. They prefer food, light, moisture, and avoid competition. The kitchen uniquely and continuously provides that right environment for bacterial growth. By analogy to the human body, the mouth is the "dirtiest" place in the human body. Where else do the three key components of growth come together so well? And this is not theoretical as about 64% of Americans aged 65 and older have moderate or severe gum (periodontal) disease, according to the US Centers for Disease Control and Prevention.

There is a general understanding about the connection between oral health and whole body health, but it is not appreciated as a major cause of system-wide (systemic) diseases. *Scientific American* documented many of the diseases caused by oral disease in their separate publication titled, *Oral and Whole Body Health* [89]. Through the efforts of a colleague, this valuable publication is now available for free on the Internet. Here are a couple of lesser-known manifestations of periodontal disease:

1. Loss of physical height with age. In this case, the periodontal derived bacteria enter the nervous system through the roots of the teeth. It eventually migrates into the spinal fluid and causes inflammation of the spinal discs resulting in loss of structural integrity, loss of height, and collapse, causing height loss and back pain.
2. A Brown University Medical School study found (Harvard University had a similar finding): Periodontal disease might increase the risk for pancreatic cancer [90]. Moreover,

increased levels of antibodies against specific commensal oral bacteria, which can inhibit growth of pathogenic bacteria, might reduce the risk of pancreatic cancer.

These are relatively unexpected findings, and they point out the connectivity of diseases. Who would have thought that height loss could be related to pancreatic cancer at some basic root cause level? Therefore it should be no surprise, based on all we have studied, that there is also a connection between oral infection and AD.

Before we delve into the oral/Alzheimer's connection, let's briefly look at a brilliant review titled, *Systemic Diseases Caused by Oral Infection*, published in 2000 [91]. This publication provides both perspective and insights into cause and effect (mechanisms). This work also predates the growing appreciation for the connection between oral infection and Alzheimer's but the mechanisms clearly apply.

> It has become increasingly clear that the oral cavity can act as the site of origin for dissemination of pathogenic organisms to distant body sites, especially in immunocompromised hosts such as patients suffering from malignancies, diabetes, or rheumatoid arthritis or having corticosteroid or other immunosuppressive treatment. A number of epidemiological studies have suggested that oral infection, especially marginal and apical periodontitis, may be a risk factor for systemic diseases.
> Bacteremia (the presence of bacteria in the blood) was observed in 100% of the patients after dental extraction, in 70% after dental scaling, in 55% after third-molar surgery, in 20% after endodontic treatment, and in 55% after bilateral tonsillectomy. All root-canals contained anaerobic bacteria
> Another study involving 735 children undergoing treatment for extensive dental decay found that 9% of the children had detectable bacteremias before the start of dental treatment.

Would you like to guess which children are susceptible to diabetes and obesity? If you guessed the 9%, you are probably correct. These are inflammatory conditions either triggered or exacerbated by infection, in this case from the oral cavity.

> In a recent review article [92], Page proposed that periodontitis may affect the host's susceptibility to systemic disease in three ways: by shared risk factors, by subgingival biofilms acting as reservoirs of gram-negative bacteria, and through the periodontium acting as a reservoir of inflammatory mediators.

Consider reading the review article quoted earlier. You will find that all the key factors attributed to AD are connected to periodontal/oral infection.

Dr. Charles Mayo, a founder of the Mayo Clinic, attempted to alert the world as to the severe adverse health consequences of poor oral health. Mayo believed in the "focal infection" theory of disease, something so archaic that today almost no one has heard of it. The theory basically states that an oral infection can influence the health of the entire body. Addressing the Chicago Dental Society in 1913, Mayo said, "The next great step in preventative medicine must come from the dentists."

Mayo appointed Dr. Edward C. Rosenhow to head a team of researchers dedicated to focal infection theory. From 1902 to 1958, Rosenhow conducted experiments and published more than 300 papers, 38 of which appeared in the *Journal of the American Medical Association*. During the same period, Weston A. Price, founder of the research institute of the National Dental Association, published his findings indicating that dental and oral infections were often the primary cause of disease.

These two medical pioneers established a simple but profound fact. If you pull an infected tooth, the patient will often recover from disease, even serious disease, from chronic fatigue to cancer, from dermatitis to diabetes, from hemorrhoids to heart disease. Drs. Rosenhow and

Price theorized that disease often originated from infections in the mouth that entered the bloodstream and eventually caused major problems in some part of the body. The evidence they amassed and published is staggering, yet the next great step Dr. Mayo hoped for did not come, and their work is largely forgotten today.

The Alzheimer's/Periodontal Diseases connection was reviewed in 2008 by a team from New York University. The article is titled, *Inflammation and Alzheimer's disease: Possible role of periodontal diseases* [93]. The abstract of the paper explains the thesis and conclusion with superb adequacy:

> The molecular and cellular mechanisms responsible for the etiology and pathogenesis of Alzheimer's disease (AD) have not been defined; however, inflammation within the brain is thought to play a pivotal role. Studies suggest that peripheral infection/inflammation might affect the inflammatory state of the central nervous system. Chronic periodontitis is a prevalent peripheral infection that is associated with gram-negative anaerobic bacteria and the elevation of serum inflammatory markers including C-reactive protein. Recently, chronic periodontitis has been associated with several systemic diseases including AD. In this article we review the pathogenesis of chronic periodontitis and the role of inflammation in AD. In addition, we propose several potential mechanisms through which chronic periodontitis can possibly contribute to the clinical onset and progression of AD. Because chronic periodontitis is a treatable infection, it might be a readily modifiable risk factor for AD.

The lesson learned is that infection can cause a myriad of diseases. What is not discussed is that some infections respond to treatments better compared to others. Periodontal infection, for the most part, is treatable. However, prevention is always preferable to treatment. The concept of prevention being preferable to treatment is part of the medical Hippocratic Oath. Not all infectious materials that could lead to or facilitate AD are presented in this book. McCully points out that over 50 bacteria (at least) are associated with cardiovascular diseases. The same number is likely tied to AD. Q-fever is a bacteria that is particularly difficult to treat. Thus, AD caused by Q-fever may not respond to treatment, whereas AD caused by *C. pneumoniae* may respond reasonably well.

In a follow-on study prompted by the 2008 review cited earlier, the NYU team looked at both periodontal infection and tooth loss [94]. They showed a clear connection between infection, tooth loss, and decay in cognitive function. A key finding was subjects with periodontal inflammation were nine times more likely to test in the lower range of the DST (cognitive test) compared to subjects with little or no periodontal inflammation.

A recent study at the University of Florida analyzed brain samples from 10 people with Alzheimer's and 10 people without the brain disease and found gum disease-related bacteria in the brain samples from 4 of the 10 Alzheimer's patients [95]. No such bacteria were found in the brain samples from people without Alzheimer's. This is interesting evidence that there is no one single cause or set of causes for Alzheimer's, at least when it comes to infection.

MICROBES AND MENTAL ILLNESS

This section is included in this book because the author of this information is the director of the National Institute of Mental Health—Tom Insel, MD. We are now learning that depression is a precursor to AD. This doesn't mean that if you are depressed, the you are doomed. Our bodies become ill then go into remission for many diseases all the time. It is key that if you are depressed, that you take action to ameliorate your condition. The solution does not involve taking the psychoactive drugs that treat symptoms—or rather—just dope you so you don't feel anything. Thus part of a differential diagnosis should include looking at toxicity, especially toxicity caused by microbes, that is infectious species. But, just like we stated regarding Alzheimer's at the beginning of this chapter, microbes are not the cause of depression (and all the other neuropsychological disorders), but they can exacerbate your condition and establish a viscous cycle of proliferation which makes you continually worse.

Here is a reproduction of Dr. Thomas Insel's "Directors Blog: Microbes and Mental Illness," from August 13, 2010 [96].

Hints that some mental illness may be linked to infectious agents and/or autoimmune processes date back to at least the early 20th Century. In the 21st Century, the field of microbiomics, which is mapping the microbial environment of the human organism, may transform the way we think about human physical and mental development. It is already clear that 90% of "our DNA" is microbial, not human. "We" are, in fact, "super-organisms" made up of thousands of species, many of which are being identified for the first time. And there are persistent individual differences in our microbial ecology established early in life.

Insights from microbiomics have proven important for understanding obesity and Type 1 diabetes, but microbiomics has not yet been a focus for research on mental illness. Yet, there are many clues linking microbiology and mental disorders, such as epidemiologic evidence of increased risk for schizophrenia associated with prenatal exposure to influenza. Probably the most compelling case for such involvement is children who develop obsessive compulsive disorder (OCD) and/or tic disorders "overnight," following a strep infection. Despite continuing debate over its parameters, evidence is mounting in support of Pediatric Autoimmune Neuropsychiatric Disorders Associated with Streptococcal Infections (PANDAS) — or at least a syndrome modeled on it.

Last month, the NIMH Pediatric Developmental Neuroscience Branch convened dozens of experts from the field — including prominent PANDAS critics — to update the science and attempt to achieve consensus on criteria defining the syndrome. The mere fact that the conference took place signals a change in the scientific climate. Until now, whether a child presenting with sudden onset of OCD and/or tic symptoms gets checked for possible involvement of strep has varied—often depending on which medical journals a practitioner happens to read. I am hopeful that will begin to change in light of the new evidence.

Interest in PANDAS has also been spurred by an increasingly vocal network of affected families and the clinicians who are treating their often severely-impaired children. Conference participants heard reports from the front lines by some of these clinicians, who largely corroborated key features of the syndrome, originally identified by NIMH's Dr. Susan Swedo in the mid-1990s. These include sudden onset of mood swings, impulsivity, anxiety, impaired attention and poor handwriting in addition to obsessions, compulsions and tics. Dr. Swedo's studies have identified brain mechanisms through which strep antibodies act. They have also demonstrated that cleansing the blood of the antibodies, via plasma exchange or intravenous immunoglobulin, significantly diminishes the symptoms.

Impetus for the July conference came, in part, from publication of two independent studies within the past year that lend new credence to the PANDAS concept.

In the first, Columbia University researchers demonstrated, for the first time, that strep-triggered antibodies alone are necessary and sufficient to trigger a PANDAS-like syndrome in mice. In an autoimmune-disease susceptible strain of mice, exposure to strep triggered OCD-like repetitive behaviors and antibodies that attacked specific molecules in the brain. PANDAS-like behaviors also emerged in naïve mice after they received antibodies from such PANDAS mice. These included impaired learning and memory and social interaction. As in humans with PANDAS, these impairments were more common in males than females.

In the second study, a Yale University research team reported that OCD and Tourette Syndrome (tic) symptoms worsened slightly following a strep infection in some affected children. Moreover, the strep infection triggered the worsened symptoms by increasing the impact of psycho-social stress. The findings suggest that a subset of children with these disorders may be at increased risk of strep infection, which could interact with stress to exacerbate the course, as is seen in other infectious and autoimmune diseases.

Granted, these new findings are still preliminary and need to be replicated. However, the data relating to PANDAS is compelling enough to warrant following up such leads. NIMH is preparing to launch a new trial of intravenous immunoglobulin (IVIG) treatment for PANDAS this Fall, with support from a NIH Clinical Center "Bench to Bedside" award. The intramural NIMH will provide the clinical care, while data analysis will be carried out by independent teams of investigators at Yale University and the Oklahoma University Health Sciences Center. Dr. Swedo and her team are hoping to recruit 50 children with clear-cut PANDAS. They are predicting that IVIG treatment will produce striking benefits for OCD and other neuropsychiatric symptoms, and will be most effective for those children who start out with the highest levels of strep-triggered antibodies that go astray and attack parts of the brain. Moreover, monoclonal antibodies derived from these patients will be used to develop animal models of OCD that could lead to improved treatments.

Do infectious agents influence the development of autism, anxiety, or mood disorders? This remains a frontier area for NIMH research. The increasing evidence linking strep infection to OCD in children suggests that microbiomics may prove an important research area for understanding and treating mental disorders.

As with all research-types, they end with the obligatory statement about more research. Our recommendation is "take more blood." What that simply means is anyone with some chronic brain issues should have blood tests. What blood tests in particular? Those we discussed in Chapter 7 on inflammation and should also include an evaluation for the range of microbes starting with CP, toxoplasmosis, babesia, other lyme-based organism, rickesial microbes, fungi, and viruses as a starting point.

Do we really need more research to properly diagnose and treat the mentally impacted? I believe the answer is NO. We need to perform a more robust diagnosis! Parents with "labeled" children, your only option to address mental health issues is to see a functional or integrative doctor. Mainstream medicine is all about symptom suppression—sadly.

JUDITH MIKLOSSY—SPIROCHETES AND ALZHEIMER'S DISEASE

This chapter begins and ends with the pioneering work of Dr. Judith Micklossy. Dr. Miklossy from Switzerland has been most persistent at promoting the hypothesis of infection and AD, through her brilliant research. Her landmark publication from 1993 is titled, *Alzheimer's disease—A spirochetosis?* [15] The 40+ publications of Dr. Miklossy are included as references.

On her website, under the heading "The Emerging Role of Infection in Alzheimer's Disease and Stroke," Dr. Miklossy makes the following statement:

> Here we describe the line of research we have followed during the last 15 years with respect to the involvement of spirochetes in Alzheimer's disease and in cerebral infarcts. This line of research represents a panel of experiments, listed below, which are linked to each other. The goal is to answer the question, whether several types of spirochetes, including Borrelia burgdorferi, various periodontal pathogen spirochetes, intestinal spirochetes etc., may be involved in Alzheimer disease and stroke.

Definition of Infarct an area of tissue that undergoes necrosis as a result of obstruction of local blood supply, as by a thrombus or embolus.

Notice how Dr. Miklossy talks about both Alzheimer's and stroke in the same sentence.

Let's take a brief look at the "book ends" of her work by examining her papers from 1993 and 2011: *Alzheimer's disease—A spirochetosis?* [15], and *Alzheimer's disease—A neurospirochetosis. Analysis of the evidence following Koch's and Hill's criteria* [97]. In 1993 she wrote the following:

> Dementia associated with cortical atrophy and microgliosis has been observed in the late stages of two spirochetal diseases: Lyme disease, a late stage of neuroborreliosis caused by Borrelia burgdorferi (Burgdorfer et al., 1982; Pachner et al., 1989), and general paresis, tertiary stage of neurosyphilis caused by Treponema pallidum. Two cases of concurrent neocortical borreliosis and Alzheimer disease (AD) have been reported (MacDonald and Miranda, 1987; MacDonald, 1988): immunostaining showed Borrelia burgdorferi in brain tissue and the spirochetes were cultured from cerebral cortex. A careful study of 18 AD cases, using several methodological approaches, failed to support an association between Borrelia burgdorferi and AD, but the authors did not rule out the possibility that another spirochete, not detectable by their methods, may be responsible for AD (Pappolla et al., 1989).

Key word: "Associated"
In 2011, she explains:

> It is established that chronic spirochetal infection can cause slowly progressive dementia, brain atrophy, and amyloid deposition in late neurosyphilis. Recently it has been suggested that various types of spirochetes, in an analogous way to Treponema pallidum, could cause dementia and may be involved in the pathogenesis of Alzheimer's disease (AD). Here, we review all data available in the literature on the detection of spirochetes in AD and critically analyze the association and causal relationship between spirochetes and AD following established criteria of Koch and Hill. The results show a statistically significant association between spirochetes and AD (P = 1.5 × 10-17, OR = 20, 95% CI = 8-60, N = 247). When neutral techniques recognizing all types of spirochetes were used, or the highly prevalent periodontal pathogen Treponemas were analyzed, spirochetes were observed in the brain in more than 90% of AD cases. Borrelia burgdorferi was detected in the brain in 25.3% of AD cases analyzed and was 13 times more frequent in AD compared to controls. Periodontal pathogen Treponemas (T. pectinovorum, T. amylovorum, T. lecithinolyticum, T. maltophilum, T. medium, T. socranskii) and Borrelia burgdorferi were detected using species specific PCR and antibodies. Importantly, co-infection with several spirochetes occurs in AD. The pathological and biological hallmarks of AD were reproduced in vitro by exposure of mammalian cells to spirochetes. The analysis of reviewed data following Koch's and Hill's postulates shows a probable causal relationship between neurospirochetosis and AD. Persisting inflammation and amyloid deposition initiated and sustained by chronic spirochetal infection form together with the various hypotheses suggested to play a role in the pathogenesis of AD a comprehensive entity. As suggested by Hill, once the probability of a causal relationship is established prompt action is needed. Support and attention should be given to this field of AD research. Spirochetal infection occurs years or decades before the manifestation of dementia. As adequate antibiotic and anti-inflammatory therapies are available, as in syphilis, one might prevent and eradicate dementia.

Key word: "Established."
She has come a long way in providing appropriate proof for the cause/effect of AD.

Dr. Miklossy provides a brief and elegant, yet detailed summary of the connection between AD and spirochete infection under the following headings on her website:

- AD
- Chronic inflammation and AD
- Bacteria, including spirochetes, are powerful stimulators of inflammation and are amyloidogenic
- Chronic bacterial infection can cause dementia
- Spirochetes
- Our contribution to this emerging field of research

Her concluding remarks are:

> The results of all these studies when taken together allow us to conclude that Borrelia burgdorferi and oral spirochetes can persist in the brain and in analogy to Treponema pallidum, can cause dementia, cortical atrophy and amyloid deposition. Exposure to bacteria or to their toxic products, host responses similar in nature to those observed in AD may be induced. Bacteria and/or their degradation products may initiate a cascade of events leading to cell death, neurodegeneration and amyloid deposition in Alzheimer's disease.

MICROBES AND ALZHEIMER'S—INCONTROVERTIBLE PROOF

In a recent editorial in the Journal of Alzheimer's Disease, 31 scientists and clinicians wrote about incontrovertible evidence that Alzheimer's and infection are connected at one of the key roots of the disease [98]. The transcript of the editorial is reproduced here as text only without references. We encourage you to read the original editorial and all the supporting references.

> We are researchers and clinicians working on Alzheimer's disease (AD) or related topics, and we write to express our concern that one particular aspect of the disease has been neglected, even though treatment based on it might slow or arrest AD progression. We refer to the many studies, mainly on humans, implicating specific microbes in the elderly brain, notably herpes simplex virus type 1 (HSV1), Chlamydia pneumoniae, and several types of spirochaete, in the etiology of AD. Fungal infection of AD brain has also been described, as well as abnormal microbiota in AD patient blood. The first observations of HSV1 in AD brain were reported almost three decades ago. The ever-increasing number of these studies (now about 100 on HSV1 alone) warrants reevaluation of the infection and AD concept.
>
> AD is associated with neuronal loss and progressive synaptic dysfunction, accompanied by the deposition of amyloid-beta peptide, a cleavage product of the amyloid-beta protein precursor (AbetaPP), and abnormal forms of tau protein, markers that have been used as diagnostic criteria for the disease. These constitute the hallmarks of AD, but whether they are causes of AD or consequences is unknown. We suggest that these are indicators of an infectious etiology. In the case of AD, it is often not realized that microbes can cause chronic as well as acute diseases; that some microbes can remain latent in the body with the potential for reactivation, the effects of which might occur years after initial infection; and that people can be infected but not necessarily affected, such that 'controls', even if infected, are asymptomatic.

Evidence for an Infectious/Immune Component

Viruses and other microbes are present in the brain of most elderly people. Although usually dormant, reactivation can occur after stress and immunosuppression; for example, HSV1 DNA is amplified in the brain of immunosuppressed patients.

1. Herpes simplex encephalitis (HSE) produces damage in localized regions of the CNS related to the limbic system, which are associated with memory, cognitive and affective processes, as well as personality (the same as those affected in AD).
2. In brain of AD patients, pathogen signatures (e.g., HSV1 DNA) specifically colocalize with AD pathology.
3. HSV infection, as revealed by seropositivity, is significantly associated with development of AD.
4. AD has long been known to have a prominent inflammatory component characteristic of infection.

5. Polymorphisms in the apolipoprotein E gene, APOE, that modulate immune function and susceptibility to infectious disease, also govern AD risk. Genome-wide association studies reveal that other immune system components, including virus receptor genes, are further AD risk factors.
6. Features of AD pathology are transmissible by inoculation of AD brain to primates and mice.

Evidence for Causation

1. In humans, brain infection (e.g., by HIV, herpesvirus, measles) is known to be associated with AD-like pathology. Historical evidence shows that the clinical and pathological hallmarks of AD occur also in syphilitic dementia, caused by a spirochaete.
2. In mice and in cell culture, A deposition and tau abnormalities typical of AD are observed after infection with HSV1 or bacteria; a direct interaction between A PP and HSV1 has been reported. Antivirals, including acyclovir, in vitro block HSV1-induced A and tau pathology.
3. Olfactory dysfunction is an early symptom of AD. The olfactory nerve, which leads to the lateral entorhinal cortex, the initial site from where characteristic AD pathology subsequently spreads through the brain, is a likely portal of entry of HSV1 and other viruses, as well as *C. pneumoniae*, into the brain, implicating such agents in damage to this region. Further, brainstem areas that harbor latent HSV directly irrigate these brain regions: brainstem virus reactivation would thus disrupt the same tissues as those affected in AD.

Growing Evidence for Mechanism: Role of beta-amyloid

1. The gene encoding cholesterol 25-hydroxylase (CH25H) is selectively upregulated by virus infection, and its enzymatic product (25-hydroxycholesterol, 25OHC) induces innate antiviral immunity.
2. Polymorphisms in human CH25H govern both AD susceptibility and beta-amyloid deposition, arguing that beta-amyloid induction is likely to be among the targets of 25OHC, providing a potential mechanistic link between infection and beta-amyloid production.
3. Beta-amyloid is an antimicrobial peptide with potent activity against multiple bacteria and yeast. Beta-amyloid also has antiviral activity.
4. Another antimicrobial peptide (beta-defensin 1) is upregulated in AD brain.

Regarding HSV1, about 100 publications by many groups indicate directly or indirectly that this virus is a major factor in the disease. They include studies suggesting that the virus confers risk of the disease when present in brain of carriers of the 4 allele of APOE, an established susceptibility factor for AD (APOE 4 determines susceptibility in several disorders of infectious origin, including herpes labialis, caused usually by HSV1). The only opposing reports, two not detecting HSV1 DNA in elderly brains and another not finding an HSV1–APOE association, were published over a decade ago. However, despite all the supportive evidence, the topic is often dismissed as "controversial." One recalls the widespread opposition initially to data showing that viruses cause some types of cancer, and that a bacterium causes stomach ulcers.

In summary, we propose that infectious agents, including HSV1, *C. pneumoniae*, and spirochetes, reach the CNS and remain there in latent form. These agents can undergo reactivation in the brain during aging, as the immune system declines, and during different types of stress (which similarly reactivate HSV1 in the periphery). The consequent neuronal damage—caused by direct viral action and by virus induced inflammation—occurs recurrently, leading to (or acting as a cofactor for) progressive synaptic dysfunction, neuronal loss, and ultimately AD. Such damage includes the induction of A which, initially, appears to be only a defense mechanism.

AD causes great emotional and physical harm to sufferers and their careers, as well as having enormously damaging economic consequences. Given the failure of the 413 trials of other types of therapy for AD carried out in the period 2002–2012, antiviral/antimicrobial treatment of AD patients, notably those who are APOE 4 carriers, could rectify the "no drug works" impasse. We propose that further research on the role of infectious agents in AD causation, including prospective trials of antimicrobial therapy, is now justified.

EMERGING CONNECTION BETWEEN INFECTION AND DISEASE

Here is an interesting excerpt from the Atlantic Monthly from February, 1999 [99]

The catalogue of suspected chronic diseases caused by infection, according to David A. Relman, an assistant Professor of Medicine, Microbiology, and Immunology at Stanford University, now includes, "sarcoidosis, various forms of inflammatory bowel disease, rheumatoid arthritis, systemic lupus erythematosus, Wegener's granulomatosis, diabetes mellitus, primary biliary cirrhosis, tropical sprue, and Kawasaki disease." Ewald and Cochran's list of likely suspects would include all of the above plus many forms of heart disease, arteriosclerosis, Alzheimer's disease, many if not most forms of cancer, multiple sclerosis, most major psychiatric diseases, Hashimoto's thyroiditis, cerebral palsy, polycystic ovary disease, and perhaps obesity and certain eating disorders. From an evolutionary perspective, Cochran says, anorexia is strikingly inimical to the survival principle. "I mean, not to eat—what would cause that?

References

[1] Tuberculosis Fact sheet No. 104. World Health Organization; 2010.
[2] World Health Organization. Epidemiology. Global tuberculosis control: epidemiology, strategy, financing, 2009. pp. 6–33.
[3] Kumar V, Abbas AK, Fausto N, Mitchell RN. Robbins basic pathology. 8th ed. Philadelphia, PA: Saunders Elsevier; 2007. pp. 516–522.
[4] Raoult D, Roux V. Rickettsioses as paradigms of new or emerging infectious diseases. Clin Microbiol Rev 1997;10(4):694–719.
[5] Fischer O. Miliare Nekrosen mit drusigen Wucherungen der Neurofibrillen, eine regelmässige Veränderung der Hirnrinde bei seniler Demenz. Mschr Psychiat Neurol 1907;22:361–72.
[6] Alzheimer A. Über eigenartige Krankheitsfälle des späteren Alters. Z Ges Neurol Psychiat 1911;4:356–85.
[7] Miklossy J, et al. Beta-amyloid deposition and Alzheimer's type changes induced by Borrelia spirochetes. Neurobiol Aging 2006;27(2):228–36.
[8] Ohanian SH, Schwab JH. Persistence of group A streptococcal cell walls related to chronic inflammation of rabbit dermal connective tissue. J Exp Med 1967;125(6):1137–48.
[9] Fox ALVIN. Role of bacterial debris in inflammatory diseases of the joint and eye. Apmis 1990;98(7–12):957–68.
[10] Stanek G, Reiter M. The expanding Lyme Borrelia complex—clinical significance of genomic species? Clin Microbiol Infect 2011;17(4):487–93.

[11] Ang CW, et al. Large differences between test strategies for the detection of anti-Borrelia antibodies are revealed by comparing eight ELISAs and five immunoblots. Eur J Clin Microbiol Infect Dis 2011;30(8):1027–32.

[12] Schmitz JL, Powell CS, Folds JD. Comparison of seven commercial kits for detection of antibodies to Borrelia burgdorferi. Eur J Clin Microbiol Infect Dis 1993;12(6):419–24.

[13] Esmarch F, Jessen W. Syphilis und Geistesstörung. Allg Zeitschr Psychiat 1857;14:20–32.

[14] Armstrong H. On some clinical manifestations of congenital syphilis. BMJ 1914;1:958–61.

[15] Miklossy J. Alzheimer's disease—A spirochetosis? NeuroReport 1993;4:841–8.

[16] Bonfiglio F. Di speciali reperti in uno caso di probabile sifilide cerebrale. Rev Sper Freniatr 1908;34:196–206.

[17] Fischer O. Die presbyophrene Demenz, deren anatomische Grundlage und klinische Abgrenzung. Z ges Neurol Psychiat 1910;3:371–471.

[18] Letemendia F, Pampiglione G. Clinical and electroencephalographic observations in Alzheimer's disease. J Neurol Neurosurg Psychiatry 1958;21(3):167.

[19] Wisniewski HM. Possible viral etiology of neurofibrillary changes and neuritic plaques. In: Katzman R, Terry RD, Bick KL, editors. Alzheimer's disease: Senile dementia and related disorders. Aging, vol. 7. New York: Raven Press; 1978. p. 555–7.

[20] Renvoize EB, Hambling MH. Cytomegalovirus infection and Alzheimer's disease. Age Ageing 1984;13(4):205–9.

[21] Tateishi J. Recent topics on Alzheimer's disease. The possibility of Alzheimer's disease as a delayed-type infection—study on prions of proteinaceous infective particles. Nihon Rinsho 1988;46(7):1607.

[22] Fallon BA, Nields JA. Lyme disease: a neuropsychiatric illness. Am J Psychiatry 1994;151(11):1571–83.

[23] Ravnskov U, McCully KS. Vulnerable plaque formation from obstruction of vasa vasorum by homocysteinylated and oxidized lipoprotein aggregates complexed with microbial remnants and LDL autoantibodies. Ann Clin Lab Sci 2009;39(1):3–16.

[24] Ott SJ, et al. Detection of diverse bacterial signatures in atherosclerotic lesions of patients with coronary heart disease. Circulation 2006;113(7):929–37.

[25] Marshall BJ, Warren RJ. Unidentified curved bacilli in the stomach of patients with gastritis and peptic ulceration. Lancet 1984;323(8390):1311–5.

[26] Marshall BJ, et al. Prospective double-blind study of supplementary antibiotic therapy for duodenal ulcer associated with Campylobacter pyloridis infection. Dig Dis Sci 1986;31:889.

[27] Marshall BJ, et al. Prospective double-blind trial of duodenal ulcer relapse after eradication of Campylobacter pylori. Lancet 1988;332(8626):1437–42.

[28] Marshall BJ, et al. Rapid urease test in the management of Campylobacter pyloridis-associated gastritis. Am J Gastroenterol 1987;82(3):200–10.

[29] Goodwin CS, et al. Enzyme-linked immunosorbent assay for Campylobacter pyloridis: correlation with presence of C. pyloridis in the gastric mucosa. J Infect Dis 1987;155(3):488–94.

[30] Thagard, P. Ulcers and bacteria I: discovery and acceptance; 1997.

[31] Saikku P, et al. Chronic Chlamydia pneumoniae infection as a risk factor for coronary heart disease in the Helsinki Heart Study. Ann Internal Med 1992;116(4):273–8.

[32] Grayston JT. Antibiotic treatment of atherosclerotic cardiovascular disease. Circulation 2003;107(9):1228–30.

[33] Grayston JT, Jackson LA, Kennedy WJ, et al. Secondary prevention trials for coronary artery disease with antibiotic treatment for Chlamydia pneumoniae: design issues. Am Heart J 1999;138:S545–9.

[34] Grayston JT. Antibiotic treatment of C. pneumoniae for secondary prevention of cardiovascular events. Circulation 1998;97:1669–70.

[35] Uriarte SM, et al. Effects of fluoroquinolones on the migration of human phagocytes through Chlamydia pneumoniae-infected and tumor necrosis factor alpha-stimulated endothelial cells. Antimicrob Agents Chemother 2004;48(7):2538–43.

[36] MacIntyre A, et al. Chlamydia pneumoniae infection promotes the transmigration of monocytes through human brain endothelial cells. J Neurosci Res 2003;71(5):740–50.

[37] Brundin P, Melki R, Kopito R. Prion-like transmission of protein aggregates in neurodegenerative diseases. Nat Rev Mol Cell Biol 2010;11(4):301–7.

[38] Hardy J, Revesz T. The spread of neurodegenerative disease. N Engl J Med 2012;366(22):2126–8.

[39] Balin BJ, et al. Identification and localization of Chlamydia pneumoniae in the Alzheimer's brain. Med Microbiol Immunol 1998;187(1):23–42.

[40] Hudson AP, et al. Chlamydia pneumoniae, APOE genotype, and Alzheimer's disease. Chlamydia pneumoniae and chronic diseases. Berlin Heidelberg: Springer; 2000. 121–36.

[41] Stratton CW, et al. Association of Chlamydia pneumoniae with multiple sclerosis: protocol for detection of C. pneumoniae in the CSF and summary of preliminary results. Chlamydia pneumoniae and chronic diseases. Berlin Heidelberg: Springer; 2000. 137–52.

[42] Hammerschlag MR, et al. Is Chlamydia pneumoniae present in brain lesions of patients with multiple sclerosis? J Clin Microbiol 2000;38(11):4274–6.

[43] Gieffers J, et al. Failure to detect Chlamydia pneumoniae in brain sections of Alzheimer's disease patients. J Clin Microbiol 2000;38(2):881–2.

[44] Gerard HC, et al. Chlamydophila (Chlamydia) pneumoniae in the Alzheimer's brain. FEMS Immunol Med Microbiol 2006;48(3):355–66.

[45] Balin BJ, et al. Chlamydophila pneumoniae and the etiology of late-onset Alzheimer's disease. J Alzheimer's Dis 2008;13(4):371–80.

[46] Arking EJ. Chlamydia pneumoniae: an emerging pathogen in alzheimer's disease. Diss. Allegheny University of the Health Sciences, MCP Hahnemann School of Medicine, Department of Pathology and Laboratory Medicine, Clinical Pathology Program; 1998.

[47] Arking EJ, et al. Ultrastructural analysis of Chlamydia pneumoniae in the Alzheimer's Brain. Pathogenesis (Amst) 1999;1(3):201–11.

[48] Mahony J, et al. Chlamydia pneumoniae in the Alzheimer's brain—Is DNA detection hampered by low copy number. Proceedings of the fourth meeting of the European Society for Chlamydia Research. Helsinki-Finland; Clinical Diseases 2000.

[49] Mahony JB, et al. Identification of Chlamydia pneumoniae in the Alzheimer's brain. Neurobiol Aging 2000;21:245.

[50] Hudson AP. What is the evidence for a relationship between Chlamydia pneumoniae and late-onset Alzheimer's disease? Lab Med 2001;32(11):680–5.

[51] MacIntyre A. Chlamydia pneumoniae infection in human monocytes and brain endothelial cell: initiating factors in the development of Alzheimer's disease. Philadelphia: Diss. MCP Hahnemann University; 2002.

[52] Little S, et al. Induction of Alzheimer's disease-like pathology in the brains of young, non-transgenic, BALB/c mice following infection with chlamydia pneumoniae: a model for late onset/sporadic Alzheimer's disease. Neurobiol Aging 2002;23(1):S23.

[53] Jaremko M, et al. Chlamydia pneumoniae in cerebrospinal fluid of patients with Alzheimer's Disease and in atherosclerotic plaques. Przegl Epidemiol 2003;56(Suppl.):107–14.

[54] Balin BJ, et al. Chlamydia pneumoniae in the pathogenesis of Alzheimer's disease. Chlamydia pneumoniae infection and disease. United States: Springer; 2004. p. 211–26.

[55] Robertson M. Is Chlamydia associated with Alzheimer's? Drug Discov Today 2004;9(11):469.

[56] Appelt DM, et al. P3-189 The effects of the chlamydia pneumoniae infectious process on neuronal cells: implications for Alzheimer's disease. Neurobiol Aging 2004;25:S409.

[57] Little CS, et al. P2-014 The induction of Alzheimer's disease-like pathology and recovery of viable organism from the brains of BALB/c mice following infection with chlamydia pneumoniae. Neurobiol Aging 2004;25:S227–8.

[58] Gérard HC, et al. Chlamydia pneumoniae as a potential etiologic agent in sporadic Alzheimer's disease. London, United Kingdom: Taylor and Francis Medical Books; 2005.

[59] Gérard HC, et al. The load of Chlamydia pneumoniae in the Alzheimer's brain varies with APOE genotype. Microb Pathog 2005;39(1):19–26.

[60] Way DS, et al. P3-223: Caspase activity is inhibited in neuronal cells infected with Chlamydia pneumoniae: implications for apoptosis in Alzheimer's disease. Alzheimers Dement 2006;2(3):S441.

[61] Hudson AP, Dreses-Werringloer U, Whittum-Hudson JA. P1-176: Chlamydia pneumoniae isolates from the Alzheimer's brain are poorly related to other strains of this organism. Alzheimers Dement 2006;2(3):S147.

[62] Gérard HC, et al. Chlamydophila (Chlamydia) pneumoniae in the Alzheimer's brain. FEMS Immunol Med Microbiol 2006;48(3):355–66.

[63] Dreses-Werringloer U, et al. Chlamydophila (Chlamydia) pneumoniae infection of human astrocytes and microglia in culture displays an active, rather than a persistent, phenotype. Am J Med Sci 2006;332(4):168–74.

[64] Hudson AP, et al. P1-165: Isolates of Chlamydia pneumoniae from the Alzheimer's brain infecting cultured human astrocytes and microglia display an active, not a persistent, growth phenotype. Alzheimers Dement 2006;2(3):S144.

[65] Paradowski B, et al. Evaluation of CSF-Chlamydia pneumoniae, CSF-tau, and CSF-Abeta42 in Alzheimer's disease and vascular dementia. J Neurol 2007;254(2):154–9.

[66] Balin B, et al. Method of diagnosing Alzheimer's disease by detection of chlamydia pneumoniae. US Patent Application 11/623,796.

[67] Appelt DM, et al. P4-059: induction of amyloid processing in astrocytes and neuronal cells coinfected with herpes simplex virus 1 and Chlamydophila (Chlamydia) pneumoniae: linkage of an infectious process to Alzheimer's disease. Alzheimers Dement 2008;4(4):T685.

[68] Szczerba S M. The use of magnetic resonance imaging to evaluate Chlamydia as an aetiological agent in Alzheimer's disease; 2009.

[69] Balin BJ, et al. Analysis of Chlamydia pneumoniae-infected monocytes following incubation with a novel peptide, acALY18: a potential treatment for infection in Alzheimer's disease. Alzheimers Dement 2010;6(4):S268.

[70] Shima K, Kuhlenbäumer G, Rupp J. Chlamydia pneumoniae infection and Alzheimer's disease: a connection to remember? Med Microbiol Immunol 2010;199(4):283–9.

[71] Hammond CJ, et al. Immunohistological detection of Chlamydia pneumoniae in the Alzheimer's disease brain. BMC Neurosci 2010;11(1):121.

[72] Balin B, et al. Infection of human monocytes, olfactory neuroepithelia, and neuronal cells with Chlamydia pneumoniae alters expression of genes associated with Alzheimer's disease. Alzheimers Dement 2011;7(4):S561–2.

[73] Anzman JM, Christine JH, Brian JB. Analysis of inflammasome gene regulation following chlamydia pneumoniae infection of thp1 monocytes: implications for Alzheimer's disease; 2012.

[74] Williams K, et al. Infection with chlamydia pneumoniae alters calcium-associated gene regulation and processes in neuronal cells and monocytes: implications for Alzheimer's disease; 2012.

[75] Cappelini C, Denah A. Chlamydia pneumoniae infection of neuronal cells induces changes in calcium-associated gene expression consistent with Alzheimer's disease; 2013.

[76] Bailin B. Analysis of autophagy and inflammasome regulation in neuronal cells and monocytes infected with Chlamydia pneumoniae: Implications for Alzheimer's disease; 2013.

[77] Brandt J, et al. Electron microscopy studies elucidate morphological forms of Chlamydia pneumoniae in blood samples from patients diagnosed with mild cognitive impairment (MCI) and Alzheimer's disease (AD)(LB81). FASEB J 2014;28(Suppl. 1):LB81.

[78] Mann DMA, Tinkler AM, Yates PO. Neurological disease and herpes simplex virus. Acta Neuropathol 1983;60(1–2):24–8.

[79] Wozniak MA, Frost AL, Itzhaki RF. Alzheimer's disease-specific tau phosphorylation is induced by herpes simplex virus type 1. J Alzheimers Dis 2009;16(2):341–50.

[80] Mims CA. Viral aetiology of diseases of obscure origin. Br Med Bull 1985;41(1):63–9.

[81] Koch R. Die Aetiologie der Tuberculose. Mitt Kaiser Gesundh 1884;2:1–88.

[82] Patterson PH. Cytokines in Alzheimer's disease and multiple sclerosis. Curr Opin Neurobiol 1995;5(5):642–6.

[83] Kohler J, Heilmeyer-Kohler H. Dementia in multiple sclerosis: is it Alzheimer's or a new entity? Alzheimers Dement 2009;5(4):P504–5.

[84] Chidiac C, Braun E. Atherosclerosis, multiple sclerosis, and Alzheimer's disease: what role for Herpesviridae? Pathol Biol (Paris) 2002;50(7):463–8.

[85] Chapman J, et al. APOE genotype is a major predictor of long-term progression of disability in MS. Neurology 2001;56(3):312–6.

[86] Dal Bianco A, et al. Multiple sclerosis and Alzheimer's disease. Ann Neurol 2008;63(2):174–83.

[87] Lassmann H. Mechanisms of neurodegeneration shared between multiple sclerosis and Alzheimer's disease. J Neural Trans 2011;118(5):747–52.

[88] Stratton CW, Wheldon DB. Multiple sclerosis: an infectious syndrome involving Chlamydophila pneumoniae. Trends Microbiol 2006;14(11):474–9.

[89] Abbate J, Guynup S, Genco R. Scientific American presents: oral and whole body health; 2006. p. 6–49.

[90] Michaud DS. Role of bacterial infections in pancreatic cancer. Carcinogenesis 2013;34(10):2193–7.

[91] Li, Xiaojing, et al. Systemic diseases caused by oral infection. Clin Microbiol Rev 2000;13(4):547–58.

[92] Page RC. The pathobiology of periodontal diseases may affect systemic diseases: inversion of a paradigm. Ann Periodontol 1998;3:108–20.

[93] Kamer AR, et al. Inflammation and Alzheimer's disease: possible role of periodontal diseases. Alzheimers Dement 2008;4(4):242–50.

[94] Kamer AR, et al. Periodontal inflammation in relation to cognitive function in an older adult Danish population. J Alzheimers Dis 2012;28(3):613–24.

[95] Poole S, et al. Determining the presence of periodontopathic virulence factors in short-term postmortem Alzheimer's disease brain tissue. J Alzheimer's Dis 2013;36(4):665–77.

[96] Post by Former NIMH Director Thomas Insel: Microbes and Mental Illness. Transforming the understanding and treatment of mental illnesses. NIH, Aug 13, 2010. Available from: http://www.nimh.nih.gov/about/director/2010/microbes-and-mental-illness.shtml

[97] Miklossy J. Alzheimer's disease—a neurospirochetosis. Analysis of the evidence following Koch's and Hill's criteria. J Neuroinflammation 2011;8:90.

[98] Hill JM, et al. Pathogenic microbes, the microbiome, and Alzheimer's disease (AD). Front Aging Neurosci 2014;6:127.

[99] Hooper, JA. New Germ Theory. The Atlantic Monthly, 1999. Available from: https://www.theatlantic.com/magazine/archive/1999/02/a-new-germ-theory/377430/

Image Credit: C. K. Magee's emulsion of pure cod liver oil; J. A. Magee & Co., ca. 1870–1900. Available at: https://www. digitalcommonwealth.org/search/commonwealth:h415pf47m

CHAPTER

10

Alzheimer's Disease Prevention

America's health care system is in crisis precisely because we systematically neglect wellness and prevention.

—*Tom Harkin (United States Senator, IA)*

Why is the illustration earlier appropriate? Alzheimer's will be eliminated, one person at a time, through developing good habits in our youth and not by treating the elderly with severe disease. Did your grandmother force you to take cod liver oil? If you are now elderly and free from dementia or other chronic diseases, say a prayer of thanks to her.

People with Alzheimer's disease (AD) have a serious health problem that is not hopeless. This concept is contrary to the establishment that says there is no way to slow or reverse this disease. This book illustrates that science and medicine does have a solid understanding of many of the aspects of the disease. With this knowledge, a differential diagnosis and targeted treatments can attack the disease at its root. Even under these circumstances, reversal of Alzheimer's is challenging at best. What is the real solution to Alzheimer's? Prevention!

The places to start investigating prevention strategies are the written works of Paul Clayton and by taking a history lesson on Claude Bernard. Both of these scientists offer real and practical solutions to AD disease prevention.

Prevention strategies are best designed on the foundation of knowledge about what causes AD. Can you base your strategy on the potential formation of beta-amyloid? This concept is difficult to understand, let alone use to devise an avoidance strategy. And, as explained in Chapter 2, the best and brightest minds in research and medicine have failed to help Alzheimer's sufferers with any type of beta-amyloid approach. Evading disease caused by inflammation and/or infection is a much more manageable approach.

Remember, if AD onset is pushed back by a mere 5 years, the number of afflicted will be reduced by about half. Those are good odds, and there are ways to hold AD at bay, as opposed to what we hear in the news about this disease. What is the approach to preventing, or at least pushing Alzheimer's way into our future? The answer is to understand and address the true causes of the disease and enhance the health of your immune system. Both of these approaches will reduce inflammation, chronic diseases in general, and AD.

In this chapter, we do not focus on prevention strategies published by the major thought leaders including the Alzheimer's Association, WebMD, and the Mayo Clinic. Please browse through those sites as they speak to general measures, such as exercising more,

The End of Alzheimer's. http://dx.doi.org/10.1016/B978-0-12-812112-2.00010-0

curbing smoking, and taking care of your mental health. Here we mainly focus on factors that have a profound impact on immune system function including vital micronutrients that are often deficient or in excess and cause immune system dysfunction due to their discordant balance.

In their Alzheimer's prevention section the Mayo Clinic states:

"According to a statement from the National Institutes of Health (NIH), a number of factors could play a role in whether you develop AD. However, more research is needed before modification of any of these factors can be proved to prevent AD [1]."

The Mayo Clinic and the other pundits are raising the white flag over AD. This is a major disservice to you and your family because they are ignoring bona fide research in favor of promoting a cry for more research funding. For example, as you will read in this chapter, the Framingham Study shows conclusively that fish oil consumption reduces the future risk of Alzheimer's by as much as 75%. Why is that not a headline on every AD website? Some do state the value of omega-3s but only in very guarded terms.

PREVENTION THROUGH INTERNAL BALANCE

Most of this book has been a projection of researchers' voices from around the globe and through time. This section on prevention is no exception as the Trillion Dollar Conundrum—that is, the wealth of research available on science and medicine, holds the answers. Start with a keyword search on "Alzheimer's" and "prevention." A title search returns over 700 articles that is equal to a lot of nighttime reading. Now focus on those that you, with a little personal responsibility, are able to incorporate into your life. For now, limit your reading to those that include immune health and inflammation control in some form.

An article by a group of Australians titled, *The molecular basis of the prevention of Alzheimer's disease through healthy nutrition*, pops up as interesting [2]. The abstract to their paper is included here. Read it with caution because, as discussed in Chapter 7, inflammation is a treasure and is there to protect us.

> The Alzheimer's disease (AD) brain shows numerous pathological phenomena, including amyloid plaques, neurofibrillary tangles, elevated levels of advanced glycation end products and their receptor, oxidative damage and inflammation, all of which contribute to neurodegeneration. In this review, we consider these neuropathologies associated with AD and propose that inflammation and oxidative stress play major pathogenic roles throughout disease progression. It is believed that oxidative stress and inflammation not only play major roles early in the disease, but that they act in a reinforcing cycle, amplifying their damaging effects. Therefore, epidemiological studies indicate that anti-inflammatory, and neuroprotective agents including those from medicinal plants and health promoting foods may protect against AD. This concept is further supported by evidence that certain diets (such as a Mediterranean diet) have been associated with a lower incidence of AD. This review highlights specific foods and diets thought to lower the risk of developing AD and discusses the potential of healthy nutrition in disease prevention.

What is an antiinflammation strategy?

1. Antiinflammation is a strategy to build immune health so when our immune system acts on our behalf, it does so decisively and then quickly settles back down to normal levels.

2. Antiinflammation is a strategy to create an internal balance that then suppresses the growth of otherwise opportunistic pathogens or other "insults" that impact our health.
3. Antiinflammation is not a strategy to suppress inflammation (with NSAIDs and other immune system depressants). This may be a proper strategy to quell short-term inflammation that causes pain but, it is inappropriate for the management of chronic inflammation.

Be careful not to confuse the terms antioxidant and antiinflammation. They are not the same. In fact, they may be diametrically in opposition. The innate immune system is primarily an oxidative process, with white blood cells as our first line of defense. They identify invaders then kill them with peroxide—a highly oxidizing substance. Oxidative stress is frequently cited as a cause of Alzheimer's and other inflammatory diseases, but this is likely from downstream effects of the inflammatory process and is not primary. David Sinclair from Harvard Medical School was a founder of Sirtris Pharmaceuticals, the science of which was based on the super antioxidant resveratrol. After the sale of Sirtris, Sinclair stated during an NPR interview, "Antioxidants have shown disappointing results in the area of anti-aging medicine."

The Australians are on the right track because they are not proponents of supplements; rather, they endorse whole foods. This is the right approach for building a healthy immune system and avoiding chronic disease. No single diet type provides the balance upon which our physiology is based. The concept of a "diet" infers a restriction of some type. The mid-Victorians did not "diet" per se; they struggled to take in sufficient calories to compensate for their toils and labors. Fortunately for them, their foods were natural and unprocessed. That is the key. The best diets, from an antiinflammation perspective, appear to be the Paleo and Mediterranean. Both these nutritional concepts steer clear of processed foods and excessive carbohydrate consumption.

Past experience does tell us what to avoid, or at least control, in our quest for proper internal balance. *High carbohydrate diets and Alzheimer's disease*, a scientific article produced by the University of Colorado, does well explaining the current surge in Alzheimer's and other inflammatory diseases [3]. "Evolutionarily discordant high carbohydrate diets are proposed to be the primary cause of AD by two general mechanisms. (1) Disturbances in lipid metabolism within the central nervous system inhibit the function of membrane proteins, such as glucose transporters and the amyloid precursor protein. (2) Prolonged excessive insulin/IGF signaling accelerates cellular damage in cerebral neurons. These two factors ultimately lead to the clinical and pathological course of AD. This hypothesis also suggests several preventative and treatment strategies. A change in diet emphasizing decreasing dietary carbohydrates and increasing essential fatty acids (EFA) may effectively prevent AD." "Effectively prevent" may be language that is too strong. However, this advice about carbohydrates and essential fatty acids may certainly play a role in the prevention of AD.

This study poses a conundrum almost as big as the previously discussed Trillion Dollar Conundrum. Specifically, if carbohydrates are limited, what is the replacement source of calories? Madison Avenue has brainwashed the United States and the world into believing that high fat intake equals obesity. Yet our store shelves are filled with low fat (high carbohydrate) alternatives that are making us fat. The answer to the carbohydrate/fat conundrum is to increase the intake of healthy fats that maintain a longer feeling satiation and quell inflammation. The book *Eat Fat Lose Fat* by Dr. Mary Enig and Sally Fallon explains how to reduce carbohydrates, increase healthy fat intake, and improve overall health [4]. A simple strategy to employ that helps avoid carbohydrate overload is to (1) cook and eat a home, and (2) grocery shop along the outside isle of the store but avoid the deli counter. By skipping the inner isles, your cart will contain significantly fewer carbohydrates and processed foods.

Foods today are somewhat stripped of their nutritional value due to over farming of land, processing, genetic modifications, and a host of other reasons. The elderly are particularly at risk of deficiencies because they do not absorb minerals well, compared to younger people. Many people do supplement in an attempt to be healthy, and this is probably important for both the active and the sedentary because even those with sedentary bodies have metabolically active and hungry brains. Some vitamins, nutrients, and supplements may be harmful in excess and should be purchased and taken based on knowledge, not just based on the latest headline in *Self*, by that "Oz" headline machine, or *Men's Health*. Calcium, which will be discussed in great detail later in this chapter, is a mineral that is the poster child for a well-intended message that may have gone terribly wrong. There are extensive research compilations on the "calcium hypothesis of dementia." Therefore, don't assume that if it is good for one specific thing that it is healthy for you long-term or is essential, as a supplement, for your good health. The Harvard School of Public Health teaches, "The dose makes the poison." Both oxygen and water are toxic at either extremes of dose.

INFLAMMATION STARTS IN THE GUT

The Microbiome may be the hottest topic in medicine today, particularly with progressive practitioners. Microbiome refers generally to the gut and specifically to the beneficial microorganisms that work symbiotically with our bodies to produce immunity, provide nutrients, and manage waste. Dysbiosis refers to a condition with microbial imbalances on or within our bodies. Bysbiosis is most prominent in the digestive tract or on the skin, but can also occur on any exposed surface or mucous membrane, such as the vagina, lungs, mouth, nose, sinuses, ears, nails, or eyes.

Researchers are discovering a causal relationship between an imbalanced gut microbiome (gut dysbiosis) and a growing number of conditions and diseases—for example, acne, allergies, asthma, celiac disease, chronic Lyme disease, Crohn's disease, diabetes, Graves disease, gum and tooth disease, irritable bowel, lichen planus, lupus, multiple sclerosis (MS), psoriasis, rheumatoid arthritis, UTIs, all the other 80–100 autoimmune diseases, and some cancers. See Autoimmune Disorders for more information.

Serious memory loss, as with Alzheimer's, has also been found to be related to gut bacterial imbalances. How does this happen? When the microorganisms living in the gut become out of balance, inflammation develops—and chronic inflammation is the hallmark of chronic diseases including AD. We now appreciate that chronic inflammation is our leads to opportunistic pathogen proliferation. Apparently in dysbiosis, when beneficial bacteria are lost, antagonistic pathogens are able to grow. And their development leads to a general decline in immunity so other latent and harmful species can also thrive, leading to inflammation, neuroinflammation, and AD.

Here is the conclusion from an article titled, "Obesity and Gut's Dsbiosis Promote Neuroinflammation, Cognitive Impairment, and Vulnerability to Alzheimer's disease: New Directions and Therapeutic Implications [5]."

> Systemic inflammation occurs due to LPS (bacterial remnants) efflux from the gut; this up-regulates neuroinflammation including that in the hippocampus and cerebellum. Brain pro-inflammatory cytokine generation and synthesis, i.e. neuroinflammation promotes amyloid deposition and tau hyperphosphorylation that enhance hypofunction/dysfunction in key brain regions, including the hippocampus and cerebellum. This cascade of events promotes neuronal injury/apoptosis and degeneration, leading to cognitive impairment and vulnerability to Alzheimer's dementia.

Your solution to prevention of this significant problem is actually quite simple but you have to start young, in fact very young. However, it is never too late to protect your health.

1. Expecting mothers need to develop good gut health following some of the suggestions later.
2. Have a natural birth and drench your newborn in all your beneficial bacteria. This is their first and vital exposure.
3. Allow your 0–3 year old to play on clear dirt. Soil is the earth's gut and is full of beneficial microbes and minerals. Why do our very young always touch things and reach for their mouths? In nature, very few things are toxic so this action is not due to a death wish. They are naturally building their immunity.
4. Avoid GMO and processed food from very early in life (prebirth for mothers) until your last days.
5. Take a multimineral supplement. This supports the function of our enzymes. And healthy bacteria have enzymes too. This will promote gut health.
6. Eat fermented foods as they contain beneficial food-processing microbes.
7. Take broad-spectrum probiotics and, more importantly take prebiotics. Too much of a good thing is also bad so be careful when choosing a probiotic. Another approach is to switch brands frequently.
8. Read what follows as there are many more suggestions to create good health and prevent AD.

MAGNESIUM

Do you know that wheat has 3 times as many genes as a human? How is this relevant to AD you might ask? Wheat has many more genes compared to us because it does not have the capacity to move to find proper nutrition to grow. It must have, within its gene pool, adequate diversity to accommodate life and growth regardless of the quality of soil upon which its seeds fall. Crafty humans are aware of this and are able to modify the genes of wheat and other plants so they can grow in nutrient deficient soil. It is less expensive to produce seeds that can grow in magnesium deficient soils than it is to fertilize the fields with magnesium. The consumer thus loses a source of valuable nutrients. This is one significant concern regarding GMO seeds, crops, and our good health.

Magnesium is a special, necessary, yet all-to-often overlooked mineral. Sixty percentage of the body's magnesium is found in bone yet our focus is only calcium intake. The majority of magnesium in muscle is found in the mitochondria, where it plays a key role in metabolism and is believed to be involved in the permeability of the outer membrane [6].

Dietary intake of magnesium has gone down dramatically over the past 100 years. It is estimated that 68%–80% of Americans are magnesium deficient [7–9]. In places where water is harder, levels of magnesium are higher, and the incidence of coronary artery disease is lower. Magnesium deficiency apparently leads to early and sudden mortality by cardiovascular diseases. Almost 8 million deaths from sudden cardiac failure occurred in the United States between 1940 and 1994 that were largely attributed to magnesium deficiency [10–13].

Evaluation of your homeostasis for magnesium is not straight forward, as serum levels (in the blood) are a poor indicator of magnesium status. Most of the magnesium in our bodies resides inside of cells contrary to calcium that resides outside of cell membranes. Heart muscle levels are almost 20 times higher than serum levels. It turns out that measuring white blood cell levels is a more sensitive test. The best test is ionic magnesium measurement or elemental X-ray analysis. However, none of these methods is definitive [14,15]. Many factors regulate magnesium absorption, including vitamin D levels [16]. As calcium levels go down, magnesium absorption increases. High intakes of calcium, protein, vitamin D, and alcohol all increase magnesium requirements. Without adequate magnesium, bones will be dense, but some will have poor integrity. Northern European countries, where the calcium to magnesium ratio is 4:1, have the highest rates of osteoporosis [17].

In lieu of tests, the best way to insure adequate magnesium is simply through supplementation or, preferably, through dietary modifications, as magnesium has a very favorable safety profile. Abundant magnesium leads to a loose stool and then to diarrhea. The best dietary sources of magnesium are whole grains, nuts, and fruits. These include buckwheat flour, tofu, figs, cashews, avocado, millet, and brewer's yeast. All green plants contain magnesium, as this metal is at the center of the chlorophyll molecule.

Magnesium has an effect of relaxing smooth muscle and is therefore useful in conditions, such as hypertension, dysmenorrhea, constipation, asthma, angina, stroke, heart attack, and AD. It decreases coagulation and acts as a calcium channel blocker, helping the heart to pump more effectively and regulating blood pressure [18–20]. Magnesium is involved in the function of more than 300 enzymes, as well as in regulating muscle contractility and nerve impulses. Virtually all body systems also rely on magnesium for at least some of their metabolic functions [21–23]. Do you find it interesting that drug companies invest billions of dollars in drugs that are involved in the action of a couple of enzymes yet we take their drugs while being magnesium deficient? This is quite a paradox.

Magnesium deficiency is insidious because it can mimic many other disorders. These include fatigue, poor nail growth, irritability, weakness, dysmenorrhea, muscle spasms or tightness, cardiomyopathy, anorexia, sugar cravings, hypertension, and anxiety [24–28]. Does magnesium deficiency mimic these diseases or is the deficiency the disease? Deficiency can result from kidney disease and intake of diuretics, and it can cause depletion of potassium intracellularly and affect muscles and bones. Magnesium deficiency can be caused by poor absorption or high metabolic use, as is likely the case for hyperthyroidism, kwashiorkor, diabetes mellitus, alcoholism, pancreatitis, parathyroid disorders, high dietary phytic acid, and diarrhea [29–32].

The clinical use of magnesium can be applied to a variety of conditions. These include constipation, muscle cramping, torticollis, acute angina following a myocardial infarct or stroke, asthma, kidney stone prevention (especially when given with vitamin B6), and dysmenorrhea. Other candidates for magnesium supplementation are GI spasms or cramping, eclampsia, heart disease (especially cardiomyopathy) [33,34], diabetes mellitus, nocturnal muscle cramps, mitral valve prolapse, toxemia of pregnancy, fibromyalgia, migraine headaches, lead toxicity, general fatigue, anxiety, and irritability. Isn't it interesting to see the interconnectedness of many of these diseases as shown simply and elegantly through magnesium deficiency?

General dosing of daily magnesium should be approximately 450 mg, considering all sources. However, for GI cramping, asthma, constipation, and heart disease, it is recommended to take magnesium to improve bowel tolerance (until the bowel movements become "loose"). Fortunately, the toxicity of magnesium is fairly low, with diarrhea being the biggest problem. Reduce the dose until bowel movements return to normal to prevent other possible symptoms of magnesium toxicity, such as calcium deficiency, hypotension, depletion of potassium, and respiratory depression [35].

MAGNESIUM AND INFLAMMATION

In two large observation studies (the Women's Health Initiative [36] and Nurses Health Study [37]) greater magnesium (Mg) intake was associated with lower levels of inflammation as measured by CRP, IL-6, and TNF-α receptor, a measure of TNF-α activity. Data from the Multi-Ethnic Study of Atherosclerosis (MESA) failed to find significant differences in IL-6 or CRP levels between individuals with the highest and lowest magnesium intakes, but did find a significant association between greater dietary magnesium and the lower levels of the inflammation-associated proteins homocysteine and fibrinogen [38]. Magnesium was rated as the most anti-inflammatory dietary factor in the Dietary Inflammatory Index, which rated 42 common dietary constituents on their ability to reduce CRP levels based on human and animal experimental data. This did not include foods that are often mistaken as supplements. For example, fish oil has a more profound impact on lowering CRP compared to magnesium.

"Memory functions decline with age, and severely deteriorate during AD. Several studies suggest that dietary/environmental factors can reduce the prevalence of AD in humans. Magnesium is essential for maintaining normal body and brain functions." This is according to Chinese researchers as presented at a Shanghai conference in 2012 [39]. Magnesium deficiency is common in the elderly and is an important factor to consider in the prevention and management of dementias including Alzheimer's.

MAGNESIUM AND ALZHEIMER'S

Stepping back to 1990, *Magnesium depletion and pathogenesis of Alzheimer's disease*, by a French researcher, presented evidence indicating dementias are associated with a relative insufficiency of magnesium (Mg) in the brain [40]. Such insufficiency may be attributable to low intake or retention of Mg; high intake of a neurotoxic metal, such as aluminum (Al), which inhibits activity of Mg-requiring enzymes; or impaired transport of Mg and/or enhanced transport of the neurotoxic metal into brain tissue.

Finally, a recent paper portends what may be an important part of the future of Alzheimer's treatment [41]. This paper was written by German researchers and has a clear thesis and conclusion. Part of the abstract is reproduced here:

> The cholinergic deficit in Alzheimer's disease (AD) remains the cornerstone for the understanding of chemical signal transfer. Hypofunctions of cholinergic systems are significantly involved in the signs and symptoms of senile dementia of the Alzheimer type... Magnesium is directly involved in numerous important biochemical reactions and is particularly a necessary cofactor in more than 300 enzymatic reactions and specifically in all those processes involving the utilization and transfer of adenosine triphosphate. A study in patients with different diagnoses showed low enzyme activity of choline esterase in erythrocytes (red blood cells). Administration of magnesium resulted in normal catalytic activity of choline esterase. The measurement of the enzyme activity of choline esterase is a possibility to prove magnesium deficiency and to verify the efficacy of magnesium administration. Magnesium deficiency, resulting from low magnesium dietary intake, is more common and may be corrected by magnesium supplementation.

Take your magnesium—it is such a simple health enhancement measure—and discard your calcium.

CALCIUM

With our newly found understanding about the connection between Alzheimer's and cardiovascular diseases, let's look at the impact of calcium supplementation on the latter. The medical literature is full of research on this topic as a scholar.google search yields 285 articles, just between 2010 and 2013, using the keywords "calcium" and "cardiovascular" in a title-only search. The first one that comes up paints a vivid picture. The title is, *Effect of calcium supplements on risk of myocardial infarction and cardiovascular events: meta-analysis* [42]. This team from New Zealand and the United States conclude:

> Calcium supplements (without co-administered vitamin D) are associated with an increased risk of myocardial infarction (heart attack). As calcium supplements are widely used these modest increases in risk of cardiovascular disease might translate into a large burden of disease in the population. A reassessment of the role of calcium supplements in the management of osteoporosis is warranted.

A reassessment of the role of calcium supplement indeed! The National Osteoporosis Foundation, on their website states, under "Debunking the Myths."

> Myth #9: Taking extra calcium supplements can help prevent osteoporosis.
> Taking more calcium than you need does not provide any extra benefits. Estimate the amount of calcium you get from foods on a typical day to determine whether a supplement is right for you.

You probably get enough calcium in your foods to prevent osteoporosis and supplementation is not worth the risk. What you do not have at sufficient levels are vitamin D and magnesium. There are no appreciable side effects from taking these essential components compared to calcium.

"Evidence from limited data suggests that vitamin D supplements at moderate to high doses may reduce CVD risk, whereas calcium supplements seem to have minimal to doubtful cardiovascular effects. Further research is needed to elucidate the role of these supplements in CVD prevention." This is the conclusion by another set of researchers [43]. Although the researchers did not find any negative contribution from calcium (or did they, but the vitamin D was protective?), they showed the importance of vitamin D. Based on current data, treating 1,000 people with calcium supplements for 5 years would prevent only 26 fractures but would cause an additional 14 heart attacks. Both these numbers would most likely decrease if 1,000 people had sufficient vitamin D and magnesium in their systems.

The New York Academy of Sciences assembled a book titled, *The Calcium Hypothesis of Dementia*, in early 1990s based on a summit on the topic. A classic paper included in the books is titled, *The calcium rationale in aging and Alzheimer's disease* [44]. The full original article is available in electronic form on the Internet. Some key points from this article are reproduced here:

- Calcium is required for the function of all cells in the body, including neurons.
- Calcium is intimately involved in a variety of "plastic" changes in the brain.
- Calcium thus is likely to have key roles in the cellular processes underlying aging-related changes in the brain, including normal age-associated memory impairments as well as more severe dementias, including AD.
- The pivotal role of calcium in so many neuronal processes dictates the need for precise regulation of its intracellular levels. Any dysregulation, however subtle, could lead to dramatic changes in normal neuronal function.
- The calcium hypothesis, which posits that in the aging brain, transient or sustained increases in the average concentration of intracellular free calcium contribute to impaired function, eventually leading to cell death.
- The hypothesis suggests that the final common pathway that may contribute to cognitive deterioration of aging vertebrates, including persons with AD or other aging-related dementias, is increased free calcium within neurons. The functional impairment that characterizes a patient at a particular time in the aging-related disease process may be relieved by reducing excessive calcium influx.

There are plenty of updates to this important research. A google.scholar search over a 4 year period starting in 2009 yields 20 articles with Alzheimer's and calcium in the title. The key term appears to be "dysregulation" of calcium. Whatever the scientific jargon, please consider taking the advice already provided, and do not upset your calcium balance with supplements.

VITAMIN D

The Miracle Vitamin, by Paula Dranov, states, "new evidence shows that getting enough D may be the most important thing you can do for your health [45]." This is a true statement, and the preferred way to get vitamin D is through sun exposure and by taking cod liver oil, as it contains the key fat-soluble vitamins A and D. Most importantly, this natural source contains all the variations (isomers) of the vitamins.

The health benefits of vitamin D are prominently highlighted in the New York Well Blog. Key recent headlines concerning vitamin D include:

2010: What do you lack? Probably vitamin D. [46]
2013: Low vitamin D tied to aging problems. [47]
2014: Low vitamin D tied to a pregnancy risk. [48]
2014: Vitamin D may lower cholesterol. [49]
2014: Low vitamin D levels linked to disease in two big studies. [50]
2014: Low vitamin D tied to premature death. [51]

Clearly, insufficient vitamin D impacts our health from birth to death.

But it turns out that the term "vitamin" is a misnomer for vitamin D. It is really a hormone. The word "vitamin" means something our body needs that it can't make, so must be obtained from food. "D hormone" (vitamin D) is instead an essential substance that we make on our skin from sun exposure. It is a hormone like thyroid, estrogen, or testosterone. Using the proper word "hormone" reminds us that it affects multiple parts of the body and that it is essential to every cell in the body.

From what molecule does vitamin D come when light hits our skin? Cholesterol [52]. Yes, that same "evil" substance that the drug companies claims causes so much harm. Maybe we are learning that cholesterol is not so evil and even important for our protection against AD.

Vitamin D, the fat-soluble hormone is naturally present in very few foods, added to others, and available as a dietary supplement. It is also produced when ultraviolet rays from sunlight strike the cholesterol in the skin and trigger vitamin D synthesis. Vitamin D obtained from sun exposure, food, and supplements is biologically inert and must undergo a chemical reaction (hydroxylation) in the body for activation. One reaction occurs in the liver and converts vitamin D to 25-hydroxy vitamin D, also known as 25 vitamin D, vitamin D3, or simply vitamin D. Under normal conditions, another reaction occurs primarily in the kidney to form the physiologically active 1,25-dihydroxyvitamin D, also known as calcitriol.

Importantly, activation of 25-hydroxy vitamin D to the 1,25-dihydroxyvitamin D occurs in inflamed tissue. Here, the activated form of vitamin D is working in concert with our immune system to deal with the inflammation. This "activation" process is often the cause for the failure of ingested vitamin D supplements to raise the serum vitamin D levels in patients. In this capacity, a measurement of blood vitamin D levels, for those under supplementation, may reveal a disease process in progress. Those patients with low vitamin D levels, but who appear to have adequate intakes of the substance should be tested for the activated (1,25-dihydroxy) form of vitamin D.

The activated form of vitamin D may positively impact AD. A Canadian group carried out long-term treatment of mice with activated 1,25-dihydroxy vitamin D reduced beta-amyloid plaque formation, importantly of both the soluble and insoluble type. Of particular importance, the amyloid reduction occurred in the hippocampus region of the brain. This led to improvement in conditioned fear memory. The data suggest that the vitamin D receptor and treatment with vitamin D or its activated form is important therapeutically for the prevention and treatment of AD [53].

Vitamin D promotes calcium absorption in the gut and maintains adequate (and balanced) serum calcium and phosphate concentrations to enable normal mineralization of bone and to prevent low calcium concentrations. It is also needed for bone growth and bone remodeling by osteoblasts and osteoclasts. Without sufficient vitamin D, bones can become thin, brittle, or misshapen. Vitamin D sufficiency prevents rickets in children and osteomalacia in adults. Vitamin D also helps protect older adults from osteoporosis.

Vitamin D has other roles in the body, including modulation of cell growth, neuromuscular and immune function, and reduction of inflammation. Many genes encoding proteins that regulate cell proliferation, differentiation, and apoptosis are modulated in part by vitamin D. Many cells have vitamin D receptors, and some convert the diol to the triol of vitamin D. Serum concentration of the diol of vitamin D is the best indicator of vitamin D status. It reflects vitamin D produced by sunlight and that obtained from food and supplements and has a fairly long circulating half-life of 15 days. Vitamin D functions as a biomarker of exposure, but it is not clear to what extent vitamin D levels also serve as a biomarker of effect. Serum vitamin D levels do not indicate the amount of vitamin D stored in body tissues. Vitamin D, although not synthesized by sunlight in the winter in the northern hemisphere, is available to the body by storage in fat throughout the year, assuming adequate exposure to sunlight during summer months.

There is considerable discussion about the serum concentrations of vitamin D associated with deficiency (e.g., rickets and other degenerative diseases), a scientific consensus process has not developed adequacy for bone health and optimal overall health. Based on its review of data of vitamin D needs, a committee of the Institute of Medicine concluded that persons are at risk of vitamin D deficiency at serum vitamin D concentrations <30 nmol/L. Some are potentially at risk for inadequacy at levels ranging from 30–50 nmol/L. Practically all people are sufficient at levels ≥50 nmol/L; the committee stated that 50 nmol/L is the serum vitamin D level that covers the needs of 97.5% of the population. Serum concentrations >150 nmol/L are associated with potential adverse effects. These adverse affects, however, are relatively mild. There has never been an incident of serum concentration >125 nmol/L from sun exposure alone. It is nearly impossible to achieve a level of 150 nmol/L through supplementation.

According to research from the United Kingdom and Canada, "Vitamin D was initially thought to play a restricted role in calcium homeostasis, but the pleiotropic (multi-factorial) actions of vitamin D in biology and their clinical significance are only now becoming apparent [54]." In their publication, the researchers found 2,776 binding sites for the vitamin D receptor along the length of the genome. These were unusually concentrated near a number of genes associated with susceptibility to autoimmune conditions, such as MS, Crohn's disease, systemic lupus erythematosus (or 'lupus'), rheumatoid arthritis, and to cancers, such as chronic lymphocytic leukemia and colorectal cancer. They also showed that vitamin D had a significant effect on the activity of 229 genes. "Vitamin D status is potentially one of the most powerful selective pressures on the genome in relatively recent times." As with magnesium, the action of vitamin D is the envy of the drug companies.

Vitamin D appears to exert antiinflammatory activity by the suppression of proinflammatory prostaglandins and inhibition of the inflammatory mediator NF-κβ [55]. Although intervention studies of its antiinflammatory activity in humans are lacking, several observational studies suggest vitamin D deficiency may promote inflammation. Vitamin D deficiencies are more common among patients with inflammatory diseases (including rheumatoid arthritis, inflammatory bowel disease, systemic lupus erythematosus, AD, and diabetes) than in healthy individuals [56–63]. They also occur more frequently in populations that are prone to low-level inflammation, such as obese individuals and the elderly. Vitamin D levels can drop following surgery (a condition associated with acute inflammation), with a concomitant rise in CRP. Low vitamin D status was associated with elevated CRP in a study of 548 heart failure patients and with increases in IL-6 and NF-κβ in a group of 46 middle-aged men with endothelial dysfunction [64].

VITAMIN D AND ALZHEIMER'S

Elizabeth Pogge from the Midwestern University College of Pharmacy crafted a paper titled, *Vitamin D and Alzheimer's Disease: Is there a Link?* [65] Dr. Pogge wrote, "The current observational studies seem to identify a link between vitamin D and dementia, particularly AD. Before this evidence can be used to make a recommendation for routine supplementation in elderly patients to prevent AD, more prospective trials with a longer follow-up period are needed to show a causality relationship."

The following is a recommended rewrite to the conclusion earlier, replacing the "need more research," with something helpful to humanity. "The link between Alzheimer's and vitamin D is very interesting especially considering that most of modern society is deficient in vitamin D. We therefore recommend that all people consider having vitamin D levels measured and supplement as necessary to ensure everyone have levels widely considered sufficient. In lieu of measurement, consider supplementing with vitamin D because this vitamin has been proven safe, even at high levels, over years of study and clinical experience." Which recommendation will save more lives and reduce human suffering?

Let's circle back to that statement about vitamin D being the most important thing you can do for your health. Here is part of an abstract from a United Kingdom and Canadian 2013 research article [66]:

> This review highlights the epidemiological, neuropathological, experimental, and molecular genetic evidence implicating vitamin D as a candidate in influencing susceptibility to a number of psychiatric and neurological diseases. The strength of evidence varies for schizophrenia, autism, Parkinson's disease, amyotrophic lateral sclerosis, and Alzheimer's disease, and is especially strong for multiple sclerosis.

Vitamin D was discovered as the cure for rickets, and those on vitamin D therapy were found to have lower incidences of infectious diseases, such as Tuberculosis. The 1903 Nobel Prize in Medicine or Physiology was awarded to Professor Niels Finsen in recognition of his work on the treatment of diseases, and in particular the treatment of lupus vulgaris by means of concentrated light. Lupus vulgaris is, as we now know, a form of tuberculosis, with localized lesions on the skin, especially that of the face, such as the nose, eyelids, lips, and cheeks. What do you suppose the light therapy produced on the exposed skin—vitamin D maybe?

Modern research shows that high doses of vitamin D help tuberculosis patients recover more quickly. Vitamin D, given in addition to antibiotic treatment, appears to help patients through likely antibiotic action of its own, by dampening down the body's inflammatory response to infection, and enabling tissue to recover more quickly and with less damage. "We found that a large number of these inflammatory markers fell further and faster in patients receiving vitamin D," according to researchers of a paper titled, *Vitamin D accelerates resolution of inflammatory responses during tuberculosis treatment* [67].

A large group of researchers from across the pond and the United States combined to determine whether low vitamin D concentrations are associated with an increased risk of incident all-cause dementia and AD [68]. In this study, they evaluated 1,658 ambulatory adults free from dementia including AD. Vitamin D levels were measured, and the patients were tested subsequently for an average of 5.6 years. Dementia occurred in 171 participants including 102 cases of AD. Individuals considered deficient in vitamin D were 2.25 times more susceptible to dementia and AD; those considered insufficient in vitamin D were more than 1.5 times more susceptible compared to those who were at sufficient levels of vitamin D. The 14 MDs and PhDs. concluded:

"Our results confirm that vitamin D deficiency is associated with a substantially increased risk of all-cause dementia and AD. This adds to the ongoing debate about the role of vitamin D in non-skeletal conditions."

These researchers are not alone in their claims about the value of vitamin D and Alzheimer's. Just within the past year, the following titles listed provide further validity to this important relationship.

2012: Vitamin D, cognition, and dementia: a systematic review and metaanalysis. [69]
2013: Low serum vitamin D concentrations in Alzheimer's disease: a systematic review and meta-analysis. [70]
2013: Meta-analysis of memory and executive dysfunctions in relation to vitamin D. [71]

One little vitamin (hormone) yet so much power to promote good health, and you can obtain it naturally, at no cost, through some prudent exposure to the sun.

VITAMIN E

The jury has cast a unanimous verdict against antioxidants in diseases of inflammation. Our own clinical experience, particularly with advanced macular degeneration, has been negative. That is, patients report an increase in symptoms, especially bleeding in the wet form of age-related macular degeneration (AMD), when on vitamin E supplementation. However, popular wisdom is that gobbling up free radicals is critical for maintaining health. Proceed with caution. A Cambridge-based biotech company sold a resveratrol concept (a powerful antioxidant that you no doubt have seen make its way onto drug store shelves recently) to GSK pharma for $728,000,000 on the premise that mice fed their antioxidant compound(s) lived twice as long. GSK abandoned the project after they couldn't reproduce the results. We have yet to learn from Ponce de Leon. Antioxidants are not turning out to be the fountain of youth.

A United Kingdom group wrote, *Vitamin E for Alzheimer's disease and mild cognitive impairment*, in 2008 [72]. The authors stated that vitamin E is a dietary compound with antioxidant properties involved in scavenging free radicals. Laboratory and animal studies have pointed toward a possible role for vitamin E in the prevention and management of cognitive impairment. To date, only one randomized controlled trial has assessed the efficacy of vitamin E in the treatment of AD patients and only one assessed the role of vitamin E in patients with mild cognitive impairment (MCI). In the vitamin E study, for moderately severe AD patients, a lower number of those taking vitamin E declined to incapacity over a 2-year period compared with the placebo group. However, AD patients taking vitamin E experienced a greater number of falls. In the MCI study, vitamin E 2000 IU daily produced no significant difference in the rate of progression to AD compared to the placebo group.

What is your call to action? Be careful about supplementing with vitamin E until further notice. Also, not all supplements are created equally. There are several forms (isomers) of the basic compound of vitamin E. Any supplementation should include that blend of isomers in their naturally occurring ratios. In a 2008 study, the authors said, "the combination of [alpha-tocopherol] and [gamma-tocopherol] supplementation appears to be superior to either supplementation alone on biomarkers of oxidative stress and inflammation and needs to be tested in prospective clinical trials [73]."

ANTIOXIDANTS (GENERAL)

Be careful about supplementation with antioxidants. This includes vitamin C. The best use of antioxidant supplements is "situationally." For example, take vitamin C in high doses when you feel the onset of a cold. Then, after the cold, stop and get your vitamin C from natural sources.

Often a negative result provides us as much or more information than a positive result. In this case, the recent findings on the adverse effects of antioxidants helps explain a lot about health and disease. Particularly,

- it helps explain the mechanism of action of our immune system to protect us;
- it substantiates the involvement of pathogens in disease, and illustrates the interconnectedness of disease; and
- it shows how antioxidant research may be misinterpreted and our "healthy use" recommendations.

We are told to take antioxidants to overcome free radicals. What are free radicals? They are those dreaded reactive oxygen species (ROS) associated with oxidative stress and aging that we are told we must eliminate from our bodies to stay young, healthy, and beautiful. But our immune system produces and uses free radicals, such as superoxide and hydrogen peroxide, to preserve our health. Somewhere there is a contradiction!

According to Harvard Medical School, "When you were a kid your mom poured it (hydrogen peroxide) on your scraped finger to stave off infection. When you got older you might have even used it to bleach your hair. Now there's another possible function for this over-the-counter colorless liquid: your body might be using (producing) hydrogen peroxide as an envoy that marshals troops of healing cells to wounded tissue [74]."

Antioxidants scavenge (remove) ROS, including those produced from hydrogen peroxide. Since the ROS are often fighting against unwanted invaders in our body, this antioxidant action is working against our body and our good health. Let's go back in time to when scientists were first discovered how human immunity worked against bacteria and toxins. Here is an article from 1973 that reveals the essence of immunity. It is titled, "Biological defense mechanisms. The production by leukocytes of superoxide, a potential bactericidal agent. [75]."

For those interested in reading the technical "punch line," look further later. But for the casual reader, here is a simplified interpretation of this work that, in 1973, was ground-breaking in understanding how our immune system actually works.

Simplified explanation: White blood cells, which are the first line of defense against outside invaders. These cells take up oxygen and convert it to a very powerful and destructive substance called superoxide. This substance is produced to kill pathogens and alter other substances that do not belong in our bodies and are responsible for disease. Thus white blood cells and the ROS they produce are part of our defense against sickness.

From the article (detailed explanation): "It has recently become apparent that biological systems are able to convert oxygen into a compound of great reactivity. This compound, the superoxide anion, is an extremely powerful oxidation-reduction reagent, capable of undergoing either oxidation to oxygen or reduction to hydrogen peroxide with the liberation of large amounts of energy. The production of this compound by the one-electron reduction of oxygen in biological systems, together with its reactivity, suggested it as a possible killing agent in leukocytes (white blood cells). In the following communication, we report evidence for the production of superoxide by phagocytizing (to envelop and destroy bacteria and other foreign material) leukocytes.

ROS can help protect us against heavy metals and other toxins too. Certain chemical substances that we ingest or take in through our skin or the air are toxic to our bodies. Examples include lead from paint and benzene from gasoline. Often, our bodies cannot simply remove these substances because of their physical properties, such as solubility. Superoxide (free radicals) can interact with these harmful substances and do one or more things to the toxin. First, it can alter its structure through oxidation. The new structure may have less toxicity compared to the original compound—making it less harmful to the body. Second, oxidized substances tend to be more soluble in water. Water soluble substances are more easily removed from our bodies because our circulatory and lymphatic systems are water-based.

Since Alzheimer's is a disease the mainly strikes our elders, we must pay close attention to the special needs of the geriatric cohort. Older people have a higher baseline of inflammation, and this happens for a reason. As the innate and adaptive immune systems decline, the body supports life with secondary, tertiary, and quaternary immunity. Cholesterol is one such component of alternative immunity, amyloid, to some degree is another, and inflammation—white cells, T cells, cytokines, are yet another. They patrol the body in its defense. Antioxidants can reduce their efficacy by extracting free radicals that they produce intentionally to protect us.

We have seen, in our clinics, that chronic antioxidant supplementation has adverse effects on our older populations. The results on patients do not lie—and are preferable to even double-blind studies—because they are occurring real-time before our very eyes.

IRON

Clearly, we need iron in our bodies to survive, and plenty of it. Iron is at the center of the heme molecule of hemoglobin that is responsible for the uptake and transport of oxygen to tissue in need, through our circulatory system. However, iron dysregulation is receiving substantial consideration as a factor contributing to AD. The critical question, which is yet to be fully answered, is: can limiting or removing "excess" iron prevent or curtail Alzheimer's?

Researchers at UCLA provide some clarity about iron, iron metabolism, and the potential for excess iron to cause or exacerbate AD. "It is difficult to measure iron in tissue when the tissue is already damaged. But the MRI technology we used in this study allowed us to determine that the increase in iron is occurring together with the tissue damage (of Alzheimer's). We found that the amount of iron is increased in the hippocampus and is associated with tissue damage in patients with Alzheimer's but not in the healthy older individuals—or in the thalamus. So the results suggest that iron accumulation may indeed contribute to the cause of AD [76,77]."

The researchers indicate that the findings are not all bad news. "The accumulation of iron in the brain may be influenced by modifying environmental factors, such as how much red meat and iron dietary supplements we consume and, in women, having hysterectomies before menopause [76]."

Most of the basic building blocks in our bodies can be deleterious when deficient or in excess. This is true for some of the most fundamental substances like water and salt. This issue to be solved, always, is the root cause. The solution for dehydration is obvious, take in water. An iron dysregulation in part of the brain impacted by Alzheimer's may not have such an obvious answer. Controlling your iron intake, especially by limiting iron supplementation that is not recommended by a doctor you trust, may be prudent based on the UCLA and other studies. Women clearly have less of a need to reduce iron intake. Several male doctors we know, who are aware of the general oxidative stress that excess iron can create, give blood at least once each year.

ZINC

Yes, there are many magnificent papers on zinc and AD including one by Brits in 2010 titled, *The Role of Zinc in Alzheimer's Disease* [78]. The authors inform us that zinc, the most abundant trace metal in the brain, has numerous functions, both in health and in disease. "Zinc has multifactorial functions in AD. Zinc is critical in the enzymatic non-amyloidogenic processing of the amyloid precursor protein (APP) and in the enzymatic degradation of the amyloid-β (Aβ) peptide. Zinc binds to beta-amyloid promoting its aggregation into neurotoxic species, and disruption of zinc homeostasis in the brain results in synaptic and memory deficits. Thus, zinc dyshomeostasis (variable concentrations) may have a critical role to play in the pathogenesis of AD."

From the perspective of inflammation, zinc-containing antioxidant proteins reduce reactive oxygen species (free radicals), which indirectly inhibits nuclear factor kappa beta (NF-κβ) activity and prevents the production of several inflammatory enzymes and cytokines. NFkB is a protein that acts as a switch to turn inflammation on and off in the body. It is sometimes described as a "smoke sensor" that detects dangerous threats like free radicals and infectious agents. In response to these threats, NFkB "turns on" the genes that produce inflammation. As we age, NFkB expression in the body increases, as does chronic inflammation sets the stage for aging body's defense against diseases ranging from atherosclerosis and diabetes to Alzheimer's. Zinc can also inhibit nuclear factor kappa beta NF-κβ in a more direct manner. Zinc supplementation is associated with decreases in inflammation in populations that are prone to zinc deficiency, such as children and the elderly. Low level inflammation and circulating proinflammatory factors (CRP, TNF-α, IL-6, and IL-8) were reduced in elderly subjects by moderate zinc supplementation in several studies. Like zinc, selenium deficiencies are common in chronic inflammatory states associated with disease, where selenium supplementation has been associated with reductions in inflammation and better patient outcomes.

Mass General researchers reviewed zinc through an article titled, *Zinc takes the center stage: its paradoxical role in Alzheimer's disease* [79]. They indicated, "Zinc in human nutrition is undoubtedly essential," and "... the protective effect of zinc against beta-amyloid cytotoxicity, coupled with anecdotal results from a few zinc supplementation studies warrant further research."

The Age-Related Eye Disease Study (AREDS), explored in Chapter 6, set out to investigate if vitamins and/or minerals could slow the progression of AMD. A conclusion from one of the many AREDS-generated research articles reads, "The AREDS trial results suggest that antioxidants and zinc, either alone or in combination, were modestly effective for category 3 and 4 patients with AMD. The trials leaves unanswered the question of supplementation for category 1 and 2 patients as well as the long-term safety of the agents. Due to the morbidity of the visual loss associated with AMD and the lack of treatments, it may be reasonable to use supplementation in the selected high-risk group [80]." These results do not marginalize the value of zinc but suggest that the high doses used in the study do not significantly alter the impact that the proper physiological levels of zinc already exert.

What should you do regarding zinc? Stay tuned as more information about its connection to AD emerges. Identify foods high in zinc that are found on the outer isle of your grocer's shelves including: liver, certain mushrooms, asparagus, chard, scallops, lamb, beef, maple syrup, shrimp, green peas, yogurt, oats, pumpkin seeds, sesame seeds, turkey, miso, and spelt. Consider taking zinc supplementation up to daily recommendations by the USDA. Never take high doses of any supplement, as this is not natural to our history, with few exceptions. Vitamin D is a notable exception where we store the vitamin in our adipose tissue for consumption in the winter as few foods provide it naturally.

VITAMIN K

Vitamin K is a fat-soluble vitamin. The "K" is derived from the German word "koagulation." Coagulation refers to the process of blood clot formation. Vitamin K is essential for the functioning of several proteins involved in blood clotting. There are two naturally occurring forms of vitamin K. Vitamin K_1 also known as phylloquinone, is synthesized by plants and is the predominant form in the diet. Vitamin K_{12} comes from animal sources and synthesis by intestinal bacteria.

In the Rotterdam Heart Study, people eating lots of Edam and Gouda cheese had higher levels of vitamin K_2 and less artery calcification [81]. Higher levels of Vitamin K_2 are also associated with lower risk of prostate cancer. By keeping calcium out of the brain, K_2 may also help prevent AD. There is a clear pattern: too much calcium in the wrong places cause trouble. Calcium in the right place (i.e., bone) is a good thing. Vitamin K_2 appears to make sure calcium goes into the right places in the right amounts.

A relative deficiency of vitamin K is common in aging men and women. The concentration of vitamin K is lower in the circulating blood of APOE4 carriers, the gene that, to some degree, predisposes a person to AD. Evidence is accumulating that vitamin K has important functions in the brain. It is now proposed that vitamin K deficiency contributes to the process of AD and that its supplementation may have a beneficial effect in preventing or treating the disease. Vitamin K may also reduce neuronal damage associated with cardiovascular disease.

CURCUMIN

Extensive in vitro (outside the body) and animal studies have examined the effects of curcumin on experimentally induced inflammatory diseases (atherosclerosis, arthritis, diabetes, liver disease, gastrointestinal disorders, and cancers) and disease markers (lipoxygenase, cyclooxygenase, TNF-α, IL-1β, NF-$\kappa\beta$, and others). Fewer human studies have examined curcumin's effects on patient-oriented outcomes in inflammatory diseases, but most of the small, randomized controlled trials of curcumin have consistently shown patient improvements in several inflammatory diseases, including psoriasis, irritable bowel syndrome, rheumatoid arthritis, and inflammatory eye disease.

Curcumin is undergoing studies in AD, and initial results show no benefit from the compound. A combined Japanese and German team wrote, *Curcumin and Alzheimer's Disease,* and reported a lack of effect [82]. Their abstract is provided here because it is very educational. Curcumin is involved in the inhibition of so many mechanisms thought to be important to AD; yet early results are disappointing.

> Curcumin has a long history of use as a traditional remedy and food in Asia. Many studies have reported that curcumin has various beneficial properties, such as antioxidant, anti-inflammatory, and antitumor. Because of the reported effects of curcumin on tumors, many clinical trials have been performed to elucidate curcumin's effects on cancers. Recent reports have suggested therapeutic potential of curcumin in the pathophysiology of Alzheimer's disease (AD). In in vitro studies, curcumin has been reported to inhibit amyloid-β-protein (Aβ) aggregation and Aβ-induced inflammation, as well as the activities of β-secretase and acetylcholinesterase.
>
> In in-vivo studies, oral administration of curcumin has resulted in the inhibition of Aβ deposition, Aβ oligomerization, and tau phosphorylation in the brains of AD animal models, and improvements in behavioral impairment in animal models. These findings suggest that curcumin might be one of the most promising compounds for the development of AD therapies.
>
> At present, four clinical trials concerning the effects of curcumin on AD have been conducted. Two of them that were performed in China and USA have reported no significant differences in changes in cognitive function between placebo and curcumin groups, and no results have been reported from two other clinical studies. Additional trials are necessary to determine the clinical usefulness of curcumin in the prevention and treatment of AD.

Rule of thumb: If you are studying the literature for information about disease, completely ignore "animal only" studies. They are probably right 50% of the time when translated to humans, so consider flipping a coin instead. The lack of detailed studies with curcumin and its antioxidant properties suggests that you should enjoy curry, but avoid supplementation.

For those of you hungry for more information, there is plenty on curcumin and AD, provided by UCLA at the following web address:

http://Alzheimer.neurology.ucla.edu/Curcumin.html

DHEA

DHEA is a hormone that is naturally made by the human body. DHEA is an adrenal steroid hormone, the precursor to the sex steroids testosterone and estrogen. DHEA is abundant in youth, but steady declines with advancing age and may be partially responsible for age-related decreases in sex steroids. In cell culture and animal models, DHEA can suppress inflammatory cytokine activity, in some cases more effectively than either testosterone or estrogen. Chronic inflammation itself may reduce DHEA levels. DHEA supplementation in elderly volunteers (50 mg/day for 2 years) significantly decreased TNF-α and IL-6 levels, as well as lowered visceral fat mass and improved glucose tolerance (both associated with inflammation) in a small study [83]. Remember from the previous chapter that blocking IL-6 restores the production of neurons.

In a sample of newly diagnosed Alzheimer's patients, researchers did not find significant association between presence of Alzheimer's or impairment in cognitive domains and DHEA levels. This result was confirmed by scientists from Oregon who also explain the pitfalls of rodent models for measuring specific effects of DHEA relevant to humans [84].

ESTROGEN (AND OTHER HORMONES)

Estrogen is important to the building and maintenance of nerve networks in the brain from early on in life. Several studies are now pointing to the fact that estrogen may offer protection against AD in postmenopausal women. One study conducted on almost 90,000 postmenopausal women found that those taking estrogen had a significantly longer life, and by the time of their deaths, the women on estrogen had a 40% lower incidence of AD [85].

Estrogen docking sites are present in several regions of the brain, including those involved in memory (such as the hippocampus). When activated by estrogen, these sites, in turn, activate processes that are beneficial to the brain. In addition, estrogen may in effect raise levels of certain brain chemicals (neurotransmitters). These include the neurotransmitters acetylcholine (implicated in memory), serotonin (implicated in mood), noradrenaline (implicated in mood and other autonomic functions), and dopamine (implicated in motor coordination). Thus, estrogen facilitates networking between nerve cells, promoting their ability to "talk to" one another.

The medical literature is full of articles discussing (mainly) the benefits of estrogen toward AD. A "title only" search of scholar.google returns 350 articles! The first of these articles was published in 1986 [86]. The abstract from a 1994 paper from the USC School of Medicine provides a very nice summary of the cause/effect of estrogen replacement therapy [87].

The authors explored the possibility that estrogen loss associated with menopause may contribute to the development of Alzheimer's disease by using a case-control study nested within a prospective cohort study. The Leisure World Cohort includes 8,877 female residents of Leisure World Laguna Hills, a retirement community in southern California, who were first mailed a health survey in 1981. From the 2,529 female cohort members who died between 1981 and 1992, the authors identified 138 with Alzheimer's disease or other dementia diagnoses likely to represent Alzheimer's disease (senile dementia, dementia, or senility) mentioned on the death certificate. Four controls were individually matched by birth date (±1 year) and death date (+1 year) to each case. The risk of Alzheimer's disease and related dementia was less in estrogen users relative to nonusers (odds ratio = 0.69, 95%; confidence interval 0.46 = 1.03). The risk decreased significantly with increasing estrogen dose and with increasing duration of estrogen use. Risk was also associated with variables related to endogenous estrogen levels, it increased with increasing age at menarche and (although not statistically significant) decreased with increasing weight. This study suggests that the increased incidence of Alzheimer's disease in older women may be due to estrogen deficiency and that estrogen replacement therapy may be useful for preventing or delaying the onset of this dementia.

Do not forget that women account for two-thirds of all Alzheimer's cases.

Another 1994 paper and one from 2010 provide hope to Alzheimer's sufferers in that "estrogen replacement may improve cognitive performance of women with this (AD) illness [88,89]." Recent studies are helping us fine-tune what constitutes most beneficial estrogen replacement therapy.

> Previous studies in postmenopausal women have reported that estrogen treatment (ET) modulates the risk for developing Alzheimer's disease (AD). It has recently been hypothesized that there may be a 'critical period' around the time of menopause during which the prescription of ET may reduce the risk of developing AD in later life. This effect may be most significant in women under 49 years old. Furthermore, prescription of ET after this point may have a neutral or negative effect, particularly when initiated in women over 60–65 years old. In this paper, we review recent studies that use in vivo techniques to analyze the neurobiological mechanisms that might underpin estrogen's effects on the brain post menopause. Consistent with the 'critical period' hypothesis, these studies suggest that the positive effects of estrogen are most robust in young women and in older women who had initiated ET around the time of menopause.

FISH OIL IN CARDIOVASCULAR DISEASES

Eat plenty of fish, especially cold-water fish. Supplement with fish oil or, preferably, cod liver oil. Do not make any excuses or be influenced by media. Fish oil is not a vitamin, mineral, or other supplement; it is simply a food that is deficient in the diet of many of those with chronic diseases. Along with vitamin D, fish oil has a panoply of benefits to health including those suffering with AD. Fish oil is the best source of the omega-3 fatty acids eicosapentaenoic acid (EPA) and docosahexaenoic acid (DHA) that can only be synthesized to a limited extent in humans. These fatty acids are essential constituent of the membranes of neurons. Thus fish oil supplementation, or consuming fish, is critical. Omega-3 fatty acids have been well studied for their prevention of cardiovascular disease and mortality in tens of thousands of patients. The antiinflammatory effects of omega-3s contribute to this activity. They have also proven successful at improving patient outcomes in scores of studies of inflammatory diseases, particularly asthma, inflammatory bowel disease, cardiovascular diseases, AD, and rheumatoid arthritis. Emerging studies show fish oil conferring a wealth of benefits in guarding against depression, cancer, osteoporosis, and other crippling diseases of aging.

Fish oil lowers triglycerides. Dangerously high levels of triglycerides have become more common as Americans develop metabolic syndrome, cardiovascular disease, and other nutrition-related illnesses. Elevated triglycerides greatly increase risk for heart disease. The right dose of omega-3s can significantly reduce triglyceride levels and help correct other cardiac risk factors that accompany metabolic syndrome [87,90,91]. In fact, omega-3 fatty acids derived from fish oils are now available in the form of a prescription drug called Omacor, which has been approved specifically for the treatment of elevated triglycerides [92]. Do not get in the habit of succumbing to the prescription pad. Instead, eat plenty of fish and take fish oil or cod liver oil in large amounts to compensate for a suboptimal diet.

Silent inflammation triggers a chain of events leading to heart disease and other illnesses. Omega-3 fatty acids suppress multiple steps in this inflammatory process, inhibiting the production of inflammatory cytokines and prostaglandins probably through a variety of mechanisms that includes immune system augmentation. Furthermore, omega-3 fats boost production of antiinflammatory compounds. These antiinflammatory effects may have important implications for fighting heart disease and numerous other disease processes associated with excessive inflammation [93–99]. People who consume a greater amount of omega-3 fatty acids demonstrate lower levels of C-reactive protein, a cardiovascular risk factor, suggesting that omega-3 supplementation might help prevent cardiovascular disease by, for example, reducing infectious burden, thus lowering inflammation [100]. Omega-3s from fish oil may help to modify the structure of atherosclerotic plaque in ways that make it less dangerous. In fact, studies show that omega-3s can slow the rate of atherosclerotic plaque growth. Fish oil rich in omega-3 fatty acids contributes to healthy vascular function by increasing the production of an important blood vessel-dilating substance in the endothelial cells [101,102].

Omega-3 supplementation actually changed the composition of unstable atherosclerotic plaque, making it less likely to rupture and thus less dangerous. Subjects who had severe carotid plaque and were scheduled to have it surgically removed received either fish oil or sunflower oil prior to surgery. When the plaque was removed at surgery and examined, researchers found that those who took fish oil had less plaque inflammation as well as more stable plaque. By contrast, those who took sunflower oil had more unstable, rupture-prone plaque [103]. Fish oil also helps reduce certain proteins that promote abnormal blood clotting and inhibit platelet aggregation, two effects that reduce the likelihood of clot formation on active, ruptured coronary plaque that could result in a heart attack [104]. One of the most dramatic benefits of fish oil is its ability to prevent sudden death, particularly sudden cardiac death. Scientists believe that omega-3 fatty acids from cold-water fish may help prevent these sudden deaths by reducing potentially fatal abnormal heart rhythms, or arrhythmias.

Australian and Chinese scientists studied the results of 11 trials and uncovered a significant reduction in plasma homocysteine in association with greater intake of omega-3 polyunsaturated fatty acids [105]. Treatment periods ranged from 6 to 48 weeks, and doses varied between 0.2 and 6 g per day. The analysis confirmed a reduction in plasma homocysteine levels in association with omega-3 fatty acid supplementation, with an average decrease of 1.59 μm/L experienced by those who supplemented compared to those who received a placebo. "Our systematic review provides, to our knowledge, the most comprehensive assessment to date of the effects of omega-3 polyunsaturated fatty acids on plasma homocysteine," the authors announced.

Dr. Kilmer McCully, the pioneer of the homocysteine theory showed, way back in 1993, that fish oil lowered plasma homocysteine in subjects with elevated cholesterol (hyperlipidemia) [106].

FISH OILS IN ALZHEIMER'S AND DEMENTIA

Since vascular diseases are tied to Alzheimer's and other dementias, there should be no surprise that fish oils are helpful for Alzheimer's sufferers and also play a key role in prevention. The literature on this topic is rich and extends back into the 1990s. The Japanese took the early lead through publications and patents. One such patent from 1994 is titled, *Brain function ameliorant composition, learning capacity enhancer, mnemonic agent, dementia preventive, dementia curative, or functional food with brain function ameliorant effect* [107]. Here is a part of the abstract from that patent:

> The invention aims at ameliorating brain functions to thereby effect learning capacity enhancement, memory enhancement, and prevention and cure of senile dementia, and to provide a functional food having a brain function ameliorant effect. The invention composition comprises at least one member selected from among n-3 unsaturated fatty acids, i.e., docosahexaenoic acid, eicosapentaenoic acid...

Two articles that bracket 16 years of research from 1997 to 2013 are examined here. The earlier one is titled, *Polyunsaturated Fatty Acids, Antioxidants, and Cognitive Function in Very Old Men*, (it sounds more like a new sitcom) [108]. This team from the Netherlands says, "This study raises the possibility that high linoleic acid intake is positively associated with cognitive impairment and high fish consumption inversely associated with cognitive impairment." The 2013 article was likely chosen from the most conservative group, the Alzheimer's Drug Discovery Foundation [109]. Polyunsaturated fatty acids (fish oil) are not really drug candidates; they are foods, although there are a couple on the market. The title of their article includes the key word "prevention." They state, "Of particular relevance, epidemiology indicates a higher risk of cognitive decline in people in the lower quartile of n-3 LC-PUFA intake or blood levels."

Higher blood levels of EPA are associated with a lower risk of dementia and depression in elderly persons in a French study [110]. EPA is an omega-3 polyunsaturated fatty acid found in certain fish that may decrease the risk of dementia and AD. The study included 1,214 French persons aged 65 years or older who were examined for dementia and blood levels of fatty acids over four years. Depression was also assessed because it has been related to both low EPA and dementia. By 4 years, 65 patients had developed dementia. A higher level of EPA was associated with a lower likelihood of dementia, even after accounting for depression and other patient characteristics. An association between depression and dementia was also confirmed. The authors concluded, "Because depression and dementia share common vascular risk factors, the vascular properties of EPA could contribute to decrease depression and dementia risk simultaneously."

Researchers at the Rush Institute of Healthy Aging conducted a study to see if consuming fish and different omega-3 fatty acids protect against Alzheimer's [111]. Over 800 participants unaffected by AD (between the ages of 65 and 94) were monitored from 1993 to 2000 and then followed-up for a 4-year period to see if they developed Alzheimer's. Researchers discovered patients who ate fish once or more per week or increased the amount of omega-3 fatty acids in their diet had a 60% lower risk of developing AD. While these are only the results of one single study, they are encouraging.

Mentally return to Chapter 2 for a moment to put this "60% lower risk" into perspective based on the new efforts of big pharma and those who live and die with the Amyloid Cascade Hypothesis. All therapeutics lowering beta-amyloid failed to improve cognitive functioning and other aspects of AD. In fact, many of the patients enrolled in these trials got worse compared to a sugar pill. Desperate for these drugs to work, spokespeople for big pharma say that they will "soldier on" and test their drugs on people who do not have Alzheimer's or mild cognitive impairment. They argue that in patients with symptoms, the disease has progressed too far, and the therapy did not stand a chance (this is not correct, their therapy is based on bad science). Basically, they will reproduce the study by the researchers at Rush Institute. What chance does their beta-amyloid therapy have at outperforming omega-3 fatty acids? How likely is it that the FDA will ask big pharma to compare their results to omega-3 fatty acids rather than a sugar pill?

Omega-3s may reduce symptoms in mild AD. A study published in the October 2006 issue of the *Archives of Neurology* suggests fish oil supplements may slow cognitive decline in patients with very mild Alzheimer's [112]. This double-blind, placebo-controlled study had 174 Alzheimer's patients receive either a placebo or 430 mg of DHA along with 150 mg EPA, 4 times a day. The randomized treatment lasted for 6 months, and then all subjects received the fish oil supplements for another 6 months. Patients who took omega-3 fatty acids experienced less change in their rate of cognitive decline compared to those who took the placebo. A smaller group of 32 patients with mild AD experienced a significant reduction in their rate of cognitive decline compared to the placebo group. And surprisingly enough, when the placebo-group patients took the omega-3 supplements for 6 months post-placebo trial, they also experienced the same reduction in mental decline.

PUFA 3, PUFA 6, AND PUFA 6/3 RATIO

Omega-3 fats are also known as PUFA 3 (polyunsaturated fatty acids). The key clinical omega-3 fats are EPA and DHA, which are found largely in cold-water fish. The PUFA 6/3 ratio is also important as PUFA 6 tends to promote proinflammatory molecules while PUFA 3 does just the opposite. High ratios (more PUFA 6) (>5) are associated with chronic silent inflammation. The following is a very concise abstract that discusses the history of our diets with respect to PUFAs in general and provides information on diseases impacted by an "imbalance" of PUFAs. The paper was written by Artemis P. Simopoulos from The Center for Genetics, Nutrition and Health, Washington, DC, and is titled, *The Importance of the Omega-6/Omega-3 Fatty Acid Ratio in Cardiovascular Disease and Other Chronic Diseases* [113].

Several sources of information suggest that human beings evolved on a diet with a ratio of omega-6 to omega-3 essential fatty acids (EFA) of ~1 whereas in Western diets the ratio is 15/1–16.7/1. Western diets are deficient in omega-3 fatty acids, and have excessive amounts of omega-6 fatty acids compared with the diet on which human beings evolved and their genetic patterns were established. Excessive amounts of omega-6 polyunsaturated fatty acids (PUFA) and a very high omega-6/omega-3 ratio, as is found in today's Western diets, promote the pathogenesis of many diseases, including cardiovascular disease, cancer, and inflammatory and autoimmune diseases, whereas increased levels of omega-3 PUFA (a lower omega-6/omega-3 ratio), exert suppressive effects.

In the secondary prevention of cardiovascular disease, a ratio of 4/1 was associated with a 70% decrease in total mortality. A ratio of 2.5/1 reduced rectal cell proliferation in patients with colorectal cancer, whereas a ratio of 4/1 with the same amount of omega-3 PUFA had no effect. The lower omega-6/omega-3 ratio in women with breast cancer was associated with decreased risk. A ratio of 2–3/1 suppressed inflammation in patients with rheumatoid arthritis, and a ratio of 5/1 had a beneficial effect on patients with asthma, whereas a ratio of 10/1 had adverse consequences.

These studies indicate that the optimal ratio may vary with the disease under consideration. This is consistent with the fact that chronic diseases are multigenic and multifactorial. Therefore, it is quite possible that the therapeutic dose of omega-3 fatty acids will depend on the degree of severity of disease resulting from the genetic predisposition. A lower ratio of omega-6/omega-3 fatty acids is more desirable in reducing the risk of many of the chronic diseases of high prevalence in Western societies, as well as in the developing countries. Thankfully, the solution to your good health is simple—eat more fish and supplement with fish oils. And you can improve your chances of good health by decreasing your intake of the inflammation-creating omega-6 (PUFA-6) fats. The books by Barry Sears on the "Zone" diet provide what you need to know about avoidance of omega-6s [114,115].

Excess PUFA 6 interferes with the health benefits of PUFA 3, in part because they compete for the same rate-limiting enzymes. A high proportion of 6/3 fat in the diet shifts the physiological state in the tissues toward many diseases that involve blood clotting, inflammation, and vessel constriction. Chronic excessive production of $n-6$ eicosanoids derived from PUFA 6 is associated with heart attacks, thrombotic stroke, arrhythmia, arthritis, osteoporosis, inflammation, mood disorders, obesity, and cancer. Medications used to treat and manage these conditions work by blocking the effects of the PUFA 6 known as arachidonic acid. Many steps in formation and action of n-6 hormones from $n-6$ arachidonic acid proceed more vigorously than the corresponding competitive steps in formation and action of n-3 hormones from $n-3$ compounds.

The PUFA 6:3 ratio plays a role in dementia and Alzheimer's, too. A 2013 German review article delved into the science behind the PUFA ratio and dementia [116]. The abstract is reproduced here:

> It has been suggested that the intake of certain fatty acids may influence the risk of dementia. However, current reviews have focused only on the therapeutic effects of omega-3 fatty acids, mostly as supplements. To date, the evidence for the relevance of the omega-6/omega-3 ratio has been neglected. Therefore, we searched the databases Alois, Medline, Biosis, Embase, Cochrane Central Register of Controlled Trials, and The Cochrane Database of Systematic Reviews for 'essential fatty acids' and 'dementia' and aimed to conduct a comprehensive review across study types.
>
> All studies that reported on the association between the n-6/n-3 ratio and dementia or cognitive decline were selected. In the 13 animal studies we examined, the dietary n-6/n-3 ratio was shown to affect brain composition, Alzheimer's disease pathology, and behavior. Our review of the 14 studies in humans that fulfilled the selection criteria (7 prospective studies, 3 cross-sectional studies, 1 controlled trial, 3 case-control studies) provided evidence, albeit limited, supporting an association between the n-6/n-3 ratio, cognitive decline, and incidence of dementia. This review supports growing evidence of a positive association between the dietary n-6/n-3 ratio and the risk of Alzheimer's disease.

BERBERINE

Berberine, an isoquinoline alkaloid isolated from medicinal herbs frequently used in traditional Eastern medicine, has multiple therapeutic effects for metabolic disorders, microbial infection, neoplasms, and inflammation [117]. Increasing interest has focused on its antiinflammatory effects. In microglia, the main immune system of the brain, berberine suppresses neuroinflammatory responses and attenuates the production of inflammatory mediators.

Substantial evidence also shows that berberine exerts neuroprotection in cerebral ischemia (blood flow loss to regions of the brain) and AD [118].

There are a variety of other supplements and food intake modifications that can impact your likelihood of developing chronic diseases like Alzheimer's. Google David Wheldon of England to see his list of recommendations for MS, as MS and AD are both neurodegenerative processes, and there is logical overlap.

PREVENTION—IN YOUR CONTROL

AD is preventable. Claude Bernard, Alois Alzheimer's, and other doctors and scientists of history who documented the dearth of "Alzheimer's" disease even 100 years ago present that proof. Paul Clayton, a modern man, indicates that chronic degenerative diseases were 10% of what we experience today, during the mid-Victorian era (1870). What is this pall that has overcome our people and inflicted us with rampant and expanding poor health? I think we all know the answers.

The comforts, convenience, and economics of modern society have superseded our good sense. Everything we do on a day-by-day and minute-by-minute basis perpetuates poor health. It starts with a bowl of sugar-laden cereal with fat-free milk, after a restless night's sleep. Next we jump into an automobile and struggle through a commute to get to a climate-controlled office. We rush through lunch prepared of more processed foods that our bodies rapidly convert to sugars and fats. Now back to our climate-controlled bubble. We are rewarded at the end of the workday with a ride in our well-deserved luxury car that doesn't even require a finger to open the trunk. We settle into a nice Stouffer's dinner, because our spouse worked to afford his or her car, too. Finally, some free quality time with the boob tube, some carbonated beverage, coffee, dessert, a cigar, and a nightcap. Off we go for another restless night's sleep.

Surprise…

You know what to do (sort of—except you think low fat is good—it's just the opposite, especially for your brain). But you don't do it anyway because the circle of life doesn't provide the luxury of time to truly care for yourself.

In the spirit of keeping it simple, focus on increasing your intake of vitamin D, fish oils, magnesium, and zinc while avoiding calcium and omega-6s. A key issue is dose. Most of the time, when researchers find that these supplements are ineffectual, the reason is likely the low dosing level. The PUFA 6/3 ratio helps us understand this conundrum. If your PUFA ratio is >15/1 (plenty of Americans have a 40/1 ratio or higher), then supplementing with 500 mg/day of PUFA 3 is like giving someone who is dying of thirst a shot glass of water. The amount is insufficient, and the impact on chronic health will not be measurable. USRDAs are established to be very conservative numbers and are not appropriate guidelines for people who are significantly out of balance.

Look upon foods differently compared to supplements. Fish oil (and cod liver oil) is food, not a supplement. Our family takes 15 grams of cod liver oil daily! Come on, it's not that bad. It is not the same thing your grandmother gave you. Here is how we take cod liver oil. We take a shot glass and fill it half full (half empty if you prefer) with cod liver oil. Drink some orange juice, then shoot the cod liver oil to the back of your throat and quickly swallow. Next, drink some more orange juice. You won't know what hit you, in a good way. Do this in the evening just prior to bed, as this will reduce or eliminate any upset caused by the cod liver oil. Also, if you are new to this, start with one-eighth of a shot glass and work your way up.

Focus on preventative measures, and you will have no need to read Chapter 11.

> The doctor of the future will give no medicine, but will instruct his patient in the care of the human frame, in diet and in the cause and prevention of disease. – *Thomas Alva Edison.*

Take your cod liver oil.
Here's to your health!

References

[1] Graff-Radford J. Alzheimer's prevention: does it exist? Mayo Clinic. Available from: http://www.mayoclinic. org/diseases-conditions/Alzheimers-disease/expert-answers/Alzheimers-prevention/faq-20058140; 2014.

[2] Steele M, Stuchbury G, Münch G. The molecular basis of the prevention of Alzheimer's disease through healthy nutrition. Exp. Gerontol 2007;42(1):28–36.

[3] Henderson ST. High carbohydrate diets and Alzheimer's disease. Med Hypotheses 2004;62(5):689–700.

[4] Enig M.G, Fallon S. Eat fat, lose fat; 2005.

[5] Daulatzai MA. Obesity and gut's dysbiosis promote neuroinflammation, cognitive impairment, and vulnerability to Alzheimer's disease: new directions and therapeutic implications. J Mol Genet Med 2014;S1(01):005.

[6] Jung DW, Brierley GP. Matrix magnesium and the permeability of heart mitochondria to potassium ion. J Biol Chem 1986;261:6408–15.

[7] King DE, Mainous AG, et al. Dietary magnesium and C-reactive protein levels. J Amer Col Nutrition 2005;24:166–71.

[8] Subar AF, Krebs-Smith SM. Dietary sources of nutrients among US adults, 1989 to 1991. J Amer Dietetic Assoc 1998;98:537–47.

[9] Ford ES, Mokdad AH. Dietary magnesium intake in a national sample of U.S. adults. J Nutr 2003;133:2879–82.

[10] Turlapaty PD, Altura BM. Magnesium deficiency produces spasms of coronary arteries: relationship to etiology of sudden death ischemic heart disease. Science 1980;208:198–200.

[11] Caspi J, Rudis E, et al. Effects of magnesium on myocardial function after coronary artery bypass grafting. Ann Thorac Surg 1995;59:942–7.

[12] Eisenberg MJ. Magnesium deficiency and sudden death. Am Heart J 1992;124:544–9.

[13] Rude RK, Singer FR. Magnesium deficiency and excess. Ann Rev Med 1981;32:245–59.

[14] Ralston MA, Murnane MR, et al. Magnesium content of serum, circulating mononuclear cells, skeletal muscle, and myocardium in congestive heart failure. Circulation 1989;80:573–80.

[15] Newhouse IJ, Johnson KP, et al. Variability within individuals of plasma ionic magnesium concentrations. BMC Physiol 2002;2:6–13.

[16] Brannan PG, Vergne-Marini P, et al. Magnesium absorption in the human small intestine. Results in normal subjects, patients with chronic renal disease, and patients with absorptive hypercalciuria. J Clin Invest 1976;57:1412–8.

[17] Nieves JW. Osteoporosis: the role of micronutrients. Am J Clin Nutr 2005;81:1232S–9S.

[18] Murphy E, Freudrich CC, Lieberman M. Cellular magnesium and Na/Mg exchange in heart cells. Ann Rev Physiol 1991;53:273–87.

[19] Shechter M, Sharir M, et al. Oral magnesium therapy improves endothelial function in patients with coronary artery disease. Circulation 2000;102:2353–8.

[20] Maier JA, Malpuech-Bruegère C, et al. Low magnesium promotes endothelial cell dysfunction: implications for atherosclerosis, inflammation and thrombosis. Biochim Biophys Acta 2004;1689:13–21.

[21] Galland L. Magnesium and immune function: an overview. Magnesium 1988;7:290–9.

[22] Wolf FI, Cittadini A. Magnesium in cell proliferation and differentiation. Front Biosci 1999;4:607–17.

[23] Reid JD, Hunter CN. Current understanding of the function of magnesium chelatase. Biochem Soc Trans 2002;30:643–5.

[24] Thung-Shenq L, Slaughter TF, et al. Regulation of human tissue tranglutaminase function by magnesium nucleotide complexes. J Biol Chem 1998;273:1776–81.

[25] Rassmussen HH, Mortensen PB, Jensen IW. Depression and magnesium deficiency. Int J Psychiatry Med 1989;19:57–63.

[26] Morris ME. Brain and CSF magnesium concentrations during magnesium deficit in animals and humans: neurological symptoms. Magnes Res 1992;5:303–13.

[27] Iseri LT, Freed J, Bures AR. Magnesium deficiency and cardiac disorders. Am J Med 1975;58:837–46.

[28] Al-Ghamdi SM, Cameron EC, Sutton RA. Magnesium deficiency: pathophysiologic and clinical overview. Am J Kidney Dis 1994;24:737–52.

[29] Rude RK. Magnesium deficiency and diabetes mellitus: causes and effects. Postgrad Med 1992;92:222–4.

[30] Whang R, Whang DD, Ryan MP. Refractory potassium repletion: a consequence of magnesium deficiency. Arch Intern Med 1992;152:40–5.

[31] Flink EB. Magnesium deficiency: etiology and clinical spectrum. Acta Med Scand Suppl 1998;209:125–37.

[32] Rink EB. Magnesium deficiency in alcoholism. Alcohol Clin Exp Res 1986;10:590–4.

[33] Hogue CW, Hyder ML. Atrial fibrillation after cardiac operation: risks, mechanisms, and treatment. Ann Thorac Surg 2000;69:300–6.

[34] Miller S, Crystal E, Garfinkle M, et al. Effects of magnesium on atrial fibrillation after cardiac surgery: a metaanalysis. Heart 2005;91:618–23.

[35] Fassler CA, Rodriguez RM, Badesch DB, et al. Manesium toxicity as a cause of hypotension and hypoventilation. Archives Intern Med 1985;145:1604–6.

[36] Patterson Ruth E, et al. Measurement characteristics of the Women's Health Initiative food frequency questionnaire. Ann Epidemiol 1999;9(3):178–87.

[37] Kim HJ, et al. Longitudinal and secular trends in dietary supplement use: nurses' health study and health professionals follow-up study, 1986-2006. J Acad Nutr Diet 2014;114(3):436–43.

[38] Burnett-Hartman AN, et al. Supplement use contributes to meeting recommended dietary intakes for calcium, magnesium, and vitamin C in four ethnicities of middle-aged and older Americans: the Multi-Ethnic Study of Atherosclerosis. J Am Diet Assoc 2009;109(3):422–9.

[39] Liu G. Prevention of cognitive deficits in Alzheimer's mouse model by elevating brain magnesium. Mol Neurodegener 2012;7(Suppl. 1):L24.

[40] Durlach J. Magnesium depletion and pathogenesis of Alzheimer's disease. Magnesium research: official organ of the International Society for the Development of Research on Magnesium 1990;3(3):217.

[41] Heinitz, Matthias F. Magnesium and Alzheimer's disease: the cholinergic hypothesis. Schweiz Zschr Ganzheitsmed 2012;24(6):371–4.

[42] Bolland, Mark J, et al. Effect of calcium supplements on risk of myocardial infarction and cardiovascular events: meta-analysis. BMJ 2010;341:c3691.

[43] Wang L, et al. Systematic review: vitamin D and calcium supplementation in prevention of cardiovascular events. Ann Int Med 2010;152(5):315–23.

[44] Disterhoft JF, Moyer JR, Thompson LT. The calcium rationale in aging and Alzheimer's disease. Ann N Y Acad Sci 1994;747(1):382–406.

[45] Dranov, P. The miracle vitamin. 2006:162–9.

[46] Brody JE. What do you lack? Probably vitamin D. The New York Times, July 26, 2010. Available from: http://www.nytimes.com/2010/07/27/health/27brod.html

[47] Bakalar N. Low vitamin D tied to aging problems. The New York Times, Jul 19, 2013. Available from: http://well.blogs.nytimes.com/2013/07/19/low-vitamin-d-tied-to-aging-problems/?emc=eta1

[48] Bakalar N. Low vitamin D tied to a pregnancy risk. The New York Times, Jan 30, 2014. Available from: http://well.blogs.nytimes.com/2014/01/30/low-vitamin-d-tied-to-a-pregnancy-risk/

[49] Bakalar N. Vitamin D may lower cholesterol. The New York Times, Mar 13, 2014. Available from: http://well.blogs.nytimes.com/2014/03/13/vitamin-d-may-lower-cholesterol/

[50] O'Connor A. Low vitamin D levels linked to disease in two big studies. The New York Times, Apr 1, 2014. Available from: http://well.blogs.nytimes.com/2014/04/01/low-vitamin-d-levels-linked-to-disease-in-two-big-studies/

[51] Bakalar N. Low vitamin D tied to premature death. The New York Times, Jun 12, 2014. Available from: http://well.blogs.nytimes.com/2014/06/12/low-vitamin-d-tied-to-premature-death/

[52] Berg JM, Tymoczko JL, Stryer L. Biochemistry. Section 26.4, Important derivatives of cholesterol include bile salts and steroid hormones. 5th ed. New York: W H Freeman; 2002. Available from: http://www.ncbi.nlm.nih.gov/books/NBK22339/

[53] Durk MR, et al. 1α, 25-Dihydroxyvitamin D3 reduces cerebral amyloid-β accumulation and improves cognition in mouse models of Alzheimer's disease. J Neurosci 2014;34(21):7091–101.

[54] Ramagopalan SV, et al. A ChIP-seq defined genome-wide map of vitamin D receptor binding: associations with disease and evolution. Genome Res 2010;20(10):1352–60.

[55] Krishnan AV, Feldman D. Molecular pathways mediating the anti-inflammatory effects of calcitriol: implications for prostate cancer chemoprevention and treatment. Endocr Relat Cancer 2010;17(1):R19–38.

[56] Guillot X, et al. Vitamin D and inflammation. Joint Bone Spine 2010;77(6):552–7.

[57] Sato Y, Asoh T, Oizumi K. High prevalence of vitamin D deficiency and reduced bone mass in elderly women with Alzheimer's disease. Bone 1998;23(6):555–7.

[58] Annweiler C, et al. Association of vitamin D deficiency with cognitive impairment in older women; cross-sectional study. Neurology 2010;74(1):27–32.

[59] Evatt ML, et al. Prevalence of vitamin d insufficiency in patients with Parkinson disease and Alzheimer disease. Arch Neurol 2008;65(10):1348.

[60] Oudshoorn C, et al. Higher serum vitamin D3 levels are associated with better cognitive test performance in patients with Alzheimer's disease. Dement Geriatr Cogn Disord 2008;25(6):539–43.

[61] Wilkins CH, et al. Vitamin D deficiency is associated with worse cognitive performance and lower bone density in older African Americans. J Natl Med Assoc 2009;101(4):349.

[62] Annweiler C, et al. Serum vitamin D deficiency as a predictor of incident non-Alzheimer dementias: a 7-year longitudinal study. Dement Geriatr Cogn Disord 2012;32(4):273–8.

[63] Buell JS, et al. 25-Hydroxyvitamin D, dementia, and cerebrovascular pathology in elders receiving home services. Neurology 2010;74(1):18–26.

[64] Jablonski KL, et al. 25-Hydroxyvitamin D deficiency is associated with inflammation-linked vascular endothelial dysfunction in middle-aged and older adults. Hypertension 2011;57(1):63–9.

[65] Pogge E. Vitamin D and Alzheimer's disease: is there a link? Consult Pharm 2010;25(7):440–50.

[66] Deluca GC, et al. The role of vitamin D in nervous system health and disease. Neuropathol Appl Neurobiol 2013;39(5):458–84.

[67] Coussens AK, et al. Vitamin D accelerates resolution of inflammatory responses during tuberculosis treatment. Proc Natl Acad Sci 2012;109(38):15449–54.

[68] Littlejohns TJ, et al. Vitamin D and the risk of dementia and Alzheimer disease. Neurology 2014;83:920–8.

[69] Balion C, Griffith LE, Strifler L, et al. Vitamin D, cognition, and dementia: a systematic review and meta-analysis. Neurology 2012;79:1397–405.

[70] Annweiler C, Llewellyn DJ, Beauchet O. Low serum vitamin D concentrations in Alzheimer's disease: a systematic review and meta-analysis. J Alzheimers Dis 2013;33:659–74.

[71] Annweiler C, Montero-Odasso M, Llewellyn DJ, Richard-Devantoy S, Duque G, Beauchet O. Meta-analysis of memory and executive dysfunctions in relation to vitamin D. J Alzheimers Dis 2013;37:147–71.

[72] MGEKN I, Quinn R, Tabet N. Vitamin E for Alzheimer's disease and mild cognitive impairment. Cochrane Database Syst Rev 2008;3:CD002854.

[73] Devaraj S, et al. Gamma-tocopherol supplementation alone and in combination with alpha-tocopherol alters biomarkers of oxidative stress and inflammation in subjects with metabolic syndrome. Free Radic Biol Med 2008;44(6):1203–8.

[74] Hydrogen peroxide marshals immune system. News Alert; Harvard Medical School Office of Communications & External Relations, Sept 20, 2016. Available from: http://web.med.harvard.edu/sites/RELEASES/html/060309_mitchison.html

[75] Babior BM, Kipnes RS, Curnutte JT. Biological defense mechanisms. The production by leukocytes of superoxide, a potential bactericidal agent. J Clin Invest 1973;52(3):741.

[76] UCLA study suggests iron is at core of Alzheimer's disease. UCLA Newsroom, Aug 20, 2013. Available from: http://newsroom.ucla.edu/releases/ucla-study-suggests-that-iron-247864

[77] Raven, Erika P, et al. Increased iron levels and decreased tissue integrity in hippocampus of Alzheimer's disease detected in vivo with magnetic resonance imaging. J Alzheimer's Dis 2013;37(1):127–36.

[78] Watt NT, Whitehouse IJ, Hooper NM. The role of zinc in Alzheimer's disease. Int J Alzheimer's Dis 2010;2011:971021.

[79] Cuajungco MP, Fagét KY. Zinc takes the center stage: its paradoxical role in Alzheimer's disease. Brain Res Rev 2003;41(1):44–56.

[80] Age-Related Eye Disease Study Research Group. A randomized, placebo-controlled, clinical trial of high-dose supplementation with vitamins C and E, beta carotene, and zinc for age-related macular degeneration and vision loss: AREDS report no. 8. Arch Ophthalmol 2001;119(10):1417–36.

[81] Engberink MF, et al. Inverse association between dairy intake and hypertension: the Rotterdam study. Am J Clin Nutr 2009;89(6):1877–83.

[82] Hamaguchi T, Ono K, Yamada M. Review: curcumin and Alzheimer's disease. CNS Neurosci Ther 2010;16(5):285–97.

[83] Araghiniknam M, et al. Antioxidant activity of dioscorea and dehydroepiandrosterone (DHEA) in older humans. Life Sci 1996;59(11):PL147–57.

[84] Sorwell KG, Urbanski HF. Dehydroepiandrosterone and age-related cognitive decline. Age 2010;32(1):61–7.

[85] Lafferty FW, Fiske ME. Postmenopausal estrogen replacement: a long-term cohort study. Am J Med 1994;97(1):66–77.

[86] Fillit HM, et al. Estrogen levels in postmenopausal women with senile dementia-Alzheimer's type (SDAT) are significantly lower than matched controls. Soc Neurosci Abstr 1986;12:A59.11.

[87] Paganini-Hill A, Henderson VW. Estrogen deficiency and risk of Alzheimer's disease in women. Am J Epidemiol 1994;140(3):256–61.

[88] Henderson VW, et al. Estrogen replacement therapy in older women: comparisons between Alzheimer's disease cases and nondemented control subjects. Arch Neurol 1994;51(9):896.

[89] Craig MC, Murphy DGM. Estrogen therapy and Alzheimer's dementia. Ann N Y Acad Sci 2010;1205(1):245–53.

[90] Menuet R, Lavie CJ, Milani RV. Importance and management of dyslipidemia in the metabolic syndrome. Am J Med Sci 2005;330(6):295–302.

[91] Balk EM, Lichtenstein Ah, Chung M, Kupelnick B, Chew P, Lau J. Effects of omega-3 fatty acids on serum markers of cardiovascular disease risk: a systematic review. Atherosclerosis 2006;189(1):19–30.

[92] Harris WS, Ginsberg HN, Arunkul N, et al. Safety and efficacy of Omacor in severe hypertriglyceridemia. J Cardiovasc Risk 1997;4(5–6):385–91.

[93] Watkins BA, Li Y, Lippman HE, Seifert MF. Omega-3 polyunsaturated fatty acids and skeletal health. Exp Biol Med 2001;226(6):485–97.

[94] Ciubotaru I, Lee YS, Wander RC. Dietary fish oil decreases C-reactive protein, interleukin-6, and triacylglycerol to HDL cholesterol ratio in postmenopausal women on HRT. J Nutr Biochem 2003;14(9):513–21.

[95] Adam O, Beringer C, Kless T, et al. Anti-inflammatory effects of a low arachidonic acid diet and fish oil in patients with rheumatoid arthritis. Rheumatol Int 2003;23(1):27–36.

[96] Solomon DH. Selective cyclooxygenase 2 inhibitors and cardiovascular events. Arthritis Rheum 2005;52(7):1968–78.

[97] Hippisley-Cox J, Coupland C. Risk of myocardial infarction in patients taking cyclo-oxygenase-2 inhibitors or conventional non-steroidal anti-inflammatory drugs: population-based nested case-control analysis. BMJ. 2005;330(7504):1366.

[98] Cleland LG, James MJ. Marine oils for anti-inflammatory effect—time to take stock. J Rheumatol 2006;33(2):207–9.

[99] Leaf A. On the reanalysis of the GISSI-Prevenzione. Circulation 2002;105(16):1874–5.

[100] Lopez-Garcia E, Schulze MB, Manson JE, et al. Consumption of (n-3) fatty acids is related to plasma biomarkers of inflammation and endothelial activation in women. J Nutr 2004;134(7):1806–11.

[101] Connor SL, Connor WE. Are fish oils beneficial in the prevention and treatment of coronary artery disease? Am J Clin Nutr 1997;66(4 Suppl.):1020S–31S.

[102] Abeywardena MY, Head RJ. Longchain n-3 polyunsaturated fatty acids and blood vessel function. Cardiovasc Res 2001;52(3):361–71.

[103] L69: Thies F, Garry JM, Yaqoob P, et al. Association of n-3 polyunsaturated fatty acids with stability of atherosclerotic plaques: a randomized controlled trial. Lancet 2003;361(9356):477–85.

[104] Vanschoonbeek K, Feijge MAH, Paquay M, et al. Variable hypocoagulant effect of fish oil intake in humans: modulation of fibrinogen level and thrombin generation. Arterioscler Thromb Vasc Biol 2004;24(9):1734–40.

[105] Huang T, et al. High consumption of Ω-3 polyunsaturated fatty acids decrease plasma homocysteine: a meta-analysis of randomized, placebo-controlled trials. Nutrition 2011;27(9):863–7.

[106] Olszewski AJ, McCully KS. Fish oil decreases serum homocysteine in hyperlipemic men. Coron Artery Dis 1993;4(1):53–60.

[107] Nisikawa M, Shoji K, Kazuaki M. Brain Function Ameliorant Composition, Learning Capacity Enhancer, Mnemonic Agent, Dementia Preventive, Dementia Curative, Or Functional Food With Brain Function Ameliorant Effect. Wipo Patent No. 1994005319. 18 Mar. 1994.

[108] Kalmijn S, et al. Polyunsaturated fatty acids, antioxidants, and cognitive function in very old men. Am J Epidemiol 1997;145(1):33–41.

[109] Dacks PA, Shineman DW, Fillit HM. Current evidence for the clinical use of long-chain polyunsaturated N-3 fatty acids to prevent age-related cognitive decline and Alzheimer's disease. J Nutr Health Aging 2013;17(3):240–51.

[110] Samieri C, et al. Low plasma eicosapentaenoic acid and depressive symptomatology are independent predictors of dementia risk. Am J Clin Nutr 2008;88(3):714–21.

[111] Morris MC, et al. Consumption of fish and n-3 fatty acids and risk of incident Alzheimer disease. Arch Neurol 2003;60(7):940.

[112] Freund-Levi Y, et al. omega-3 fatty acid treatment in 174 patients with mild to moderate Alzheimer disease: OmegAD study: a randomized double-blind trial. Arch Neurol 2006;63(10):1402–8.

[113] Simopoulos AP. The importance of the omega-6/omega-3 fatty acid ratio in cardiovascular disease and other chronic diseases. Exp Biol Med 2008;233(6):674–88.

[114] Sears B. The anti-inflammation zone. New York: HarperCollins; 2009.

[115] Sears B. The omega Rx zone. New York: HarperCollins; 2009.

[116] Loef M, Walach H. The omega-6/omega-3 ratio and dementia or cognitive decline: a systematic review on human studies and biological evidence. J Nutr Gerontol Geriatr 2013;32(1):1–23.

[117] Kulkarni SK, Dhir A. Berberine: a plant alkaloid with therapeutic potential for central nervous system disorders. Phytother Res 2010;24:317–24.

[118] Durairajan SS, Liu LF, Lu JH, Chen LL, Yuan Q, et al. Berberine ameliorates beta-amyloid pathology, gliosis, and cognitive impairment in an Alzheimer's disease transgenic mouse model. Neurobiol Aging 2012;33:2903–19.

CHAPTER

11

Differential Diagnosis Toward a Cure for Alzheimer's

It must be remembered that physicians of today are trained to treat the sick, and they must learn how to examine so-called well persons to prevent them from getting sick.

—Dr. Charles Mayo

A NEW DIAGNOSTIC LANGUAGE—RISKS AND CAUSES

As stated by Dr. Joseph Martin (former Dean of the Harvard Medical School), "There is a great need to find a way to prevent Alzheimer's disease, delay its onset, or retard its progress; an effective treatment that delayed its onset by an average of five years would reduce the cost to society by nearly 50% which adds up to hundreds of billions annually and save untold human suffering [1].

Because Alzheimer's disease (AD) has such a long asymptomatic phase, people must be treated when they have mild symptoms of early dementia or even before symptoms appear. Medicine now has inexpensive technologies to diagnose the early signs of AD 20 years before a patient becomes aware of any cognitive problems. This is the best time to treat Alzheimer's. Dr. Charles Mayo (the founder of the world famous Mayo Clinic) told us 100 years ago that: "It must be remembered that physicians of today are trained to treat the sick, and they must learn how to examine so-called well persons to prevent them from getting sick [2]."

Modern medicine cannot accurately diagnose AD, even in those who have severe dementia. This is the consensus of essentially every Alzheimer's pundit. It is generally agreed that definite specific confirmation of AD requires autopsy. Here pathologists are able to differentiate Alzheimer's from other types of dementia. However, in general, there is rather poor correlation between the degree of brain atrophy, including Alzheimer's hallmarks, and the functionality of the patient. If medicine admits they cannot diagnose advanced Alzheimer's, what can standard-of-care medicine do for the early (asymptomatic) diagnosis?

If we admit that diagnosis is difficult or impossible then what? Is it truly important to you, a patient, whether you have any of the many types of dementia or Alzheimer's type dementia? Will treatment vary if you have Lewy Body Dementia, vascular dementia, or AD? No! The treatments are all the same within the standard of care. All these diseases are treated essentially in the same way.

The End of Alzheimer's. http://dx.doi.org/10.1016/B978-0-12-812112-2.00011-2

Why then is there such a focus on diagnosing Alzheimer's type dementia with great accuracy? The answer lies in the treatments targeted for AD, that is antiamyloid therapy. However, as Chapter 2 clearly demonstrates, antiamyloid therapies of all types fail to change the course of Alzheimer's and more than likely make the disease worse.

You as the "Alzheimer's" patient or an individual concerned about your potential to develop Alzheimer's are best served by focusing on your risk for the disease or its acceleration. Assessing risk is best achieved through a differential diagnosis. The concept of a differential diagnosis is really the evaluation of multiple measures across multiple medical disciplines some of which indicate disease risks while others indicate disease causes. When you establish your specific risk factors and causes, now you have something actionable. That is, now you have targets that may change the course of the disease in your favor.

We state that there are two parts to a differential diagnosis, namely risks and causes. This is very empowering to you, a person concerned about or afflicted with a neurodegenerative process like Alzheimer's. With a good understanding of risks, you will be able to "diagnose" Alzheimer's yourself and work with your healthcare professionals to help you stop or minimize the impact of these diseases.

You might ask, is it worth worrying about my risks because there is no (reported) cure for AD? This is a valid question, and the answer is: indeed it is worth worrying about risks because these risks are not Alzheimer's, per se. The risks are associated with causes that manifest in a disease that medicine happens to name Alzheimer's, for example. Causes are preventable or "curable" and when you attack causes, that will help you prevent or stop Alzheimer's.

A recent medical research paper concluded that around half of the AD cases may be attributable to potentially modifiable factors. And that number is based on a short set of risk factors including: diabetes, midlife hypertension, midlife obesity, physical inactivity, smoking, depression, and educational attainment [3]. Here we discuss many more risk factors, some of which are modifiable and others that are not. However, all tolled, the number of AD cases we believe are attributable to potentially modifiable risk factors far exceeds 50%. This means controlling or preventing AD is substantially in your own hands. (Genetic factors play a minimal role in the great majority of Alzheimer's cases; your internal environment is the most important determinant of your potential for AD).

Let's review what we have learned from Chapters 1–10 that will help you understand your Alzheimer's risk and guide you on a path of either prevention or effective treatments.

KEEP ASKING YOUR DOCTOR "WHY," NOT "WHAT"

Do not allow your doctor to give you "ten-cent-word" diagnoses. Many are meaningless and simply mask a lack of understanding as to the cause of your ailment. A classic example is "Essential Hypertension." It is a fancy word that means elevated blood pressure of unknown cause. But you, when you receive such an elegant diagnosis, are convinced your doctor is on top of your condition, after all, you were prescribed a drug. But why is your blood pressure elevated? Is lowering it indiscriminately the right thing to do?

The problem of diagnostic nomenclature extends back probably since medicine was first practiced. Consider this excerpt from a 1908 Journal of the American Medical Association article By Dr. Wiel. [4] We pick up the dialog after the author discusses a variety of meaning-less diagnoses:

> There are many more meaningless diagnoses than these, and it may be that each section of the country has its own particular foibles of this nature, but should not all conscientious physicians discard the use of these terms, and if there are none better to be found in the present terminology, why use terminology at all? It is self-evident that the expression "I don't know" is better than "biliousness" and the rest of the category, and brings some comfort that there is after all much doubt left, in the clearing of which we can find use for our years.
>
> Our distance from Utopia in medicine, as in all things, is vast, and though we shall never attain the ideal, we make one step toward it when we face our ignorance when we find it, and we make still another when we try to overcome it. Rather than call things by false names or meaningless names, let us call them by no names at all, and so, for the love of Æsculapius and Hippocrates, let us hear little more of "biliousness," "ptomaine poisoning" and the like.

Patients in the 21st century seeking medical treatment frequently insist on or even demand a diagnosis to explain their set of symptoms. Once a name has been provided, they can direct their energy to potential treatment to resolve, relieve, or at least lessen their symptoms. Dr. Wiel eloquently stated in 1908, "the indispensible and essential action predicating rational treatment of disease is establishing a diagnosis." He goes on to say some diagnoses are used to "cover our ignorance and to pander to the desire of the patient" seeking an answer.

Today, although technology and diagnostic testing are vastly improved, different factors now affect the diagnostic capabilities of the physician. Busy lifestyles, high patient volumes, hypochondriacal patients, and office schedules which allot a fixed amount of time per patient tempt even the most ethical practitioner to provide a diagnosis not on evidence-based medicine, but rather a nonverifiable opinion designed to pacify a patient desperately seeking an answer.

Physicians have now coined such scientific sounding names, such as fibromyalgia, chronic fatigue syndrome, sick building syndrome, repetitive strain injury, multiple chemical sensitivity, and myofascial pain. Patients desperate for a diagnostic explanation frequently glom onto these names, feel relieved when provided, and seem anxious to share their new "diagnosis" with friends and family. Support groups now are fashionable which reinforce their belief of an unproven diagnosis and give the new "name" further credibility. These groups provide enormous comfort, empathy, reassurance, and embellishment as our patient shares symptoms and stories with similarly affected individuals.

Our current insurance and billing system mandates a diagnosis, even one that is fabricated, to allow appropriate reimbursement for the provider. In the slotted area on the insurance form, the "new diagnosis" is entered and a properly assigned diagnostic code now confirms and validates the diagnosis. No diagnostic code has been yet established to reimburse the physician who writes, "I don't know," or "aches and pains" in the diagnostic slot.

ALZHEIMER'S RISKS—CONVENTIONAL WISDOM

Organizations like the Mayo Clinic, WebMD, and the Alzheimer's Association publish risk factors for Alzheimer's. These conservative organizations rely on medical research studies that show bona fide cause/effect relationships. Many of these risk factors may seem obvious and may not be critically important because they are mostly based on common sense, yet AD rates are increasing exponentially. However, if you truly work to avoid these risks, you will absolutely be less likely to progress to the clinical stage of the disease. Here is a basic list of risk factors:

Age: Okay, this risk factor doesn't help you much. However, many Alzheimer's sufferers outlive the national average, so in some ways, these people were able to maintain good health before AD took over. However, in essentially all cases these older sufferers had subclinical systemic inflammation for years before being afflicted with AD. The way their health decayed prior to the onset of AD is often obvious but is not recognized as a disease or important. For example, sudden weight loss, balance issues, and deterioration in strength is a clear sign of accelerated aging and possibly Alzheimer's. Many of our "pre-Alzheimer's" patients report losing significant distance off their golf drive. Indeed this can be part of normal aging, but when the change is abrupt, it usually is indicative of disease.

Family History: This is often lumped in with genetics. Clearly people in the same family have similar genes, so if family members have the disease, then you may assume that you, too, are predisposed. However, the medical research asserts that the common "environment" among family members is a much stronger influence on your future potential for AD compared to common genes. If you have family members with AD, evaluate their environment (nutrition, exercise, chemical exposure) in comparison to yours and make changes based on consideration of risk factors. In my family (TJL), my dad came down with Alzheimer's at the age of 82. When I review the family environment, fish was clearly missing from our diet. My father all but refused to eat fish except during Lent. All my family members have modified our "environment" to include frequent meals featuring oily, cold-water fishes.

Sex: Women are substantially more likely than men to develop AD. Indeed women live longer, but there is something more profound than just longer life. Hormones and low-fat diets are likely contributions to the difference in AD rates for men and women.

Mild cognitive impairment: People with mild cognitive impairment (MCI) have memory problems or other symptoms of cognitive decline that are worse than might be expected for their age, but not severe enough to be diagnosed as dementia. Those with MCI have an increased risk—but not a certainty—of later developing dementia. Also, people with accelerated memory decay are systemically sick, and many die before the disease progresses into dementia. Taking action to ameliorate risk factors at this stage may help delay or prevent the progression to dementia.

Head trauma: People who've had a severe head trauma or repeated head trauma appear to have a greater risk of AD. What do you do if you have had concussions in the past? Understand and control all the other risks that are within your control.

Heart health: Do you find it interesting that all the Alzheimer's pundits consider heart health a very important risk factor yet modern medicine only focuses on the brain? If you have Alzheimer's, you will only have a neurologist who pays little attention to the condition of your heart. Many of the items listed by the major-medical websites give a list of heart health risk factors that are critical to you controlling or preventing Alzheimer's. One of the most important is homocysteine levels. High homocysteine is highly correlated to Alzheimer's risk. However, the homocysteine-lowering strategy can profoundly impact your outcome. As discussed in Chapter 7, simply lowering homocysteine levels with B-vitamins does not work. Challenge your doctors to find the cause of the elevated homocysteine (do not accept B vitamin deficiency as the answer). The doctor who knows how to modulate your physiology to lower homocysteine will go a long way toward preventing or abating AD. Again, do not disregard the connectivity of diseases. When you take healthy measures that result homocysteine levels going down, many chronic diseases may be avoided.

Heart health warning: Cardiovascular disease management is becoming a very controversial area, particularly for the elderly. How you treat a 55-year-old male with severe cardiovascular disease is quite different compared to a 75-year-old woman with high blood pressure. Clearly exercise, good nutrition, and ceasing smoking will benefit the heart and reduce your odds of AD regardless of your age or sex. However, the Mayo site states high blood pressure, high blood cholesterol, and poorly controlled diabetes as risk factors for Alzheimer's. Remember a very simple truism, your brain is the control center for your body.

Blood pressure: It is often elevated in the elderly due to calcified and stiff vessels. The brain, being up hill from the heart, suffers when arteries can deliver less blood. What does the brain do? It signals for more blood flow the only way it can, by increasing blood pressure. Many studies point to the need for elderly patients to have higher blood pressure compared to healthy middle-aged people. A blood pressure of 150/95 is not a risk for Alzheimer's and instead is probably protective. However, many seniors on blood pressure lowering medications have blood pressure that are too low for the health of their kidneys and brain. Often seniors will take their blood pressure medication at night because it makes them light headed if they take it in the morning. This is a dangerous practice, especially for seniors inclined to make a bathroom run during the night.

Cholesterol: About 25% of all the cholesterol in your body is in your brain. Since the brain is only 2%–3% of the mass of your body, the brain has 10 times as much cholesterol compared to other bodily tissues. Study after study shows that there is essentially no correlation between cholesterol intake and blood cholesterol levels. The liver produces cholesterol in response to signals by the brain. People, especially women, with low cholesterol levels after the age of 70 have much higher mortality rates and more AD. The low-fat, anticholesterol campaign has failed as illustrated by the massive increases in obesity and chronic degenerative diseases prevalent in our society today.

Poorly controlled diabetes (part 1): This piece of conventional wisdom is dangerously wrong. "Controlled" diabetes generally implies keeping blood glucose levels below 170 mg/dL. The ACCORD (Action to Control Cardiovascular Risk in Diabetes) study is the largest multicentered randomized study ever done on diabetes. It clearly shows that diabetics on "tight" glucose control have significantly higher death rates compared to those not "tightly" controlled. Why? The brain, again, tells the story. The brain consumes 20%–25% of all the oxygen we inhale. The brain is 10 times more metabolically active compared to the average tissue in the body. Under tight glucose control, there is less glucose available as an energy supply. Adding to the problem is that diabetics are insulin resistant, meaning their bodies cannot make efficient use of the available glucose and it has difficulty entering the brain. Because the brain is so highly metabolic, it must be allowed to control the output of glucose from the liver so it can obtain the energy it requires to function.

Poorly controlled diabetes (part 2): The ACCORD MIND study is looking at the impact …of glucose control on cognitive function. University of Edinburgh researchers wanted to know why type 2 diabetes patients are at increased risk of cognitive decline [5]. They suspected hypoglycemia (most likely caused by insulin administration [6]) as a "candidate risk factor." They concluded, "The relationship between cognitive impairment and hypoglycemia appeared complex, with severe hypoglycemia associated with both poorer initial cognitive ability and accelerated cognitive decline." Here is our request to medicine: Stop killing people (cardiovascular deaths) and destroying their brains. Find the cause of insulin resistance and treat it. The various medical websites provided very different information on a "hypoglycemia" search. The Mayo Clinic boldly states, "Hypoglycemia is most common among people who take insulin, but it can also occur if you are taking oral diabetes medications." The American Diabetes Association did not make any reference to the connection between taking insulin and hypoglycemia. Why? Do they have your best health interests in mind? A search of "ACCORD" on their site brings you to a page titled, "Mixed Results From ACCORD." Deeply imbedded in the article is how tight control of blood sugar increases mortality. Shouldn't "tight control with insulin kills you faster" be the headline on the American Diabetes Association home page?

Type 2 diabetes is an inflammatory disease that results in elevation of the blood glucose in response to the inflammation. With inflammation, even increased serum insulin levels are not able to overcome the insulin resistance. Higher than normal levels of glucose in the circulatory system is a "last ditch" effort of the brain to protect itself from an energy crisis.

Lifelong learning and social engagement: Studies have found an association between lifelong involvement in mentally and socially stimulating activities and reduced risk of AD. Factors that may reduce your risk of Alzheimer's include:

- higher levels of formal education;
- a stimulating job;
- mentally challenging leisure activities, such as reading, playing games or playing a musical instrument; and
- frequent social interactions.

ALZHEIMER'S RISKS THAT YOU CAN DETERMINE

We believe that AD can be prevented or reversed if the disease is recognized early and proper treatment is started. You personally hold the key to early recognition of Alzheimer's and other dementias in yourself and in those whom you love. Medicine is too hung up on the nuance of diagnoses between pure Alzheimer's and other types of dementias. That differentiation will not help you, but understanding and acting upon Alzheimer's risks that are gaining new appreciation will help you keep Alzheimer's at bay.

Many of the emerging risks associated with AD are actually diseases themselves. Others are biomarkers for degenerative processes associated with diseases. Here is a listing and explanation of some of the most important "risks" associated with the current or future development of AD. After you understand these risk factors, you will be in a position to determine your risk for Alzheimer's more accurately than Alzheimer's specialists who languish in mostly brain-only studies.

Diet

Shockingly, neither the Mayo Clinic nor the Alzheimer's Association list diet as a risk factor for AD. Diet and exercise have the most profound impact on our bodies' immune system. When our bodies are under attack, the immune system is activated, and we note inflammation—the consequence of an activated immune system. Most Alzheimer's experts agree that inflammation is a major part of the disease, yet they are not bridging the gap between nutrition, inflammation, and Alzheimer's.

Fortunately, many studies show the connection between AD and diet. A very significant finding comes from the famous Framingham Heart Study. This is a longest ongoing study of human health extending back to 1948 and is a cooperative project between the National Heart, Lung, and Blood Institute and Boston University. The study has looked beyond the heart because they have gathered systemic data on patients for over 60 years. Here is a significant finding of the Framingham Heart Study regarding AD:

> A recent report from the Framingham Heart Study showed that persons with plasma phosphatidylcholine DHA in the top quartile of values had a significantly (47%) lower risk of developing all-cause dementia than did those in the bottom quartile [7]. Significantly (P = 0.04) greater protection was obtained from consuming 2.9 fish meals per week than from consuming 1.3 fish meals per week [8].

Why does eating fish curb cardiovascular disease and Alzheimer's? Fig. 11.1 sheds light on the matter.

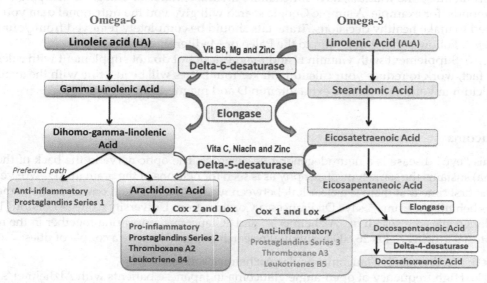

FIGURE 11.1 **How inflammation is downregulated by** $n-3$ **fatty acid.** *From Verma OP. Flaxseed—Miraculous Anti-ageing Divine Food. Available at: http://flaxindia.blogspot.in/2013/04/flaxseed-miraculous-anti-ageing-divine.html*

The authors of the study on the benefits of fish oils ($n-3$ fatty acids) state, "Accumulating evidence in both humans and animal models clearly indicates that a group of very-long-chain polyunsaturated fatty acids, the $n-3$ fatty acids (or omega-3), have distinct and important bioactive properties compared with other groups of fatty acids. $n-3$ Fatty acids are known to reduce many risk factors associated with several diseases, such as cardiovascular diseases, diabetes, and cancer. Modulation of specific genes by $n-3$ fatty acids and cross-talk between these genes are responsible for many effects of $n-3$ fatty acids."

Fish oils have strong antiinflammatory properties for many reasons including boosting your immune system function. If you are interested to learn more about the role of fish oil on inflammation perform a scholar.google.com search for Neuroprotectin D1 and Resolvin. They are a few of the break down products of fish oil that play important roles in the protection of the brain and the eye and in helping to terminate (resolve) the inflammatory process. Alzheimer's and the other major chronic diseases that plague us are diseases of inflammation. We discussed nutrition and supplements as part of Chapter 10. However, it is worth reviewing the key points that are provided here as action steps.

- *Step 1*. Evaluate your diet and replace empty calories with inflammation-suppressing foods like fish.
- *Step 2*. Reduce intake of sugary foods and other foods that contribute to diabetes. The book *Grain Brain* by Dr. David Perlmutter provides appropriate guidance [9]. A simple actionable step is to avoid foods and soft drinks that contain fructose, particularly corn syrup.
- *Step 3*. Avoid low-fat diets. Higher fat diets have more calories per gram but also contain a much higher nutrient content, especially for the brain. Fats that contain more omega-3s and fewer omega-6s are better for brain health. This includes fish of all type and nuts. Certain nuts, like walnuts, have much more favorable ratios of omega-3s compared to almonds, for example. A simple Google search will give you the nutritional data you need to make healthy decisions. Trans fats should be completely removed from your diet.
- *Step 4*. Follow a diet that is akin to the Paleo or Mediterranean diets.
- *Step 5*. Supplement with vitamin D and magnesium but do not supplement with calcium. In fact, work to reduce your calcium intake. Your bones will be just fine with the available calcium in your diet and the extra vitamin D and magnesium.

Glaucoma

This "eye" disease is a neurodegenerative disorder. The optic nerve in the back of the eye (retina) sustains the same type of atrophy as is seen for neurons of the brain in AD. "We've seen for the first time that there is a clear link between what causes AD and one of the basic mechanisms behind glaucoma," says Dr. Francesca Cordeiro from University College London [10]. A scholar.google.com search of "Alzheimer's" and "Glaucoma" appearing together in the medical journal articles yields 26,600 individual research papers. Here are a couple of titles:

2003: McKinnon SJ. Glaucoma: ocular Alzheimer's disease? [11]
2006: High frequency of open-angle glaucoma in Japanese patients with Alzheimer's disease. [12]

2012: Glaucoma and Alzheimer disease: age-related neurodegenerative diseases with shared mechanisms? [13]

Glaucoma is clearly more than a disease the eye and brain. The medical review article titled, *A sick eye in a sick body? Systemic findings in patients with primary open-angle glaucoma* demonstrates that glaucoma is a whole-body disease [14]. Eye diseases seldom occur in isolation from the rest of the body. If the body is sick in glaucoma and glaucoma and Alzheimer's have overlapping causes, then it only makes sense that Alzheimer's is a disease of the whole body that manifests in the brain.

Glaucoma is an early warning sign for future AD. There may be cases in which Alzheimer's precedes glaucoma, the diseases occur at the same time, or AD occurs after glaucoma develops. However, an optometrist or ophthalmologist can look for glaucoma routinely, noninvasively, and at a low cost. Even if glaucoma and Alzheimer's develop at the same time, you cannot "see" Alzheimer's in the brain but you can see glaucoma in the eye. Recall from previous chapters that Alzheimer's incubates and develops over decades. By the time Alzheimer's is clinically noticeable in a patient, their brain has suffered significant atrophy. Watching out for the earliest stages of glaucoma is a very good way to evaluate your risk for Alzheimer's in the future. What do you do if you are diagnosed with glaucoma and are concerned about the future possibility of Alzheimer's?

- *Step 1.* Reread Chapters 6–10 and continue reading this chapter so you have the power of knowledge to help you find (and guide) a doctor who can help you.
- *Step 2.* Find that doctor who regards glaucoma as a systemic disorder and not an eye-only disease.
- *Step 3.* Institute measures outlined throughout this book to support immune system health and carefully follow the treatment regiment prescribed by your (new?) doctor who understands the root causes of glaucoma (and Alzheimer's).

> If your doctor investigates and treats glaucoma properly, as a systemic inflammatory disorder, you are on the right path to either prevent or treat Alzheimer's.
>
> —*Clement L. Trempe, MD*

Macular Degeneration

This eye disease is also a systemic chronic inflammatory neurodegenerative disease related to aging that is proceeded by 10–20 years by mild elevation of blood markers of inflammation. The causes of inflammation and disease are the exact same factors that lead to Alzheimer's, cardiovascular disease, and type 2 diabetes. Thus macular degeneration is, and a biomarker of, both neurodegeneration and cardiovascular disease. Dr. Don Anderson of the University of California has spent the better part of 25 years characterizing deposits in the retina called drusen that often form early in macular disease. These deposits contain the same misfolded protein associated with AD. His work culminated in many research papers and this press release from UC Santa Barbara, "Scientists at the Center for the Study of Macular Degeneration at the Neuroscience Research Institute of the University of California, Santa Barbara have found a link between the brain plaques that form in AD and the deposits in the retina that are associated with age-related macular degeneration (AMD) [15]."

Recently the work of Frost has confirmed and extended the work of Dr. Anderson [16]. Frost has developed a method to highlight the amyloid deposits in drusen and make detection of these formations in the eye clearer. However, any eye doctor with standard optometric equipment is capable of detecting drusen formations very early along the disease process. Identifying the earliest stages of macular degeneration, by way of drusen, is arguably the earliest way to assess future risk of AD.

As with glaucoma, macular degeneration is not an "eye-only" disease. Your steps to protect yourself from Alzheimer's using macular degeneration as a biomarker/diagnostic are the same as for glaucoma.

- *Step 1.* Find a doctor who treats macular degeneration as a systemic disorder. The current eye injections of inhibitors of the vascular endothelial growth factor will not help with Alzheimer's. In fact they are much more likely to accelerate and worsen the condition because it is known that these eye injections cause patients to have serious systemic "adverse events" in up to 40% of patients treated within two years. "Adverse events" is a convenient medical term that means more heart attacks and stroke. The intraocular injection of VEGF inhibiting drugs (the drugs for macular degeneration and some other eye diseases of the retina) cause a significant decrease in the circulating level of VEGF for up to one month. VEGF is essential for the rejuvenation of all the vascular endothelium (blood vessel linings) in the body.
- *Step 2.* Reread Chapters 6–10 to gain a better understanding of the causes of drusen formation, macular degeneration, and AD. Armed with this information, you may be able to find a doctor willing and able to help you prevent and battle AD.
- *Step 3.* Institute measures outlined throughout this book to support immune system health and carefully follow the treatment regiment prescribed by your doctor who understands the root causes of macular degeneration (and Alzheimer's).

Cortical Cataract

Harvard Medical School elegantly showed that deceased Alzheimer's patients they studied had beta-amyloid plaques in their brains [17]. They also showed that these same patients had "supranuclear" cataracts—also known as cortical cataracts. Their work showed that the amount of beta-amyloid in the brain corresponded with the amount found in the eye, in the cortical cataract. These cortical cataracts generally form long before a patient has clinical AD. In medicine today, cortical cataract are not removed at early stages and usually just observed. The reason is these structures form on the edge of the lens where lens stem cells are most prevalent, and have very little impact on vision for years. Few doctors recognize these formations for their rich diagnostic information.

Cortical cataracts are also an early warning sign for Alzheimer's, but what is an early warning sign for cataracts? They do not just suddenly appear but rather form over time. The process looks something like this:

First the top cortical layer of the lens periphery starts on increase in thickness before any of the beta amyloid fibrils (the actual cataract) becomes visible. This swelling produces accurately measurable optical changes, an effect known as coma aberration. At this stage the cortex (edge) of the lens has a milky appearance probably caused by accumulation of beta amyloid precursor (monomers and oligomers). Eventually the milky deposit coalesces and form clearly visible fibrils called cortical cataracts that Harvard Medical School showed contains the same beta amyloid as is found in Alzheimer's brains.

Drusen found in the retina is a very early indicator of AD. However, coma aberration attributable to lens swelling may appear even earlier. The challenge for you, the patient, is to find an optometrist or ophthalmologist willing to investigate you, who most likely is otherwise healthy, for these Alzheimer's biomarkers. And, once identified, the greater challenge is to find a doctor who understands the fundamental processes that lead to the deposition of these misfolded proteins (cortical cataracts and their precursors).

- *Step 1*. Find an optometrist who specialized in "pathology." Many optometrists will claim they understand eye pathology and indeed they do. However, few include looking at eye pathology in their routine practice. These eye doctors and their technicians are skilled at measuring pathologies like cataract, glaucoma, macular degeneration, dry eye, and retinal nerve physiology.
- *Step 2*. If you have evidence of early cortical cataract formation or any evidence of early AMD or glaucoma, you must find a doctor who appreciates that those eye diseases are early visual biomarkers for chronic systemic inflammatory disease process related to accelerated aging. They are not simply eye diseases as the great majority of eye doctor view and treat these conditions. The doctor you find must be interested and willing to do a thorough appropriate medical history and appropriate blood tests to get at the root of your disease.
- *Step 3*. Appreciate and try to institute measures outlined throughout this book to support immune system health and carefully follow the treatment regiment prescribed by your doctor who understands the root causes of cortical cataracts (and Alzheimer's).

All chronic diseases, including cortical cataract formation and the diseases with which it is associated are multifactorial. That is, more than one thing is contributing to the over activation of the normal chronic systemic inflammatory process associated with aging. The tricky part of managing chronic diseases properly is that inflammation is part of the normal aging process. As our immune system slowly deteriorates (Immunosenescence) our body need to adapt by using the inflammatory process to protect us from disease. Your goal is to locate a doctor who will take the time to find out if you have an over activated immune system and work with you to control the contributing factors. Those factors that contribute to disease are usually different from one person to another and the way that patients adhere to and respond to treatment is also very different from one person to another. The modern 10 min office visit of the standard-of-care does not provide adequate time for your doctor to personalize your diagnosis and treatment.

Retinal Nerve Fiber Layer Thinning

Retinal nerve fiber layer (RNFL) thinning is early concrete evidence that you could have glaucoma especially if the trend continues. Measurement of RNFL thinning is mainly used to study the progression of glaucoma. Multiple sclerosis (MS) is another neurodegenerative disease, usually diagnosed in younger individuals, which is often correlated to RNFL thinning. Chapter 6 delved into the RNFL thinning/Alzheimer's connection. Many studies show a strong correlation between RNFL thinning/atrophy (glaucoma) and brain atrophy as measured using MRI. Measuring RNFL thickness is an important way to measure and study a person's neurodegenerative disease process. The equipment used to measure this parameter, optical coherence tomography (OCT) is an instrument similar to a CAT scan except that a weak beam of light is use instead of an X-ray beam. These instruments are extremely accurate and precise. Thus, while the eye diseases discussed earlier are excellent ways to evaluate your AD risk, OCT measurements of the RNFL provide a more objective way to follow the course of the disease. A recent research article corroborated this statement as follows, "In our study, the thickness of RNFL in patient with AD was lower than that of controls. This suggests that OCT has the potential to be used in the early diagnosis of AD as well as in the study of therapeutic agents [18]."

How do you go about determining your RNFL thickness and if a degenerative disease process impacts it?

- *Step 1.* Find an optometrist or ophthalmologist who specialized in "pathology." Many optometrists will claim they understand eye pathology and indeed they do. However, few include looking at eye pathology in their routine practice. These eye doctors and their technicians are skilled at measuring pathologies like cataract, glaucoma, macular degeneration, dry eye, and retinal nerve physiology.
- *Step 2.* If you have RNFL thinning or atrophy you must find a doctor who appreciates that this is a systemic disorder and is willing to do appropriate system-wide tests (blood tests for example) to get at the root of your disease.
- *Step 3.* Institute measures outlined throughout this book to support immune system health and carefully follow the treatment regiment prescribed by your (new?) doctor who understands the root causes of RNFL thinning and/or atrophy (and Alzheimer's).

Cardiovascular Disease

A recent article in the top medical journal, the New England Journal of Medicine confirms 25 years of effort by Jack C. de la Torre. Dr. de la Torre's work was highlighted in Chapter 8. The New England Journal of Medicine article is titled, "Proteotoxicity and cardiac dysfunction – Alzheimer's disease of the heart?" [19] There is sufficient evidence linking Alzheimer's and other dementias to inflammation and vascular problems. The challenge you will have, if you want to effectively be treated for cardiovascular disease to prevent future Alzheimer's, is that the current cardiovascular treatments do not delve into true root causes. In fact, some of the major treatments for cardiovascular disease including statins and blood pressure lowering drugs may actually exacerbate AD and other dementias.

AD desperately needs new science and new approaches to control and reverse the disease. Unfortunately, cardiovascular disease also needs new approaches. The cholesterol management approach is not working. Statins are presumed to be miracle drugs, but other cardiovascular drugs are actually much more important for controlling (not curing) the disease. These include beta blockers, ACE inhibitors, and Angiotensin receptor blockers. These latter drugs do appear to have positive impacts on Alzheimer's while statins do not (despite desperate attempts by that industry to prove otherwise).

Here is an action plan for those of you with diagnosed cardiovascular disease who are now concerned about your potential to develop AD.

- *Step 1.* Reread Chapter 8 to fully understand the AD/cardiovascular disease connection.
- *Step 2.* Find a doctor who will measure your inflammation burden and will also check for intracellular infection and other causes of elevated inflammation.
- *Step 3.* Find a doctor who knows how to treat the causes of inflammation from the root. Remember, inflammation is (generally) a treasure. It is your immune system working on behalf of your good health. Appropriate diagnostics delve into the causes of inflammation, and appropriate treatment attacks these causes.

Diabetes

We need to go no further than the article titled, *Alzheimer's Disease is Type 3 Diabetes – Evidence Reviewed.* [20] The authors emphatically conclude, "the term 'type 3 diabetes' accurately reflects the fact that AD represents a form of diabetes that selectively involves the brain and has molecular and biochemical features that overlap with both type 1 diabetes mellitus and type 2 diabetes mellitus." A study on type 2 diabetes as a risk factor for Alzheimer's suggests the process is connected to "cerebrovascular pathology." That is, the vascular disease caused by diabetes is a major contributor to AD. [21] This makes sense because diabetes and vascular disease are interwoven as is AD and vascular disease. The take-home lesson for you, concerned about future Alzheimer's, is that systemic sugar metabolism dysfunction is a severe risk factor for Alzheimer's.

The good news is that you now know that managing diabetes will help curb Alzheimer's. However, we already cited the ACCORD study that showed how tight control of sugar worsens outcomes. People on tight glucose control are sicker and die sooner. Also, the ACCORD MIND study shows that hypoglycemia (mostly caused by insulin treatment in diabetics) contributes to and accelerates cognitive decline. What are your action items for preventing or treating Alzheimer's if you have diabetes?

- *Step 1.* Recognize the link between diabetes and cardiovascular disease. The point of intersection is inflammation; therefore, review Chapter 7 on inflammation.
- *Step 2.* Locate a doctor who does not use the old, tired, and deadly way for managing diabetes—that is, controlling glucose levels with insulin injections. This method treats symptoms, not causes. The causes of type 2 diabetes (and that of cardiovascular and AD) are those factors that increase your inflammation burden. The key is your environment, particularly your diet that can either promote or mitigate inflammation. The second is infection that increases the inflammatory burden associated with aging above normal levels. Pathogens proliferate opportunistically when our immune system weakens. A good way to assess your susceptibility is to monitor your baseline inflammation. You will achieve better health if your baseline levels are lowered consistently, over time.

Periodontal Disease

Bruce Paster is the Senior Member of the Staff and Chair of the Department of Microbiology at the Forsyth Institute, part of the Harvard Medical School. He told us unequivocally that everyone who is 55 years or older has some level of periodontal infection. What about a connection between gum disease and Alzheimer's? The medical literature is loaded with research confirming the connection. Here is one of thousands, *Emerging evidence for associations between periodontitis and the development of Alzheimer's disease* [22] The root of the AD/gum disease connection is inflammation as discussed in detail in the following journal article, *The Microbiome and Disease: Reviewing the Links between the Oral Microbiome, Aging, and Alzheimer's Disease* [23].

There is more to oral health than just brushing and flossing regularly. Cosmetic and functional surgery and implants that consider only the health of the mouth often have severe system-wide health consequences.

Root Canals: Each of your teeth is a vital (living) organ. Like your liver, lungs, kidneys, and other organs, teeth sometimes become diseased and fail. In that case, a dentist may recommend root canal treatment, which starts with removing the dentinal pulpal complex—the tooth's "guts," so to speak, which are rich in nerves, blood vessels, and delicate connective tissue. After cleaning the chamber, the dentist fills it with rubbery putty and caps the remaining tooth structure with a restoration. Establishment dentists call this "saving" the tooth, which is odd since, by definition, a root canal tooth is a dead tooth. All the living stuff—what once kept it alive—is gone. But this doesn't mean that nothing is happening inside. The absence of the pulp has major consequences. Because there is no blood supply, pathogenic bacteria that were formerly held in check by your immune system multiply. Due to the oral cavities' proximity to the blood supply and the central nervous system (CNS), the bacteria can invade both systems, with health damaging consequences. For a comprehensive yet understandable explanation of the impact of oral bacteria, please read *Scientific American Presents: Oral and Whole Body Health* [24].

Dental Implants: Implants are a foreign body in your mouth, despite the best efforts of the dental community to use substances that the body accepts. Also, as in the case of root canals, the implant is not living and thus is not graced with blood flow and protective immunity the blood supplies. These foreign bodies harbor bacteria and proliferate inflammation. If you are not a believer, try this simple experiment. Place a small splinter under your skin. You will undoubtedly note that the area will soon become inflamed. Why? It is becoming infected, and your body's immune system is going to work for you. Your immune system will likely lose the battle unless the splinter is removed. Even antibiotics will not permanently relieve your symptoms of inflammation, unless the source (the splinter) is removed.

Proper oral hygiene is much more than preventative dentistry—it is preventative disease! One of the diseases that proper dental care can help prevent is Alzheimer's. Here are the steps:

- *Step 1.* Brush after every meal.
- *Step 2.* Floss after every meal.
- *Step 3.* See your dentist at least 2 times a year but preferably 4 times, regardless of your age.
- *Step 4.* Do not obtain root canals or dental implants.

- *Step 5.* If you have periodontal disease, have it treated aggressively.
- *Step 6.* Remove teeth that are infected immediately.
- *Step 7.* Remove dental implants and teeth subject to root canal surgery.

An article in *Business Insider* states, "We've mentioned before that not flossing daily can shave years off your lifespan. Here's more proof that poor dental hygiene—which allows harmful bacteria to enter the bloodstream—leads to a host of medical problems, including heart disease, respiratory problems, and diabetes." [25] They are missing the connection to Alzheimer's, but diabetes and heart disease should be all the motivation you need to take action.

The connection between oral hygiene, periodontal disease, and chronic disease is not new. The doctor who tried to bring this concept to prominence was Charles Mayo, the founder of the Mayo Clinic. Here is an excerpt about Dr. Mayo's concept on oral and whole body disease. [26]

Dr. Charles Mayo, founder of the famous Mayo Clinic, believed in the 'focal infection' theory of disease, something so archaic that today almost no one has heard of it. The theory basically states that an oral infection can influence the health of the entire body. Addressing the Chicago Dental Society in 1913, Mayo said, 'The next great step in preventative medicine must come from the dentists.'

Mayo appointed Dr. Edward C. Rosenhow to head a team of researchers dedicated to focal infection theory. From 1902 to 1958, Rosenhow conducted experiments and published more than 300 papers, 38 of which appeared in the *Journal of the American Medical Association*. During the same period, Weston A. Price, founder of the research institute of the National Dental Association, published his findings indicating that dental and oral infections were often the primary cause of disease.

These two medical pioneers established a simple but profound fact. If you pull an infected tooth, the patient will often recover from disease—serious disease, from chronic fatigue to cancer, from dermatitis to diabetes, from hemorrhoids to heart disease. Drs. Rosenhow and Price theorized that disease often originated from infections in the mouth that entered the bloodstream and eventually caused major problems in some part of the body. The evidence they amassed and published is staggering, yet the next great step Dr. Mayo hoped for did not come, and their work is largely forgotten today.

Dr. Mayo brought a couple of brilliant ideas to light, none of which are currently being practiced by the clinic that bears his name, or any other major medical institution for that matter. One of those ideas is the concept of focal infection discussed above and the other is "grand rounds." In grand rounds, all the specialists in medicine come together to evaluate the patient and bounce ideas off each other. Today, specialists see patients in isolation of other specialists. Your PCP, who is supposedly managing your care, has neither the time nor expertise to supplant the diagnostic and treatment power of the grand rounds concept. Why have grand rounds gone out of favor? If we had to guess, it is the influence of the drug companies who have developed specific drugs for each of the myriad of medical specialties. Many of these drugs would fall into obscurity in the grand rounds environment.

The concept of grand rounds was discussed in a recent *Medpage Today* article [27]. The topic was not Alzheimer's; however, the findings about the need for grand rounds to help avert serious disease hold true for all the major chronic diseases. The authors provide the following insights. "It takes a village to successfully treat and manage patients... but those villages are few and far between." "About 84% of doctors believe that coordinated care among healthcare professionals is important but only 33% believe that coordinated care is currently adequate in their respective countries." "About 60% of the doctors in France said there were adequate professionals outside the hospital setting to help coordinate care. Among physicians in other countries, however, only about 30% to 40% said coordinated outpatient care was feasible. The United States came in at the low end, with about 27% of the surveyed doctors indicating there were adequate settings for coordinated care in their country."

Other Alzheimer's Predetermining Factors

The items mentioned earlier provide you with a guide to evaluate your future risk of AD. There is vast and compelling evidence to support a strong link between any of those items and Alzheimer's. The items listed here are also important for your Alzheimer's risk self-diagnosis but probably have many confounding health elements compared to those discussed earlier.

Chronic stomach problems: Any stomach problems or family history of stomach problems could imply *helicobacter pylori* infection. This bacterium is considered both synergistic and pathogenic and, the latter "condition" is linked to AD. Helicobacter pylori infection is a cause of stomach ulcers, and the doctors who made that discovery were awarded the Nobel Prize in Medicine in 2005.

History of positive tuberculosis skin test or night sweats: Tuberculosis is making a comeback and people positive for tuberculosis have more AD. It may not be the tuberculosis itself that is causing AD, but active tuberculosis is an indication that your immune system may be failing.

History of or current Lyme disease: This is very similar to tuberculosis. It may not be a cause but an indication of your immune system health.

History of leg cramps: This may infer a magnesium deficiency.

Raw or very rare meat: Toxoplasmosis is an infectious species obtained from undercooked meat. Toxo is a cause of a number of neuropsychiatric disorders and may be a contributor to AD.

Weight loss: If you are a senior, have you experienced sudden weight loss not attributable to a diet program? "Epidemiologic studies have shown that weight loss is commonly associated with AD and is a manifestation of the disease itself. The etiology of weight loss in AD appears multifactorial. Hypotheses to explain the weight loss have been suggested (e.g., atrophy of the mesial temporal cortex, biological disturbances, and higher energy expenditure); however, none have been proven [28]."

Trips and falls: According to a WebMD article, "Falls may be an early sign of AD, researchers report. In a study of 125 older adults who appeared physically and cognitively healthy, two-thirds of those with large deposits of Alzheimer's-associated plaque in their brains suffered falls. In contrast, only one-third of those with little or no plaque experienced falls [29]."

Function: Changes in memory and general functioning is a late sign of pending and advancing dementia. The specific issue for which to watch include: forgetfulness; challenges in planning or solving problems; difficulty completing familiar tasks at home; confusion about time and/or place; being challenged by special relationships; language problems; impaired decision making, withdrawal from social activities; and mood and personality changes.

Sudden loss of physical strength: This condition may be a result of lost brain control or due to an overlapping disease called Inclusion Body Myositis that could be considered AD of the muscle.

DIAGNOSING AND TREATING THE "WELL" TO PREVENT DISEASE

Dr. Mayo commented over 100 years ago about the need to treat the "well" to prevent them from becoming ill. Little has changed in the investigation and treatment of so-called well persons to prevent them from getting sick. This is true for all the chronic inflammatory diseases related to aging including Alzheimer's. The statistics bear out the need for "well person" treatment as these chronic diseases are increasing exponentially. If you wait for your doctor to make a diagnosis of (chronic) disease, it will be far more advanced and much more difficult, if not impossible, to control.

The problem a doctor faces in treating the "well" is he or she needs a diagnosis to get paid and be justified to order blood tests. Let's consider the case of a homocysteine test. You read, in Chapter 7, that elevated homocysteine is highly correlated to the future risk of Alzheimer's. The Mayo Clinic website states that "Elevated Homocysteine" is an Alzheimer's risk factor [30]. Please let us know what happens if you are a "well person" and you go to the Mayo Clinic and ask for a homocysteine blood test because of your concern about future Alzheimer's. Even if you already have a diagnosis for a disease that predisposes you to develop AD in the next 10 years, such as AMD, glaucoma, type 2 diabetes, or cardiovascular disease, your doctor is often not permitted (or at least, you will not be reimbursed for the test) by these diagnoses to order blood tests or other diagnostic procedures, such as MRI or special tests of the eye to document the early changes related to AD. You may get these tests if you insist and are willing to pay cash.

Many articles are written about unnecessary tests and costs associated with diagnosing and treating "healthy" people. But our society is overwhelmed with slowly developing chronic diseases. We must find cost-effective ways to examine the so-called "well person." Proof of the need to better manage the health of the well comes from the medical literature and is summarized in the *New York Times* [31].

> After a routine test of her blood sugar eight years ago, Randi Sue Baker, a seriously overweight 64-year-old, learned that type 2 diabetes was bearing down.
>
> With that test result, she joined the 79 million Americans over the age of 20 who have prediabetes. Up to 70 percent of them will go on to develop diabetes, but 90 percent don't even know they are at risk. In fact, as many as 28 percent of adults with full-blown diabetes don't know they have it, according to Edward W. Gregg, a senior epidemiologist at the Centers for Disease Control and Prevention.

In this *New York Times* article, the author Jane Brody goes on to explain that diabetes accounted for $245 billion in health care expenses in 2012, about one in six health care dollars. You now know that people with this condition often progress into Alzheimer's, if they survive diabetes.

The main goal of the examination of an apparently well person is to eliminate the probability of future chronic diseases. This patient is in good health with no complaints (no diagnosis).

The very early diagnosis of AD is often very simple. Hundreds of millions of research dollars are spent to find early biomarker AD before it becomes symptomatic and very challenging to treat. And, as Chapter 2 illuminated, more than 200 studies have been performed on early dementia and have failed miserably.

Does it make more sense to spend that research money in the real world—that is, on people like you? Wouldn't we curb much more disease by screening our population based on known risk factors, biomarkers, and overlapping disease? A homocysteine test is relatively inexpensive and, as explained in Chapter 7, may be the single most important measure of your future chronic disease health. What would the return on investment be for a homocysteine test (in combination with other simple and highly correlated tests that are already known)? Compare this to the Pittsburgh dye [32] that is toxic, followed by an expensive and nerve-racking PET scan? Priceless.

DIFFERENTIAL DIAGNOSIS FOR THE WELL AND THE AFFLICTED

> You never change things by fighting the existing reality. To change something, build a new model that makes the existing model obsolete.
>
> —*Richard Buckminster Fuller*

The steps to brain health, Alzheimer's prevention, and an understanding of the cause(s) of AD for the purposes of effective treatment include six components:

1. Screening
2. Baseline measurement
3. Comprehensive assessment
4. In depth root cause(s) analysis
5. Intervention(s)
6. Repeat steps 1–5 periodically to measure the efficacy of the treatment on the patient and to make adjustments to improve the treatment protocol and clinical results.

In dentistry, these same steps are followed. The dentist and hygienist screen your oral health with simple methods (visually). They acquire baseline information with X-rays and closely note recession, gum pockets, and other markers of oral health and disease. A comprehensive assessment is conducted, which may involve a specialist looking at cavities and extraction needs. Occasionally, your dentist may look into a more in-depth root cause analysis, in the case of, for example, periodontal disease, and refer you to a specialist for evaluation and treatment. Interventions include action items determined from the previous steps. Of course, we all understand the concept of preventative dentistry, and many of us return to the dentist on a regular basis to keep ahead of oral disease and decay.

The steps 1–5 are a logical progression of activities that one should consider, regardless of health or age. A baseline assessment is a fairly broad evaluation. However, for AD, what is proposed here is far different compared to a routine physical performed within the standard-of-care. The eye-screening portion of this diagnostic protocol is far more predictive of your future chance of having a serious health problem than what any doctor can determine through a routine physical examination. Chapter 6 contains the evidence. Why our major health pundits, including the National Institutes of Health (who funded some of the research on the eye) are keeping this information from doctors and the public is only a guess.

Before embarking on the journey through the subsequent pages on a differential diagnosis toward a treatment let's briefly consider what tests provide the earliest diagnosis of a person with Alzheimer's or other chronic diseases. First, no single test or single measurement of that test is every truly adequate to characterize disease. Medicine tries to accommodate our busy lifestyles and the need for profit by creating science around a single test like cholesterol. If you ignore the hype, that type of approach is an abject failure. Thus how can one be sure of a diagnosis? We discussed this in Chapter 3 and summarize the ideas here.

Demonstrated reproducibility or trends: repeat tests at defined intervals. The rule of thumb scientists use in evaluating laboratory data is to obtain three tests and then use the standard deviation to determine the statistically significant value. In medicine, tests are expensive, and often blood needs to be drawn. Thus it would be impractical to draw three vials of blood at the same time and send the samples to three different laboratories.

Since we are mainly concerned with chronic diseases with long incubation times, it makes sense to get the "key" tests done regularly, or at least annually. Increase the frequency to quarterly for any tests that show slightly abnormal values. Certain tests for chronic diseases also are markers for acute events, such as trauma. An example is C-reactive protein (CRP). If your CRP is high, it does not infer chronic disease. However, if you retest in 1–3 months and it is still high, then there may be something brewing in your body. A third test in another 1–3 months will provide the certainty needed to determine if you are "silently" ill.

Get a range of tests: Alzheimer's researchers are desperately seeking the definitive test for beta-amyloid plaques in sufferer's brain. However, science we presented in Chapter 2 shows that some Alzheimer's patients are free of beta-amyloid while some healthy people have burdens of amyloid that researchers thought were sufficient for the subjects to have disease. Having a single test, like cholesterol, is expedient and profitable, but it does not weave a story about your health. Alzheimer's is a story, not a moment in time.

Which tests provide early disease detection? The tests that provide the earliest indication of disease generally go in the following order, from earliest to latest.

- **Blood tests:** When we are sick, our immune system reacts almost immediately. The adaptive immune system is slow, taking about 5 days to respond to new threats. In acute (immediate) disease, blood markers of immune activity relax back to normal levels once the illness passes. In chronic disease, inflammatory markers elevate and, although they may fluctuate, they tend to stay higher than normal baseline levels.
- **Eye tests:** The eye is a tremendous biomarker for diseases of itself but also for diseases of the whole body. Our bodies are robust and often overcome illnesses, particularly low-grade illnesses. When left unchecked, the causes of these diseases can proliferate, leading to slow degradation of tissue. The eye is an early biomarker for a couple of reasons. First, it is transparent, and disease development at its earliest stages is seen optically. The lens at the front of the eye is a magnifying "glass" that helps doctors see disease markers developing in the back of the eye. Second, the eye is less prone to develop an autoimmune reaction to mild antigenic stimulation. The eye is "immune privileged" and an outstanding witness to what is happening in the rest of our body.
- **Brain test:** We are blessed with brains that build significant redundancies as we develop and grow. Noticeable or even measurable loss of brain function occurs only after

significant disease and atrophy. Another exacerbating feature of brain measurement is the integrity of the protective skull and blood/brain barrier that somewhat isolates the brain.

Summary: Although the eye is a fantastic marker for Alzheimer's and neurodegenerative diseases in general, if you want to be diagnosed as early as possible, when you have the best chance of preventing or stopping the disease, obtain blood tests. Some of the right tests are listed in the section titled, "Baseline Screening" (later).

Routine Screening

Routine screening, by definition, should be conducted at regular intervals to look for initial signs and symptoms of disease, preferably before clinical symptoms start to appear. This screening involves the evaluation of possible and absolute biomarkers that are either known or thought to precede disease that we have already covered in this chapter. By design, routine screening is as noninvasive, cost-effective, and expedient as possible. Medical technicians and other lower-level healthcare members may conduct routine screening. Modern standard-of-care does not perform routine screening despite claims to the contrary. The physical exam is an outdated test still used by modern medicine, and does little to diagnose chronic diseases like AD at their early stages.

The standard physical exam has hardly changed during the past 100 years. It is not the process of a preventative check-up that is failing us; rather, it is the tests that are offered. These tests are oblivious to important targets for disease. And the real problem is that only the limited tests offered are covered by health insurance.

> The experimenter who does not know what he is looking for will not understand what he finds.
> —*Claude Bernard*

A strong case exists for the connection between the eye, the brain, the body, morbidity (disease), and mortality (early death). This is very true for biomarkers and other indicators of AD. A routine screening, which is very efficient, cost-effective, and noninvasive, may therefore start and end with a comprehensive eye exam as discussed in Chapter 6. The exam must focus on: specific markers for general systemic diseases; specific markers for cardiovascular diseases; formations associated with neurodegeneration; and the hallmarks of AD including beta-amyloid formations and neurofibrillary tangles.

With the eye exam, doctors are able to search for the two major hallmarks of AD and indicators of other "comorbid" diseases, such as cardiovascular disease and diabetes. The key tests are:

- Slit lamp microscopy for lens evaluation,
- Fundus camera for retina evaluation,
- Optical coherence tomography (OCT) to analyze the retinal nerve fiber layer,
- Scanning Laser Polarimeter (SLP) to measure the retinal nerve fiber layer and microtubule health, and
- VER/VEP to measure brain signal strength and response.

What we are recommending here and what you will get from a routine exam with your eye doctor are not quite the same. As part of every routine eye examination your eye doctor will measure your visual acuity, test your eye movements, and measure your eye pressure. He or she will likely use the following optical instrument to examine your eyes, but only for eye evaluation purpose and not for preventative screening:

Slit lamp microscope: This device looks at the front part (anterior segment) of the eye, particularly the lens and the cornea to detect chronic and aging changes. Nuclear cataracts are readily seen and are known to correlate with the presence of cardiovascular disease. Chapter 6 discusses the Age-Related Eye Disease Study that shows how visual impairment, caused by cataract formation, is a predictor or early mortality due to some type of vascular disease. Cortical cataracts are also easily detected and are highly correlated to AD. Evidence of this is provided by a Harvard Medical School study. Pseudo-exfoliation is believed to be the early formations of cortical cataracts, thus the earliest potential indicator for future AD. Pseudo-exfoliation has been considered an "amyloid" material since 1973.

The American Optometric Association (and others) publishes a grading scale for cataracts that correlates with disease. A slit lamp with an attached camera can be employed to accurately track the progression or regression of cataract. The Oculus Pentacam, a tomography instrument, can be used to augment the slit lamp information as it performs very precise and accurate measurements of lens structures. However, most optometric and ophthalmic offices do not have this device. The slit lamp microscope is adequate for the initial screening phase.

Fundus camera and ophthalmoscope: These are used to examine the back of the eye (posterior chamber or "Fundus"). The Fundus camera is capable of recording detailed photographs of the retina, the retinal blood vessels, the retinal pigment epithelium under the transparent retina, and the optic nerve. Neurodegenerative disease is inferred by a pale or cupping optic nerve. Drusen is an amyloid formation and is indicative of a latent systemic and neurodegenerative disease process. In addition, the Fundus enables the observation of the integrity of the microvessels in the back of the eye. Quantifiable irregularities noted are indicative of local and systemic inflammation and cardiovascular disease. This system is also useful for the detection and the measurement of diabetic retinopathy. The Optos version of the fundus camera obtains a wide-angle photo of the entire posterior and mid periphery of the eye in a single photo. This instrument allows the biomarkers and eye features to be qualified and quantified to accurately follow disease. Finally, glaucoma is observed via the fundus camera and other means. Many recent studies indicate that glaucoma is AD of the eye and that AD is glaucoma of the brain.

Other instruments that should be used for preventative health screening purposes, but are seldom part of a routine eye examination include:

Optical coherence tomography (OCT): This instrument is used to measure macular edema and macular volume, which is related to brain volume and thus brain neurons, and provides information similar to a brain MRI. Critical is the measurement of the nerve fiber layer. The layers in the RNFL connect and correspond to different areas in the brain. There is a strong correlation between the RNFL and glia, the white matter in the brain that is a measure of neurotangle density (intelligence). There is also a strong correlation between reduced macula volume and MS in young people and AD in older people. A weaker but emerging connection exists between Parkinson's disease (PD) and the RNFL.

Macular degeneration is also observed with OCT. OCT is useful for the detection and measurement of diabetic retinopathy. It provides an objective digital measure of many parameters that can be quantified and tracked to evaluate patient whole-body health and changes to their health status.

Scanning laser polarimeter (SLP): The SLP looks beyond thickness and density of the RNFL (measured by OCT) to underlying structural organization that is key to RNFL Integrity. This system studies microtubules of the axons by measuring the polarization of light. A decrease in polarization infers unhealthy axons, thus a neurodegenerative process. Recent studies suggest that RNFL microstructures undergo changes in orientation and density before RNFL anatomical thickness changes become apparent. The SLP measurement depends on both RNFL thickness and the cumulative level of organization of its microstructures. Alteration of these microstructures is potentially tied to the two hallmarks of AD, beta-amyloid and neurofibrillary tangles. The latter is caused by phosphorylation of tau protein. This is quickly becoming the "new" therapeutic target for AD within pharmaceutical companies.

These tests require approximately 10 minutes of clinic time each. Some of these tests require eye dilation. All these tests produce digital data that can be compared against data obtained at each subsequent visit. Thus, any changes in underlying health status may be accurately tracked. Also, standards are well-known, so these tests provide a good assessment of the type and the extent of any deterioration or disease as compared to a large population of age-mates.

For the purposes of routine screening, a cognitive functioning test (the MMSE or equivalent, see Chapter 3) should be recorded to establish a baseline for an individual. These tests, although reported by some researcher as useful for early detection, are clearly late-stage tests. They need to be performed because they have become somewhat of a standard for assessing the degree of existing (symptomatic) disease. Neurological function does not "fall off a cliff." The inflammatory degradation associated with dementias is a slow and progressive process. Therefore, it is critical to establish "ground level" functioning for perfectly healthy people. A one-time test (one data point) may be corrupted by any number of factors including sleep, medications, and stress. Similar to high blood pressure, the cognitive tests should be recorded regularly, taking into consideration that a high frequency may lead to false readings as the person taking the test "learns" the expectations.

The Mini-Mental State Exam (MMSE): This test is a widely used test for cognitive function among the elderly; it includes tests of orientation, attention, memory, language, and visual-spatial skills. In general, participants scoring below education-adjusted cut-off scores on the MMSE may be cognitively impaired. There are arguably many more, better tests for cognitive impairment that can be used in a screening mode. Many of these are mentioned in Chapter 3.

When should you have your first routine screening? Immediately. How often should you have a routine screening? For many of us, these tests are already done if you have seen you optometrist of ophthalmologist recently. However, there a few, if any, eye doctors willing to share with their patients the association between eye pathology and diseases like Alzheimer's. Your doctor can best determine when your next screening should be conducted. As a rule-of-thumb, screening should be performed starting at age 40 then every 5 years thereafter until the age of 60 when the frequency should be every 2 years.

Baseline Determination

We are all unique individuals. You want to be compared to yourself, not some amorphous statistic, don't you? So often in medicine, the patient is judged "by the numbers," rather than by looking at the person. This "baseline" step can be comprehensive or limited based on the judgment of your doctor and particularly on his or her evaluation of your phenotype. Age may determine the depth and breadth of this assessment. In baseline determination, a patient will go through a wide range of tests to establish the base or current status with regard to brain health, cognitive function, and the factors that may impact brain health. Baseline tests include ocular tests, blood testing, and neurological and psychological assessment. Baseline testing also includes any or all of the tests in "Routine Screening."

Advanced Ocular Tests

These tests are in addition to those in routine screening. Some of these tests are very obscure while others do not have adequate reimbursement for doctors to even consider doing. However, they all provide valuable information about your health status, particularly with regard to neurodegenerative disease. Many of these tests are potentially useful to drug development companies who are looking for that next blockbuster drug to defeat AD.

Optical path deviation (OPD): This is a research and clinical tool used in refractive surgery that very accurately measures the optical path aberration in the cornea and the lens (Cataract). In the case of refractive surgery, the corneal aberration can be corrected based on those measurements (Lasik surgery). The same instrument can also accurately measure changes in the lens before the cataract is visible. Often this lens swelling is due to the accumulation of a misfolded protein precursors associated with Alzheimer's. Slight changes in the accumulation and coalescing of these proteins in the eye can cause a significant change in aberration, thus this tool can be used very early on to potentially diagnose disease and measure treatment efficacy. The OPD produces a quantifiable digital report that can be used in a health report card and also facilitates tracking of patient progress.

Visual evoked response (VER): This simple device measures the transmission of signals from the retina to the back of the brain across four synapses and between five neurons. The signal generated is highly correlated to acute and chronic neurodegenerative processes, and it is highly reproducible, quantifiable, and very useful in tracking progression/regression of brain response.

Micro perimeter assessment: This device has both an infrared and a color fundus camera, and a real-time tracking system that allows for a full automatic assessment of the eye's fixation stability. It measures the stability of the eye fixation that is controlled by the six muscles surrounding each eye. The extent that a subject cannot focus, or maintain a stable focus, is indicative of a neurodegenerative process. This device offers a simple, yet semiobjective and quantifiable way to characterize the extent of disease and evaluate treatments. This instrument also measures visual fields.

Visual field: Our peripheral visual fields (peripheral vision) gradually constricts as a normal process of aging. More rapid or advanced visual field contraction indicates accelerated aging. This device also allows the study of side vision. Inflammation leading to tissue disease is a potential cause of peripheral visual field changes as are brain tumors. The typical system provides both qualitative and quantitative results that can be used in a health report card and to track patient progress.

Pentacam Scheimpflug (imaging of the cornea and the crystalline lens): This instrument takes over 25,000 very accurate measurements in less than 1 s of the anterior (front) structures of the eye. This camera has five different functions; hence its name "penta" (five) camera. Two of those functions are useful to study the changes in the eye associated with AD and aging seen in the lens and the cornea.

- **Lens tomography:** The Pentacam performs very precise and reproducible measurements of the lens. Information obtained includes; precise measurements of cataracts including density, and average lens opacity (related to cataracts and inflammation). The accurate lens measurement could be useful to follow the changes caused by the accumulation of beta amyloid protein (one AD hallmark biomarker) in the lens.
- **Corneal thickness and volume:** Like skin thickness, corneal thickness diminishes with age. Premature thinning is a sign of accelerated aging. This instrument imparts more objectivity compared to the slit lamp microscope and provides information about the cornea and the lens not obtained in a slit-lamp measurement. The data obtained is very objective and classifiable, and thus is useful in a health report card and for monitoring patient progress with respect to both neurodegenerative and cardiovascular diseases. Formations like nuclear and cortical cataracts are accurately mapped through tomography.

Retinal laser Doppler instrument: Laser Doppler systems provide imaging to measure blood vessel size and speed of blood flow. Patients with Alzheimer's and other diseases of inflammation show a significant narrowing of the venous blood column diameter compared with the control subjects and a significantly reduced venous blood flow rate compared with the control subjects. This test saw considerable buzz as an Alzheimer's "diagnostic" technique at the International Alzheimer's Association meeting in Paris, 2011. However, this is a rather nonspecific test that is worth recording as part of a differential diagnosis.

Blood Tests

AD is now recognized as a disease of the brain and of the general cardiovascular system. Markers in the system-wide (systemic) blood may give clues as to the origin and progression of the disease. During baseline testing, some key blood tests should be considered. They are listed in order from most-to-least important based on the author's judgment, evaluation of the literature, and relevant clinical experience. Some consideration in placement was given to the ability of the patient to modify that risk factor in a favorable way and thus to prevent or slow AD.

Homocysteine: This is a metabolic by-product of methionine metabolism. Progressively elevated blood levels of homocysteine are a documented risk marker for cardiovascular events, AMD, glaucoma, and AD.

C-reactive protein: CRP is the single best measure of your "chronic disease temperature." CRP is one of a number of acute phase reactant proteins that increases in response to inflammatory stimuli. In large epidemiologic studies, elevated levels of CRP have been shown to be a strong indicator of CVD. The plasma cytokine interleukin-6 (IL-6) plays an important role in mediating inflammation and is a central stimulus for the acute-phase response. In particular, IL-6 induces the hepatic (liver) synthesis of CRP. IL-6 is implicated as one of the important inflammatory substance that halts generation of new neurons (neurogenesis). In fact, IL-6 is the only known cytokine capable of inducing all acute-phase proteins involved in the inflammatory response. [33] For more information read "Hope for the Afflicted" later in this chapter.

Complete blood count with differential (CBC): Parameters obtained in this test tell a lot about the level of immune system activity by measuring white blood cell counts and other factors that help determine why the immune system is active. The most useful data obtained from the CBC with differential includes:

White blood cell (WBC, leukocyte) count: White blood cells are part of your immune system. They protect the body against infection. If an infection develops, white blood cells attack and destroy the bacteria, virus, or other organisms causing it. White blood cells are bigger than red blood cells but fewer in number. When a person has a bacterial infection, the number of white cells rises very quickly. The number of white blood cells is routinely used to determine if our body is responding to infection, to see how the body is dealing with cancer treatment, or if an immune system disorder exists.

- Neutrophil granulocytes are generally referred to as either neutrophils or polymorphonuclear neutrophils (or PMNs) and are subdivided into segmented neutrophils (or segs) and banded neutrophils (or bands). Neutrophils are the most abundant type of white blood cells in mammals and form an essential part of the innate immune system. Neutrophils are recruited to the site of injury within minutes following trauma and are the hallmark of acute inflammation. These cells also protect the body against infection by destroying bacteria.

- Eosinophil granulocytes, usually called eosinophils or eosinophiles (or, less commonly, acidophils), are white blood cells that are one of the immune system components responsible for combating multicellular parasites and certain infections. Along with mast cells, they also control mechanisms associated with allergy and asthma. They are granulocytes that develop during hematopoiesis in the bone marrow before migrating into blood.

- Lymphocytes and natural killer (NK) cells: NK cells are a part of innate immune system and play a major role in defending the body from both tumors and virally infected cells. NK cells distinguish infected cells and tumors from normal and uninfected cells by recognizing changes of a surface. NK cells are activated in response to a family of cytokines called interferons. Activated NK cells release cytotoxic (cell-killing) granules that then destroy the altered cells. They were named "natural killer cells" because of the initial notion that they do not require prior activation in order to kill cells.

- Basophils appear in many specific kinds of inflammatory reactions, particularly those that cause allergic symptoms. Basophils contain anticoagulant heparin, which prevents blood from clotting too quickly. They also contain the vasodilator histamine, which promotes blood flow to tissues. They can be found in unusually high numbers at sites of ectoparasite infection (e.g., ticks). Like eosinophils, basophils play a role in both parasitic infections and allergies.

Vitamin D: This hormone is a "neurosteroids hormone" as such is a potential biomarker of AD. [34] Patients with AD have a lower serum vitamin D compared to matched controls. In a prospective (forward looking) study it has been demonstrated that low levels of vitamin were associated with substantial cognitive decline in the elderly population [35]. In fact, it has been shown that supplementation with vitamin D improves cognition in patients with AD. More commonly vitamin D is known to promote calcium absorption in the gut and maintains adequate serum calcium and phosphate concentrations to enable normal mineralization of bone. Vitamin D has other roles in the body, including modulation of cell growth, neuromuscular and immune function, and reduction of inflammation. Many genes encoding proteins that regulate cell proliferation, differentiation, and apoptosis are modulated in part by vitamin D. Vitamin D3 is neuroprotective and therapeutic in attenuating iron-induced neurotoxicity in the CNS. The toxic effects of iron on the brain is important in AD and it is even more important in PD (see discussion on Iron and Ferritin in Chapter 10). Iron is now added to the flour and as we get older we all accumulate iron that is stored as ferritin in our body. If your serum ferritin is high you should give blood on a regular basis and make sure that your serum vitamin D level is optimal.

Please take your vitamin D supplements, get sensible sun and have your vitamin D level check annually. Be aware that if you take vitamin D (25-hydroxy vitamin D) and your serum level does not go up to the optimal level of 40–80, you should see a doctor and have your 1,25 dihydroxy vitamin D measured. If it is elevated, you have an inflammatory disease process that is causing your 25-hydroxy vitamin D to be activated. Do not take higher amount of vitamin D if your 1,25 dihydroxy vitamin D is elevated. See a doctor familiar with the causes of vitamin D activation and is willing to perform root-cause tests. In Dr. Trempe's experience only a very few doctors measure 1,25 dihydroy vitamin D.

Omege-3 fatty acids: These critical substances are also known as polyunsaturated fatty acids or PUFAs, particularly PUFA 3s. The key clinical omega-3 fats are EPA and DHA, which are found largely in cold-water fish. It is well established in current literature that a higher blood level of these important fats may help to reduce the risk of AD, heart disease, and stroke.

PUFA 6/3 ratio: High levels (>5) are associated with chronic silent inflammation.

Glucose: It is a type of sugar that the body uses for energy. An abnormal glucose level in your blood may be a sign of diabetes. For some blood glucose tests, you have to fast before your blood is drawn. Other blood glucose tests are done after a meal or at any time with no preparation. Hemoglobin A1c (HbA1c, A1c) is a more dependable way to determine your average blood glucose level. The red blood cells that circulate in the body live for about 3 months before they die. When sugar sticks to these cells, it gives us an idea of how much sugar has been around for the preceding 3 months. In most laboratories, the normal range is considered 4%–5.9%. In diabetes, it's 7.5% or above, and in prediabetics it's around 7.0%. The benefits of measuring A1c is that it gives a more reasonable view of what's happening over the course of time (three months), and the value does not bounce as much as finger stick blood sugar measurements.

Insulin: This hormone is associated with the characterization of the Atherogenic Lipid Profile and Metabolic Syndrome. Abnormal fasting insulin, especially when combined with other risk factors, identifies patients at significantly higher risk for CVD and Alzheimer's.

Tumor necrosis factor alpha (TNF-α): TNF-α is a growth factor for immune cells and osteoclasts, the cells that break down bone. It has well-known proinflammatory functions and may be elevated in chronic infections, certain cancers, and hepatitis C. TNF-α is being (incorrectly, we believe) considered a direct therapeutic target for AD [36]. Indeed, antiinflammatory strategies may have short-term benefits of relieving symptoms, but long term, suppressing immune function (inflammation) always fails. Thus TNF-α is another important marker of inflammation, and elevated levels should be used for diagnostic purposes and a reason to dig deeper and find a cause for its elevation. The section in this chapter titled, "Hope for the Afflicted" should shed light on the shortcoming of just suppressing inflammation.

Fibrinogen: This is a plasma glycoprotein that can be transformed by thrombin into a fibrin clot in response to injury. The combination of elevated fibrinogen with other CVD risk factors can substantially increase disease potential. "Fibrinogen is normally found circulating in blood, but in AD it deposits with Aβ in the brain parenchyma and the cerebral blood vessels. We found that Aβ and fibrin(ogen) interact, and their binding leads to increased fibrinogen aggregation, Aβ fibrillization, and the formation of degradation-resistant fibrin clots. Decreasing fibrinogen levels not only lessens cerebral amyloid angiopathy and BBB permeability, but it also reduces microglial activation and improves cognitive performance in AD mouse models [37]." As with elevated tumor necrosis factor alpha, this does not imply that fibrinogen is a therapeutic target. That will be determined through more research. However, fibrinogen is a bona fide diagnostic parameter for AD.

Interleukin-6 (IL-6): Elevated levels may occur in different conditions including chronic infections, autoimmune disorders, certain cancers, and AD. This test may be redundant because of the connection between IL 6 and CRP.

Ceramides: "Serum ceramides increase the risk of AD," according to a Mayo Clinic group [38]. They may not increase the risk on their own, but their presence at a high level surely is an indicator of increased risk. "Compared to the lowest tertile, the middle and highest tertiles of ceramide were associated with a 10-fold and 7.6-fold increased risk of AD respectively. Total and high-density lipoprotein cholesterol and triglycerides were not associated with dementia or AD." "Results from this preliminary study suggest that particular species of serum ceramides are associated with incident AD." A routine blood test for ceramides is not available in the standard-of-care, but the technology is available for such tests.

Sex hormones:

- Testosterone is a hormone made by your body and is responsible for the normal growth and development of the male sex organs and for maintenance of other sexual characteristics. In men, testosterone is produced in the testes, the reproductive glands that also produce sperm. The amount of testosterone produced by the testes is regulated by the hypothalamus and the pituitary gland. Testosterone deficiency can also lead to a number of disturbing symptoms, including loss of stamina and lean muscle mass, reduced libido, anxiety, depression, and cognitive decline.

- Estrogen is probably the most widely known and discussed of all hormones. The term "estrogen" actually refers to any of a group of chemically similar hormones; estrogenic hormones are sometimes mistakenly referred to as exclusively female hormones when in fact both men and women produce them. Estrogens act on the CNS through genomic mechanisms, modulating synthesis, release and metabolism of neurotransmitters, neuropeptides and neurosteroids, and nongenomic mechanisms, influencing electrical excitability, synaptic function, and morphological features. Therefore, estrogen's neuroactive effects are multifaceted and encompass a system that ranges from the chemical to the biochemical to the genomic mechanisms, protecting against a wide range of neurotoxic insults. Many biological mechanisms support the hypothesis that estrogens might protect against AD by influencing neurotransmission, increasing cerebral blood flow, modulating growth proteins associated with axonal elongation, and blunting the neurotoxic effects of β-amyloid.

- Pregnenolone, testosterone, estrogen, cortisol, and DHEA are members of a family of natural hormones that are essential for human survival. Scientists have discovered that pregnenolone also can be manufactured in the brain from cholesterol instead of being transported through the blood-brain barrier from other parts of the body. This supports recent findings showing that pregnenolone is involved in a variety of brain-related functions, such as memory, concentration, and mood [39].

Apolipoprotein E (apoE): This is an inherited trait. The apoE genotype predicts lipid abnormalities and responsiveness to different dietary fat intake. The e4 version of the apoE gene indicates an individual's increased risk for developing late-onset AD. People who inherit one copy of the APOE e4 allele have an increased chance of developing the disease; those who inherit two copies of the allele are at even greater risk. The apoE e4 allele may also be associated with an earlier onset of memory loss and other symptoms.

Magnesium: This mineral plays many vital roles in preventing heart disease, controlling blood pressure, and maintaining healthy cholesterol levels. This test is placed low on the list because serum levels probably do not accurately portray the true balance of magnesium. Are you taking magnesium supplements?

Adiponectin: This substance is a protein hormone that modulates a number of metabolic processes, including glucose regulation and fatty acid catabolism. Adiponectin is exclusively secreted from adipose tissue into the bloodstream and is very abundant in plasma relative to many hormones. Korean researchers studied the literature and concluded that there is a correlation between adiponectin and AD. "Adiponectin is an adipocytokine released by the adipose tissue and has multiple roles in the immune system and in the metabolic syndromes such as cardiovascular disease, type 2 diabetes, obesity, and also in the neurodegenerative disorders including AD. Adiponectin regulates the sensitivity of insulin, fatty acid catabolism, glucose homeostasis and anti-inflammatory system through various mechanisms. Previous studies demonstrated that adiponectin modulates memory and cognitive impairment and contributes to the deregulated glucose metabolism and mitochondrial dysfunction observed in AD. Here, we aim to summarize recent studies that suggest the potential correlation between adiponectin and AD [40]."

Cholesterol/HDL ratio: Many studies have sought to show the benefit of high HDL levels and the results remain mixed. However, the ratio of total cholesterol to HDL is physiologically important. A ratio of <4 is preferred.

Calcium: This mineral is an important mineral in the body. Abnormal calcium levels in the blood may be a sign of kidney problems, bone disease, thyroid disease, cancer, malnutrition, or another disorder. Excess calcium is connected with hardening of the arteries and dementia. The "calcium hypothesis of dementia" is an emerging theory on one of the potential root causes of accelerated brain aging.

β2 microglobulin: This is a measure of the activity of the acquired immune system and can provide information about infection and inflammation. Tests for β2 microglobulin are being evaluated for Alzheimer's with favorable results [41]. A *Daily Mail* article discusses this test as viable for AD, however a spinal tap is required [42].

Urinary albumin: Studies have shown that elevated levels of urinary albumin in people with diabetes or hypertension are associated with increased risk of developing cardiovascular disease (CVD).

Kidneys: Blood tests for kidney function measure levels of blood urea nitrogen (BUN) and creatinine. Both of these are waste products that the kidneys filter out of the body. Abnormal BUN and creatinine levels may be signs of a kidney disease or other disorders.

Myeloperoxidase: Recent studies have reported an association between myeloperoxidase levels and the severity of coronary artery disease. It has been suggested that myeloperoxidase plays a significant role in the development of the atherosclerotic lesion and rendering plaques unstable. Researchers continue to link cardiovascular and AD, and the marker myeloperoxidase (MPO) is not providing exception to this emerging rule. Research led scientists to make the following statement: "AD patients showed significantly increased plasma levels of MPO, which could be an important molecular link between atherosclerosis and AD [43]."

N-terminal pro-brain natriuretic peptide (NT-proBNP): This is a progressive CVD risk marker with powerful, independent prognostic value for detection of clinical and subclinical cardiac dysfunction [44]. Elevated levels indicate the presence of ongoing myocardial stress and potentially an underlying cardiac disorder.

Lp-PLA2: It is a marker for vascular-specific inflammation and also plays a causal role in the vascular inflammatory process, leading to the formation of vulnerable, rupture-prone plaque. Elevated levels have been shown to be powerful predictors of ischemic stroke and heart attack risk [45]. This marker is now becoming important in Alzheimer's evaluation. Researchers from the Netherlands make the following claim: "This is the first study to our knowledge that shows that Lp-PLA2 is associated with the risk of dementia independent of cardiovascular and inflammatory factors and provides evidence for a potential role of Lp-PLA2 in identifying subjects at risk for dementia [46]."

Lipoprotein(a) [Lp(a)]: It is an inherited abnormal protein attached to LDL. Lp(a) increases coagulation and triples cardiovascular disease risk. Medical research shows the connection between AD and CVD through the following statement: "It is suggested that increased Lp(a) serum concentrations, by increasing the risk for cerebrovascular disease, may have a role in determining clinical AD [47].

ESR or SED rate: This is a nonspecific test used to detect chronic inflammation associated with infections, autoimmune disorders, and cancer. This test has been used in trials that evaluated differential parameters for evaluation of dementia [48].

F2-Isoprostanes (F2-IsoPs): These are the "gold-standard" for quantifying oxidative stress. Increased free radical-mediated injury to the brain is proposed to be an integral component of several neurodegenerative diseases, including AD. Lipid peroxidation is a major outcome of free radicals' mediated injury to brain, where it directly damages membranes and generates a number of oxidized products. F2-Isoprostanes (F2-IsoPs), one group of lipid peroxidation products derived from arachidonic acid (omega-6 fatty acid), are especially useful as in vivo biomarkers of lipid peroxidation. F2-IsoP concentration is selectively increased in diseased regions of the brain from patients who died from advanced AD, where pathologic changes include beta-amyloid, neurofibrillary tangle formation, and extensive neuron death [49].

Ferritin: Checking your iron levels is done through a simple blood test called a serum ferritin test. The study of iron in the human brain is particularly important in the context of AD. Iron is both essential for healthy brain function and is implicated as a factor in neurodegeneration. The chemical form of the iron is particularly critical, as this affects its toxicity, and disrupted iron metabolism is linked to regional iron accumulation and pathological hallmarks, such as senile plaques and neurofibrillary tangles [50].

Haptoglobin: Plasma haptoglobin is oxidized in mild cognitive impairment and AD. Oxidation of haptoglobin contributes to amyloid fibril formation. Extracellular chaperones are impaired in AD of which haptoglobin is one. Haptoglobin may be considered a putative marker of AD progression [51].

Uric acid: It is a risk factor of CVD, as well as a major natural antioxidant, prohibiting the occurrence of cellular damage. According to some research, "Notwithstanding the associated increased risk of cardiovascular disease, higher levels of uric acid are associated with a decreased risk of dementia and better cognitive function later in life [52]." However, other research suggests that the correlation does not exist [53]. Uric acid remains an important test that should be performed routinely to measure health and health trends.

Vitamin B6: Low circulating vitamin B6 is highly correlated to markers of inflammation that contribute to AD [54]. This marker, taken together with other inflammation markers, helps strengthen the case for inflammation, when all are pointing at an increase in inflammatory body burden.

We have just listed over two-dozen tests, most of which involve drawing blood. Indeed, having these tests done raises the usual issues of insurance reimbursement, doctors catching up with emerging data, interpretations, and interventions based on the results. The biggest hurdle you face is that the best use of these tests is on the apparently "well person," as most of these tests should be used in a prevention/early detection mode.

We have not listed these tests in any particular order because every test provides useful information. The more tests you can obtain, the better. However, the order of listing of these tests is not completely coincidental, at least for the first half dozen tests.

Diseases Comorbid with Alzheimer's Disease

The diagnosis of any of the following diseases is thus a likely diagnostic for AD potential. Several of these were already explained earlier in this chapter, under the heading "Alzheimer's Risks That You Can Determine."

- Mild cognitive impairment
- Dementia
- Glaucoma
- Macular degeneration
- Lyme Disease
- Infectious disease(s) (chronic)
- Diabetes Type II
- Cardiovascular disease(s)
- Amyloid-based diseases
- Autoimmune diseases
- Rheumatoid arthritis
- Other inflammatory diseases

Neurological and Psychological Tests

First, being honest, no one enjoys undergoing the "challenge" or tests that measure your cognitive function. They are humiliating. However, the standard-of-care uses these tests and they provide a benchmark common to the neurological profession. The MMSE is the most commonly prescribed test to characterize the state of cognitive function. Other tests have emerged that claim to provide better performance in terms of accuracy and reproducibility. The best solution is to obtain measurements from more than one of these tests. This will give a broader assessment of cognitive function and also likely allow repeat testing to be less corrupted by memorization by the person taking the tests. The following list includes the more commonly administered tests:

- Wisconsin Card Sorting Test
- The Rey-Osterrieth Test
- ADAS-Cog (Alzheimer's Disease Assessment Scale-Cognitive)
- Wechsler Test of Adult Reading (WTAR)
- Dementia Rating Scale-Second Edition (DRS-2).

Yet another cognitive test for "early" detection is in development. This is smartly named the AQ or "Alzheimer's Questionnaire." This quiz is reported to be 90% accurate in detecting signs of memory loss. The AQ should not be used as a definitive guide for diagnosing AD or mild cognitive impairment (MCI). However, it is a quick and simple-to-use indicator that may help physicians determine which individuals should be referred for more extensive memory testing. In the scheme of early diagnosis, people who "fail" the AQ are late in the disease process compared to diagnoses associated with peripheral blood and eye tests.

Comprehensive Assessment

The comprehensive assessment is carried out by your doctor and includes all the elements of a detailed health assessment but must go well beyond the standard-of-care physical. It starts with a review and evaluation of all the testing and evaluation from the routine screening and baseline determination steps. These tests should be current, within 1 year of the comprehensive assessment. This assessment should include, at a minimum, the following elements besides those already stated here:

History review

- Family history.
- Personal history.
- Travel history.
- Work history.
- Nutritional history and current nutritional habits.
- Concussions or other previous trauma.
- History of previous infections or viruses.
- Surgeries and any anesthesia.
- Current and previous medications.
- Current health profile with an emphasis on blood pressure, diabetes, cardiovascular disease, neuropathy, sudden weight changes, ocular conditions, and any inflammatory conditions including asthma, arthritis, and other joint pain.
- Current medications, especially those that impact blood pressure or have any known side effects that impact brain health.
- Oral health currently and historically including any periodontal disease and surgeries. If your current doctor does not ask you about your oral health and hygiene, find another doctor because he or she will not have the overall understanding of disease etiology (causes) and pathology to help you with AD.
- General review of immune system function and health.

In-Depth Root Cause(s) Analysis

AD is substantially preventable. How do we know that? It is really a modern disease. The case presented by Dr. Alzheimer in 1907 was relatively rare a scant 100+ years ago. Indeed we are long lived but, to a large degree, we are living less healthy lives later in life due to chronic diseases. The May 2013 issue of *National Geographic Magazine* indicates that today, those with a life expectancy of 80 suffer 19 years of degrading health. Why are those of us who are relatively long-lived so unhealthy? Diet is a big part of the puzzle according to Paul Clayton, who studied the mid-Victorian and post mid-Victorian diet of England in the 1870s and later. The post mid-Victorian workers did not change what they ate by all that much. At that time, food processing and mass production of food was emerging. Clayton determined that the processing of food led to more empty calories and less general nutritional value, especially in the lower cost foods that poorer workers could afford.

Does nutritional consideration impact us as much today? Yes. Even more than in the past, we are all continually exposed to processed foods, a plethora of artificial additives, and other ingredients to win our buying patterns by creating desirable taste. The measurement of PUFA (polyunsaturated fatty acid also known as omegas) ratios is a good indicator of just how far out of balance our diets have become. From an evolutionary perspective, the PUFA 6 to PUFA 3 ratio is best at about 3 to 1. The American diet is 15 to 1 or more.

Plenty of people live long healthy lives and don't come down with chronic disease like Alzheimer's. You know some of them, and you also know how they have comported themselves over the years. You can do it as well, but it requires more commitment and vigilance than ever, because we are long lived, and there are much more unhealthy "temptations" available to all of us. Not to mention we are subjected to a barrage of media touting the latest fad, miracle prevention, or diet.

The good news is that chronic diseases like Alzheimer's are very preventable. Question everything. Seek multiple sources of corroborating evidence. Get your ideas from the most credible sources. Consider reading the *Zone* books by Dr. Barry Sears for good ideas on nutrition. Dr. Liponis from Canyon Ranch has published some very useful books including *Ultraprevention* [55]. This book gives some very good suggestions on how to maintain your health throughout your life.

Inflammation is the Diagnostic Clue

The decades of blaming cholesterol for chronic disease is rapidly coming to an end. Yes, there are 21,000,000 Americans on statin therapy, and these numbers will not drop precipitously. How can they? You cannot expect your doctor to just say, "Um, Mr. or Mrs. Patient, we were wrong about cholesterol, and I'm concerned about statin side effects, so I'm going to remove you from this drug immediately." A colleague was just prescribed Lipitor by Mass General Hospital! Mass General has raised the white flag when it comes to treating causes of cardiovascular disease.

The new "next big thing" appears to be a combination of inflammation and homocysteine. It looks like medicine has it right this time. However, the major emphasis has been on inflammation controlling or homocysteine controlling therapies. Again, the true root cause(s) of elevated levels of homocysteine and inflammation are taking a back burner. The standard-of-care has a model, like a romance novel writer, for diagnosing and treating disease. For example, cholesterol is high, and the patient is sick, therefore lower cholesterol. The next big thing looks something like this: inflammation is high, and the patient is sick, therefore lower the inflammation. Substitute the word "inflammation" with any other marker and that is what you can expect out of clinical medicine in the future.

Inflammation is a treasure, and it is our body's response to some type of imbalance. It is our immune system response. To keep it simple, let's consider (again) what has been known for centuries about the activity of our immune system. The following information is an excerpt from the www.pfizerpro.com website. There is a certain irony to this source because Pfizer markets Lipitor, the best-selling drug of all time, which just happens to be a statin.

Immunology Refresher

Immune System: A coordinated system of cells, tissues, and soluble molecules that constitute the body's defense against invasion by nonself entities, including infectious and inert agents and tumor cells.

The immune system has four key tasks:

1. **Recognition:** Detect infection or harm
2. **Effector function:** Contain and eliminate infection
3. **Regulation:** Control activity to avoid damage to the body
4. **Memory:** Remember exposure; react immediately and strongly upon re-exposure

The Pfizer site shows how the two aspects of the immune system, innate and adaptive, are different:

- Innate Immunity

  - Nonspecific
  - Present at all times
  - Immediate but general protection
  - Activates adaptive immune response
  - Does not improve with repeated exposure to a pathogen.

  Adaptive Immunity

  - Develops in response to infection
  - Protective against specific pathogens
  - Leverages components of the innate response
  - Develops memory, which may provide lifelong immunity to reinfection with the same pathogen.

  The recurring theme is that pathogens or infectious species trigger an immune response, thus inflammation.

Diagnose for Causes of Inflammation: Pathogens

Antibody titers: Bacteria are not measured directly, but rather are determined by measuring "antibodies" that are targeted to a specific "antigen."

Definition of Antigen Any substance that causes the immune system to produce antibodies against it. The substance may be from the environment or formed within the body. The immune system will kill or neutralize any antigen that is recognized as a foreign and potentially harmful invader.

A sample is taken from the patient blood and is then challenged with known antigens to detect the presence of antibodies to these antigens. A titer is a measure of how much a sample can be diluted before antibodies can no longer be detected. Titers are usually expressed as ratios, such as 1:256, meaning that one part serum to 256 parts saline solution (dilutant) results in no antibodies remaining detectable in the sample. A titer of 1:8 is, therefore, an indication of lower numbers of bacteria antibodies than a 1:256 titer.

There are concerns over the interpretation of such tests, however, as a high titer does not necessarily mean that a person is infected, nor does a low titer necessarily mean that they have either a low-grade infection or none at all. Antibody tests, such as these detect only free antibodies in the sample, and those already bonded together with an antigen are not detected. As such, patients with a high number of antigen-antibody complexes may have a significant infection but this will not be borne out by testing if there are few free antibodies in their serum sample. A low titer may in fact demonstrate significant success on the part of the immune system in fighting off an infection with bacteria, whereas a high titer could show residual antibodies to a previous infection, or unsuccessful attempts to bond to the antigens in the bloodstream by the antibodies. Measuring for the infectious species implicated at the root of Alzheimer's is tricky business.

Antibodies may take up to 2 months to reach peak levels, with immunoglobulin M (IgM) appearing between 2 and 4 weeks after infection, and immunoglobulin G (IgG) taking between 4 and 6 weeks to reach detectable levels in most cases. Ensuring that the correct testing, either for IgM, IgG, or both, is carried out at the appropriate time is also important so as not to invalidate the purpose of the test.

Most of the infection related chronic inflammatory diseases are caused by intracellular (inside the cells) infections. Antibodies that are formed and circulate are able to control but not eradicate these infections because the antibodies are not able to enter the cells. From time to time, these microorganism escape the cells and there is reactivation of the immune reaction and elevation of the antibody titer. For example, about 15% of the United States general population is infected and seropositive for toxoplasmosis [56]. Those infected individual are usually asymptomatic but the the immune surveillance can fail and the infection can reactivate and result in loss of vision and neurologic illness. This chronic infection can be treated but not cured with antibiotics. When the infection reactivates the antibody titer increases and the spread of the infection halts.

In some cases, a positive result on an antibody test may be false due to potential crossover reactions for antibodies to other bacteria, such as syphilis, or viruses, such as Epstein-Barr or human immunodeficiency virus (HIV). An autoimmune response in the body may also confound the results, as antibodies to the patient's own tissues may be detected in conditions, such as Lupus or Rheumatoid Arthritis. However, in some cases, active bacteria are at the root of autoimmune diseases, a fact that is not well appreciated in medicine.

Polymerase chain reaction (PCR): Another method for detecting and measuring bacteria, especially those implicated in AD, is PCR testing. This test is used to detect the bacterial genes in a blood or tissue sample.

The traditional basis for the identification of living organisms usually requires their isolation and growth in the laboratory. Reliance on these parameters have limited our awareness of the role of bacteria in the pathophysiology of diseases. PCR tests are used to detect the genetic signature of bacteria in a blood or tissue sample. This method of bacterial detection of infectious agents is particularly useful in culture-negative samples (those that do not lead to growth by traditional methods) which is usually the case for atherosclerotic or brain tissue samples obtain at autopsy.

Broad-range PCR primers are first used to detect the evidence of all known bacteria, including the eubacteria (cells without nucleus) in tissue sample. Alternatively, PCR is also be used to detect the highly specific portions of the genetic signatures that is unique for each bacterium. The same type of genetic analysis is used in forensic medicine to determine if a tissue sample belongs to a human and to which specific individual human the sample belongs.

The presence of genetic information in a tissue sample does not mean that the detected specific information belongs to a living organism because the genetic evidence can remain in tissue for prolong periods of time after the death of the organism. A full horse genome was recently extracted from the bone of a horse that lived 735,000 years ago. In that same sample more than 12 billion additional DNA molecules were detected that belong to various bacteria that kill or contaminated the sample over that time period. The horse and the bacteria are long dead but the full genetic information of the horse and of the bacteria are still preserved.

This forensic type of bacterial information done at autopsy on the brain of AD patients has revealed a significant number of genetic bacterial markers. This does not necessarily indicate an active infection at the time of death but rather indicate that those identified bacteria could have played a role at some point in the disease process and that we should test for those bacteria in patients with early signs of AD. If a patient has a high CRP, a high homocysteine, and a high antibody titer to a disease-causing bacteria it is better to treat those patients because, in our experience, it is the best and only way to decrease both the CRP and homocysteine and notice a beneficial effect in our patients.

Dr. Kilmer McCully, in his review paper, reported on the work of Ott and Stephan et al. who identified over 50 unique bacteria genetic markers in various atheromatous plaques [57]. After studying a vast number of CVD sufferers, affected individuals were found to have between 8 and 12 distinct species in their plaques. With regard to AD, and to limit the amount of blood drawn, the "usual suspects" (those noted most in the literature) are those that should be tested for first. The following is a list of bacteria worth searching for in anyone with elevated inflammatory markers and any signs of cognitive impairment. This testing should also be considered for anyone with signs of cardiovascular diseases and the other "comorbid" diseases presented in this and other chapters.

- *Chlamydia pneumoniae*
- *Mycoplasma pneumoniae*
- Toxoplasmosis
- Rickettsia diseases (the three most common)
- Lyme disease (there are many and the standard Lyme test is for just one)
- Tuberculosis (we see many patients with untreated tuberculosis, and it can impact the brain)
- H-Pylori (there is a simple breath test)
- Q-Fever
- Periodontal bacteria

Recall that most with infections that appear to predispose people to Alzheimer's have a good immune system and are completely asymptomatic when infected. The best example is tuberculosis: Over 95% of the healthy infected people with tuberculosis are not aware that they are infected. This is the reason why all medical personnel in every hospital in the United States of America is tested for tuberculosis every year in order to detect those asymptomatic yet infected.

In the experience of Dr. Trempe, it is not rare for patients with early neurodegenerative diseases to have highly positive antibody titers against rickettsial diseases. Those infections can be controlled but not completely "cured" with treatment. These patients require repeated treatments over the course of years. The problem is that rickettsial organisms are very small obligate (depend on the host for fuel) intracellular infection that have a close evolutionary relationship and share many common genetic materials with our own mitochondria [58].

Remember the definition for Typhus:

Typhus Any of several similar diseases caused by Rickettsiae. The name comes from the Greek word 'typhos τῦφος' meaning smoky or hazy, describing the state of mind of those affected with typhus.

Treatable bacteria cause this smoky or hazy brain.

DISEASE MANAGEMENT PROGRAM

It is very important to realize that many patients that have cognitive impairment and are diagnosed as having AD do not have the disease (Chapter 1). In a paper titled, *Much of late life cognitive decline is not due to common neurodegenerative pathologies*, the authors performed analysis on 856 deceased participants from two studies of aging and dementia [59]. Their study provides information on the causes of cognitive decline. The participants had been evaluated regularly since 1994 or 1997 and had a mean age at death of 88. The authors examined the rate and timing of cognitive decline and linked these to the three age-related pathologies. Alzheimer pathology explained 22% of the decline, gross infarcts (strokes) 2%, and Lewy bodies 8%. When considered together, these were associated with a faster rate of cognitive decline, but accounted for only 41% of the variation in decline.

The Disease Management Program is all about making you well if you have true Alzheimer's or other causes of accelerated cognitive decline. Here your doctor takes into consideration the results from the screening, baseline measurements, comprehensive assessment, and in-depth root cause(s) analysis. This program is all about treatments, measuring how you respond to treatment, and then making treatment adjustments. What are the treatments? The answer to that depends upon the diagnosis. The next section provides proof that the right diagnosis and the right treatments will work. There is hope for the afflicted and even more hope for those in the earliest and asymptomatic stages of Alzheimer's. Please do get diagnosed (a differential diagnosis, that is). A proper diagnosis will ensure you the best chance to beat the processes leading to your cognitive decline, including Alzheimer's.

Hope for the Afflicted

We have discussed prevention and diagnosis at length but we want more; we want that treatment. We are programmed to ask, "What is the treatment?" It is a natural question because a diagnosis does not make us well, but the treatment might. It is not possible to suggest a treatment without a complete differential diagnosis. Hopefully you do not feel that this response avoids the tough issues about Alzheimer's.

Another way to address treatment is to address "treatability" instead. How sick, or how far progressed into the AD process is too far for a person (patient) to expect either stabilization or improvement? Within the Amyloid Cascade Hypothesis, the assumption is that anyone with the slightest clinical sign of cognitive deficiency is too far gone for the therapy to work (Chapter 2). Fortunately, this is the wrong therapeutic approach. We have many patients that have remained stable for years after a diagnosis of AD was made. Some patients show clear improvement. There is hope for those clinically afflicted with early AD.

As usual, shall we consult with some researchers to get an answer to this profound question, "Who is too far gone to treat?" It turns out that hope for the sufferers of Alzheimer's has a strange "bedfellow," a study on cranial irradiation. Why? Researchers at Stanford University studied models for the inflammation created by radiation of the brain using dead bacteria (referred to as LPS). The researchers had the savvy to appreciate that their work extended beyond inflammation created by radiation because they, and many others, show that the same inflammatory process occurs in the brain whether it is stimulated by irradiation, concussion, or chronically by inflammation and infection.

Stanford University Study Provides Hope

The work of this Stanford group over a decade ago is extremely important in AD. It essentially says that even those severely afflicted with Alzheimer's have some level of hope for recovery. This is tremendous news. Now let's get it into clinics!

What follows are excerpts from, and a blow-by-blow explanation of, the Stanford paper titled, *Inflammatory blockade restores adult hippocampal neurogenesis* [60]. This article is cited over 1,000 times, thus there is a tremendous body of literature related to this topic since its publication.

First, here is an interpretation of the title:

- Inflammatory blockade = Stopping inflammation
- Hippocampal = Part of the brain, in particular, the gray matter that has a central role in memory processes.
- Neurogenesis = Birth of neurons

The title, in laypersons terms, would be "Stopping (the cause of) inflammation leads to the birth of new neurons and the restoration of brain function and memory".

This is so exciting that the conclusion, as it relates to Alzheimer's, is provided here, first. Then this brilliant paper is reviewed in detail.

- Alzheimer's is recognized to be an inflammatory disease.
- A clear cause of the inflammation and neuroinflammation is a variety of infectious species including spirochetes, other bacteria, and viruses.
- Since the infection produces inflammation, and stopping the inflammation renews the growth of neurons, then treating the infection provides a solution (cure—read Appendix 3 for a definition of medical cure) for AD.
- Treatment is not simply dosing with antibiotics. There are many stages to creating a host (you, the patient) that is highly receptive to controlling or eradicating the infectious species implicated in AD. The degree of difficulty is related to age, the burden, and type of pathogen, and other aspects of your personal phenotype.
- Prevention is the best approach.
- For those who are afflicted with AD, review the recommendations of Dr. David Wheldon that are designed for MS but target *C. pneumonia*, which is one of the pathogens that can contribute to the neuroinflammation associated with both Alzheimer's and MS.

From the Stanford paper (1): "The birth of new neurons within the hippocampal region of the central nervous system continues throughout life, and the amount of neurogenesis correlates closely with the hippocampal functions of learning and memory [61,62]. The generation of new neurons within the hippocampus is mediated by proliferating neural stem or progenitor cells (NPC) [63–65] that are widespread within the adult brain but instructed by local signaling to produce neurons only in discrete areas [66,67]. Alterations in the microenvironment of the stem cell may allow ectopic neurogenesis to occur [68,69] or even block essential neurogenesis, leading to deficits in learning and memory [70–72] such as that observed in patients who receive therapeutic cranial radiation therapy [73]."

Interpretation: The belief that we are born with a fixed number of brain cells and that the number of brain cells diminishes with age is a misplaced belief. Paul Allen, a Microsoft cofounder, funded research at Stanford to map brain stem cells. It turns out that the brain is full of stem cells, and processes that occur in cells in the rest of the body also occur in brain cells. For example, all cells have a relatively short life, referred to as senescence. When cells die, they are replaced through the action of stem cells, and the brain is no exception. The good news is that the birth of new neurons continues throughout life. Changes to the brain, or local areas within the brain (the so-called microenvironment) may alter the behavior of brain stem cells. New neuron growth may occur in unexpected areas (ectopic) or be halted. If senescent (dying) neurons are no longer replaced, then brain atrophy will occur, leading to loss of brain function and disease labels like Alzheimer's.

From the Stanford paper (2): "In animal models, cranial irradiation ablates hippocampal neurogenesis, in part by damaging the neurogenic microenvironment, leading to a blockade of endogenous neurogenesis [72,73]. Injury induces pro-inflammatory cytokine expression both peripherally and within the central nervous system and induces stress hormones, such as glucocorticoids, that inhibit hippocampal neurogenesis [70]. The extensive microglial inflammation and release of pro-inflammatory cytokines that accompanies this irradiation-induced failure suggests that inflammatory processes may influence neural progenitor cell activity [72,74,75]."

Interpretation: Radiation therapy stops the birth of neurons. The radiation produces inflammation both locally and systemically, thus it can be detected in blood. The inflammation produced by the radiation, and not the radiation itself, is suggested to halt the growth of new neurons.

From the Stanford paper (3): "To determine the effects of inflammation on adult hippocampal neurogenesis, we injected bacterial lipopolysaccharide (LPS) into adult female rats to induce systemic inflammation. [76–79] The intraperitoneal (i.p.) administration of LPS causes a peripheral inflammatory cascade that is transduced to the brain via interleukin 1β from the cerebral vasculature [76] and causes a strong up-regulation of central pro-inflammatory cytokine production [76,79]."

Interpretation: What luck! They introduced bacteria (dead bacteria, also known as LPS) to simulate the inflammation created by radiation. Moreover, they introduced the bacteria into the body cavity, and the inflammation created there quickly migrated to the brain where more inflammation was created.

From the Stanford paper (4): "The neuroinflammation achieved in the LPS paradigm was accompanied by ... a 35% decrease in hippocampal neurogenesis."

Interpretation: The mechanism for producing new neurons to replace those that die off was reduced by 35%.

From the Stanford paper (5): "To determine whether inflammatory effects could be countered pharmacologically, animals were treated concurrently with a single dose of intraperitoneal LPS and daily doses of the nonsteroidal anti-inflammatory drug (NSAID) indomethacin [2.5 mg/kg, i.p., twice each day]. The effect of peripheral LPS exposure on neurogenesis was completely blocked by systemic treatment with indomethacin, whereas indomethacin alone had no effect on neurogenesis in control animals."

Interpretation: Administering an antiinflammatory allowed the production of new neurons to commence again and achieve levels of production that were present before the inflammation.

From the Stanford paper (6): "To determine the extent to which microglial activation might directly affect neural stem or progenitor cells, microglia were stimulated in vitro with LPS. LPS is a potent activator of microglia and up-regulates the elaboration of pro-inflammatory cytokines, including interleukin-6 (IL-6) and tumor necrosis factor–α (TNF-α). LPS-stimulated or resting microglia were then co-cultured with normal neural stem cells from the hippocampus under conditions that typically stimulate the differentiation of 30 to 40% of the progenitor cells into immature Dcx-expressing neurons. Neurogenesis in the presence of microglia was assessed as the increase or decrease in Dcx-expressing cells relative to control. Co-culture with activated but not resting microglia decreased in vitro neurogenesis to approximately half of control levels. LPS added directly to precursor cells had no effect on neurogenesis [80]."

Interpretation: The activated brain immune system cells, caused by the inflammation (microglia cells) decreased the creation of new neurons by about one-half compared to controls.

From the Stanford paper (7): "Activated microglia produce the potent pro-inflammatory cytokines IL-1β, TNF-α, interferon-γ (INF-γ), and IL-6. Progenitor cells were allowed to differentiate in the presence of each cytokine, and the relative expression of Dcx was scored after 60 hours. Exposure to recombinant IL-6 (50 ng/ml) or to TNF-α (20 ng/ml) decreased in vitro neurogenesis by approximately 50%, whereas the effects of IL-1 or INF-γ were not significant [80]. Addition of neutralizing anti–IL-6 antibody to CM from activated microglia was able to fully restore in vitro neurogenesis. This implicated IL-6 as a key inhibitor of neurogenesis in microglial CM. Although recombinant TNF-α also suppressed neurogenesis, IL-6 blockade alone appeared sufficient to restore neurogenesis in the presence of microglial CM."

Interpretation: They determined what inflammatory molecules appear to be responsible for the stoppage of neuron stem cells and the birth of new neurons. Specifically, controlling or removing IL-6 adequately restores production of new neurons. Note that IL-6 tracks with CRP. Both of these molecules were discussed in Chapter 7. We call CRP levels a measure of your chronic disease temperature. Clearly, putting out the fire (that is, reducing inflammation) helps the body and the brain recover and improve.

From the Stanford paper (8): "Chronic microglial activation and peripheral monocyte recruitment with the accompanying increase in local pro-inflammatory cytokine production, including IL-6, emerge as potent antineurogenic components of brain injury."

Interpretation: Inflammation injures your brain.

From the Stanford paper (9): "The in vitro data suggests that IL-6 inhibition of neurogenesis is primarily due to a blockade in neuronal differentiation rather than selective influences on cell death or proliferative activity."

Interpretation: Inflammation stops stem cell activity rather than killing existing cells. They die on their own in a programmed way.

From the Stanford paper (10): "Inflammatory blockade with indomethacin decreased microglial activation, accounting for part of the restorative effect of this treatment on neurogenesis after irradiation."

Interpretation: This study was performed on mice, not on Alzheimer's patients. The mice do not actually have Alzheimer's; however, they do have inflammation that blocks neuron stem cell activity. Do not interpret that indomethacin is an appropriate drug for Alzheimer's. Subsequent studies proved that patients treated with that substance did not improve or got worse. The mice do not have the disease. Inflammation was induced. People with AD have an underlying disease that causes the inflammation. That underlying cause must be treated. Remember, inflammation is actually a treasure of nature and is fighting that underlying cause of disease. If you inappropriately remove the inflammation, the disease will proliferate and ultimately create more inflammation and more disease.

From the Stanford paper (11): "In addition, the microvasculature of the hippocampus is a critical element of the neurogenic microenvironment [81–83] and both endotoxin and irradiation-induced inflammation disrupts the association of proliferating progenitor cells with microvessels [72]. The recruitment of circulating inflammatory cells is highly dependent on the endothelial status and elaboration of chemokines. One of the most robust effects of indomethacin in the present paradigm is the reduction in peripheral monocyte recruitment, suggesting that the inflammatory status of endothelial cells [e.g., expression of chemokines and/or ICAM (intercellular adhesion molecule)] may be normalized by indomethacin. Indeed, one known attribute of indomethacin treatment is the normalization of vascular permeability, which likely affects the neurogenic vascular microenvironment [84]."

Interpretation: The inflammation is in the blood, where the problem starts. Thus we can detect Alzheimer's-type inflammation through a blood test.

From the Stanford paper (12): "Neuroinflammation and microglial pathology are associated with many diseases of cognition in which memory loss features prominently, such as Alzheimer's disease, Lewy Body Dementia, and AIDS Dementia Complex. [85–87] Further, serum IL-6 levels in humans correlate with poor cognitive performance and predict risk of dementia [88]."

Interpretation: Many diseases of the brain and the body start with inflammation (and its causes). Regardless of the disease, the processes of inflammation are the same.

Hope Through Stopping Inflammation?

The Stanford team teaches us that, when inflammation is removed, regrowth of neurons commences. A team down the coast, at UCLA, showed a similar effect in a small but real group of patients with Alzheimer's [89]. They conclude:

> An increasing amount of basic science and clinical evidence implicates inflammatory processes and resulting glial activation in the pathogenesis of AD. This small, open-label pilot study suggests that inhibition of the inflammatory cytokine TNF-alpha may hold promise as a potential approach to AD treatment.

Here is an inside look into this encouraging research.

UCLA (1): Tumor necrosis factor (TNF)-alpha, a proinflammatory cytokine, has been implicated in the pathogenesis of AD.

Interpretation: (TNF)-alpha is elevated in most (maybe all) Alzheimer's patients. It is one of several well-known markers of inflammation. The UCLA team implies that this inflammatory marker is part of the cause of AD.

UCLA (2): To investigate the use of a biologic TNF-alpha inhibitor, etanercept was given by perispinal extrathecal administration for the treatment of AD.

Interpretation: Etanercept binds specifically to TNF and blocks its interaction with cell-surface TNF receptors. By avidly binding excess TNF, etanercept functions as an extraordinarily potent TNF antagonist. Because of the known role of inflammation in AD pathogenesis, etanercept has been suggested as a possible therapeutic agent for AD.

UCLA (3): We administered etanercept, 25–50 mg, once weekly by perispinal administration to 15 patients of average age 76.7. Main outcome measures included the Mini-Mental State Examination (MMSE), the Alzheimer's Disease Assessment Scale-Cognitive subscale (ADAS-Cog), and the Severe Impairment Battery (SIB). There was significant improvement with treatment, as measured by all of the primary efficacy variables, through 6 months: MMSE increased by 2.13 ± 2.23, ADAS-Cog improved (decreased) by 5.48 ± 5.08, and SIB increased by 16.6 ± 14.52.

Interpretation: These AD patients improved! This is not seen with standard-of-care medicine available today. Blocking (TNF)-alpha, thus stopping inflammation, leads to improved cognition in AD patients.

Why is this not treatment in the clinic today? Etanercept is a patented and approved drug that could be prescribed "off label" for the treatment of AD. But somehow this approach has not caught on. Eight years after the UCLA research (published in 2006), a review article on this general topic appeared titled, *Tumor Necrosis Factor Alpha: A Link between Neuroinflammation and Excitotoxicty.* [90] Not much progress has been made on therapies to stop (TNF)-alpha and help AD patients. These scientists draw the same conclusion as from the past, "modulating TNF-alpha signaling may represent a valuable target for intervention."

We hesitate to use the term "all," but all studies using antiinflammatory treatments do not improve patients with chronic inflammatory diseases. Sure, they can provide temporary relief and comfort and the etanercept example in AD is a classic example. However, if/when these studies are extended to 5 or 10 years, our guess is the patient will have deteriorated more rapidly compared to placebo. Inflammation is a treasure of our health. We need to work with, not against, inflammation. It is your immune system at work for you.

What is your hope if it is not stopping IL-6 and TNF-alpha? Find the root causes of this elevation in inflammatory cytokines and treat those causes. When the levels of these substances go back to normal naturally you, as a patient, will experience the health benefits described in this research.

Hope Conclusion

Your body's immune system is there to help, and we know it is working when inflammation is detected. You can diagnose and treat Alzheimer's effectively if you are able to find the often treatable cause(s) of inflammation. It is the cause(s) of inflammation that, when treated, will truly help Alzheimer's sufferers.

References

[1] Martin JB. Molecular basis of the neurodegenerative disorders. N Engl J Med 1999;340(25):1970–80.

[2] Roberts WC. Facts and ideas from anywhere. Proceedings (Baylor University. Medical Center) 2008;21(1):93.

[3] Norton S, et al. Potential for primary prevention of Alzheimer's disease: an analysis of population-based data. Lancet Neurol 2014;13(8):788–94.

[4] Wiel HI. Meaningless diagnoses. JAMA 1908;L(23):1889–90.

[5] Feinkohl I, et al. Severe hypoglycemia and cognitive decline in older people with type 2 diabetes: the Edinburgh Type 2 Diabetes Study. Diabetes Care 2014;37(2):507–15.

[6] Diabetic hypoglycemia. Mayo Clinic, 2012. Available from: http://www.mayoclinic.org/diseases-conditions/diabetic-hypoglycemia/basics/causes/con-20034680

[7] Schaefer EJ, Bongard V, Beiser AS, et al. Plasma phosphatidylcholine docosahexaenoic acid content and risk of dementia in Alzheimer disease. Arch Neurol 2006;63:1545–50.

[8] Connor WE, Connor SL. The importance of fish and docosahexaenoic acid in Alzheimer disease. Am J Clin Nutr 2007;85(4):929–30.

[9] Perlmutter D. MD:"grain brain." The surprising truth about wheat, carbs, and sugar-your brain's silent killers. New York: Little, Brown and Company; 2013.

[10] Wellcome Trust. Alzheimer's disease linked to Glaucoma. ScienceDaily, 7 August 2007.

[11] McKinnon SJ. Glaucoma: ocular Alzheimer's disease? Front Biosci 2003;8:s1140–56.

[12] Tamura H, et al. High frequency of open-angle glaucoma in Japanese patients with Alzheimer's disease. J Neurol Sci 2006;246(1):79–83.

[13] Yvonne O. Glaucoma and Alzheimer disease: age-related neurodegenerative diseases with shared mechanisms? J Clin Exp Ophthalmol 2012;S4:004.

[14] Pache M, Flammer J. A sick eye in a sick body? Systemic findings in patients with primary open-angle glaucoma. Surv Ophthalmol 2006;51(3):179–212.

[15] Gallessich G. Scientists at UCSB link brain plaques in Alzheimer's disease to eye disease. The Current, May 9, 2003. Available from: http://www.ia.ucsb.edu/pa/display.aspx?pkey=943

[16] Frost SM, Macaulay SL. Ocular biomarkers for neurodegenerative and systemic disease. Psychiatry 2013;3:e233.

[17] Goldstein LE, et al. Cytosolic β-amyloid deposition and supranuclear cataracts in lenses from people with Alzheimer's disease. Lancet 2003;361(9365):1258–65.

[18] Kirbas S, et al. Retinal nerve fiber layer thickness in patients with Alzheimer disease. J Neuroophthalmol 2013;33(1):58–61.

[19] Willis MS, Patterson C. Proteotoxicity and cardiac dysfunction—Alzheimer's disease of the heart? N Engl J Med 2013;368(5):455–64.

[20] Suzanne M, Wands JR. Alzheimer's disease is type 3 diabetes—evidence reviewed. J Diabetes Sci Technol 2008;2(6):1101–13.

[21] Vagelatos NT, Eslick GD. Type 2 diabetes as a risk factor for Alzheimer's disease: the confounders, interactions, and neuropathology associated with this relationship. Epidemiol Rev 2013;35:152–60.

[22] Poole S, Singhrao SK, Crean SJ. Emerging evidence for associations between periodontitis and the development of Alzheimer's disease. Faculty Dental J 2014;5(1):38–42.

[23] Shoemark DK, Allen SJ. The microbiome and disease: reviewing the links between the oral microbiome, aging, and Alzheimer's disease. J Alzheimers Dis 2014;43(3):725–38.

[24] Abbate J, Guynup S, Genco R. Scientific American presents: oral and whole body health. 2006:p. 6–49.

[25] Spector D. INFOGRAPHIC: Not Flossing Your Teeth Is Killing You. Business Insider, Apr 18, 2012. Available from: http://www.businessinsider.com/infographic-not-flossing-your-teeth-is-killing-you-2012-4

[26] Stuber J. Book says oil-pulling therapy detoxifies body. Samaritan Ministries, Sept 3, 2012. Available from: http://samaritanministries.org/book-says-oil-pulling-therapy-detoxifies-body/

[27] Susman E. Treating Afib: It Takes a Village. MedPage Today, Sept 05, 2014. Available from: http://www.medpagetoday.com/MeetingCoverage/ESC/47522

[28] Gillette-Guyonnet S, et al. Weight loss in Alzheimer disease. Am J Clin Nutr 2000;71(2):637s–42s.

[29] Laino C. Falls an early clue to Alzheimer's: falls more common in preclinical Alzheimer's disease, Study Finds. WebMD, Jul 18, 2011. Available from: http://www.webmd.com/Alzheimers/news/20110718/falls-an-early-clue-to-Alzheimers

[30] Alzheimer's disease: symptoms and causes. Mayo Clinic. Available from: http://www.mayoclinic.org/diseases-conditions/Alzheimers-disease/basics/risk-factors/con-20023871

[31] Brody JE. Averting diabetes before it takes hold. The New York Times, Sept 8, 2014. Available from: http://well. blogs.nytimes.com/2014/09/08/prediabetes-blood-sugar/?emc=eta1

[32] Kolatajune G. Promise Seen for Detection of Alzheimer's. The New York Times, Jun 23, 2010. Available from: http://www.nytimes.com/2010/06/24/health/research/24scans.html?pagewanted=all

[33] Castell JV, Gomez-Lechon MJ, David M, Andus T, Geiger T, Trullenque R, et al. Interleukin-6 is the major regulator of acute phase protein synthesis in adult human hepatocytes. FEBS Lett 1989;242:237–9.

[34] Annweiler C, Llewellyn DJ, Beauchet O. Low serum vitamin D concentrations in Alzheimer's disease: a systematic review and meta-analysis. J Alzheimers Dis 2013;33(3):659–74.

[35] Llewellyn DJ, et al. Vitamin D and risk of cognitive decline in elderly persons. Arch Int Med 2010;170(13):1135–41.

[36] Tobinick EL, Gross H. Rapid cognitive improvement in Alzheimer's disease following perispinal etanercept administration. J Neuroinflammation 2008;5(1):2.

[37] Cortes-Canteli M, et al. Fibrinogen and altered hemostasis in Alzheimer's disease. J Alzheimers Dis 2012;32(3):599–608.

[38] Mielke M, et al. Serum ceramides increase the risk of Alzheimer disease: the women's health and aging study II. Neurology 2012;79(7):633–41.

[39] Vallée M, Willy M, Moal ML. Role of pregnenolone, dehydroepiandrosterone and their sulfate esters on learning and memory in cognitive aging. Brain Res Rev 2001;37(1):301–12.

[40] Song J, Lee JE. Adiponectin as a new paradigm for approaching Alzheimer's disease. Anat Cell Biol 2013;46(4):229–34.

[41] Ward MA, et al. Evaluation of CSF cystatin C, beta-2-microglobulin, and VGF as diagnostic biomarkers of Alzheimer's disease using SRM. Alzheimers Dement 2011;7(4):S150–1.

[42] Chapman J. Warning test for Alzheimer's. Mail Online, Sept 8, 2014. Available from: http://www.dailymail.co.uk/health/article-177005/Warning-test-Alzheimers.html

[43] Tzikas S, et al. Increased myeloperoxidase plasma levels in patients with Alzheimer's disease. J Alzheimer's Dis 2014;39(3):557–64.

[44] Welsh P, et al. Do cardiac biomarkers NT-proBNP and hsTnT predict microvascular events in patients with type 2 diabetes? Results from the ADVANCE trial. Diabetes Care 2014;37(8):2202–10.

[45] Suchindran S, et al. Genome-wide association study of Lp-PLA2 activity and mass in the framingham heart study. PLoS Genet 2010;6(4):e1000928.

[46] van Oijen M, et al. Lipoprotein-associated phospholipase A2 is associated with risk of dementia. Ann Neurol 2006;59(1):139–44.

[47] Solfrizzi V, et al. Lipoprotein (a), apolipoprotein E genotype, and risk of Alzheimer's disease. J Neurol Neurosurg Psychiatry 2002;72(6):732–6.

[48] Larson EB, et al. Diagnostic tests in the evaluation of dementia: a prospective study of 200 elderly outpatients. Arch Int Med 1986;146(10):1917–22.

[49] Montine TJ, et al. F2-isoprostanes in Alzheimer and other neurodegenerative diseases. Antioxid Redox Signal 2005;7(1–2):269–75.

[50] Collingwood J, Dobson J. Mapping and characterization of iron compounds in Alzheimer's tissue. J Alzheimers Dis 2006;10(2):215–22.

[51] Cocciolo A, et al. Decreased expression and increased oxidation of plasma haptoglobin in Alzheimer disease: insights from redox proteomics. Free Radic Biol Med 2012;53(10):1868–76.

[52] Euser SM, et al. Serum uric acid and cognitive function and dementia. Brain 2009;132(2):377–82.

[53] Chen X, et al. Serum uric acid levels in patients with Alzheimer's disease: a meta-analysis. PLoS One 2014;9(4):e94084.

[54] Friso S, et al. Low circulating vitamin B6 is associated with elevation of the inflammation marker C-reactive protein independently of plasma homocysteine levels. Circulation 2001;103(23):2788–91.

[55] Hyman M, Liponis M. Ultraprevention: the 6-week plan that will make you healthy for life. New York: Simon and Schuster; 2003.

[56] Jones JL, Deanna KM, Wilson M. Toxoplasma gondii infection in the United States, 1999-2000. Emerg Infect Dis 2003;9(11):1371.

[57] Ravnskov U, McCully KS. Vulnerable plaque formation from obstruction of vasa vasorum by homocysteinylated and oxidized lipoprotein aggregates complexed with microbial remnants and LDL autoantibodies. Ann Clin Lab Sci 2009;39(1):3–16.

[58] Emelyanov VV. Evolutionary relationship of Rickettsiae and mitochondria. FEBS Lett 2001;501(1):11–8.

[59] Boyle PA, et al. Much of late life cognitive decline is not due to common neurodegenerative pathologies. Ann Neurol 2013;74(3):478–89.

[60] Monje ML, Hiroki T, Palmer TD. Inflammatory blockade restores adult hippocampal neurogenesis. Science 2003;302(5651):1760–5.

[61] Shors TJ, et al. Neurogenesis in the adult is involved in the formation of trace memories. Nature 2001;410(6826):372–6.

[62] Feng R, et al. Deficient neurogenesis in forebrain-specific presenilin-1 knockout mice is associated with reduced clearance of hippocampal memory traces. Neuron 2001;32(5):911–26.

[63] Cameron HA, McKay R. . Curr Opin Neurobiol 1998;8(677).

[64] Gage FH, Kempermann G, Palmer TD, Peterson DA, Ray J. Multipotent progenitor cells in the adult dentate gyrus. J Neurobiol 1998;36:249–66.

[65] Palmer TD, Takahashi J, Gage FH. The adult rat hippocampus contains primordial neural stem cells. Mol Cell Neurosci 1997;8:389–404.

[66] Suhonen JO, Peterson DA, Ray J, Gage FH. Differentiation of adult hippocampus-derived progenitors into olfactory neurons in vivo. Nature 1996;383:624–7.

[67] Luskin MB. Neuroblasts of the postnatal mammalian forebrain: their phenotype and fate. J Neurobiol 1998;36(2):221–33.

[68] Nakatomi H, et al. Regeneration of hippocampal pyramidal neurons after ischemic brain injury by recruitment of endogenous neural progenitors. Cell 2002;110(4):429–41.

[69] Magavi SS, Leavitt BR, Macklis JD. Induction of neurogenesis in the neocortex of adult mice. Nature 2000;405(6789):951–5.

[70] Cameron HA, Tanapat P, Gould E. Adrenal steroids and N-methyl-D-aspartate receptor activation regulate neurogenesis in the dentate gyrus of adult rats through a common pathway. Neuroscience 1998;82(2):349–54.

[71] Madsen TM, Kristjansen PE, Bolwig TG, Wortwein G. Arrested neuronal proliferation and impaired hippocampal function following fractionated brain irradiation in the adult rat. Neuroscience 2003;119(3):635–42.

[72] Monje ML, Mizumatsu S, Fike JR, Palmer TD. Irradiation induces neural precursor-cell dysfunction. Nat Med 2002;8(9):955–62.

[73] Monje ML, Palmer T. Radiation injury and neurogenesis. Curr Opin Neurol 2003;16(2):129–34.

[74] Picard-Riera N, et al. Experimental autoimmune encephalomyelitis mobilizes neural progenitors from the subventricular zone to undergo oligodendrogenesis in adult mice. Proc Natl Acad Sci USA 2002;99:13211–6.

[75] Vallieres L, Campbell IL, Gage FH, Sawchenko PE. Reduced hippocampal neurogenesis in adult transgenic mice with chronic astrocytic production of interleukin-6. J Neurosci 2002;22(2):486–92.

[76] Turrin NP, et al. Pro-inflammatory and anti-inflammatory cytokine mRNA induction in the periphery and brain following intraperitoneal administration of bacterial lipopolysaccharide. Brain Res Bull 2001;54:443–53.

[77] Shaw KN, Commins S, O'Mara SM. Lipopolysaccharide causes deficits in spatial learning in the watermaze but not in BDNF expression in the rat dentate gyrus. Behav Brain Res 2001;124:47–54.

[78] Terrazzino S, Bauleo A, Baldan A, Leon A. Peripheral LPS administrations up-regulate Fas and FasL on brain microglial cells: a brain protective or pathogenic event. J Neuroimmunol 2002;124:45–53.

[79] Vallieres L, Rivest S. Regulation of the genes encoding interleukin-6, its receptor, and gp130 in the rat brain in response to the immune activator lipopolysaccharide and the proinflammatory cytokine interleukin-1beta. J Neurochem 1997;69(4):1668–83.

[80] Deleted in review

[81] Leventhal C, Rafii S, Rafii D, Shahar A, Goldman SA. Endothelial trophic support of neuronal production and recruitment from the adult mammalian subependyma. Mol Cell Neurosci 1999;13:450–64.

[82] Louissaint A Jr, Rao S, Leventhal C, Goldman SA. Coordinated interaction of neurogenesis and angiogenesis in the adult songbird brain. Neuron 2002;34:945–60.

[83] Palmer TD, Willhoite AR, Gage FH. Vascular niche for adult hippocampal neurogenesis. J Comp Neurol 2000;425:479–94.

[84] Reichman HR, Farrell CL, Del Maestro RF. Effects of steroids and nonsteroid anti-inflammatory agents on vascular permeability in a rat glioma model. J Neurosurg 1986;65:233–7.

[85] Dickson DW. Neuropathology of Alzheimer's disease and other dementias. Clin Geriatr Med 2001;17:209–28.

[86] Mackenzie IR. Activated microglia in dementia with Lewy bodies. Neurology 2000;55:132–4.

[87] Brew BJ. AIDS dementia complex. Neurol Clin 1999;17:861–81.

[88] Weaver JD, et al. Interleukin-6 and risk of cognitive decline: MacArthur studies of successful aging. Neurology 2002;59:371–8.
[89] Tobinick E, et al. TNF-alpha modulation for treatment of Alzheimer's disease: a 6-month pilot study. Medscape Gen Med 2006;8(2):25.
[90] Olmos G, Lladó J. Tumor necrosis factor alpha: a link between neuroinflammation and excitotoxicity. Mediat Inflamm 2014;2014:861231.

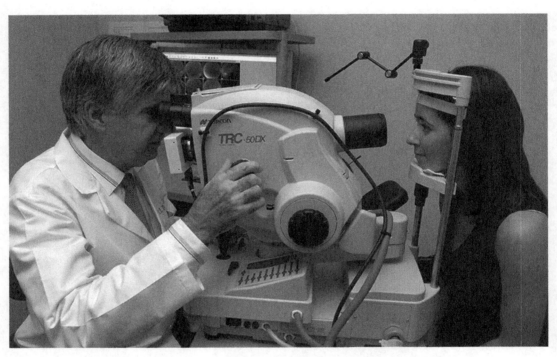

Image Credit: Ophthalmic care. © Dr Trempe, Center for Eye Care, New England College of Optometry.

Personal Stories

My father (TJL) (Papa) was a tinkerer. He enjoyed solving problems around the house, especially his children's houses. He was a navy man and engineering officer on several destroyers during the Pacific conflict of World War II. Before the navy, he had joined the merchant marines, and after the war, he went to both Northeastern University and Wentworth Institute to get his engineering degree. Upon returning to civilian life he worked at a high-rise in Boston as the head of maintenance and eventually became the superintendent in charge of all aspects of maintaining a building, including the renting of space, renovations, demolition, heating, and air conditioning. All of his children, my older brother, older sister, and I enjoyed an opportunity to work at "the building." The boys got to apprentice with carpenters, plumbers, roofers, and a variety of other skilled tradesmen. Thus the Lewis boys became handy and crafty, more like the jack-of-all-trades but really the master of none.

Papa often came to my aging, four-chimney farmhouse in southwestern New Hampshire. Where does his repair effort begin? I feared going to work because who knew what alteration he would make. The issue was that he didn't appreciate a "punch list," so he just did what he saw fit. And often his efforts and my desires didn't align. However, he ultimately always did something useful and clever, and I (hopefully) always showed appreciation.

In the fall of 1993, Papa, along with my mother "Neema" arrived and were accompanied by heavy rain and winds. Sadly the foliage took a hit and so too did the house. The house, being very old, had leaks and other problems exposed by the rain that Papa resolved to address. The issue of the day was the leaking roof over the porch. The porch didn't receive use except in the summer so there was no urgent need for repair. But this storm exposed the vulnerability of the roof and there was water on the porch floor.

Papa expressed dismay over the roof and the problem of the water on the floor. I dismissed the problem as something that required budget planning and that I would have it professionally repaired next year. Another day of work passed, and I came home to a chagrined household. "Go and check out the porch," said my daughter. I knew Papa had been up to his tricks. He would often fashion a Rube Goldberg solution when he wasn't inclined to travel to the hardware store, so I was prepared for an interesting solution to the leak. I inspected the porch roof from the outside and then the inside and saw nothing. Then I was instructed to look down. There they were—several holes were drilled through the floor. My father pronounced that the holes would let the water out and prevent the wood from rotting. This was a barely plausible explanation, but the solution was really very out of character for him.

The End of Alzheimer's. http://dx.doi.org/10.1016/B978-0-12-812112-2.00012-4

Years later, when my dad began the outward signs of suffering with clinical dementia and/or Alzheimer's disease (AD), I reflected back on this porch roof incident and now realize that the irrational behavior was the sign of a brain that was becoming diseased. However, in the early stages of disease, it appears that these departures are few and far between, so it is difficult to correlate one or two aberrant acts to disease as opposed to just a momentary lapse of judgment due to boredom or some other reason, such as a senior moment. As time passed, however, these types of behaviors started happening more frequently, from occurring every few months to more weekly occurrences. That is when we knew that Papa must be sick.

Alzheimer's is clearly a very complex disease. We don't have to look far for proof either. Here we are in the 21st century; we have sent humans to the moon and unraveled the mystery of the human genome, yet all pundits and thought leaders, from the Alzheimer's Association to the National Institutes for Health, continue to assert that there is no cure or even a way to slow the progress of the disease. Treatments are simply palliative, easing some signs of cognitive impairment temporarily. Even more confounding is that there isn't a consistent equivocal diagnosis, let alone an unequivocal one for the disease. Most medical professionals in neurology quietly agree that Alzheimer's can only be definitively diagnosed upon death.

Here we have the most dreaded and costly disease ever to face humankind from the standpoint of human dignity, emotional and financial costs to loved ones, and overall cost to society, and we are essentially defenseless to even understand the disease at its most basic level, at least "officially," despite billions of dollars in research spending. Medicine has raised the white flag.

PAPA'S STEADY DECLINE

With time Papa's memory lapsed more frequently as did his unusual actions. He was 80 years old when a specialist who took him through a further series of cognitive tests including the Mini Mental State Exam saw him. He did not score a normal "30" that even a child of 10 could pass with a perfect score. At that time, he was diagnosed definitively with advancing cognitive impairment/dementia/Alzheimer's. The advantage of an actual dementia or Alzheimer's diagnosis meant that he would be prescribed pharmaceuticals, including Aricept by Pfizer or other so-called acetyl cholinesterase inhibitors. We were told that these drugs do not change the course of his disease but might make him more functional on a daily basis. These drugs are known to stimulate healthy brain neurons and thus provide more function to those not yet impacted by the disease. These drugs, we now know, stimulate the brain similarly to nicotine. However, the effect of these drugs diminishes with time because there are fewer and fewer healthy neurons to stimulate.

The prognosis given back in 2000, 7 years after the roof incident, was that there was nothing that could be done to curb the course of the disease. That is still the mantra of standard medicine today. We were told to make appropriate plans for his care that included making sure that he had a stable, consistent, and familiar routine. This would allow him to use his longer-term memory and thus enable him to be more functional. We also received strong recommendations to quickly move him into assisted living, starting with part-time "day care" in facilities experienced with dementia patients. We were told that he would need full-time nursing home care in short order.

The slow transition from part-time dementia to full-time dementia is particularly difficult, especially in families of the 1940s and 1950s where the husband ruled supreme. My father always drove, managed the finances, and generally ruled the roost. He was not (I believe) aware of his periods of cognitive lapses, thus he was beyond reluctant to give up any of his command. One of the biggest challenges the family faced was his driving. Even as he became severely demented, it was very difficult to get him to cede to my mother when it came to driving. We all knew this created a danger to himself, my mother, and anyone unfortunate enough to be on the road when he was driving. That meant that either great risk was taken, or my mother and father stayed home, even when my mother was more than capable to drive, for example, the 2 hours to visit grandchildren. Needless to say, she could not leave him home alone, so confronting the driving issue was almost a daily struggle as life goes on and there are errands to run. Thank God, we were eventually able to wrest the driving from him, and my mother could still venture out to take care of necessities and enjoy her family and her life.

My mother is very much "old school," having lived through the depression and "the war." She married my father at the age of 24 and was completely committed to family. She taught school briefly but then devoted herself to raising three children over the next 20 years. She went back to teaching, only to help the family financially in preparation to send us kids to college, and only after I reached grade school. She never could envision abandoning him to someone else but did concede to having him go to a nearby day care type facility a couple days each week. This allowed her to catch up with chores and daily household maintenance that she had now assumed completely at the age of 77.

My mother so completely and selflessly managed my father and the home that my siblings and I were somewhat insulated from his true condition. We did learn later that he was somewhat typical in that during his severe episodes, he would lash out and become violent. She often explained bruises as being caused by her clumsiness. We later smartened up as we each educated ourselves and realized that Papa could get violent. This would happen during the height of care when she was doing simple things like dressing him or getting him to use the bathroom. We now understand that dementia sufferers have some level of awareness of what they want or should do, but cannot either do it or communicate their needs, like a very young child. Thus they apparently develop a great deal of frustration that is expressed as anger. This makes complete sense for a man like my father who was a leader of many, always in control of the helm, and fiercely independent.

Regardless of my father's behavior and the prompting of his doctors, my mother disregarded any suggestions to place him in full-time care. She had vowed, at the time of their wedding, to be there for him for better and for worse and in sickness and in health. She was not one to compromise on her promise.

Amidst tragedy, light moments can be found. Some of the things Papa would say were so off-the-wall that they were almost funny, especially to the innocent grandchildren. Here are some examples recorded in my daughter's journal:

> Papa: "Where did you get your hat?"
> Anya: "I don't have a hat on."
> Papa: "You don't?"
> Anya: "No."
> Papa: "Somebody must be crazy."
> Anya: "Are you going to bed soon Papa?"

Papa: (walks toward the window) "I'm going to go see what this one's name is."
Anya: "What?"
Papa: (lifts up curtain) "Yup, looks like this one is all the way."
Anya: "All the way what?"
After hours of trying to get papa's attention by calling him, I realized that he didn't realize that he was Papa, the name I have been calling him for the past 17 years.

—Anya Lewis

My mother continued to be his caregiver with the exception of visits to the day care center. They eventually refused him because he became too difficult to manage. At that time, we explored full-time facilities that could take him. Being a veteran, a VA facility was available to care for him, but they did not have any immediate availability. My mother used that as an excuse to manage his care herself until a bed opened up. By now he was almost impossible to care for, incontinent, and frequently violent. All the doctors told my mother that she was going to kill herself by caring for him. She said she would submit when the VA facility opened. A bed finally opened in September of 2004, and he went there. She, of course, visited him daily. I don't believe any other family members had the courage to go visit him at that facility.

On October 8th, 2004, just 1 month after going to the VA facility, we received a call from a doctor with a deep and somber voice informing us that Daniel Lewis had passed. I took the call and my mother quickly joined. She managed the conversation. She didn't cry and just accepted the news. She had done all she could humanly do for him, and now her job was completed. We buried him in a plot in Peabody, MA that they had reserved 40 years before.

We donated his brain to research. The family was pleased we made this contribution; however, we had no way to obtain feedback as to how his brain was used and how, if in any way, that donation contributed to finding either a cause or a cure.

Search for Answers

Charles Colton, British churchman and writer (1780–1832) wrote: "Body and mind, like man and wife, do not always agree to die together." We need to find out why the brain dies before the body, leading to the devastating disease now called "Alzheimer's disease."

Throughout the decade of my father's AD, my family was in relative denial. What could we do anyway; we were told there was nothing that could be done except to be supportive. With my mother being both stoic and capable, I contributed little and frankly gave little consideration to my father's conditions. Several things that make this disease so insidious exacerbated this.

First, the disease comes on gradually and often just in quiet waves. Thus, my two siblings and I did not frequently witness any big issue early on. Also, my sister, the classical caregiver of a family besides the wife, lived 5,000 miles away in Hawaii. My brother lived 1,200 miles away in Indiana and had a large tract of land, a farm, and many animals for which he provided care. This then left me to helping mom on a regular basis. During the period of my father's decline, I was involved in raising three children and all the trappings of family life.

I also yearned for a career beyond what I was doing, which at the time, was working for a mid-sized corporation. Nights and weekends were consumed with developing new opportunities. I also moonlighted consulting for a company in Pennsylvania. I eventually joined that company and wound up having a brutal travel schedule for the next few years. That company soon imploded, and I maintained a rigorous consulting schedule for the next couple of years that also involved substantial travel. While this was ongoing, I started a company from scratch with a local businessman. Because of financial issues, I continued the consulting work while working on this start-up.

All these family circumstances substantially left my mother to fend for herself. Life gets busy, and life goes on. Besides, we were told there was nothing we could do. And we children didn't have the same resolve to keep Dad out of a professional care facility. What I neglected to consider was that, since his fate was sealed, my efforts were not for my father but rather to help my mother. She was always so strong and capable, so I assumed that she could and would handle anything.

Ten years after the passing of my dad, my mom, at the age of 91, is doing heroically well. And, dedicated to my father's memory, I am on a crusade to change the course of AD with a little help from $1,000,000,000,000 of annual medical and related research.

Why did Papa become afflicted with Alzheimer's? That is a perplexing question because he had a large garden that fed the family fresh fruits and vegetables. Papa was very active but not an athlete. That means he obtained sensible exercise that is tied to better brain health. His job was not particularly stressful, and he enjoyed it immensely. His home life was also free from stress. He didn't smoke, and he drank in moderation. He developed glaucoma about 10 years prior to being diagnosed with Alzheimer's. Had we known then what we know now, we would have done an exhaustive differential diagnosis to try to find root causes of the glaucoma. Proper management of glaucoma probably would have staved off his Alzheimer's.

Papa exposed himself to one major risk factor that, taken by itself, should probably not have caused him to have his disease. He almost never ate fish.

THE STORY OF DR. LEE

Did you just read *Hope for the Afflicted* in the last chapter? The story of Dr. Lee is a real-life story of someone diagnosed with presumed Alzheimer's who was treated to remove the "inflammatory blockade." Dr. Lee improved.

I met Dr. Lee at a conference of concierge doctors in Scottsdale, AZ in 2011. He was the first concierge doctor in "the valley" of the Scottsdale area. Before he started that practice, he was one of the first doctors at the Mayo facility in Arizona and was a highly trained Mayo clinician. He is a robust man of about 6'2" who was a star at basketball during his college years. Three years prior, Dr. Lee had relinquished his medical practice because he was having severe memory problems. In a single patient visit, it was becoming an all too common event for Dr. Lee to ask a patient the same question three or more times.

When I met Dr. Lee in Scottsdale he appeared quite functional, but that was just on the surface at first glance. I asked him for his cell phone number and the stammering began. He could not come up with the number. Dr. Doug was well aware of his dementia problem and was, of course, very frightened by his prognosis. His former colleagues at Mayo diagnosed him as having AD.

Shortly after our meeting, Dr. Lee flew to Boston to be evaluated. We cannot say he doesn't have Alzheimer's or another form of dementia. However, based on the experience of a doctor colleague, what we can say is that there were clear indications of what was causing his disease with very high certainty. He had elevated markers of inflammation in his blood and multiple infectious species. He has undergone 1 year of treatment under close supervision by Dr. Bryan, who took over Dr. Lee's medical practice. Dr. Lee's mini mental score has increased by approximately 7 points, and he is clearly much more functional by all measures and according to everyone in his inner circle who interacts with him regularly. Dr. Lee would like to think he could return to medical practice. What is more important is that the MMSE score didn't decline over the year of therapy. At this stage in the disease, Dr. Lee could have experienced a 1–5 point decline in the MMSE. Thus a 7-point increase is really an 8–12 point turnaround!

The story of Dr. Lee is not a single, isolated, or coincidental event. His improvement is a direct consequence of the billions of dollars of research presented in this book. More importantly, it is the efforts of a courageous doctor willing to step beyond the standard-of-care, read the latest research, and translate that information for the benefit of his patients. This doctor is single-handedly circumventing the Trillion Dollar Conundrum.

You or your loved one could experience the same positive results if you are able to obtain a true differential diagnosis that guides your doctors to a path of curative therapy.

MR. LP HOLDING ALZHEIMER'S AT BAY

I (TJL) met Mr. LP about 20 years ago. At that time, he had sold his successful business and was interested in investing in new technology. We spent some time together evaluating some offerings by interesting entrepreneurs. Mr. LP eventually invested in one of those early-stage companies, and I didn't see much of him for quite some time.

A good friend of mine, Tom, moved next door to LP. He told me that he and his wife often went to dinner with Mr. LP and his wife. About 8 years ago, Tom told me that they stopped going to dinner because Mr. LP was developing Alzheimer's (diagnosed at a local major hospital), as he was unable to participate in normal conversation and would get very antsy before dinner was finished.

I stopped by Mr. LP's home after hearing the news about his disease. His was an elegant old New England home. He was on the back porch attended by a visiting nurse. Mr. LP did not recognize me and, after some leading questions, barely remembered the small company in which he had invested a scant 8 years before. He went to see my coauthor after long discussion with Tom, his wife, and the wife of Mr. LP.

Six years after his first clinical appointment and subsequent treatment, Mr. LP still lives at his magnificent home and dines out with his dedicated wife. Is Mr. LP cured of Alzheimer's? Most would argue that he is not technically cured; however, his disease progression has halted by all measures, and in some areas he has actually improved. He still suffers from restlessness, but his memory is clearly improved compared to when he had his first clinical visit. Mrs. LP decided that the approach a local major hospital was using (i.e., measuring Mr. LP's decay and using tired, old treatment) was not working, and she stopped those visits.

Objectively, it appears that Mr. LP's memory and overall functioning improved slightly within the first 3–6 months of treatment and has subsequently not deteriorated. The key to his treatment was a revealing differential diagnosis that directed a treatment regiment and a dedicated wife who kept him compliant with the program.

SF'S DISEASE REVERSES

According to this 75-year-old male and his wife and caregiver, he improved in many ways that impacted his ability to take on the basic activities of daily living and then some. He stated, "Dr. Trempe's unique way on treating healed my eye. Even more importantly his treatments led to my health improving on several fronts including improved muscle strength, the ability to walk without a cane, better memory, ability to do basic math, the ability to take a phone call, and vast improvement in the way my hands would shake and tremble."

MR. 84 IMPROVES DRAMATICALLY

Mr. 84 is an 84-year-old gentleman who came to my coauthor because of macular degeneration. From the very first visit, a loving daughter who oversaw his care accompanied him. He had a myriad of diseases when Dr. Trempe first saw him, including macular degeneration, dementia (Alzheimer's?), and severe muscle atrophy.

During his first clinic visit he was asked a series of simple questions, akin to those on the Mini Mental State Exam, and he could seldom find words to articulate an answer. His daughter explained that he could often describe things, like an orange, but often could not come up with the actual word "orange." Here are some of the key differences before and after a comprehensive differential diagnosis and treatment.

Before: Walked with a cane and held on to a shopping cart for stability (poor balance), struggled to reach a standing position from a sitting position (muscle strength), tremor (so bad he required a straw to drink coffee), antisocial behavior (unable to engage with people around him), poor cognitive functioning (unable to find the right words to describe objects and things), and poor central vision.

After: He shed his cane and walks independently, stands up with authority, his tremor is completely gone, he is engaging (okay, a bit of a crotchety old-time New Englander), he can identify an orange without hesitation, and his vision improved using all optometric measure.

Mr. 84, when listening to his doctors and daughter talking about his memory 7 months after initial treatment chimed in, "There is nothing wrong with my memory." No one had cause to disagree.

SOME FAMOUS PEOPLE WHO HAD ALZHEIMER'S DISEASE

Joe Adcock – Baseball player
Mabel Albertson – Actress
Dana Andrews – Actor
George Balanchine – Dancer, choreographer
Rudolph Bing – Opera impresario
James Brooks – Artist
Abe Burrows – Author
Glenn Campbell – Actor
Joyce Chen – Chef

Aaron Copeland – Composer
Willem DeKooning – Artist
Thomas Dorsey – Father of gospel music
Tom Fears – Hall of fame professional football player and coach
Louis Feraud – Prominent fashion designer
Arlene Francis – Actress
Mike Frankovich – Movie producer
John Douglas French – Physician
Barry Goldwater – Arizona Senator
Rita Hayworth – Actress
Raul Silva Henriquez – Roman Catholic cardinal, human rights advocate
Philip Klutznick – Real estate developer, adviser to five Presidents of the United States
Mervyn Leroy – Director
Daniel Lewis – My father
Jack Lord – Actor
Ross MacDonald – Author
Burgess Meredith – Actor
Iris Murdoch – Author
Edmond O'Brien – Actor
Arthur O'Connell – Actor
Marv Owen – Baseball player
Molly Picon – Actress
Otto Preminger – Director
Bill Quackenbush – Hall of fame professional hockey player
Ronald Reagan – Former President of the U.S.
Harry Ritz – Performer
Sugar Ray Robinson – Boxer
Norman Rockwell – Artist
Betty Schwartz – First woman to win an Olympic gold medal in track
Simon Scott – Actor
Irving Shulman – Screenwriter
Kay Swift – Composer
Alfred Van Vogt – Science fiction writer
E.B. White – Author
Harold Wilson – British Prime Minister

LETTER FROM PRESIDENT RONALD REAGAN TO THE AMERICAN PEOPLE: NOV. 5, 1994 [1]

My Fellow Americans, I have recently been told that I am one of the millions of Americans who will be afflicted with AD. Upon learning this news, Nancy and I had to decide whether as private citizens we would keep this a private matter or whether we would make this news known in a public way. In the past, Nancy suffered from breast cancer, and I had my cancer surgeries. We found through our open disclosures we were able to raise public awareness. We were happy that as a result, many more people underwent testing. They were treated in early stages and able to return to normal, healthy lives. So, now we feel it is important to share it with you. In opening our hearts, we hope this might promote greater awareness of this condition. Perhaps it will encourage a clearer understanding of the individuals and families who are affected by it. At the moment

I feel just fine. I intend to live the remainder of the years God gives me on this earth doing the things I have always done. I will continue to share life's journey with my beloved Nancy and my family. I plan to enjoy the great outdoors and stay in touch with my friends and supporters. Unfortunately, as AD progresses, the family often bears a heavy burden. I only wish there was some way I could spare Nancy from this painful experience. When the time comes I am confident that with your help she will face it with faith and courage. In closing, let me thank you, the American people, for giving me the great honor of allowing me to serve as your president. When the Lord calls me home, whenever that may be, I will leave with the greatest love for this country of ours and eternal optimism for its future. I now begin the journey that will lead me into the sunset of my life. I know that for America there will always be a bright dawn ahead. Thank you, my friends. May God always bless you. Sincerely, Ronald Reagan.

FINAL THOUGHT FROM A CHEMIST

Did you take chemistry in high school? How about basic chemistry in college or the dreaded "P" chem.? Physical chemistry in particular provides us with many useful ways to describe and understand the world around us. Medicine and human physiology is very much part of that world. Real-world concepts that chemistry teaches us are thermodynamics and kinetics.

Thermodynamics is a measure of the energy required for something to happen, while kinetics is a measure of the rate at which something does happen. Take the burning of gasoline, for example. When it burns, it does so violently, releasing plenty of energy in the form of heat and light. This reaction is always ready to go because the products (water and carbon dioxide) are more stable compared to the reactants (gasoline and oxygen). However, the gasoline and oxygen could stay combined indefinitely without reacting. Why? There is an "activation barrier" to overcome.

The overarching measure that determines if a reaction is even possible is the Gibbs Free Energy (ΔG). If ΔG is negative, the reaction has enough energy to go; if the kinetics is right, the activation energy can be overcome.

Aging tells us that the ΔG of the human body is negative. We are more stable by returning to our base elements compared to the marvelous and intricate structures God created that allows us to be alive. Our goal to maintain our health is a simple one, to raise the activation barrier bar as high as possible to slow the reactions to our more stable state, then return to Earth as ashes and dust. All aspects of bodily homeostasis, immune health, and emotional well-being contribute to raising the bar.

Stay Well.

Reference

[1] Post Stephen G. The moral challenge of Alzheimer disease: ethical issues from diagnosis to dying. Baltimore: JHU Press; 2000.

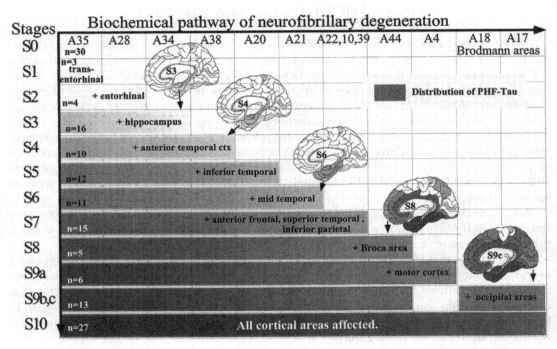

Biochemical pathway of neurofibrillary degeneration

Stages	A35	A28	A34	A38	A20	A21	A22,10,39	A44	A4	A18	A17
S0	n=30										Brodmann areas
S1	n=3 trans-entorhinal										
S2	n=4 + entorhinal							Distribution of PHF-Tau			
S3	n=16	+ hippocampus									
S4	n=10		+ anterior temporal ctx								
S5	n=12			+ inferior temporal							
S6	n=11				+ mid temporal						
S7	n=15			+ anterior frontal, superior temporal, inferior parietal							
S8	n=5						+ Broca area				
S9a	n=6							+ motor cortex			
S9b,c	n=13									+ occipital areas	
S10	n=27			All cortical areas affected.							

Image Credit: Pathway of neurofibrillary degeneration (NFD) in aging and Alzheimer's disease, from Delacourte A. et al. The biochemical pathway of neurofibrillary degeneration in aging and Alzheimer's disease. Neurology 1999;52(6):1158-65.

Stages of Alzheimer's

Patients with Alzheimer's disease (AD) generally have three kinds of symptoms. First is impairment in activities of daily living. The ability to perform activities, such as brushing, bathing, toilet habits, and dressing is lost, usually in the more advanced stages of the disease. Second is abnormal behavior in the patient. This can be very distressing; for example, patients may suspect their spouse of 50 years to be a thief, not recognize their own children, or make aggressive advances toward caregivers. Some patients even become abusive, possibly in contradiction to their past behavior. Restlessness at night is another pattern that is very troublesome to the entire family. Third, intellectual capability of the patient is lost. The ability to perform activities that they were very good at, such as arithmetic calculations, planning, giving their opinions, and making difficult decisions is lost. They seem to get confused when confronted with the slightest problem. The most striking aspect is the pattern of loss of intellectual function that follows the principle of "last learned, first lost."

In the early stages of the disease, the most complicated functions, such as those learned in a professional setting, are the first to be lost. However, basic capabilities, such as toilet habits and recognition of immediate family members, are preserved. As the disease progresses, additional functions are lost. In the more advanced stages, the patient becomes almost childlike, being completely dependent on others in terms of being fed, bathed, dressed, and exercised. The rapid progression of this disease leads to premature death, often due to causes, such as food going into the lungs or fracture due to a fall. Patients can, however, survive up to 20 years after the onset of illness.

The early symptoms of AD, such as forgetfulness, loss of olfactory sense, inability to learn new things, loss of concentration, unexplained weight loss, and difficulties in walking may be overlooked because they resemble signs of natural aging. Memory for events in the remote past is preserved in the early stages of the disease.

In healthy individuals, similar symptoms can result from fatigue, grief or depression, vision or hearing loss, the use of alcohol or certain medications, or simply the burden of too many details to remember at once. But when memory loss worsens, family and friends can perceive the existence of more serious problems. Often there are accompanying problems, such as hearing loss and a decline in reading ability, as well as general physical debility in patients of AD. These further add to their disability.

Between the inevitable consequences of aging and AD is the "gray area" in which some people suffer loss of intellectual functions that are more than mild and yet not severe enough to be considered AD. Scientists are unsure of the terminology for these patients. Many terms are in use, such as minimal cognitive impairment, mild cognitive impairment, or benign senescence of old age. This is an area of intense research to determine whether these cases

The End of Alzheimer's. http://dx.doi.org/10.1016/B978-0-12-812112-2.00013-6

will eventually progress to AD and, if so, who will progress and why, or which of these cases remain in the "grey area."

Differences between normal signs of aging and early AD:

Memory and concentration

Normal	Early signs of Alzheimer's disease
Periodic minor memory lapses or forgetfulness of part of an experience. Occasional lapses in attention or lapses in attention or concentration.	Misplacement of important items. Confusion about how to perform simple tasks. Trouble with simple arithmetic problems. Difficulty in making routine decisions. Confusion about month or season.

Mood and behavior

Normal	Early signs of Alzheimer's disease
Temporary sadness or anxiety based on appropriate and specific cause. Changing interests. Increasingly cautious behavior.	Unpredictable mood changes. Increasing loss of outside interests. Depression, anger, or confusion in response to change. Denial of symptoms.

Differences between normal and late Alzheimer's disease:
Language and speech

Normal	Late signs of Alzheimer's disease
Unimpaired language skills.	Difficulty in completing sentences or finding the right words. Reduced and/or irrelevant conversation.

Movement and coordination

Normal	Late signs of Alzheimer's disease
Increasing caution in movement. Slower reaction time.	Visibly impaired movement or coordination, including slowing of movements, halting gait, and reduced sense of balance.

STAGES OF ALZHEIMER'S

Placing all AD or dementia patients into one diagnostic category is very limiting because we may learn with time that the stages of the disease, particularly the early stages, advance treatment options more quickly. Consider the possibility that the disease progresses due to several overlapping factors and that the initial factors are more treatable than those that develop later. This thesis has plausibility because, as the disease progresses and the body weakens, other or more factors may play into its progression. Thus breaking down the disease into stages for the purpose of classification, diagnosis, and treatment is a good first step to developing solutions to this complex malady.

We know for sure that AD is a progressive chronic disease that has a long incubation or long asymptomatic period. By the time an official diagnosis is made, the person's function is usually significantly impaired, and treatment rarely helps. Doctors' groups are suggesting a redefinition of AD that would include even mild memory and behavioral symptoms. This is an increase in definition of stages for clinicians from one stage (no Alzheimer's/yes Alzheimer's) to include a mild cognitive impairment. The original criteria for AD, written in 1984, defined the disease in a single stage and assumed people who didn't have symptoms did not have the disease. This original definition addressed only later stages of the disease, when it ravages the patient's ability to function. Disease experts expect an increase in the number of patients receiving the Alzheimer's diagnosis as a result of the change in disease classification.

Clinicians need to quickly categorize patients and determine further diagnostic and treatments based on the stage of the disease. The presumption that there is no bona fide diagnosis or treatment makes the need for a broad range of categories mute. Nonclinicians and thus caregivers benefit from a broader range of stage definitions that allow them to note the status of their loved ones. This is most helpful for measuring deterioration either with or without medications. The standard medications given to Alzheimer's suffers do not change the course of the disease but provide some relief of symptoms. The broader scale helps the caregiver more accurately determine when the palliative effects of treatments are no longer there. Thus medications can be stopped at this time, saving money and avoiding any side effects of the medications.

The idea, proposed by the Alzheimer's Association and the National Institutes of Health, would define Alzheimer's as a "spectrum" disease, creating three stages ranging from lesser to greater severity in hopes that the devastating neurological condition could be detected earlier. The aim of identifying the disease earlier is to get patients in the pipeline for research for future treatment. When the disease isn't identified until later in its progression, patients are more impaired, and treatments that hopefully emerge are likely to be less effective in advanced cases compared to earlier ones.

The newly expanded stages of AD are meant to cover the full spectrum of the disease as it progresses over the years. The expectation is that for people working with Alzheimer's patients day-to-day, the new definition will not make a difference in practice. Most doctors working with Alzheimer's have already accepted that the disease is a progression. However, this is the first time it is to be codified into a standard guideline. In some sense, it's the formalization of things that people already know. The stages have been divided into preclinical AD, mild cognitive impairment, and dementia.

NEW ALZHEIMER'S ASSOCIATION THREE-STAGE DEFINITION OF ALZHEIMER'S

First Stage: Preclinical Alzheimer's Disease This stage is for research purposes only and will have no effect in a doctor's office, at least as far as the Alzheimer's Association is concerned. The idea is that patients could be developing AD even when they are free of cognitive or memory problems. This stage is to help researchers determine whether there is a biological change caused by Alzheimer's that can be detected through blood, spinal fluid test, neuroimaging, or more advanced testing. Right now, there is no single test that accurately predicts whether a person will develop AD.

Many studies show that Alzheimer's patients experience changes in the brain that include the buildup of amyloid protein tangles and nerve cell changes. It is unknown whether this means an inevitable progression to Alzheimer's dementia, as some Alzheimer's patients never show such markers upon autopsy. "Changes in their brain can be measured, but we can't predict for sure whether they're going to have the clinical disease," said Dr. Creighton Phelps, one-time director of the Alzheimer's Disease Centers Program of the National Institutes of Health. These tests are used only in research settings. Scientists are working to develop a more definitive test or a scan to determine Alzheimer's risk. The idea is to develop a comprehensive set of data of patients and follow their health as they age. Hopefully, with an accurate enough set of diagnoses, scientists will be able to establish an earlier cause and effect relationship that will aid in developing treatments.

Second Stage: Mild Cognitive Impairment (MCI) Long before a person gets an Alzheimer's diagnosis, he or she may show small changes in memory, behavior, and thinking. This is called mild cognitive impairment. While it does not cripple a person's ability to function throughout the day, these changes are known to impact some aspects of daily living, and friends and family members often notice these differences. Their doctors already observe some patients in this stage as "probable Alzheimer's." This is a gray area because not all memory problems are Alzheimer's-related. Cognitive difficulties could stem from other factors, such as a drug's side effects or vascular disease.

This stage could be used in specialized Alzheimer's clinics. A specialist might determine that underlying AD after a comprehensive exam, based on the disease process and symptoms, causes the cognitive problems, but there are no blood or medical tests available in the "standard-of-care" to confirm whether the mild cognitive impairment is because of Alzheimer's. There is emerging evidence about the connection between AD and cardiovascular disease. Tests for cardiovascular diseases and others should be considered at this stage of mild cognitive impairment.

Third Stage: Dementia Because of Alzheimer's This is the stage when memory, thinking, and behavioral symptoms have become so damaging that the patient's ability to function is hindered. The disease is not solely restricted to memory problems. The new guidelines include other symptoms, such as difficulty finding words, visual and spatial problems, impaired reasoning, and judgment. This is the stage with which people are most familiar. The patient eventually becomes unable to carry out basic daily tasks including eating and bathroom-related functions and becomes fully dependent on others for basic care. The purpose of setting out these updated stages of Alzheimer's is aimed at the future, Dr. Phelps said, to "define a research strategy for people who may be at risk for Alzheimer's."

SEVEN STAGES OF ALZHEIMER'S

The Alzheimer's Society of Canada presented the more comprehensive set of seven stages for the progression of AD in 2005, called The Global Deterioration Scale. The true value of the stage definitions is to enable those nonclinicians not knowledgeable in the disease to be able to more fully characterize the state of their loved ones when communicating to their health care providers.

Researchers and physicians developed these seven stages to describe how you or your loved one will change over time. Most health care professionals collapse the seven stages into early/middle/late or mild/moderate/severe, as proposed by the Alzheimer's Association and the National Institutes of Health. Today, the three-stage description is considered the standard for describing an Alzheimer's patient. While the three-stage system is more useful clinically to help doctors quickly determine courses of disease management, the seven-stage scale is helpful to caregivers in terms of providing more descriptions and definitions to help these folks describe the state of their loved ones.

Stage 1 (absence of impairment): There are no problems with memory, orientation, judgment, communication, or daily activities. Therefore, the patient is a normally functioning adult.

Stage 2 (minimal impairment): The person might be experiencing some lapses in memory or other cognitive problems, but neither family nor friends are able to detect any changes. A medical exam would not reveal any problems either, as they are likely intermittent (just like that car problem that never occurs for the mechanic).

Stage 3 (noticeable cognitive decline): Family members and friends recognize mild changes in memory, communication patterns, or behavior. A visit to the doctor might result in a diagnosis of MCI, early-stage or mild AD, but not always. Common symptoms in this stage include:

- problems producing people's names or the right words for objects;
- noticeable difficulty functioning in employment or social settings;
- forgetting material that has just been read;
- misplacing important objects with increasing frequency; and
- decrease in planning or organizational skills.

Stage 4 (early-stage/mild Alzheimer's): Cognitive decline is more evident. The person may become more forgetful of recent events or personal details. Other problems include impaired mathematical ability (for instance, difficulty counting backwards from 100 by 9s), a diminished ability to carry out complex tasks like throwing a party or managing finances, moodiness, and social withdrawal.

Stage 5 (middle-stage/moderate Alzheimer's): Some assistance with daily tasks is required. Problems with memory and thinking are quite noticeable, including symptoms, such as:

- an inability to recall one's own contact information or key details about one's history;
- disorientation to time and/or place; and
- decreased judgment and skills in regard to personal care.

Even though symptoms are worsening, people in this stage usually still know their own name and the names of key family members and can eat and use the bathroom without assistance.

***Stage 6 (middle-stage/moderate to late-stage/severe Alzheimer's)*:** This is often the most difficult stage for caregivers because it's characterized by personality and behavior changes. In addition, memory continues to decline, and assistance is required for most daily activities. The most common symptoms associated with this stage include:

- reduced awareness of one's surroundings and of recent events;
- problems recognizing one's spouse and other close family members, although faces are still distinguished between familiar and unfamiliar;
- sundowning, which is increased restlessness and agitation in the late afternoon and evening;
- difficulty using the bathroom independently;
- bowel and bladder incontinence;
- repetitive behavior (verbal and/or nonverbal); and
- wandering.

***Stage 7 (late-stage/severe Alzheimer's)*:** In the final stage, it is usually no longer possible to respond to the surrounding environment. The person may be able to speak words or short phrases, but communication is extremely limited. Basic functions begin to shut down, such as motor coordination and the ability to swallow. Total care is required around the clock.

Although the seven stages provide a blueprint for the progression of Alzheimer's symptoms, not everyone advances through the stages similarly. Caregivers report that their loved ones sometimes oscillate between two or more stages at once, and the rate at which people advance through the stages is highly individual. Still, the stages help us understand Alzheimer's symptoms and prepare for their accompanying changes and challenges.

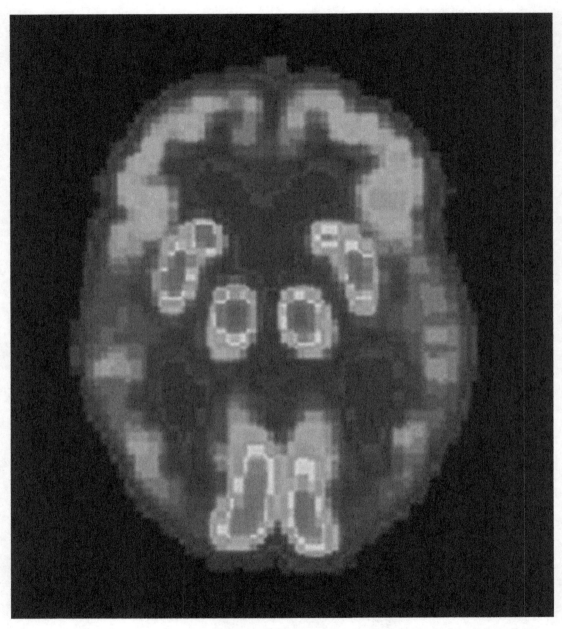

Image Credit: PET scan of a human brain with Alzheimer's disease, by US National Institute on Aging, Alzheimer's Disease Education and Referral Center. Available at Wikimedia Commons: https://commons.wikimedia.org/wiki/File%3APET_Alzheimer.jpg

Alzheimer's Diagnosis Criteria

Several groups have made recommendations as to the diagnostic tests that should be performed on an Alzheimer's suspect, based on the various standard-of-care tests that exist. The following is a review of some of these recommendations that are sold as differential diagnoses criteria. It is clear that few if any of the standard-of-care tests focus on objective measures of disease and the quest for the answer to "why." The main intent is to differentiate between the different neurological disease like Alzheimer's and dementia.

Diagnostic Criteria of Alzheimer's Disease According to ICD-10 (World Health Organization, 1993)

- Presence of dementia.
- Insidious onset and slowly progressive deterioration. It is usually difficult to detect the onset of disorders, and the patient's entourage sometimes becomes suddenly aware of the presence of deterioration. An apparent plateau may occur in the progression.
- Absence of argument, following the clinical examination and complementary investigations, to suggest that the mental state may be due to other systemic or brain disease which can induce a dementia (e.g., hypothyroidism, hypercalcemia, vitamin B12 deficiency, nicotinic acid deficiency, neurosyphilis, normal pressure hydrocephalus, or subdural hematoma.)
- Nonsudden onset and absence, at an early stage of the evolution, of neurological signs of focal damage, for example, hemiparesis, sensory loss, visual field defects or a lack of coordination (these manifestations may, however, be added secondarily).

Diagnostic Criteria of Alzheimer's Disease According to DSM-IV (American Psychiatric Association, 1994)

The development of multiple cognitive deficits manifested by memory impairment (impaired ability to learn new information or to recall previously learned information). One or more of the following cognitive disturbances:

- Aphasia (language disturbance)
- Apraxia (impaired ability to carry out motor activities despite intact motor functions)
- Agnosia (failure to recognize or identify objects despite intact sensory functions)
- Disturbances in executive functioning (i.e., planning, organizing, sequencing, abstracting)

The End of Alzheimer's. http://dx.doi.org/10.1016/B978-0-12-812112-2.00014-8

The cognitive deficits in the criteria mentioned earlier, all cause significant impairment in social or occupational functioning and represent a significant decline from a previous level of functioning.

The course is characterized by gradual onset and continuing cognitive decline.

The cognitive deficits of criteria mentioned earlier are not due to any of the following:

- Other central nervous system conditions that cause progressive deficits in memory and cognition (e.g., cerebrovascular disease, Parkinson's disease (PD), Huntington's disease, subdural hematoma, normal pressure hydrocephalus, brain tumor).
- Systemic conditions that are known to cause dementia (e.g., hypothyroidism, vitamin B12 or folate acid deficiency, niacin deficiency, hypercalcemia, neurosyphilis, HIV infection).
- Substance-induced conditions.

The deficits do not occur exclusively during the course of a delirium.

The disturbance is not better accounted for by another Axis I disorder (e.g., major depressive disorder, schizophrenia).

Diagnostic Criteria of Alzheimer's Disease According to NINCDS-ADRA (McKhann, 1984)

The NINCDS-ADRDA Alzheimer's Criteria were proposed in 1984 by the National Institute of Neurological and Communicative Disorders and Stroke, as well as the Alzheimer's Disease and Related Disorders Association (now known as the Alzheimer's Association). The criteria are among the most used in the diagnosis of Alzheimer's disease (AD). These criteria require that the presence of cognitive impairment and a suspected dementia syndrome be confirmed by neuropsychological testing for a clinical diagnosis of possible or probable AD, while they need histopathologic confirmation (microscopic examination of brain tissue) for the definitive diagnosis. They also specify eight cognitive domains that may be impaired in AD. These criteria have shown good reliability and validity.

The criteria for the clinical diagnosis of probable AD include:

- Dementia established by clinical examination and documented by the Mini-Mental Test (Folstein, 1975), the Blessed Dementia Scale (Blessed, 1968) or some similar examination, and confirmed by neuropsychological test;
- Deficits in two or more areas of cognition;
- Progressive worsening of memory and other cognitive functions
- No disturbance of consciousness;
- Onset between ages 40 and 90, most often after age 65; and
- Absence of systemic disorders or other brain diseases that in and of themselves could account for the progressive deficits in memory and cognition.

The diagnosis of probable AD is supported by:

- Progressive deterioration of specific cognitive functions, such as language (aphasia), motor skills (apraxia), and perception (Agnosia);
- Impaired activities of daily living and altered patterns of behavior;
- Family history of similar disorders, particularly if confirmed neuropathologically; and

Laboratory results of:

- Normal lumbar puncture as evaluated by standard techniques;
- Normal pattern or nonspecific changes in EEG, such as increased slow-wave activity; and
- Evidence of cerebral atrophy on CT with progression documented by serial observation.

Other clinical features consistent with the diagnosis of probable AD, after exclusion of causes of dementia other than AD, include:

- Plateaus in the course of progression of the illness;
- Associated symptoms of depression, insomnia, incontinence, delusions, illusions, hallucinations, catastrophic verbal, emotional, or physical outbursts, sexual disorders, and weight loss.

Other neurologic abnormalities in some patients, especially with more advanced disease, include:

- Motor signs, such as increased muscle tone, myoclonus, or gait disorder;
- Seizures in advanced disease; and
- CT normal for age.

Features that make the diagnosis of probable AD uncertain or unlikely include:

- Sudden, apoplectic onset;
- Focal neurologic findings, such as hemiparesis, sensory loss, visual field deficits, and incoordination early in the course of the illness; and
- Seizures or gait disturbances at the onset or very early in the course of the illness.

Clinical diagnosis of possible AD:

- May be made on the basis of the dementia syndrome in the absence of other neurologic, psychiatric, or systemic disorders sufficient to cause dementia, and in the presence of variations in the onset, in the presentation, or in the clinical course;
- May be made in the presence of a second systemic or brain disorder sufficient to produce dementia, which is not considered to be the cause of the dementia; and
- Should be used in research studies when a single, gradually progressive severe cognitive deficit is identified in the absence of other identifiable causes.

Criteria for diagnosis of definite AD are:

- The clinical criteria for probable AD and
- Histopathologic evidence obtained from a biopsy or autopsy.

Classification of AD for research purposes should specify features that may differentiate subtypes of the disorder, such as:

- Familial occurrence;
- Onset before age of 35;
- Presence of trisomy 21; and
- Coexistence of other relevant conditions such as PD.

Proposed New Diagnostic Criteria for Probable Alzheimer's Disease (Dubois et coll., 2007)

The NINCDS-ADRDA and the DSM-IV-TR criteria for Alzheimer's disease (AD) are the prevailing diagnostic standards in research; however, they have now fallen behind the unprecedented growth of scientific knowledge. Distinctive and reliable biomarkers of AD are now available through structural MRI, molecular neuroimaging with PET, and cerebrospinal fluid analyses. This progress provides the impetus for our proposal of revised diagnostic criteria for AD. Our framework was developed to capture both the earliest stages, before full-blown dementia, as well as the full spectrum of the illness. These new criteria are centered on a clinical core of early and significant episodic memory impairment. They stipulate that there must also be at least one or more abnormal biomarkers among structural neuroimaging with MRI, molecular neuroimaging with PET, and cerebrospinal fluid analysis of amyloid-beta or tau proteins. The timelines of these criteria are highlighted by the many drugs in development that are directed at changing pathogenesis, particularly at the production and clearance of amyloid-beta as well as at the hyperphosphorylation state of tau. Validation studies in existing and prospective cohorts are needed to advance these criteria and optimize their sensitivity, specificity, and accuracy.

Core diagnostic criteria:
Presence of an early and significant episodic memory impairment that includes the following features:

- Gradual and progressive change in memory function reported by patients or informants over more than 6 months;
- Objective evidence of significantly impaired episodic memory on testing: this generally consists of recall deficit that does not improve significantly or does not normalize with cueing or recognition testing and after effective encoding of information has been previously controlled; and
- The episodic memory impairment can be isolated or associated with other cognitive changes at the onset of AD or as AD advances.

Supportive features:

- Presence of medial temporal lobe atrophy
- Volume loss of hippocampi, entorhinal cortex, amygdala evidenced on MRI with qualitative ratings using visual scoring (referenced to well characterized population with age norms) or quantitative volumetry of regions of interest (referenced to well characterized population with age norms)
- Abnormal cerebrospinal fluid biomarker
- Low amyloid β1-42 concentrations, increased total tau concentrations, or increased phosphotau concentrations, or combinations of the three
- Other well validated markers to be discovered in the future
- Specific pattern on functional neuroimaging with PET
- Reduced glucose metabolism in bilateral temporal parietal regions
- Other well validated ligands, including those that foreseeably will emerge, such as Pittsburg compound B or FDDNP
- Proven AD autosomal dominant mutation within the immediate family, exclusion criteria, history

- Sudden onset
- Early occurrence of the following symptoms: gait disturbances, seizures, and behavioral changes

Clinical features:

- Focal neurological features including hemiparesis, sensory loss, visual field deficits
- Early extrapyramidal signs
- Other medical disorders severe enough to account for memory and related symptoms
- Non-AD dementia
- Major depression
- Cerebrovascular disease
- Toxic and metabolic abnormalities, all of which may require specific investigations
- MRI FLAIR or T2 signal abnormalities in the medial temporal lobe that are consistent with infectious or vascular insults

AD is considered definite if the following are present:

- Both clinical and histopathological (brain biopsy or autopsy) evidence of the disease, as required by the NIA-Reagan criteria for the postmortem diagnosis of AD; criteria must both be present
- Both clinical and genetic evidence (mutation on chromosome 1, 14, or 21) of AD; criteria must both be present

Biomarkers—Standard-of-Care

The understanding of the mechanisms implicated in the process of AD has led to the identity of biological markers of the disease. Measurements in the cerebrospinal fluid of three biomarkers are currently being explored in expert centers and specialized networks for help in diagnosing the disease:

- Increase of total tau proteins
- Hyperphosphorylated tau proteins, and
- The reduction of fragment 1–42 of amyloid-β peptide.

When the three parameters are modified, it would appear possible to predict the evolution toward dementia in patients suffering from mild cognitive disorders. But before this stage can be reached, other markers will have to be developed. A group of expert researchers recommends that research should be continued into biomarkers useful for the predictive diagnosis of AD by giving priority to those that can be measured in the peripheral blood. (Hear, hear!)

CASE REVIEW—PATIENT ED WITH MILD COGNITIVE IMPAIRMENT

Let's review a patient case managed by neurologists and neuropsychologists to see what is typically done for a patient. This is a good representation of today's standard-of-care. This case review is several pages long.

Ed is an individual who asked me to review his situation. My team had previously worked up the wife of a friend who suffers from severe dementia, similar to that of my father at his later stages. Ed included me in his evaluation because we were able to improve the condition of this lady, slowly over time. The following is a report from Ed's neuropsychologist.

Ed is a 69-year-old, right-hand dominant, Caucasian male with 16 years of education referred for a neuropsychological evaluation by his neurologist for assessment of memory loss, to assist with differential diagnosis, and to make recommendations for management and treatment.

In addition to diagnostic clinical interview, the following neuropsychological tests were administered: Wechsler test of Adult Reading (WTAR); Dementia Rating Scale-Second Edition (DRS-2); selected subtests of Wechsler Adult Intelligence Scale-Fourth Editions (WAIS-4); Trail Making Test, Parts A & B; D-KEFS Color-Word Interference Test; Controlled Oral Word Association Test (COWAT-FAS & Animals); Boston Naming Test—Second Edition (BNT-2); Benton Judgment of Line Orientation—Form H (JLO); Rey Complex Figure Test (RCFT); Logical Memory I & II subtests of the Wechsler Memory Scale – Third Edition (WMS-3); Rey Auditory Verbal Learning Test (RAVLT); Brief Visuospatial Memory Test – Revised (BVMT-R); Behavior Rating Inventory of Executive Function—Adult Version (BRIEF-A); Beck Depression Inventory—Second Edition (BDI-2); and the Beck Anxiety Inventory (BAI).

Author note: No Mini Mental State Exam (MMSE) included.

The following history was obtained through an interview with Ed and a review of available medical records.

Presenting Problem: Ed presents today for an evaluation of memory loss that started several years ago, with a gradual onset and slowly progressive course, although he describes a leveling off of this decline over the past several months. (Author note: Our staff recommended specific supplements that Ed began taking "several months ago.") He reports that he has trouble recalling details of conversations, scheduled appointments, and the names of familiar people. He reports that the information might come to him an hour later. He reports that he often enters a room, but once there cannot remember his intended purpose. He reports that he might turn on the tea, but will forget it on the stove.

He also reports difficulties with expressive language that started approximately 6 months ago. He reports that he has trouble with word-finding primarily, and that this interferes with his ability to express himself. He denies any difficulties with language comprehension. He denies recent changes in his abilities to read, spell, write, or do math. He also denies difficulties with praxis, which is the ability to plan and then execute movement.

Ed does not feel that his cognitive difficulties are interfering to any significant degree with his abilities to independently manage his ADLs and IADLs (independent activities of daily life). He denies having had any difficulties managing his medications, managing his finances, or driving.

Diagnostic testing: An MRI of the brain revealed: "There are punctate scattered foci of T2 and FLAIR white matter increased signal intensity in the left superior frontal gyrus and more posterior left frontal lobe, the right external capsule, and the left peritrigonal white matter. The remainder of the brain shows normal gray and white matter signal intensity. There is no mass, mass effect, hemorrhage, or acute infarct. Normal variant left frontal lobe development venous anomaly is noted."

Medical history: Asthma; enlarged prostate; hypertension; and neck pain. Surgical history is significant for cholecystectomy, left knee surgery, neck surgery (discectomy and fusion at C5-6 and C6-7), and prostate surgery.

Ed reports a positive family history of dementia and that his maternal grandmother who passed away 25 years ago was diagnosed with dementia in her mid-70s. He reports that his mother passed away in 2005 and was diagnosed with dementia in her 60s. He is not sure as to what type of dementia his grandmother and mother had.

Other family history is significant for cancer, diabetes, heart disease, hypertension, hyperlipidemia, and restless legs syndrome.

Current medications: Advair Diskus; aspirin; losartan potassium; ProAir HFA; cod liver oil; vitamin D3; and a multivitamin.

Author note: Our team put him on the cod liver oil; vitamin D3; and multivitamin.

Psychiatric history: Ed reports that he is currently dealing with several personal and financial stressors, and that his mood tends to fluctuate as a result. However, he denies frank symptoms of depression currently, including persistent feelings of sadness, decreased interest/motivation, or sleep disturbance. He denies frank symptoms of anxiety currently, including persistent feelings of nervousness or excessive worrying. He denies any history of depression or anxiety.

Substance use: Ed reports that he drinks two to three glasses of wine per night. He reports that he does not use illicit, nonprescribed drugs currently. He denies any history of alcohol or drug abuse/dependence. He reports that he does not use tobacco products.

Developmental/educational/vocational history: Early developmental history is reportedly unremarkable. Ed denies having had any difficulties acquiring basic academic skills in the early grades, including reading, spelling, and writing. He reports that he repeated the third grade because he had missed too many days due to illness. There is no history suggestive of an attention or learning disorder. He reports that he has 16 years of formal education, having obtained a bachelor's degree in accounting from Providence College. He says that he is currently retired.

Behavioral observations: Ed arrived promptly for his 9 a.m. appointment. He was casually and appropriately attired and well groomed. No gross motor abnormalities were noted. Rapport was easily established. Affect was full-range and appropriate to conversational topics. Mood was euthymic. Conversational speech was fluent, grammatically correct, and normal in volume, rate, and prosody. There was no gross evidence of a thought disorder or perceptual disturbance. Insight and judgment were observed to be good. During formal psychometric testing, Ed attended well to all tasks, and he did not appear to have any difficulties with comprehension. He was cooperative, socially appropriate, and engaging, and he appeared to put forth his best effort. To this extent, the results of the present evaluation are considered a valid assessment of his current level of cognitive and affective/emotional functioning.

Summary of neuropsychological testing:

Estimated premorbid intellectual functioning: on the WTAR, a measure of sight-reading considered sensitive to premorbid intellectual functioning, Ed performed at the high end of the average range (75th percentile). This index will serve as a benchmark against which other cognitive functions can be compared.

General cognitive functioning: Ed's overall performance on the DRS-2, a measure sensitive to cognitive decline in dementia, was measured in the mildly impaired range (9th percentile). This represents an approximate 2.0 standard deviation decline relative to estimated premorbid levels.

Attention and executive functions: Ed's simple attention/concentration, as measured via the DRS-2, was measured in the average range (attention, 50th percentile). His ability to initiate and maintain goal-directed behavior, as measured via the DRS-2, was measured in the average range (initiation/perseveration, 63rd percentile).

On a measure of auditory-verbal attention span and working memory, Ed performed in the average range (digit span, 37th percentile). On this measure, he was able to recite up to five digits forward (average, 25th percentile) and five digits backwards (average, 50th percentile). Performance was measured in the high average range on a measure of working memory and numeric reasoning (arithmetic, 83rd percentile). On a measure of incidental visual learning, sustained attention, and psychomotor processing speed, he performed at the high end of the average range (digit symbol-coding, 75th percentile).

On a measure of visual-motor tracking and processing speed, Ed performed in the mildly impaired range, with two errors (trails A, 48 seconds, 10th percentile). On a more complex task of visual-motor processing speed with the added demands of alternating attention and mental flexibility, he performed in the average range, without error (trails B, 74 seconds, 63rd percentile).

Ed's performance on a measure of the ease with which a person can shift his or her perceptual set to conform to changing demands and suppress a habitual response in favor of an unusual one (i.e., response inhibition) was measured in the average range (D-KEFS Color-Word Interference Test, Inhibition trial, 63rd percentile), although with nine errors (moderate to severely impaired, <1st percentile).

On the inhibition/switching trial of this test, which is a measure of one's ability to efficiently shift cognitive sets (i.e., cognitive/mental flexibility), he performed in the average range (50th percentile), and with three errors (average, 50th percentile).

Abstract reasoning: Ed's conceptualization/abstract reasoning, as measured via the DRS-2, was measured in the mildly impaired range (conceptualization, 9th percentile).

Speech/language and related functions: Verbal fluency for a phonemic category was measured in the mildly impaired range (FAS, 27 words, 10th percentile). Verbal fluency for a more easily accessible semantic category was measured in the low average range (animals, 16 words, 19th percentile). Confrontational naming was measured in the mildly impaired range (BNT-2, 50/60 correct, 8th percentile).

Ed's rapid color naming was measured in the average range (D-KEFS Color-Word Interference Test, Color Naming trial, 37th percentile). His rapid word reading was measured in the average range (D-KEFS Color-Word Interference test, Word Reading trial, 50th percentile).

Visualspatial/constructional and related functions: Ed's visualspatial/constructional ability, as measured via the DRS-2, was in the average range (construction, 50th percentile). His ability to discriminate angles, a measure of visuospatial orientation, was measured in the average range (JLO, 63rd percentile). On a complex visuoconstructional task (i.e., copying a complex figure), he performed in the low average range (RCFT-copy, 16th percentile). His approach to copying the figure was noted to be poorly planned and organized, suggesting more of an executive weakness as opposed to representing a frank deficit in visuoconstruction.

Learning and memory: Ed's verbal and visual memory, as measured via the DRS-2, was in the mild to moderately impaired range (5th percentile).

Immediate recall of verbal information presented in a meaningful context (i.e., stories) was measured in the average range (logical memory I, 50th percentile), and with an average learning curve (63rd percentile). 30-min delayed recall of the same verbal information was measured at the high end of the average range (logical memory II, 75th percentile), and with 89% retention of the originally learned information, which is considered a high average retention rate (83rd percentile).

Ed's ability to learn a list of 15 words repeated over a series of five trials on the RAVLT was measured in the average range (47th percentile), and with a superior learning curve (95th percentile). His ability to freely recall the same list of words after a short delay with distraction was measured at the low end of the average range (6/15 words, 25th percentile). His ability to freely recall the same words after a long delay (20 min) was high average (10/15 words, 83rd percentile), and with 83% retention of the originally learned information, which is considered an average retention rate (63rd percentile). Recognition memory using a yes/no format was average (25th percentile). Ed recognized 15/15 target words, and he committed six false positive errors.

Ed's ability to learn a series of six geometric shapes repeated over a series of three trials on the BVMT-R was measured in the mild to moderately impaired range (5th percentile), with a low average learning curve (19th percentile). His ability to freely recall the same shapes after a long delay (25 minutes) was mildly impaired (10th percentile), with 100% retention of the originally learned information (>16th percentile). Recognition memory using a yes/no format was average (discrimination index, >16th percentile). Ed recognized 5/6 target shapes, and he did not commit any false positive errors.

Affective/emotional functioning: On the BDI-2, Ed's responses yielded a raw score of 12, indicating the absence of clinically significant depression.

On the BAI, Ed's responses yielded a raw score of 5, indicating the absence of clinically significant anxiety.

Interpretive summary and conclusions regarding patient Ed:

Results of the current evaluation revealed overall cognitive functioning to be in the mildly impaired range (9th percentile). This represents a significant decline relative to a previously higher level (i.e., an approximate 2.0 standard deviation decline).

Ed is reporting memory loss that started several years ago, with a gradual onset slowly progressive course, although he describes a leveling off of this decline over the past several months. Results of the current evaluation revealed a memory encoding impairment of mild to moderate severity, primarily for information that he sees, with storage/consolidation and retrieval relatively well preserved. Such an encoding deficit is likely causing him to overload and miss information when a large amount of new visual information is presented to him at once, particularly when the information is complex and presented quickly. However, with repeated exposures to the to-be-learned information, he is able to increase the amount of information that he is able to learn, and he is able to hold on to the information for later recall. Since he is clearly able to increase the amount of information that he is able to learn with repeat exposures, my recommendation would be to adopt compensatory strategies that entail some aspect of repetition, such as getting into the habit of asking people to repeat important new information that he feels he did not completely understand initially, as well as paraphrasing new information so that it is more deeply encoded. Repetition helps encode information by forcing one to pay attention to it. The recommendations below should prove helpful.

His performance on measures of expressive language was characterized by mildly impaired verbal fluency for a phonemic category, low average verbal fluency for a semantic category, and mildly impaired confrontational naming. This pattern of performance suggests that he is having trouble primarily in his ability to initiate and carry out the systematic retrieval of words, an ability that is typically considered executive in nature. There is no evidence to suggest the presence of a frank acquired aphasia at this time.

Additional cognitive deficits were noted within the areas of executive control (i.e., response inhibition) and abstract reasoning.

In contrast to the earlier-delineated memory encoding, expressive language, and executive deficits, relative cognitive strengths were noted within the areas of psychomotor processing speed; simple and complex attention (i.e., focused, sustained, and divided attention; working memory); certain aspects of executive control (i.e., cognitive/mental flexibility); reading; calculations (i.e., the ability to perform simple addition, subtraction, multiplication, and division tasks); visuospatial orientation; visuoconstruction; and verbal learning and memory.

Overall, the earlier-delineated cognitive deficits are consistent with presumed neuropathology within frontal-subcortical circuits (particularly the orbitomedial prefrontal-subcortical circuit) and the left anterior temporal lobe. Recent neuroimaging revealed cerebrovascular disease within the left superior frontal gyrus and more posterior left frontal lobe. Thus, I would speculate that cerebrovascular disease is the most likely etiology of Ed's manifest cognitive impairment.

Affectively, I am not seeing any overwhelming evidence to suggest the presence of frank depression or anxiety. Thus, I do not feel that depression or anxiety, or for that matter another psychiatric disorder, can better account for Ed's manifest cognitive impairment.

In conclusion, it appears that Ed has experienced a decline in cognitive functioning relative to a previously higher level, particularly within the areas of executive control, expressive language, and memory encoding. Adaptive functioning appears to be relatively well preserved at this time, and this precludes a diagnosis of dementia. Thus, I am diagnosing Ed with Mild Cognitive Impairment (MCI), most likely related to cerebrovascular disease. At this point, I am referring Ed back to his treating neurologist for follow-up.

Diagnostic impressions:

In light of the current test results, behavioral observations, and reported history, the following ICD-9 and DSM-IV-TR diagnoses are made:

1. Mild Cognitive Impairment (MCI)—331.83

Recommendations:

1. Ed is referred back to his treating neurologist for follow-up, assistance in implementing the following recommendations, and to review laboratories and neuroimaging, as this will help rule out any metabolic, infectious, or structural causes for his manifest cognitive impairment.
2. Evaluation results will be reviewed with Ed during a feedback session.
3. Aggressive management of cerebrovascular risk factors is recommended to minimize the potential for further cognitive decline. This includes a carefully selected,

heart-healthy diet; regular exercise; actively managing stress; and routine follow-up with his PCP.

4. Individuals with memory encoding weaknesses work best when they can focus on one thing at a time, work at their own pace, and know what they will be doing before they begin a task. Transitions can be difficult for individuals with such difficulties, and taking a brief time-out starting a new task or when transitioning from one task to the next will help Ed focus his attention for the next task and plan what he will be doing next.

5. Organization is one of the bedrock concepts for improving everyday memory functioning. To assist Ed in maximizing organization of important information, I recommend that he employ checklists of activities or tasks to be performed, use of a daily planner to help him organize and remember important dates (such as deadlines and appointments), and use of a memory notebook to record important information provided to him.

6. Sustaining attentional focus and absorbing dense information at a high rate of speed become more difficult with age. To improve the likelihood that Ed will absorb and remember important information, I recommend that he gets into the habit of asking people to repeat important new information that he feels he did not completely understand initially; paraphrase new information so that it is more deeply encoded; and minimize interruptions during learning situations.

7. Repetition of new material across different contexts and at different times will result in memory traces that are more durable (e.g., during exercise, while grocery shopping). Repetition helps encode information by forcing one to pay attention to it.

8. Ed would benefit from instruction in memory compensatory strategies, such as those found in *The Memory Book* by Harry Lorayne and Jerry Lucas and *Achieving Optimal Memory* by Aaron Nelson, PhD (both are available online and at local bookstores).

9. The current testing will serve as a baseline to monitor Ed's cognitive functioning.

10. Ed is encouraged to contact me if he would like to discuss the results of the current evaluation or the proposed recommendations further. Follow-up appointments can be made available as necessary.

The foregoing information is my opinion to a reasonable degree of certainty based on the facts that were available to me at the time of the evaluation. There might be other facts that I am not aware of that could alter my opinion.

Name (Withheld); Clinical Neuropsychologist; Licensed Psychologist

Author Comment: Patient Ed Evaluation

I do hope you took the time to read and absorb the entire report. The cognitive testing was quite substantial but oddly didn't include the standard MMSE. Many patients with early cognitive impairment find these types of tests somewhat humiliating to endure but they do provide some interesting insights as to where in the brain the most damage is done. Based on the testing, a best guess as to the location of the disease was hypothesized. Let's consider some points based on statements highlighted in the report.

First, what Ed said about himself and the conclusions in the report are the same. That is, all the testing did not reveal anything new about Ed's condition but rather captured it more analytically.

Second, does any of this information shed any light on the "why?" Clearly it does tell us a bit about the "what." There is conjecture about cerebrovascular disease. However, his recommendations fall strictly into the standards-of-care. He is referring Ed to his primary care physician for evaluation. These doctors have neither the time nor the knowledge to do a detailed assessment of the vascular system that can provide Ed with meaningful information upon which a targeted treatment plan can be developed. Could Ed have been referred to a cardiovascular disease specialist? That would make sense, but since Ed does not have any cardiovascular symptoms upon which a cardiovascular specialist can diagnose, they are unable to treat him.

Ed is caught in the vicious cycle of chronic disease care in the United States. He is worried about declining into dementia, but he is not sick enough for any doctor to really delve into his condition and make a deep and broad diagnosis that can then lead to a plausible treatment program. Instead Ed is "referred back" to his neurologist. At best, Ed will be retested to show, somewhat imprecisely, how he is declining with time.

This is typical of the standard-of-care.

Image Credit: Caduceus symbol, by Rama and Eliot Lash. Available at Wikimedia Commons: https://commons.wikimedia.org/w/index.php?curid=662346

Concept of a Medical "Cure"

Our purpose is to make you diligently aware of the complexity of diagnosis.

> Genius is 10% inspiration and 90% perspiration.
>
> —*Thomas Edison*

A cure for Alzheimer's disease (AD) is 90% diagnosis and 10% treatment. Said differently by the Functional Medicine Society:

> You cannot treat what you don't measure.

Therefore, diagnosis is harder, more work, and more important than treatment. Diagnosis must be where your energies are placed. Don't search for a cure, search for an understanding of "why" through a comprehensive "differential diagnosis." When this hard work is done, a cure is at hand. There are so many clues as to cause and effect that is fleshed out through the diagnostic approaches considered here. It's extremely important to know that diseases like Alzheimer's that are complex and fester over a long period of time are "personal" because so many factors can trigger the disease.

Personalized medicine is a relatively new buzz that says our genes determine everything, including health, ability, and disease. We believe that you are not predestined by your genetic make-up as shown by the very low incidences of familial AD. You have enormous control over your fate, before and during an affliction like AD. Understanding your personal risk and thus your personal disease management and mitigation path is only achieved through a comprehensive and regular differential diagnosis.

The title of this book includes the term "cure." Just like "root cause," cure takes on many meanings. Some of the meanings do not imply a return to complete and absolutely perfect health, while others do. We believe this is the case with AD. That is, some people in the earliest stages of the disease can achieve reversal and attain "normal" cognitive functioning. Other less fortunate people have root causes that are very challenging to manage. For those people, the best that can be hoped for is a slowing of the progression of the disease. Improving prognosis is, in this context, a cure. Consider these many nuances for the definition of medical "cure."

- The treatment of any disease or of a special case.
- The successful treatment of a disease or wound.
- A system of treating diseases.
- A medicine effective in treating a disease course of therapy, a medication, a therapeutic measure, or another remedy used in treatment of a medical problem.

The End of Alzheimer's. http://dx.doi.org/10.1016/B978-0-12-812112-2.00015-X

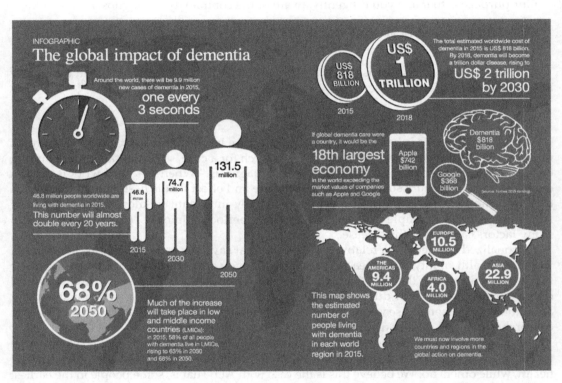

Image Credit: Infographic on the global impact of dementia, from Dementia partnerships. Available at: http://dementiapartnerships. com/wp-content/uploads/sites/2/adi-globalimpact-infographic.png

Alzheimer's Disease Statistics

PREVALENCE OF ALZHEIMER'S DISEASE AND OTHER DEMENTIAS

Study the chart above. Alzheimer's is a new and emerging epidemic. Can our genes change rapidly enough to cause the jump in Alzheimer's cases? Our shared external environment and your personal internal environment are the biggest contributors to the escalation in this disease.

An estimated 5.4 million Americans of all ages have Alzheimer's disease (AD) in 2012. This figure includes 5.2 million people age 65 and older and 200,000 individuals under age 65 who have younger-onset Alzheimer's.

- One in eight people age 65 and older (13%) has AD.
- Nearly half of people age 85 and older (45%) have AD.
- Of those with AD, an estimated 4% are under age 65, 6% are 65–74, 44% are 75–84, and 46% are 85 or older.

The estimated numbers for people over 65 come from the Chicago Health and Aging Project (CHAP), a population-based study of chronic health diseases of older people. In 2009, the National Institute on Aging (NIA) and the Alzheimer's Association convened a conference to examine discrepancies among estimates from CHAP and other studies, including the Aging, Demographics, and Memory Study (ADAMS), a nationally representative sample of older adults.

A panel of experts concluded that the discrepancies in the published estimates arose from differences in how those studies counted who had AD. When the same diagnostic criteria were applied across studies, the estimates were very similar.

National estimates of the prevalence of all forms of dementia are not available from CHAP. Based on estimates from ADAMS, 13.9% of people age 71 and older in the United States have dementia. This number would be higher using the broader diagnostic criteria of CHAP.

The estimates from CHAP and ADAMS are based on commonly accepted criteria for diagnosing AD that have been used since 1984. In 2009, the Alzheimer's Association convened an expert workgroup and the NIA to recommend updated diagnostic criteria, as described in the Overview (pages 8–9). It is unclear exactly how these new criteria, if adopted, could change the estimated prevalence of Alzheimer's. However, if AD can be detected earlier, in the preclinical stage as defined by the new criteria, the number of people reported to have AD would be larger than what is presented in this report.

The End of Alzheimer's. http://dx.doi.org/10.1016/B978-0-12-812112-2.00016-1

Prevalence studies, such as CHAP and ADAMS are designed so that all individuals with dementia are detected. But in the community, only about half of those who would meet the diagnostic criteria for AD or other dementias have been diagnosed. Because AD is under diagnosed, more than half of the 5.4 million Americans with Alzheimer's may not know they have it.

Prevalence of Alzheimer's Disease and Other Dementias in Women and Men

More women than men have AD and other dementias. Almost two-thirds of Americans with Alzheimer's are women. Of the 5.2 million people over age 65 with AD in the United States, Prevalence 2012 Alzheimer's Disease Facts and Figures 1-5 show that 3.4 million are women and 1.8 million are men. Based on estimates from ADAMS, 16% of women age 71 and older have AD or other dementias compared with 11% of men.

The larger proportion of older women who have AD or other dementias is primarily explained by the fact that women live longer on average than men. Many studies of the age specific incidence (development of new cases) of AD or any dementia have found no significant difference by gender. Thus, women are not more likely than men to develop dementia at any given age.

Prevalence of Alzheimer's Disease and Other Dementias by Years of Education

People with fewer years of education appear to be at higher risk for Alzheimer's and other dementias than those with more years of education. Prevalence and incidence studies show that having fewer years of education is associated with a greater likelihood of having dementia and a greater risk of developing dementia.

Some researchers believe that a higher level of education provides a "cognitive reserve" that enables individuals to better compensate for changes in the brain that could result in Alzheimer's or another dementia. However, others believe that the increased risk of dementia among those with lower educational attainment may be explained by other factors common to people in lower socioeconomic groups, such as increased risk for disease in general and less access to medical care.

Prevalence of Alzheimer's Disease and Other Dementias in Older Whites, African-Americans, and Hispanics

While most people in the United States living with Alzheimer's and other dementias are non-Hispanic whites, older African-Americans and Hispanics are proportionately more likely than older whites to have AD and other dementias.

Data indicate that in the United States, older African-Americans are probably about twice as likely to have Alzheimer's and other dementias as older whites, and Hispanics are about one and one-half times as likely to have Alzheimer's and other dementias as older whites.

Despite some evidence of racial differences in the influence of genetic risk factors for Alzheimer's and other dementias, genetic factors do not appear to account for these large prevalence differences across racial groups. Instead, health conditions, such as high blood pressure and diabetes that increase one's risk for AD and other dementias are more prevalent in African-American and Hispanic communities. Lower levels of education and other socio-economic characteristics in these communities may also increase risk. Some studies suggest that differences based on race and ethnicity do not persist in detailed analyses that account for these factors.

There is evidence that missed diagnoses are more common among older African-Americans and Hispanics than among older whites. For example, a 2006 study of Medicare beneficiaries found that AD or other dementias had been diagnosed in 9.6% of white beneficiaries, 12.7% of African-American beneficiaries and 14% of Hispanic beneficiaries. Although rates of diagnosis were higher among African-Americans and Hispanics than among whites, the difference was not as great as would be expected based on the estimated differences found in preva-lence studies, which are designed to detect all people who have dementia. This disparity is of increasing concern because the proportion of older Americans who are African-American and Hispanic is projected to grow in coming years. If the current racial and ethnic disparities in diagnostic rates continue, the proportion of individuals with undiagnosed dementia will increase.

INCIDENCE AND LIFETIME RISK OF ALZHEIMER'S DISEASE

While prevalence is the number of existing cases of a disease in a population at a given time, incidence is the number of new cases of a disease in a given time. The estimated annual incidence (rate of developing disease in a 1-year period) of AD appears to increase dramati-cally with age, from approximately 53 new cases per 1,000 people age 65–74, to 170 new cases per 1,000 people age 75–84, to 231 new cases per 1,000 people over age 85 (the "oldest-old"). Some studies have found that incidence levels of after age 90, but these findings are contro-versial. A recent analysis indicates that dementia incidence may continue to increase and that previous observations of a leveling off of incidence among the oldest-old may be due to sparse data for this group. Because of the increase in the number of people over 65 in the United States, the annual incidence of Alzheimer's and other dementias is projected to double by 2050.

- Every 68 seconds, someone in America develops AD.
- By mid-century, someone in America will develop the disease every 33 seconds.

Lifetime risk is the probability that someone of a given age develops a condition during his or her remaining lifespan. Data from the original Framingham Study population were used to estimate lifetime risks of AD and of any dementia. Starting in 1975, nearly 2,800 people from the Framingham Study who were age 65 and free of dementia were followed for up to 29 years. The study found that 65-year-old women without dementia had a 20% chance of developing dementia during the remainder of their lives (estimated lifetime risk), compared with a 17% chance for men. For Alzheimer's, the estimated lifetime risk was nearly one in five (17.2%) for women compared with nearly one in 10 (9.1%) for men.

The preceding statistics on Alzheimer's are from the Alzheimer's Association: ©2012 Alzheimer's Association. All rights reserved. This is an official publication of the Alzheimer's Association but may be distributed by unaffiliated organizations and individuals. Such distribution does not constitute an endorsement of these parties or their activities by the Alzheimer's Association.

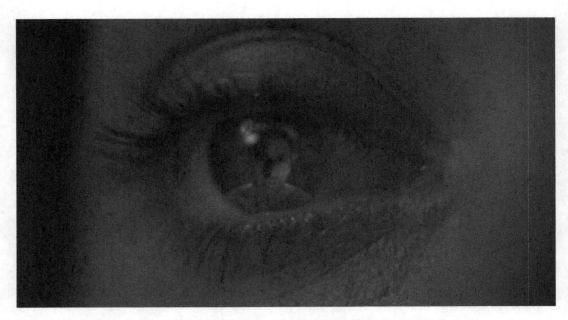
Image Credit: Available at Wikimedia Commons: https://en.wikipedia.org/wiki/File:MulderinEyeScully.jpg

Eye and Whole Body Disease

The following is a list of eye diseases that also shows up in the body indicating systemic illnesses. Notice that Alzheimer's disease is not on this list, but clearly it deserves a place of high prominence. This list is from 1990.

Acne rosacea	Emboli and thrombi	Onchocerciasis
Actinomyces	Emphysema	Osteogenesis imperfecta
Addison's disease	Encephalotrigeminal angiomatosis	Paget's disease
Albinism	Endocarditis	Pancreatic cancer
Albinism	Epidemic keratoconjunctivitis	Pancreatic disease
Albright's disease	Epidermolysis bullosa	Pemphigus
Alcoholism	Erythema multiforme	Peptic ulcer disease
Alkaptonuria	Fabry's disease	Periarteritis nodosa
Alport's syndrome	Facial deformity syndromes	Pharyngoconjunctival fever (adenovirus 3)
Amyloidosis	Galactosemia	Pneumonias
Anemia	Gaucher's disease	Pre-eclampsia
Angiomatosis retinae	Goltz-Gorlin syndrome	Pseudoxanthoma elasticum
Ankylosing spondylitis	Gonorrhea (ophthalmia neonatorum)	Psoriasis
Aortic arch syndrome	Gout	Regional enteritis or ulcerative colitis
Apert syndrome	Hay fever	Reiter's syndrome
Arterial spasm	Hemochromatosis	Relapsing polychondritis
Arteriosclerosis	Hereditary telangiectasia	Renal transplantation
Asthma	Herpes simplex	Rheumatoid arthritis
Ataxia telangiectasia (Louis-Bar syndrome)	Herpes zoster	Ring-D chromosome
Atopic dermatitis	Histiocytosis	Rubella (German measles)
Atopic dermatitis	Histoplasmosis	Rubeola (measles)
Atopic eczema	Homocystinuria	Sarcoidosis
Azotemia (acute and chronic pyelonephritis)	Human immunodeficiency virus	Schmid-Fraccaro syndrome
Behçet's disease	Hyperparathyroidism	Scleroderma
Blood cancer	Hypertension	Septicemia bacterial metastatic endophthalmitis
Breast cancer	Hyperthyroidism	Sickle cell attack
Bronchiectasis	Hypervitaminosis A, B, and D	Skin cancer
Bronchogenic carcinoma	Hypoparathyroidism	Stevens–Johnson syndrome

The End of Alzheimer's. http://dx.doi.org/10.1016/B978-0-12-812112-2.00017-3

Brucellosis
Cancer of ovary or cervix
Cancer of testis or prostate

Cancer of the genital organs
Cancer of the gut
Candida albicans
Carcinoma and sites of primary
    lesions
Cardiac myxoma
Cardiovascular diseases
Carotid artery disease
Central retinal artery occlusion
Chediak-Higashi syndrome
Cicatricial pemphigoid
Ciliopathic genetic syndromes
Coccidioidomycosis
Collagen diseases

Colon cancer
Conradi's syndrome
Cranial arteritis
Craniofacial syndromes
Cri-du chat syndrome
Crohn's disease
Cryptococcus
Cushing's disease
Cystic fibrosis of the pancreas
Cysticercosis
Cystinosis
Cytomegalic inclusion disease
Deletion of chromosome 18
Long arm chromo. 18
Dermatomyositis
Diabetes mellitus
Diabetes mellitus
Diphtheria
Echinococcosis
Ehlers-Danlos synd.

Hypothyroidism
Ichthyosis
Inclusion conjunctivitis
    (chlamydial)
Incontinentia pigmenti
Infectious mono
Influenza
Kidney cancer

Leprosy (Hansen's disease)
Leukemia
Lipidoses
Liver disease
Loiasis (Loa loa)
Lowe's syndrome
Lung cancer
Lyme disease
Lymphogranuloma venereum
    (chlamydial)
Lymphoma
Malaria
Malnutrition
Marchesani's syndrome
Marfan's syndrome
Medullary cystic disease
Melanoma
Metastatic fungal endophthalmitis
Monosomy-G syndrome
Mucopolysaccharidosis
Mumps
Muscular dystrophy disorders
Myasthenia gravis
Myxoma
Nephrotic syndrome
Neurofibromatosis
Nevus of Ota
Niemann-Pick disease
Occlusive vascular disease
Occlusive vascular disease (sudden)

Stomach cancer
Streptothrix
Syphilis

Systemic lupus erythematosus
Temporal arteritis
Thromboangiitis obl.
Thrombosis

Thyroid cancer
Toxocariasis (Toxocara)
Toxoplasmosis
Trichinosis (Trichinella)
Trisomy 13
Trisomy 18
Trisomy 21
Tuberculosis
Tuberous sclerosis (Bourneville's
    syndrome)
Tularemia
Turner's syndrome
Urticaria
Use of hormonal contraception
Vaccinia
Varicella (chickenpox)
Variola (smallpox)
Venous occlusive disease
Vernal conjunctivitis
Vitamin A deficiency
Vitamin B deficiency
Vitamin C deficiency
Vogt-Koyanagi-Harada syndrome
Wegener's granulo.
Whipple's disease
Wilms' tumor
Wilson's disease
Wyburn-Mason syndrome
Xeroderma pigmentosum

From Pavan-Langston, D. Manual of ocular diagnosis and therapy. Little, Brown and Company; 1990.

Doesn't it make sense to use the eye as a tool for disease detection and prevention?

Image Credit: Professor Alexander Fleming in his laboratory at St Mary's, Paddington, London. Available at Wikimedia Commons: https://commons.wikimedia.org/wiki/File:Synthetic_Production_of_Penicillin_TR1468.jpg

Hope, A History Lesson—Medical Pioneers

Medical knowledge continues to expand for the benefit of patients. A brief history lesson helps renew our faith in the profession. These lessons make us aware that many new cures and treatments are available if our minds are receptive to ideas that may not fit the contemporary standard-of-care.

Of the many significant medical breakthroughs, some have occurred by accident, by coincident, and by chance. Most of these have happened under humble circumstances in clinical laboratories as opposed to multimillion dollar research institutions at universities, in government laboratories, or in pharmaceutical laboratory development programs. Often discovery occurs in concert with diagnosis. The Nobel Prize in Medicine in 2005 was awarded to a pathologist who made thought-altering discovery while examining diseased tissue more thoroughly than anyone before him, on his own time. He was curious.

Malaria Cure

A South American Indian man, legend has it, unwittingly ingested quinine while suffering a malarial fever in a jungle high in the Andes. Needing desperately to quench his thirst, he drank his fill from a small, bitter-tasting pool of water that happened to be "contaminated" with quinine from the bark of an adjoining tree that was thought to be "poisonous." But when this man's fever miraculously abated, he brought news of the medicinal drink back to his tribe, which began to use a cocktail from the bark of the tree to treat malaria.

Small Pox Vaccination

In 1796, Edward Jenner, a British scientist and surgeon, when presented with a scientific curiosity, had a brainstorm that ultimately led to the development of the first vaccine. A young milkmaid had told him how people who contracted cowpox, a harmless disease easily picked up during contact with cows, never got smallpox, a deadly scourge.

X-rays

X-rays have become an important tool for medical diagnoses, but their discovery in 1895 by the German physicist Wilhelm Conrad Röntgen had little to do with medical

The End of Alzheimer's. http://dx.doi.org/10.1016/B978-0-12-812112-2.00018-5

experimentation. Röntgen wanted to determine if he could see cathode rays escaping from a glass tube completely covered with black cardboard. While performing this experiment, Röntgen noticed that a glow appeared in his darkened laboratory several feet away from his cardboard-covered glass tube. At first he thought a tear in the paper sheathing was allowing light from the high-voltage coil inside the cathode-ray tube to escape. But he soon realized he had happened upon something entirely different. Rays of invisible light were passing right through the thick paper and appearing on a fluorescent screen over a yard away.

Allergy

Charles Robert Richet, a French physiologist, made several experiments testing the reaction of dogs exposed to poison from the tentacles of sea anemones. Some of the dogs died from allergic shock, but others survived their reactions and made full recoveries. Weeks later, because the recovered dogs seemed completely normal, Richet wasted no time in reusing them for more experiments. They were given another dose of anemone poison, this time much smaller than before. The first time, the dogs' allergic symptoms, including vomiting, shock, loss of consciousness, and in some cases death, had taken several days to fully develop. But this time the dogs suffered such serious symptoms just minutes after Richet administered the poison.

Insulin

Frederick G. Banting, a young Canadian doctor, and Professor John J.R. MacLeod of the University of Toronto shared a Nobel Prize in 1923 for their isolation and clinical use of insulin against diabetes. Their work with insulin followed from the chance discovery of the link between the pancreas and blood sugar levels by two other doctors on the other side of the Atlantic decades earlier. These Nobel Prize winners had the courage and resolve to "translate" a new discovery.

PAP Smear

Dr. George Nicholas Papanicolaou's chance observation while doing a genetic study of cancer cells on a slide containing a specimen from a woman's uterus spawned the routine use of the so-called "Pap smear," a simple test that has saved millions of women from the ravages of uterine cancer.

Penicillin

The identification of penicillium mold by Dr. Alexander Fleming in 1928 is one of the best-known stories of medical discovery, not only because of its accidental nature, but also because penicillin has remained one of the most important and useful drugs in our arsenal, and its discovery triggered research into a range of other invaluable antibiotic drugs. Alexander

Fleming, a Scottish bacteriologist in London, made his discovery of penicillin by mistake when he was trying to study staphylococci bacteria. He was running experiments with the bacteria in his laboratory at London's St. Mary's Hospital and set a laboratory dish containing the bacteria near an open window. Upon returning to the experiment, he found that some mold blown in through the open window onto the dish, contaminating the bacteria. Instead of throwing away his spoiled experiment, Fleming looked closely at it under his microscope. Surprisingly, he saw not only the mold growing on the staphylococci bacteria, but a clear zone around the mold. The penicillium mold, the precursor to penicillin, was killing the harmful staphylococci bacteria.

Sterilization

Joseph Lister is the surgeon who introduced new principles of cleanliness that transformed surgical practice in the late 1800s. Widespread acceptance of Lister's procedures was rather slow, as is often the case with revolutionary new ideas. Some busy doctors were unwilling to take the time to even consider new ideas. Some found it difficult to believe in germs—living organisms that wrought havoc but were too small to see. Others tried Lister's procedures, but did so incorrectly and therefore failed to obtain the desired result. This is a great fear of those proposing complex solutions to disease and applies to Alzheimer's disease in the context of the information presented in this book.

Galileo

The Sun is the Center of our Galaxy: Galileo supported the Copernican viewpoint. It was not the church authorities that refused to look through his telescope; it was his fellow scientists! They thought that using a telescope was a waste of time, since even if they did see evidence for Galileo's claims; it could only be because Galileo had bewitched them. The "Not Invented Here" syndrome existed 400 years ago. Note: Galileo is in this section to illustrate how invention is often stifled by the "powers that be." He did improve the microscope and the telescope.

Blood Circulation

William Harvey's discovery of blood circulation caused the scientific community of the time to ostracize him. Harvey's lecture notes show that he believed in the role of the heart in circulation of blood through a closed system as early as 1615. Yet he waited 13 years, until 1628, to publish his findings in his work *Exercitatio anatomica de motu cordis et sanguinis in animalibus* or *On the Movement of the Heart and Blood in Animals*. Why did he wait so long? Galenism, or the study and practice of medicine as originally taught by Galen, was almost sacred at the time Harvey lived. No one dared to challenge the teachings of Galen. Like most physicians of his day, William Harvey was trained in the ways of Galen. Conformation was not only the norm, but was also the key to success. To rebel against the teachings of Galen could quickly end the career of any physician. Perhaps this is why he waited. Today the prescription pad, drug companies, and the standard-of-care replaced Galen, but nothing much else has changed.

For all you would-be Nobel Prize-winners and those looking for a solution to Alzheimer's disease, remember the one trait that tied all these lucky strikers together: open-mindedness. As the American physicist Joseph Henry once noted,

> The seeds of great discoveries are constantly floating around us, but they only take root in minds well prepared to receive them.

Index

blood vessels in retina, 134
eye general, 133
lens, 134
retina, 133
visual cortex, 134
visual field, 134
and eye pathology changes, 135–136
and eye structures, 185–187
microtubules, 185–187
measurement in RNFL of eye, 186–187
medical misdiagnoses
categories
doctor, 87
equivocal results, 83
false-negative, 83
false-positive, 83
healthcare professional, 86
patient, 86
specialist, 86
tests, 86
risk factors
associated with development of AD
cardiovascular disease, 358
cortical cataract, 356–357
diabetes, 359
diet, 353
glaucoma, 354
macular degeneration, 355
other predetermining factors, 362
periodontal disease, 360–361
retinal nerve fiber layer thinning, 357
type 3 diabetes, 219
Alzheimer's Disease Assessment Scale-Cognitive
(ADAS-Cog), 57
Alzheimer's drug, 117
Alzheimer's patient, 409
Alzheimer's treatment
influence of drug companies on, 119
Amalgamation, 135, 166
AMD. See Age-related macular degeneration (AMD)
American Diabetes Association, 352
American diet, 226
American Optometric Association, 367
Amgen, 116
AMP. See Antimicrobial peptide (AMP)
Amyloid
amyloidosis, 258, 264
angiopathy, 252
as biomarker, 28
buildup diseases, 259
diseases, 149
causes of, 266
fibril formation, 213
immunotherapy, 249

pathway, 34
plaques, 168
protein, 98
β-Amyloid (Aβ), 11, 12, 32, 330
as antimicrobial peptide, 36
immunotherapy, 42
protein, 195
precursor, 37
synthesis, 39
vaccination, 32, 42
Amyloid Cascade Hypothesis, 14, 27, 59, 65, 117,
203, 336
controversy related to, 27
drug companies committed to, 37–42
Bristol-Myers Squibb, 38–39
Eli Lilly, 37–38
Johnson & Johnson and Pfizer, 38
Merck, 39–42
Pfizer and Medivation, 39
evidence against, 29–34
research opposed to, 34–37
Amyloid precursor protein (APP), 29, 167, 330
Amyotrophic lateral sclerosis, 161
Andhra Pradesh Eye Disease Study, 156
Angiotensin, 358
Animalcules, 285
Animal fats, 226
Ankylosis, 264
Anti-aging medicine, 317
Antiamyloid therapies, 39, 43, 160, 348
Antibiotic drugs, 440
Antibody titers, 380
Antigen, 380
Antigenic stimulation, 380
Antiinflammation, 316
antiinflammaging recipe, 198
drugs, placebo-controlled clinical trials, 196
immunosuppressive drugs, 196
Antimicrobial peptide (AMP), 30, 31
as immunomodulators, 30
Antimicrobial therapy, 308
Antioxidants, 151, 327–329
Antiserum, 201
Aortic stenosis, 250
Aphasia, 413
Apolipoprotein E (apoE)
APOE 4 carriers, 27, 308, 332
APOE-ε4 gene, 293
gene, 307
Apoptosis, 372
APP. See Amyloid precursor protein (APP)
Apraxia, 68, 413
Arachidonic acid, 338
AREDS. See Age-related eye disease study (AREDS)

Printed in the United States
By Bookmasters